AF789612

TYING THE THREADS OF EURASIA

Sidestone Press

TYING THE THREADS OF EURASIA

Trans-regional routes and material flows in Transcaucasia, eastern Anatolia and western central Asia, c.3000-1500BC

TOBY C. WILKINSON

© 2014 T. C. Wilkinson

Published by Sidestone Press, Leiden
www.sidestone.com

ISBN 978-90-8890-244-4

Lay-out & cover design: Sidestone Press
Photograph cover: Caravan of travellers riding camels in silhouette, Nubra Valley © Phaendin | Dreamstime.com

Contents

List of Figures 11

List of Tables 19

Acknowledgements 21

1 Contexts and Frameworks of Research 23

1.1 Theoretical frameworks: trade and economy; networks and routes 23

1.2 Geographical and environmental frameworks 29
1.2.1 The landscapes of 'Transcaucasia and eastern Anatolia' 30
1.2.2 The landscapes of 'western Central Asia' 32

1.3 Archaeological frameworks: chronologies, sites and assemblages 37
1.3.1 Constructing inter-regional chronologies 38
1.3.2 Assemblages, sites and chronologies in Transcaucasia/eastern Anatolia 40
1.3.3 Assemblages, sites and chronologies in western Central Asia 43

1.4 Frameworks of movement: transportation technologies 47
1.4.1 The domestication of equids: the donkey and the horse 47
1.4.2 The domestication(s) of the camel 50
1.4.3 The first vehicles 52
1.4.4 Horse riding and the fast chariot 54
1.4.5 Water transport: the boat and the sail 55

1.5 Frameworks of interpretation: indicators for interaction 56
1.5.1 Visible and invisible flows: materials, people, language 57
1.5.2 Technology and style as indicators of interdependence 58
1.5.3 Cross-craft interaction: technology, aesthetics and skeuomorphism 60
1.5.4 Value, aesthetics and colour 'symbolism' 62

1.6 Research questions: routes and material flows 62

2 Routes: on the Trail of History and Myth 65

2.1 Introduction 65

2.2 Routes and roads from archaeological remains 67
2.2.1 Roman roads (c. 500BC - AD500) 67
2.2.2 Caravanserais and caravan routes of the Islamic era 68
2.2.3 Other road fragments: the Pasinler road 70
2.2.4 Linear hollow ways in northern Mesopotamia 71

2.3 Routes and roads in modern or recent reports 73
2.3.1 Modern road maps 73
2.3.2 Pre-automobile maps: British Naval Intelligence reports 75
2.3.3 Ethnographies: nomads, transhumance and nomadic crafts 75

2.4 Routes and roads in textual sources 76
2.4.1 Geographers, historians and itineraries 76
2.4.2 Military campaigns 79
2.4.3 Travellers' accounts 81

2.4.4 Pilgrimage routes 84
2.4.5 Legal and economic accounts 86

2.5 Critique: evaluating 'route inertia' 88
2.5.1 'Historical routes' as models for earlier ancient route systems 88
2.5.2 Routes in the archaeological imagination 89
2.5.3 Mythical routes? The 'Silk Road' as literary trope 92

2.6 Summary: route inertia and re-visioning routes 94

3 Landscape and Non-linear Networks: Finding Methods to Visualize Ancient Flow of Materials 95

3.1 Introduction 95

3.2 What is a 'route' anyway? 95
3.2.1 Types of routes: pathways and highways 96
3.2.2 Modes of representation: road-maps, distribution maps and culture areas 98

3.3 Modelling 'natural routes' with GIS 99
3.3.1 Impedance along vertices: 'network analysis' and routes 100
3.3.2 Friction over a grid: cost-surface analyses 101
3.3.3 Finding paths: least-cost-path analyses and corridors 104

3.4 Developing a GIS approach to routes 105
3.4.1 'Cost-of-passage', 'cost-surface' and 'cost-corridor' approaches 105
3.4.2 Theory: natural and anthropogenic cost factors 107
3.4.3 Praxis: realizing a working model of 'cost-of-passage' 112

3.5 Mapping material flows 113
3.5.1 Archaeotopograms: cost-of-passage to cost-surface 116
3.5.2 Some types of 'cost'-based archaeotopograms 117
3.5.3 Applying archaeotopograms 121

3.6 Summary: routes, landscape and travel density 122

4 Mapping Material Flows: Stone and Stone Objects 123

4.1 Introduction 123

4.2 Stone and its values 124

4.3 'Precious' stones 125
4.3.1 Lapis lazuli 125
4.3.2 Carnelian and etched carnelian beads 133
4.3.3 Colour, symbolism and visual arrest 136

4.4 Softstones: steatite, chlorite 137
4.4.1 'Intercultural style' vessels and 'weights' 137
4.4.2 Central Asian softstone objects: composite figurines and miniature columns 139

4.5 Chipped stone: obsidian 141
4.5.1 Obsidian use in eastern Anatolia and Transcaucasia 142
4.5.2 The lack of obsidian in central Asia 145

4.6 Weights: stone as metric mediator 146

4.7 Summary: routes and stones 151

5 Mapping Material Flows: Metals **153**
5.1 Introduction 153
5.2 Materials, geological sources and analyses of metal provenance 153
5.2.1 Social and ritual aspects of mining 155
5.2.2 Sources of copper 158
5.2.3 Sources of tin 162
5.2.4 Other metal sources: gold, silver, lead and iron 165
5.2.5 Metal circulation based on geological and archaeometallurgical analyses 167
5.3 Metal artefact types as evidence for metal flows 168
5.3.1 Metal assemblages in Transcaucasia/eastern Anatolia, 3000-1500BC 168
5.3.2 Metal assemblages in western Central Asia, 3000-1500BC 171
5.3.3 Patterns in metal assemblages 173
5.3.4 Metal artefact types 174
5.4 Regional metallurgical constellations 176
5.4.1 Chernykh's metallurgical 'foci' and 'provinces' 177
5.4.2 Internal mechanisms of metallurgical provinces? 180
5.4.3 Identifying trends in consumption: Chernykh's statistics on forms and materials 183
5.3.4 Contexts of deposition 190
5.5 Contexts of metal consumption 190
5.5.1 Metal flows and metal values in the ancient texts 190
5.5.2 Metal deposition: conspicuous sacrifice, identity and systems of valuation 194
5.5.3 Materiality of metals: liquidity, partibility and origins of exchange value 198
5.6 Metal flows traced through other materials 199
5.6.1 Weighing systems and metals? 200
5.6.2 Seals and sealing practices? 200
5.6.3 Metallschock: metal skeuomorphism 201
5.6.4 Metallschock in western Central Asia 209
5.6.5 Metallschock in Eastern Anatolia? 217
5.7 Summary: metal flows and cultural-economic trajectories 222

6 Mapping Material Flows: Textiles and Patterns **225**
6.1 Introduction 225
6.1.1 A definition of textiles and its relationship to similar materials 225
6.2 Textual evidence for textiles and dress 227
6.2.1 The Ebla archive (late 3rd millennium BC) 227
6.2.2 Neo-Sumerian/Ur III/Third Dynasty of Ur texts (c. 2100-2000BC) 228
6.2.3 Old Assyrian texts: Cappadocian/Kültepe tablets (c. 1950-1750BC) 230
6.2.4 The Mari archives (c. 1800-1750BC) 233
6.2.5 Textual evidence and its limitations 234

6.3 Direct textile evidence: fragments and impressions 235
6.3.1 Preservation, contexts and materials 235
6.3.2 Textile patterns and colour 239
6.3.3 Comments: textile remains and textile routes 243
6.4 Production of textiles: evidence for materials and technologies 244
6.4.1 Distribution of raw materials: fibres 244
6.4.2 Distribution of raw materials: dyes and colours 248
6.4.3 Spinning and spindle whorls 249
6.4.4 Loom types and loom-weights 253
6.4.5 Specialized tools and their distribution 256
6.4.6 Comments: raw materials, production tools and identities 256
6.5 Indirect evidence: depictions of dress and clothing 257
6.5.1 Figural representations of dress from Mesopotamia, Egypt and the Aegean 257
6.5.2 Figural representations of dress in Anatolia 260
6.5.3 Figural representations of dress in Central Asia 261
6.5.4 Flat figurines in Central Asia and Anatolia: civilization, dress and nakedness 262
6.5.5 Comments: clothing styles and movement 266
6.6 Indirect evidence: textile-related patterns in other media 267
6.6.1 Textile patterns and motifs in the Aegean, Egypt and Mesopotamia 269
6.6.2 Patterns and motifs in Transcaucasia/eastern Anatolia 274
6.6.3 Patterns and motifs in western Central Asia 277
6.6.4 Comments: patterns as stamps of personhood and individual identity 277
6.7 Indirect evidence: dress accessories 281
6.7.1 Pins and clothing styles 283
6.7.2 Beads, patterns and beaded dress 283
6.8 Summary: motivations and means of textile-based interaction 285

7 Discussion: Tying the Threads **289**
7.1 Introduction 289
7.2 Shifting material flows 289
7.2.1 3200-2900BC 289
7.2.2 2900-2600BC 292
7.2.3 2600-2300BC 295
7.2.4 2300-2000BC 299
7.2.6 2000-1700BC 301
7.2.7 1700-1400BC 304
7.2.8 Subsequent trends after 1400BC 306
7.3 Patterns and processes 308
7.3.1 Transcaucasia: the Kura-Arax phenomenon 309
7.3.2 Western Central Asia: 'urbanism' and the BMAC phenomenon 314
7.3.4 Linking eastern Anatolia/Transcaucasia and western Central Asia? 315
7.3.5 Systemic (im)balances: the centrality of metals in the Bronze Age? 317

7.3.6 Cross-craft indicators and the flow of metals and textiles 319
7.3.7 Documenting inequality? The mobility of value 322
7.3.8 Flows and boundaries 323
7.4 Summary of discussion 324

8 Conclusions **325**
8.1 Introduction 325
8.2 Reconstructing and re-visualizing routes 325
8.2.1 Route inertia and route dynamism 325
8.2.2 Visualizing threads: future directions for GIS and the study of ancient interactions 326
8.3 Direct and indirect evidence for interaction 328
8.3.1 Materiality, mobility and the performance of power and identity 328
8.3.2 Cross-craft interaction: 'epiaesthetic' skeuomorphism? 329
8.4 Cross-cultural flows and local development 330
8.5 Final note: whither a Bronze Age 'Silk Road'? 331

Bibliography **333**

Appendix A: Cost-of-surface models and Archaeotopograms **379**

Appendix B: Database of Sites **389**

Appendix C: Databases of Object Types **397**

Appendix D: Databases of Raw Material Sources **399**

Appendix E: Summary Statistics on Metal Trends **403**

List of Figures

1.1.	Map of south-west Asia showing the wider Near East and the broad location of the two case-study areas.	29
1.2.	Geographical and physical features of *Transcaucasia/eastern Anatolia* (base imagery: NASA's BlueMarble).	30
1.3.	Geographical and physical features of *western Central Asia* (base imagery: NASA's BlueMarble).	30
1.4.	The geographical and modern political landscape of Transcaucasia/eastern Anatolia.	34
1.5.	The geographical and modern political landscape of western Central Asia.	35
1.6.	The oscillations of the Caspian Sea.	36
1.7.	Unified chronological framework for Near East and adjacent regions, 3500-1000BC.	38
1.8.	Major archaeological sites of the 3rd and 2nd millennium BC in Transcaucasia and eastern Anatolia.	43
1.9.	Major archaeological sites of the 3rd and 2nd millennium BC in western Central Asia.	44
1.10.	A terracotta model of wagon with camel head from Altyn Depe.	51
1.11.	Drawing of a terracotta model of a cart from Arich.	52
1.12.	Terracotta models of carts and wheels from Altyn Depe.	53

2.1.	Roman roads in eastern Anatolia based on data from the Barrington Atlas.	66
2.2.	Distribution of caravanserais/(k)hans across the Near East (13th to 19th century AD).	68
2.3.	Caravanserais in eastern Anatolia and environs (13th to 19th century AD).	68
2.4.	Caravanserais in western Central Asia and environs (13th to 19th century AD).	68
2.5.	The location of the Pasinler valley near Erzurum, relative to 'eastern Anatolia'.	71
2.6.	The location of recognized linear hollow ways of the Near East, relative to 'eastern Anatolia'.	73
2.7.	Major roads in eastern Anatolia according to British Naval Intelligence in 1942.	75
2.8.	Major roads in eastern Iran according to British Naval Intelligence in 1945.	75
2.9.	Routes of seasonal migration for pasture according to British Naval Intelligence in 1942.	77

2.10.	The Persian Royal Road.	79
2.11.	Ottoman military campaigns in Eastern Anatolia during the 16th century.	81
2.12.	Alexander the Great's campaign from the Aegean to central Asia.	81
2.13.	The travels of Ibn Battuta.	83
2.14.	The supposed travels of Marco Polo.	83
2.15.	Pilgrimage to Mecca.	84
2.16.	Trade routes and settlements of the Old Assyrian Period in Central Anatolia.	86
2.17.	Combination of all reconstructed routes and itineraries in eastern Anatolia.	90
2.18.	Combination of all reconstructed routes and itineraries in western Central Asia.	90
2.19.	The evolution of route systems in eastern Anatolia.	91
3.1.	'Cost-of-passage' raster – Model 2.	114
3.2.	'Cost-of-passage' raster – Model 1.	115
3.3.	'Cost-of-passage' raster – Model 3.	115
3.4.	Relative distance (site accessibility) from Shengavit (Armenia), based on an archaeotopogram 'type A1'.	118
3.5.	Relative distance (site accessibility) from Gonur Depe (Turkmenistan), based on an archaeotopogram 'type A1'.	118
3.6.	Relative distance (site accessibility) from Ur (Iraq), based on an archaeotopogram 'type A1'.	119
3.7.	Relative distance (site accessibility) from Harappa (Pakistan), based on an archaeotopogram 'type A1'.	119
3.8.	Relative distance (site accessibility) from Arslantepe (Turkey) based on an archaeotopogram 'type A1'.	120
4.1.	Selection of objects dating to the 3rd millennium BC made from lapis lazuli.	127
4.2.	Distribution of known lapis lazuli objects and regions of intense consumption in relation to the material's sources.	128
4.3.	Possible alternative distribution routes for lapis lazuli during the 3rd and 2nd millennium BC.	130
4.4.	The least-cost-path from Ur to the source of lapis lazuli at Badakhshan.	132
4.5.	Selection of (orange-red) carnelian objects.	134

4.6.	Distribution of etched carnelian beads with a representation of the relative distances from carnelian sources.	135
4.7.	A range of 'inter-cultural' style vessels and related objects.	138
4.8.	Distribution of 'inter-cultural' style vessels of known provenance.	139
4.9.	Distribution of some major obsidian sources across Near East with broad zones.	142
4.10.	Obsidian bowl from the grave of Queen Puabi, Royal Cemetery of Ur.	143
4.11.	Examples of obsidian artefacts from 3rd millennium levels at Sos Höyük.	143
4.12.	Selection of objects confirmed as weights in recognizable weighing systems.	147
4.13.	Distribution of different weighing systems.	148
4.14.	Summary of distribution data on lapis lazuli, carnelian, 'intercultural-style' objects and weighing systems for the 3rd millennium BC.	150
5.1.	Sources of copper ore, including Ergani, Veshnoveh, Omani, Cypriot and Rajasthani regions.	157
5.2.	Archaeotopogram 'type A2' showing relative distance from copper ore sources around Eastern Anatolia.	159
5.3.	Archaeotopogram 'type A2' showing relative distance from copper ore sources around western Central Asia.	159
5.4.	Sources of tin, including Deh Hosein, Kestel and Karnab regions.	161
5.5.	Archaeotopogram 'type A2' showing relative distance from tin sources in and around greater Anatolia.	163
5.6.	Archaeotopogram 'type A2' showing relative distance from tin sources in and around Central Asia.	163
5.7.	Regions with sources of gold in Iran and Afghanistan.	166
5.8.	Regions with deposits of silver in Turkey and Iran.	166
5.9.	Regions with sources of iron in Turkey and Iran.	167
5.10.	Summary of the broad direction of circulation of metals, in 3rd and 2nd millennium BC.	169
5.11.	Prediction for centres of early tin-bronze experimentation showing sum of relative distance from copper and tin sources.	169
5.12.	Examples of the contents of metal assemblages of eastern Anatolian sites from the 3rd millennium BC.	170
5.13.	Examples of the contents of metal assemblages of eastern Anatolian sites of the very late 3rd or early 2nd millennium BC.	171
5.14.	Examples of the contents of metal assemblages of western Central Asian sites of the early to mid 3rd millennium BC.	172

5.15.	Examples of the contents of metal assemblages of western Central Asian sites of the late 3rd and early 2nd millennium BC.	173
5.16.	An example of a 'Bactrian axe'.	174
5.17.	Distribution of 'BMAC' objects outside Bactria and Margiana heartland, including 'Bactrian axes'.	174
5.18.	Examples and distribution of figure-decorated metal vessels, c. 2250-1750BC.	176
5.19.	Metallurgical Provinces, 'Chalcolithic period', c. 5500-3500BC.	178
5.20.	Metallurgical Provinces, 'Early Bronze period', c. 3500-2500BC.	178
5.21.	Metallurgical Provinces, 'Middle Bronze period', c. 2500-1750BC.	179
5.22.	Metallurgical Provinces, 'Late Bronze period', c. 1750-1250BC.	179
5.23.	Chart showing relative proportion of items classified 'tools' vs. 'ornaments' in Chernykh's database.	182
5.24.	Relative proportion of copper, copper-alloys, gold, silver and lead objects during 'Chalcolithic period' from Chernykh's database.	184
5.25.	Relative proportion of copper, copper-alloys, gold, silver and lead objects during 'Early Bronze period'.	184
5.26.	Relative proportion of copper, copper-alloys, gold, silver and lead objects during 'Middle Bronze period'.	186
5.27.	Relative proportion of pure-, arsenic- or tin- alloys used in copper objects during 'Chalcolithic period'.	186
5.28.	Relative proportion of pure-, arsenic- or tin- alloys used in copper objects during 'Early Bronze period'.	187
5.29.	Relative proportion of pure-, arsenic- or tin- alloys used in copper objects during 'Middle Bronze period'.	187
5.30.	Relative proportion of pure-, arsenic- or tin- alloys used in copper objects during 'Late Bronze period'.	188
5.31.	Metal objects classified by context of deposition, 'Early Bronze period'.	188
5.32.	Metal objects classified by context of deposition, 'Middle Bronze period'.	191
5.33.	The directionality of metal exchange according to Syro-Mesopotamian texts.	191
5.34.	Conspicuous sacrifice: metal hoarding zones.	192
5.35.	Hoarding, metal burial, 'Royal' cemeteries or graves in 3rd and 2nd millennia BC.	192
5.36.	Major conspicuous deployment of metals in funerary contexts in Caucasia and Anatolia and location of major copper sources.	197
5.37.	Regional adoption of sealing practices (cylinder/stamp).	197
5.38.	Examples of distinctive/defining forms of the 'Anatolian Trade Network'.	201

5.39. Distribution of 'depas amphikypellon' drinking vessels and apparent limits of circulation. 203

5.40. Distribution of 'Syrian flask' bottles and apparent limits of circulation. 203

5.41. Distribution of '*Metallische Ware*'. 205

5.42. Distribution of 'Metallic ware'. 205

5.43. Namazga pottery sequence schematic. 210

5.44. Examples of Namazga V ceramics from Altyn Depe. 211

5.45. Distribution of 'Namazga V' style pottery in southern Turkmenia and north east Iran. 213

5.46. Metallic vessels from the 'Royal Necropolis' of Gonur Depe, Turkmenistan. 213

5.47. Hypothetical geographical role of BMAC communities in the circulation of tin in late 3rd and early 2nd millennia BC. 215

5.48. Distribution of Hissar IIB-type burnished grey wares ('Eastern Grey Wares') and relationship with Namazga III and IV, and Kura-Arax pottery assemblages. 215

5.49. Examples of Kura-Arax/Early Transcaucasian Culture (ETC)/Khirbet-Kerak ceramics. 217

5.50. Distribution of Kura-Arax related wares. 219

5.51. The relationship between Kura-Arax communities and the metals trade between Anatolia, the Caucasus and north Syria, including Tell Mozan. 220

5.52. Summary of distribution data on metals. 223

6.1. Some identifiable places named in the Ebla archives that are associated with the distribution of different styles of textiles. 228

6.2. Maximum extent of Ur III political control, with sites that have yielded substantial Ur III textual archives. 228

6.3. Places or place names mentioned in the Kültepe tablets. 233

6.4. Kültepe textile types and their geographic origins or associations. 233

6.5. Places named in the Mari archives. 233

6.6a. Distribution of published textile fragments, impressions and threads dated from the 4th to the 2nd millennium BC. 236

6.6b. Broad earliest dates for published textile fragments, impressions and threads. 237

6.7. Examples of textile remains. 238

6.8. Distribution of plain and twill weaves 4th-2nd millennium BC. 240

6.9. Location of sites yielding textile fragments with identifiable distinctive patterning. 240

6.10.	Reproduction and highlighted enlargement of 'carpet'-like impression found in the burial pit of Kurgan 9 at Tri Brata near Elista.	242
6.11.	Detail of the patterned edge of the so-called 'Syrian-tunic' found among the grave offerings of Tutankhamun's tomb.	242
6.12.	The origins of flax and the extent of its cultivation by the 3rd millennium BC.	244
6.13.	Summary showing the likely geographical variations in the use of different textile fibres during the 3rd millennium BC.	244
6.14.	Regions likely to have been producing murex-based dyes during 3rd and 2nd millennium BC.	251
6.15.	Distribution of textiles whose spin-direction (Z- or S-) has been reported.	251
6.16.	Distribution of 'hollow-shaped' spindle whorls with approximate earliest possible date.	252
6.17.	Distribution of different loom-types during 3rd millennium BC and after.	252
6.18.	Distribution of crescent-shaped loom weights of the early 2nd millennium BC.	253
6.19.	Depiction of twill fabrics (as 'herringbone pattern') on 'Anatolian-style' seal impressions from Old Assyrian Kültepe.	254
6.20.	Khlopin's 'pile-carpet' knife.	257
6.21.	Clothing styles in Mesopotamia.	258
6.22.	'Foreigners' in Egyptian depictions wall-painting from a tomb at Beni Hassan, Egypt.	258
6.23.	Silver figurine with gold head from Hasanoğlu, in north-central Anatolia.	259
6.24.	Depictions of various textile fabrics and clothing in seal impressions, from Kültepe, showing a wide variety of fabric and styles.	259
6.25.	Depictions of the Central Asian/Iranian kaunakes dress.	263
6.26.	Examples of Bronze Age 'flat figurines' of variant dates.	263
6.27.	A selection of flat-figurine types from the 3rd and early 2nd millennia BC.	265
6.28.	Detail of lead figurine, one of a cache uncovered at Tell al-Judeidah.	266
6.29.	Geographical distribution of classes of evidence indicating clothing during the 3rd millennium BC.	268
6.30.	Geographical distribution of classes of evidence indicating clothing during the 2nd millennium BC.	268
6.31.	Painted patterns from the Aegean: Late Minoan I examples on ceramics.	269
6.32.	Painted patterns from wall-paintings uncovered in Cour 106 of Zimri Lim's palace at Mari.	270
6.33.	Painted ceiling of Egyptian Tomb of Hepzefa at Assiut.	270
6.34.	Painted patterns from Nuzi on wall-paintings and ceramics.	271

6.35.	Patterned door-threshold in frit from Mari that may have been based on a rug design.	272
6.36.	Distribution of emerging painted pottery traditions in late 3rd and early 2nd millennium BC, including 'Transcaucasian Painted Wares'.	273
6.37.	The schematic figural wall-paintings from Uruk period Arslantepe, with strong resemblance to patterns produced by textile designs.	275
6.38.	Examples of Aras, Yayla or 'Transcaucasian Painted Wares'-style painted pottery.	275
6.39.	Transcaucasian Painted Wares cultural-geographic zone.	278
6.40.	Wall-decoration from Ulug Depe (formerly Yassi Depe) with patterns reminiscent of contemporary Namazga I pottery designs.	279
6.41.	Examples of Namazga IV-style painted pottery.	279
6.42.	Distribution of Namazga IV-style painted ceramics in Turkmenistan and north-east Iran plus Namazga IV cultural-geographic zone.	280
6.43.	Central Asian seals.	280
6.44.	Concentric cross-pattern in various media.	281
6.45.	Distribution of Huot's 'type 1' double spiral-headed pins.	283
6.46.	Distribution of all double spiral-headed pins.	283
6.47.	Distribution of flat circular beads with tubes.	285
6.48.	Distribution of quadruple-spiral beads.	285
6.49.	An example of a 'beaded dress', found at Qau.	286
6.50.	Beads used in complex beaded 'collars' of the 18th dynasty.	286
6.51.	The remains of a fragment of beaded cloth with gold thread found at Acemhöyük.	286
6.52.	Summary of distribution data and of the directionality of textile flows.	287

7.1.	Summary of flows of stones, metals and textiles 3200-2900BC.	290
7.2.	Summary of flows of stones, metals and textiles 2900-2600BC.	293
7.3.	Summary of flows of stones, metals and textiles 2600-2300BC.	296
7.4.	Summary of flows of stones, metals and textiles 2300-2000BC.	300
7.5.	Summary of flows of stones, metals and textiles 2000-1700BC.	302
7.6.	Summary of flows of stones, metals and textiles 1700-1400BC.	305
7.7.	Summary of in flows of stones, metals and textiles 1400-1100BC.	307

7.8. The relationship between Kura-Arax assemblages (at their greatest extent) and the accessibility to copper sources known to modern geology. 312

7.9. The relationship between BMAC/Namazga VI-related material culture, the central BMAC zone and areas of high accessibility to tin sources. 313

B.1. All sites of eastern Anatolia in the database. 391

B.2. All sites of western Central Asia in the database. 392

B.3. Graphical representation of the 'confidence' value for each site in the database. *395*

List of Tables

1.1. The terminological mismatch between Transcaucasia/eastern Anatolian and western Central Asian 'Bronze Ages'. *38*

1.2. Broad chronological periods in western versus eastern Anatolia/Transcaucasia. *43*

1.3. Broad chronological periods in Central Asia (following Kohl 1984; Hiebert 1994a). *45*

3.1. Accumulative cost surface programs (basic/isotropic). *102*

3.2. Accumulative cost surface programs (complex/anisotropic). *102*

3.3. Least-cost-path programs. *103*

3.4. Least-cost corridors programs. *103*

4.1. Table showing common multiples allowing inter-conversion of 3rd millennium weighing systems. *148*

Acknowledgements

This book represents the fruits of research I undertook as doctoral student at the Department of Archaeology of the University of Sheffield, UK. The text is an edited and updated version of the resulting thesis completed by August 2012 and accepted December 2012. The research for that thesis was supported financially by a studentship from the University of Sheffield, with additional grants for extended travel and study in Ankara and Istanbul from the British Institute at Ankara and Koç University's Research Center for Anatolian Civilizations. The editing for this book was completed during a postdoctoral fellowship based at the department of archaeology of İstanbul Üniversitesi with the kind academic support of Prof. Dr. Şevket Dönmez and through financial sponsorship from TÜBITAK, the Turkish Academy of Sciences (BİDEB-2216). I am very grateful to all these institutions and their supporting staff.

Various individuals and institutions have given their permission for the reproduction of images appearing in this book. I would like to thank both individuals: Thomas Zimmermann (5.38c, 5.4), Agnete Wisti Lassen (6.18), Jason Ur (2.6), Elena Tsareva (6.7a), Hakop Simonyan (6.28b), Maria Shaw (6.33), Vasıf Şahoğlu (5.38a, b, e, d; 5.39 inset), Viktor Sarianidi (4.7f, 4.12, 5.15a, b, c, d, h, 5.16e, 5.18b, 5.46, 6.25b, c, d, 6.43b), Antonio Sagona (4.11, 5.49b), Lorenz Rahmstorf (4.12a, b), Marina Puturudze (5.13b), Aynur Özfırat (6.38), Kai Kaniuth (5.15g, 5.15f), MaryFran Heinsch (6.7b), Fredrik T. Hiebert (6.40), Marcella Frangipane (6.37), Stephanie Dalley (6.35), Evgenij N. Chernykh (5.12b, c, 5.13a, 5.15e), Charles Burney (5.49a, 5.50b); and institutions: the University of Pennsylvania Museum for Anthropology and Archaeology (4.1b, c, 4.5a), State Museum of History of Uzbekistan, Tashkent (4.7c), Staatliche Museen zu Berlin (6.50a), Russian Academy of Sciences (1.10, 1.12, 5.12a, 5.14 b, c, d, 5.18e, 5.44, 5.49c, 6.20 inset, 6.26c, d, 6.27[6, 7], 6.41, 6.43a), the Petrie Museum (6.49, 6.50b, d), the Oriental Institute, Chicago (5.49d, 6.27[4], 6.28a, b), Museum of Fine Arts Boston (4.12c), the Metropolitan Museum of Art (4.12f, 6.22), Instituto Nacional de Antropología e Historia, Mexico/Ebrahim Khadem Bayat (6.25a), Institut Français d'Archéologie Orientale, Le Caire (6.50c), the British Museum (4.10, 5.16, 6.50e) and Antiquity journal (5.43). A number of other individuals were very helpful either in providing advice or facilitating these permissions, to them I am also very grateful: Otabek Aripdjanov, Elizabeth Barber, Joshua Donovan, Nadezhda Dubova, Burçak Delikan, Aftandil Erkanov, Fokke Gerritsen, Alejandra Gómez Colorado, Philip Kohl, Nevine Kamal, Hajime Inagaki, Aleksandr Musin, Carolien van Zoest. Further information about these images and rights appears in the caption of the relevant figure. The creation of a large proportion of the maps presented in this book were dependent on publically-available data too frequent to mention repeatedly, including: NASA's Shuttle Radar Topography Mission (for elevation data); BlueMarble (for coloured base maps) and HydroSHEDS (for rivers and watersheds). I hereby acknowledge and thank this organization and its staff for their tremendous scientific achievements and generous policies toward public data sharing.

Fortunately on this particular research 'journey' I have rarely travelled alone. I am extremely grateful to those individuals with whom I have had the chance to discuss ideas in person or by correspondence; those who have offered their published and unpublished works and data freely and openly; those who have helped me gain access to libraries and material which I would not otherwise have done; those who helped facilitate my trips and extended stay in Turkmenistan and Turkey; and those who have read and commented on various drafts and lecture presentations of mine on the topics contained here. Among those who have helped me – each in different ways – I would like to thank most heartily: in Sheffield, Kate Lantzas, Tudur Davies, Clare Burke, Angeliki Karagianni, Vasiliki Tzevelekidi, Sarah Viner, Paul Halstead, Deborah Harlan, Rocio Fernandez Mendez, Nicky de Battista, Tim Godden, Kirsten Marsh, Müge Ergün; at UCL and in Turkmenistan, Steve Markofsky, Tim Williams, Paul Wordsworth, Gaigysyz Joraev, Barbara Cesaretti; from Berlin, Regina Attula, Sieglinde Slawisch; from Paris, Pierre Le Bigre; from Hampshire, Ian Hylands, Shaun Britton, Shirley Wilkinson, Pete Wilkinson, Rory Wilkinson; in Ankara, Ben Claasz Coocksen, Geoffrey Summers, Lutgarde Vandeput, Yaprak Eran, Gülgün Girdevan and Yiğit Erbil; in Istanbul, Giulio Palumbi, David Meiggs, Marie-Claude Boileau, Duygu Kızılaslan Paçalı, Gülten Yıldız, Gudrun Walter, Aslıhan Yener, Scott Redford; across the internet, Agnete Wisti Lassen, T. Matthew Ciolek, Aynur Özfırat; and finally, but certainly not least, Claudia Glatz, Jonathan Harle and Michele Massa. Thanks must also go to my examiner, Cameron Petrie, who suggested various improvements to the work and encouragement to question further.

Throughout the course of the research I have had constant encouragement and inspiration – both academic and personal – from my supervisor, Sue Sherratt, and my advisor, John Bennet, to whom I am extremely grateful. The presence of Andrew Sherratt was and is sorely missed as mentor and friend, but his voice continues to guide my research directions and interpretations. Finally Anja, Matilda and Alfred deserve thanks for many, many things, but particularly for their patience and forbearing in the final stages of writing when it would have been much more fun doing other things... This book is dedicated to my parents, Peter and Shirley Wilkinson who encouraged my interests in geography and history from an early age, and to whom I owe many debts – most of them the non-financial and unrepayable kind.

TCW, Istanbul, January 2014.

Chapter 1

Contexts and Frameworks of Research

We travel not for trafficking alone;
By hotter winds our fiery hearts are fanned:
For lust of knowing what should not be known,
We take the Golden Road to Samarkand
James Elroy Flecker, *Hassan*

This passage from the early twentieth century poet and playwright James Elroy Flecker sets a stage for this book, not only in the tone of mystery, excitement and romanticism from which point the pursuit of archaeological research must start but also in the specific content as it evokes the cultural motivations which drive the creation of routes, roads and long-distance trade. The origins of this book lie in a fascination with the idea represented by the term *die Seidenstraße* (the 'Silk Road'), termed thus by von Richthofen in a book on the history and geography of eastern China in the late 19th century. Though the meaning may have changed since his coinage and the geographical extent of the region with which it is associated has enlarged substantially (such that its analytical value is questionable), the phrase has gained currency as a shorthand way to indicate the contribution of ancient trade routes to the complexity, hybridity and fluidity of cultures across Eurasia during the 1st and 2nd millennia AD. In 2007 I set out to explore and map the possibility of earlier prehistoric precursors to the 'historical' silk roads to assess the antiquity of trans-regional and trans-continental cultural interconnections. Along the road it was, of course, necessary to temper romanticism with cool empiricism. But given that the available evidence is far from complete, it became very clear that our reconstructions of the ancient world must rely as much on imaginative and comparative perspectives as on raw textual, archaeological or environmental data, weaving fragmentary threads of data into a never-finished tapestry. In the following introductory chapter I set out the background to the research presented here including theoretical and methodological perspectives and the archaeological, environmental and technological contexts of the specific case-studies examined in this book.

1.1 Theoretical frameworks: trade and economy; networks and routes

Any attempt to make sense of the 'grand narrative' of human development, in Eurasia and Western Asia in particular, cannot help but face the deep interconnectivity of the ancient world. What Andrew Sherratt (2006) termed the 'Trans-Eurasian exchange' between west and east Eurasia involved the movement of materials, traditions and technologies (and presumably genes) over gigantic distances and may have begun even before the origins of farming. The overall trend of human development in the Old World over the last 8000 years, it seems generally agreed, has been one of an increasing intensity of exchange and of 'economic' interdependence through time, with material evidence pointing to ever more distant connections. The work represented in this book emerges from an interest in trying to understand, at the macro-scale, the development and consequences of these long-distance interconnections across Eurasia, primarily through the material record. As such, it attempts to address questions that attend to what has often in archaeological studies been lumped under the theoretical categories of 'trade', 'import/export' and 'trade routes'. These terms are, as has been recognized for some time (*e.g.* Appadurai 1986), clunky and anachronistic simplifications of the full process of human interaction in the past, indeed as they are for the present. They rely on an assumption that 'international trade' exists as an external relation to the village, tribe, city or state, as though there are clear boundaries between one such 'cultural entity' and another. This perspective must be a result, in part, of the origin of our standard economic outlook and theories, most of which derive precisely from that formative period of the 18th and 19th centuries in which the nation state was just emerging and where the essentialist project of national boundaries (geographic, ethnic and social) was being inscribed in intellectual, technical and administrative structures. In archaeology, this epistemological view of trade 'as external' has been carried through into its most influential theoretical paradigms: 'processual' archaeology placed trade as an external input to otherwise self-contained, geographically-constrained, locally-inspired cultures; whilst 'post-processual' archaeologies have tended to demote the study of long-distance trade for different reasons, mostly to focus instead on localized 'meanings' of material culture (including those which result from 'trade' and exchange) albeit using different models of human-material relationships.

Suspicion of trade and of merchants has, in any case, a long heritage in European scholarly discourse: the Aristotelian critique and demotion of commercial life ('chrematistiki') as the antithesis of 'the good life' and the morality of self-sufficiency ('oikonomia') is widely-known (*e.g.* Finley 1970, 15; Booth 1994), and its kernel has been diffracted through various romantic movements against commerce and capital (including of course moral or political movements such as communism). This view – placing trade as external to (or corrupting in) the development of communities and cultural life – has the advantage of being one which is also often taken by small-scale communities typical of much of the human past. Traders and travellers are frequently treated differently to other community members: sometimes with respect, and sometimes with fear (compare, for example, attitudes to city traders versus smugglers, and the different value assigned to their role in different social contexts and social climates). But the uncritical adoption in archaeological theory of such a defensive perspective toward trade fails to engage sociologically with the cause of such community views: traders, travellers and migrants are treated with special suspicion or respect, precisely because of the power their culturally liminal position. Their role in shifting and exchanging materials and the necessarily hybrid identities they must adopt to fulfil this role provides opportunities for 'magical transformation' of things, people and their relationships. Put in terms of 'transformations', we can compare traders with other community specialists and their treatment: ethnographic evidence shows blacksmiths, for example, are frequently relegated to the margin of a community (see *e.g.* McNaughton 1988), since their unique powers to fashion iron (the 'magical transformation' of one material into another) engender many forms of social danger. Traders transform a local material into another exotic material through social engagement and negotiation with distant groups – a special form of 'magic' – and hence generate socio-economic value. In contrast to the 'trade as external' thesis, I would argue that the trader or traveller's engagement with other peoples, with changing environments, and changing cultural milieu, lies precisely at the source of many forms of cultural, economic, and linguistic change. Not only do they bring 'foreign' objects whose manufacture or source may appear mysterious, but they bring new stories, experiences and behaviours that may differ radically from the vertically-inherited traditions of those individuals whose entire life is spent in one small area. It would also be a mistake to separate too strongly the social from the biological: travellers are carriers of viruses and genes, of revolutions and religions. As such, this book starts from the assumption that whilst there are many important 'local' processes of interaction (between individuals, between humans and animals/plants or the environment or between local corporate groups), systemic long-distance interaction (including external 'trade') between groups was usually at the source of most kinds of cultural innovation and transformation.

Prehistorians who have been influenced by I. Wallerstein's work on the origins and of what he called the 'modern world-system' (Wallerstein 1974) or more general Marxian explanations of – to quote A. G. Frank's compelling phrase – 'the development of underdevelopment' (Frank 1970), are those who have most often been prepared to study long-distance interaction. As many who have applied a 'world-systems' or 'centre-periphery' perspective admit, however (Rowlands, Larsen and Kristiansen 1987; Chase-Dunn and Hall 1997; A. Sherratt 1993; Stein 1999; Kohl 2011), and as Sue Sherratt has recently forcefully re-iterated (S. Sherratt 2010), a successful archaeological application should probably bear little semblance to the original formulation of 'World-Systems' as described by Wallerstein. Indeed if a label is necessary at all, and if 'world-systems' did not have such easy recognizability, it would be more useful to re-name such approaches, perhaps 'structural-interactionist' or 'systemic interactionist' (see the variety of contributions in Wilkinson, Sherratt and Bennet 2011). What all of these approaches have in come is the assumption that 'trade' and other forms of interaction were not epiphenomenal to cultural systems but central to how and why they were created: proponents attempt to explain synchronic cultural change in, and long-term economic trajectories of, different regions through the characterization of *systemic* and *structural* socio-economic relationships between regions or cultural zones. Over the *longue durée,* from the Palaeolithic to the present, a world-systems perspective allows us to recognize the general trend towards an ever increasing complexity of exchange networks, a concomitant extension of their reach and an increasing level of economic inter-dependency and regional specialization: such that goods, people and ideas travel further, and similar types of material culture and practices can be found over ever larger distances (Frank and Gills 1993). Today we might label the continuation of this process one of 'globalization' (though care should be taken with such a slippery modern catch-word). The smaller-scale origins of systemic interaction between regions with different resource bases may well be found in the cultural configurations of Near East as early as the 7th or 6th millennia BC, documented by the distribution of materials such as obsidian (see *e.g.* Healey 2007).

Such 'world-systems' approaches have tended to depend to a large degree on 'zonal' models of economy (and associated culture) between – at the risk of an unfair characterization – an expansionist and dynamic core and a shrinking and static periphery. This echoes the still-current if oft-critiqued macro-economic characterization of the modern world into two-tiers: 'first world' and 'third world'[1]. The transition to a terminology of 'developed' and 'developing' countries respectively, as part of the postcolonial critique of macro-economics, poorly masks the continued normalizing assumptions of some parts of 'development economics'[2]. Likewise efforts to rehabilitate the passive periphery of archaeological accounts into an active partner and source of innovations (G. Stein 2002), stemming from the same noble urge, or post-colonial guilt, to re-empower disenfranchised peoples – this time of the distant past – risks the same re-inscription of 'colonial'-like boundaries. The danger of such zonal models is that they depend on the simplification of an extremely complex geographic scene, material record and presumed social reality. The intertwined tentacles of alternative and competing economic systems – which may overlap and crisscross each other geographically – are simply ignored. This idea of alternative parallel economies is something we will return to in a discussion of metal consumption (Chapter 5).

It is worth making a brief exegesis on the word 'economy' and the way in which it will be used in this book. As with 'trade', I prefer to view 'economy' not as a discrete or separate activity, but as an integral part of the movement of objects and obligations between individuals and groups (including, of course, animals and plants) in space and time. An 'economic system' could be productively characterized as simply an emergent pattern of behaviour revealing values attributed to materials and services with rules (either codified or unwritten) that dictate the appropriate contexts for, and morality of, exchanges (see Bloch and Parry 1989). If this seems a somewhat abstract definition, what it means in practice is that all human interaction is in some sense 'economic' even if the rules of interaction are very different from those we consider within the 'economic sphere' in the modern world. In this sense, 'economic' structures and practices cannot be de-linked from, say, ostensibly religious structures or practices: a sacrifice to the gods is at once both 'ritual/religious' and 'economic'. The parties affected by such transactions of course include the giver and receiver (in this case, for example, priest and god), but also those not 'directly' involved. Those who are witness to the event (either directly or through third-party reports) or those who are unable to make a sacrifice (whether by poverty or boundaries of class or caste) are also in some sense 'transformed' in relative status terms.

As a system of value, 'economy' also cannot be decoupled from 'aesthetics'. Since Kant, the meaning of the word 'aesthetics' has tended to diverge from the study of the senses toward the study of 'art' as sublime beauty, which cannot be valued in economic, or at least monetary, terms[3]. The role of human senses (true 'aesthetics') in the construction of value systems becomes obvious, however, when we consider the origin of acts of consumption: most valued commodities have a direct impact on the senses of the consumer, whether that is on the taste-buds (*e.g.* in the case of chocolate, spices or coffee), or the eye (*e.g.* paintings, statues,

1 Curiously, following the collapse of the USSR and its satellites, the 'second world' and redistributive economic models have now mostly been forgotten and only occasionally invoked to try to explain, for example, Mesopotamian temple-economies.

2 There is no guarantee that the future trajectories of modern 'underdeveloped' economies will converge towards a free-market social-democratic European ideal: or that this is seen as universally desirable in such states.

3 For a critique of this Western/Kantian 'anaesthetic' approach to art, material culture and the human culture from the recent field of 'anthropology of the senses', see *e.g.* Howes (2003).

clothing). Even those modern commodities (such as crude oil) whose direct connection to the human body is tenuous, can be argued to have an ultimately sensuous aim: holiday-makers from northern Europe spending carbon on flying to sunnier climes are searching for the feel of sun, sand and sea on the skin – and 'economic migrants' travelling in the opposite direction seek the culturally-idealized bodily comfort of homes they cannot experience in their place of origin. In an archaeological context, most sensuous experiences are hard to reconstruct, especially where many of the original commodities were perishable. Often only the structure and effects of 'visual economies' (*i.e.* what is traditionally termed visual arts) are accessible. Nonetheless, this connection, between 'aesthetic' and 'economic' values should be re-iterated, since, as I will try to argue through the examples of material interaction recorded in the archaeological record, it is the sensual nature of materials (and the social aims to which these sensuous effects were directed), which drove their consumption and the incentive to exchange with particular groups in particular geographical locales, and hence transformed the 'cultures' and 'economies' of participant groups over the long-term. This perspective informs many parts of this book, including a sustained concern with the significance of colour 'symbolism' and aesthetic intermediality or what is more commonly called 'cross-craft interaction'.

One point the reader should note from this exegesis is that our terminological tools (words like 'economy', 'religion', 'politics' *etc.*) are woefully insufficient to model the complexity and variety of human experience. It is worth, therefore, making reference to an emerging set of theoretical perspectives that are precisely concerned with cultural interconnectivity, namely 'network-based' approaches. Perhaps the most sustained theoretical formulation of this approach in archaeology is that outlined recently by Carl Knappett (2011), and applied predominantly to the Aegean in the 2nd millennium BC (*e.g.* Knappett, Evans and Rivers 2011), though others have adopted similar methods and theoretical language. Such approaches are predominantly sourced from works from outside archaeology: these include Bruno Latour's 'actor network theory/ANT' (Latour 2005), Alfred Gell's agency of art (Gell 1998), and Marilyn Strathern's re-conceptualization of personhood (Strathern 1990). These works emphasize the network as a way to model complex interrelations between multiple agents, objects, landscapes, technologies and behaviours. In its favour, an agent-led 'network-approach' allows the assumed differences between human bodies, material culture and environments to become usefully ambiguous in a web of causality: humans transform objects and environments, objects and environments transform people. But there are dangers in this ambiguity if the specificity of media (what is sometimes called 'materiality') is ignored or the empirical evidence is bypassed, as explanations may become repetitive and non-specific. The rise of 'network' theory in academic discourse is probably not coincidental with the rise of digital 'networks' both in the physical sense (*i.e.* the internet, mobile phones) and for specific social applications (*i.e.* 'social networks' like Facebook, Twitter *etc.*). The textile origin of the metaphor, 'net-work', is also rarely acknowledged but the fact that a term for a concrete material object is required to represent such an ultimately abstract idea about interconnections is perhaps unsurprising. The term clearly offers a powerful metaphor for the current era – but one whose basis remains obscure. As a modern trope we should therefore treat it with caution to avoid anachronism.

These caveats aside, the term 'network' when applied to an archaeological context remains useful because it allows us to model groups of human communities who were connected to each other through sustained exchange and interaction. The advantage of the term is that it can subsume both those phenomena traditionally labelled long-distance or localized 'trade and exchange' and also 'networks' of movement behind phenomena that are discussed in more ambiguous and sometimes problematic terms such as 'diffusion', 'influence', 'cultural interaction', 'culture areas', and 'migration'. Such interaction acts at many different levels – cultural and commercial, physical and ideological – and our modern categorization of the interaction can sometimes hinder as well as help our understanding of past human societies. It is this multi-scalar aspect of 'networks' that I try to pursue in this book: a loosely-defined network-approach enables us to re-engage with unfashionable macro-concepts such as 'cultures' and 'diffusion' in a more dynamic way, mapping linkages or flows of materials (or ideas or people) more realistically, without necessarily presupposing the procedures of exchange. In an archaeological context we must accept that the low resolution and incomplete nature of the archaeological record mean that we cannot directly perceive nor fully reconstruct the 'networks' in which we are interested. What we *can* see is the accumulated material debris that was transported by these networks, whose location gives us clues (of varying detail and intensity) as to their geographical location and temporal transformations. The substantive parts of this book are devoted to exactly this: identifying examples of accumulated material debris ('distributions') which may help us to reveal mechanisms of interaction and movement in the past.

'Network'-based approaches deal with multiple *linkages* between entities. 'Systemic-interactionist' or 'world-systems' approaches deal with unequal relations between economic *zones*. What has the potential to unite both approaches is geography and space. Only the barest threads of past exchange networks may be still reconstructable, through the often accidental deposition of particular objects. But the flow of exchange, the movement that it requires, is always spatially constrained by particular landscapes both natural and human. It is for this reason that the study of 'routes' is important to understanding cultural evolution through time on both a micro- and macro-scale. The alternative meanings of 'routes', and a review of routes as they have been studied in the past will be discussed further in the following chapters (Chapter 2), but it is worth stating here the fundamental definition of routes which the content of this book aims to address: namely as both 'geographical corridors' (*i.e.* routes as *possible* corridors for travel) and 'flows of objects' (*i.e.* routes as hierarchical or differential *flows* as they developed and changed in particular times and spaces). An analogy might be between a valley (the route as corridor) and a river (the route as flow): the two are clearly interrelated, but to talk about a river only in terms of valley dimensions tells only a partial story. This definition will help explain, I hope, the sustained emphasis in Chapters 4 to 6 on material culture and its economic, aesthetic and ritual values in geographic context as essential to the understanding of 'routes' and of the economic trajectories of Old World societies – and not purely on cataloguing physical corridors.

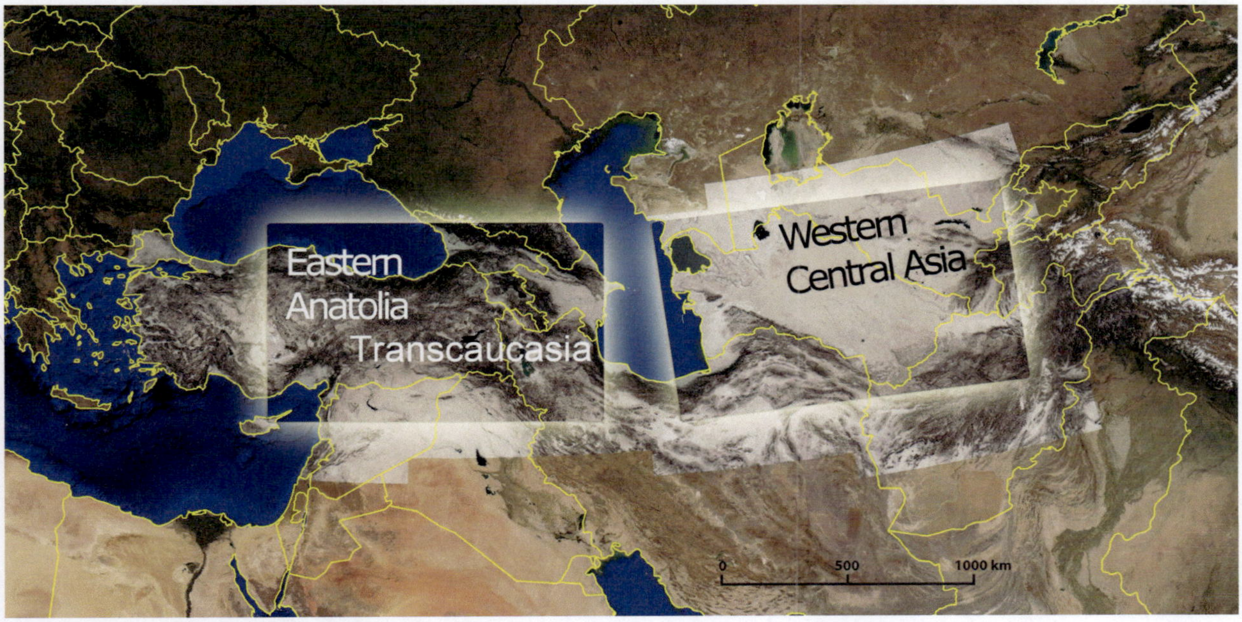

Figure 1.1. Map of south-west Asia showing the wider Near East and the broad location of the two case-study areas which will be termed 'Transcaucasia'/'eastern Anatolia' and 'western Central Asia' for the purposes of this study (base imagery: NASA's BlueMarble).

1.2 Geographical and environmental frameworks

The full geographical context of the research presented here extends over a large central block of Eurasia and includes the whole of what is traditionally called the 'Near East' or Western Asia (see Figure 1.1). As will become evident, this wide spatial field is a necessary result of the focus on inter-regional exchange because the distribution of archaeological evidence of interaction (such as precious stones, metals and textiles) forces us to look at such large scales. Of course it is impossible to treat all regions within this dizzyingly-large zone in equal detail, even if they had been equally well-investigated archaeologically (which they have not). For this reason, two smaller regions are taken as focal 'case-studies', through which larger questions and datasets are filtered to provide more manageable comparisons. In some instances absence of evidence will be as important as presence: features or materials not found inside the case-study areas, but which in the wider context have been recognized as indicators of long-distance interaction, are thus noted and discussed (for example, the discussion of lapis lazuli or weighing systems in Section 4.3.1).

The two case-study regions will generally be referred to throughout this book as 'Transcaucasia/eastern Anatolia' and 'western Central Asia'. Each is loosely defined to reflect the fact that the changing politico-cultural boundaries of the period of study can only be dimly perceived from our time, and to take account of different intensities of archaeological investigation. These areas were initially selected because, in the first instance, both regions are believed to have formed pivotal sections of the 'silk routes' during the 1st and 2nd millennia AD – albeit of very different kinds. As we will see in Chapter 2, at this later time an extensive number of caravanserais were built in both regions to facilitate various trades in exotic goods between China and the Mediterranean. By contrast, both regions are portrayed as peripheral players during the 3rd and 2nd millennia BC when compared to urban 'centres' of the Indus, Egypt and especially Mesopotamia. Kohl described Transcaucasia and Soviet Central Asia (which roughly equate to the regions described here) as forming the 'northern frontier' of the Near East (Kohl 1988). Additionally a comparison between these two smaller landscapes – very different in topography and environment – allows us to search for smaller-scale networks that could have provided the foundation for (later) inter-continental exchanges.

1.2.1 The landscapes of 'Transcaucasia and eastern Anatolia'

For the purposes of this analysis the first case-study region will be taken as a flexible area extending from the Central Anatolian Plain in Turkey in the west (roughly a line of 32 degrees of longitude) to the highlands of north-western Iran in the east (around 49 degrees of longitude); the Pontic mountains and interface with the Black Sea, in the north (41 degrees of latitude), to the beginnings of the northern Syrian plains in the south (around 36.5 degrees of latitude). These bounds, shown in Figure 1.2, include the highland Caucacus and eastern Taurus, the Armenian plateau, Kura and Arax river basins, the upper Euphrates, eastern Central Anatolia and the south-western Caspian steppes. Flexible boundaries are required, not simply because of the shifts of cultural boundaries in the past, but also because of the gaps in the history of archaeological investigation. Politically the area is divided by the modern states of Turkey, Armenia, Georgia, Azerbaijan, Iran, Russia, Iraq and Syria and there is no convenient or politically-neutral term for the whole area so defined[4]. A large proportion of the zone is known geographically as the 'Armenian plateau' but this label does not take into a account the lowland south-, west- or northern edges of our window. The term 'eastern Anatolia' (a translated modern Turkish term) is really used only for areas within the state of Turkey primarily for political reasons. 'Transcaucasia' (a translated Russian term) is perhaps the most neutral and flexible term for the overall area, though in practice its use is often restricted to the former Soviet republics of Armenia, Georgia and Azerbaijan and somewhat artificially excludes similar landscapes west of the Euphrates. In this book the terms 'Transcaucasia' and 'eastern Anatolia' should be taken to refer to the north-eastern and south-western sides of what is essentially same geographic zone.

Figure 1.2. Geographical and physical features of Transcaucasia/eastern Anatolia (base imagery: NASA's BlueMarble)

The full region is characterized by a mixture of high mountain ranges, fertile plains and river valley corridors: within this predominantly mountainous region lie the water catchments for a number of large rivers: southwards to the Persian Gulf are the Euphrates and the Tigris; northwards to the Black Sea the Kızılırmak (ancient Halys), the Yeşilırmak (ancient Iris) and Kelkit (ancient Lycus) rivers; and eastwards to the Caspian the Aras (also Arax, Araz and Araks, ancient Araxes), Kura (also Mt'k'vari, Kur, ancient Cyrus) and Kizil Üzen (also Qizil Uzun, Qizil Owzan, Kyzyl Uzen, Sefid Rud, Sepidrood, ancient Amardus) (Figure 1.4e). In the last half-century, the environment and structure of some of these valleys have been radically changed by huge dam building projects and irrigation works (such as the Ataturk, Keban and Karakaya dams on the Euphrates; or lake manipulation in highland Armenia). A considerable portion of the area has an elevation of over 1000 metres above sea-level (Figure 1.4a). This is significant for climate and agricultural potential, and seems to have encouraged various transhumant

Figure 1.3. Geographical and physical features of western Central Asia. (base imagery: NASA's BlueMarble)

4 As so often, this 'frontier' region has been defined and cut-up by competing states and empires from at least the Achaemenid era, who have each assigned their preferred externally-created name to the area.

Black Sea
Russian Federation
Northern
Caucasia
Caspian
Sea
Colchis
Daghestan
Georgia
Black Sea littoral (south)
Armenia
Azerbaijan
Turkey
Highland
Transcaucasia
Kelkit
Kura-Aras
plain
Central Anatolia
Highland
Eastern Anatolia
Nakhchevan
(Az)
Upper
Euphrates
Cappadocia
Konya
plain
Van
Urmia
Cilicia
Assyria
(Jezirah/North Mesopotamia)
Kurdistan
North-west
Iranian plateau
Amuq
Upper
Tigris
Middle
Euphrates
Cyprus
Levant
(North)
Iraq
Iran
Babylonia
(Central Mesopotamia)
Syria
Mediterranean
0 62.5 125 250 375 500
Km

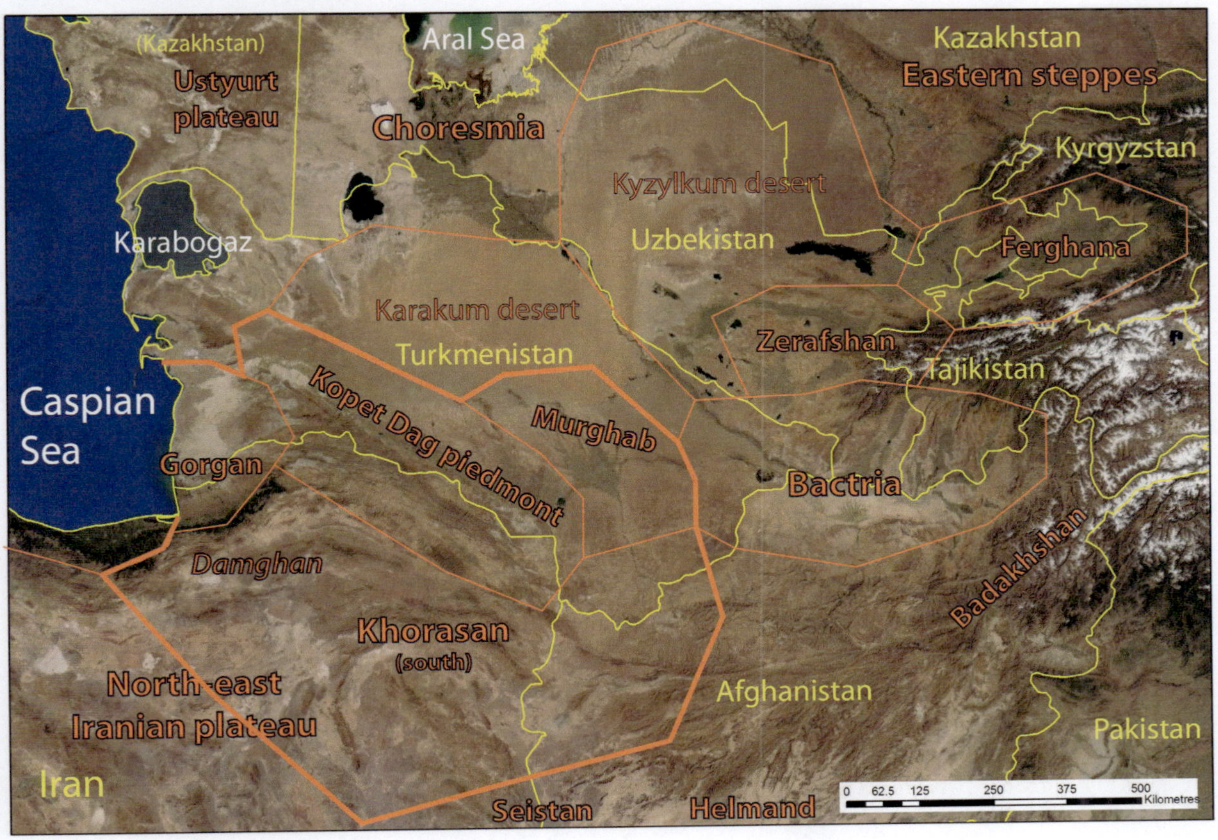
(Kazakhstan)
Ustyurt
plateau
Aral Sea
Kazakhstan
Eastern steppes
Choresmia
Kyrgyzstan
Kyzylkum desert
Karabogaz
Uzbekistan
Ferghana
Karakum desert
Zerafshan
Turkmenistan
Tajikistan
Kopet Dag piedmont
Caspian
Sea
Murghab
Gorgan
Bactria
Damghan
Badakhshan
Khorasan
(south)
North-east
Iranian plateau
Afghanistan
Pakistan
Iran
Seistan
Helmand
0 62.5 125 250 375 500
Kilometres

or pastoral agricultural practices over intensive cultivation in many periods. The general climate is continental, with warm to hot summers (Figure 1.4b), and very cold winters (Figure 1.4c). Except along the Black Sea and Mediterranean coasts, the climate is relatively dry, especially during the summer, and, in higher altitudes especially, this means relatively comfortable summers. To the visitor, used to travelling in this region only during the summer months, it is sometimes surprising to be reminded that much of the area is buried under snow for months on end, which has been a source of everyday hardship for residents, both now and in the past. Towards the low elevation Syro-Mesopotamian plain the temperatures are often very high, reaching into the 40 degrees Celsius in summer.

1.2.2 The landscapes of 'western Central Asia'

The second case-study area (as Figure 1.3) extends from the Caspian littoral in the west (54 degrees of longitude), to the Bactrian foothills of the Hindu Kush in the east (around 67 degrees east), from the relatively empty expanse of the Karakum desert in the north (somewhat arbitrarily 40 degrees north) to the Kopet Dag mountains of the south (around 35 degrees south). Divided between the modern states of Turkmenistan, Iran, Afghanistan, Uzbekistan, Kirghistan and Tajikistan (whose boundaries are mostly arteficial constructs of the 19th or 20th centuries) this area is also characterized by competing geographical labels. Early 20th or 19th century accounts would have comfortably described the area as 'western Turkestan' (from a northern or Russian perspective) or 'Khorasan' (from an Iranian perspective). More recently, the term 'Soviet Central Asia' was used within the boundaries of the USSR, until 1990 when this was prepended unsatisfactorily with 'former-' or else truncated simply to 'Central Asia'. Here I have coined the term 'western Central Asia' to indicate the geographical continuous zone between north-west Iran, northern Afghanistan and the states of 'former Soviet Central Asia', but as distinct to those areas of Central Asia lying to the east of the Pamirs in modern China.

The region so defined is dominated to a large degree by the huge plain which contains the Karakum desert (*Garagum,* in modern Turkmen, meaning 'black sands'), which are bordered in the west by the Caspian Sea and in the south and east by impressive mountain chains (Figure 1.5a). In the south, the Kopet Dag mountains mark the beginning of the Iranian plateau, after which lies the true geographic barrier, the Dasht-e Kavir desert, while to the south-east and east are the massive tectonic complexes of the Hindu Kush and Pamirs, that divide the region from the plains of the Indian subcontinent and constrain eastward movement into the Tarim Basin (and thus to China beyond). The topography (Figure 1.5a) is radically different from Transcaucasia/eastern Anatolia, and it is important to consider how these different landforms and barriers to movement affected past peoples and their cultural development.

The climate is generally much drier than Transcaucasia and Anatolia, and there are only a few large rivers. Over the last century, water distribution has been dramatically modified by Soviet interventions, such as the oft-renamed Karakum canal (previously known as the 'Lenin', then 'Communism' canal), diverting water from a point on the Amu Darya (ancient Oxus) river in the east, across modern Turkmenistan, for the purpose of industrial-scale irrigation agriculture (particularly for growing cotton – which was the Soviet assigned industry for the country).

The climate may be broadly described as extreme continental: very hot and dry in summer (Figure 1.5b), and very cold in winter (Figure 1.5c) with snowfall. Notably the climatic regime has more in common with an extreme continental or 'Near Eastern' system – winter rainfall and dry summers – than a monsoon or 'South Asian' one. This has had consequences for the kind of plants and animals which thrive, and could be exploited agriculturally. There are indicators that the climate of this region has varied considerably through time, in particular that the last three to four millennia have witnessed successive periods of aridization (Marcolongo and Mozzi 1998; Marcolongi 2002; Markofsky 2010, 26-28).

Given low precipitation, rivers are an important factor in settlement pattern structuring (Figure 1.5e): areas available for cultivation are limited without large-scale irrigation, and early settlements were thus often restricted to chains of oases along the piedmont zones. The Tedjen and Murghab rivers, alongside a few smaller streams, reach northwards from the Kopet Dag to evaporate into the desert to form fertile deltas; the Amu Darya (the ancient Oxus) comes from the east before crossing north-west to the much depleted Aral Sea. At times (*e.g.* during the Pleistocene) the river seems to have fed instead to the Caspian though precise data on these changes remains patchy (Kohl 1981: xi-xii, n. 27). The Atrak and Gorgon rivers start from the western and central Kopet Dag and flow westwards into the Caspian. This small basin, under the influence of the Caspian, is significantly more humid than nearby regions to the east. Meanwhile, in the shifting Murghab delta, the dating of archaeological sites superficially appears to show a distribution of slowly southward/upstream migration of settlement, which has often been attributed to the process of desertification mentioned above or to seismic-induced shifts in river flows (Cremaschi 1998). However, alluvial deposits may simply be deeper further up the river, and this could be masking earlier occupation levels in the south.

The history of the Caspian Sea – which lies between our two case-study regions – is fairly poorly understood. The level of this landlocked lake appears to have risen and fallen substantially through time. Over the last century the level has fluctuated between 25 to 29 metres *below* global sea-level due to differential inflow from contributory rivers (both natural and man-made), and extraction through evaporation (Arpe and Leroy 2007). To what extent variation over a longer time frame has played a part in the nature of the lake's exploitation, or indeed in the visibility of archaeological evidence, is difficult to assess. There have been some attempts to model the changing level of Caspian sea-levels over the longer-term (Kosarev and Yablonskaya 1987), which suggest significant oscillations over the last 10,000 years between a maximum of around –11m and a minimum of around –35m (Figure 1.6b), with the lowest level dating to around 3300-2900BC. Such sea-level changes would have had a more dramatic effect on the northern and south-eastern sides of the Caspian – which are much shallower (Figure 1.6a) – and these changes could mask archaeological evidence.[5] Even more difficult to model accurately are the effects of alluviation from the major sediment-carrying rivers which feed into the Sea. This includes the Kura and the Kizil Uzen/Sefid Rud in the south-west and the Volga in the north.

5 Given this data, however, the claims that archaeological sites dating to the Neolithic that are associated with wave platforms at 75m above sea-level were previously on the coast, are difficult to believe, unless we also posit significant and widespread tectonic activity.

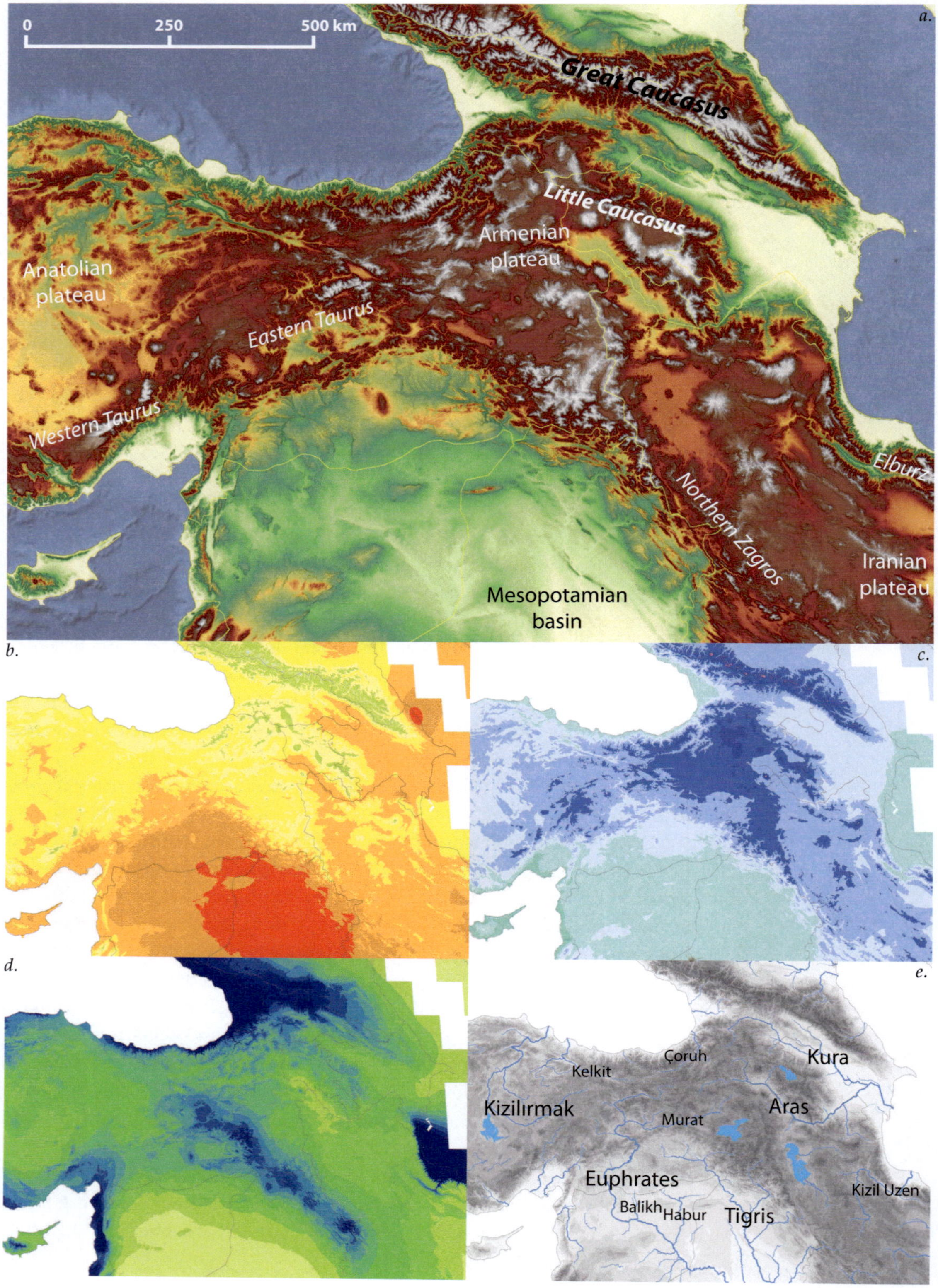

Figure 1.4. The geographical and modern political landscape of Transcaucasia/eastern Anatolia. (a) Elevation and topography; (b) Annual high temperature; (c) Annual low temperature; (d) Annual precipitation; (e) Major rivers. (Based on data from: a. NASA Shuttle Radar Topography Mission; b. c. d. WorldClim.org bioenvironmental database; e. HydroSHEDS).

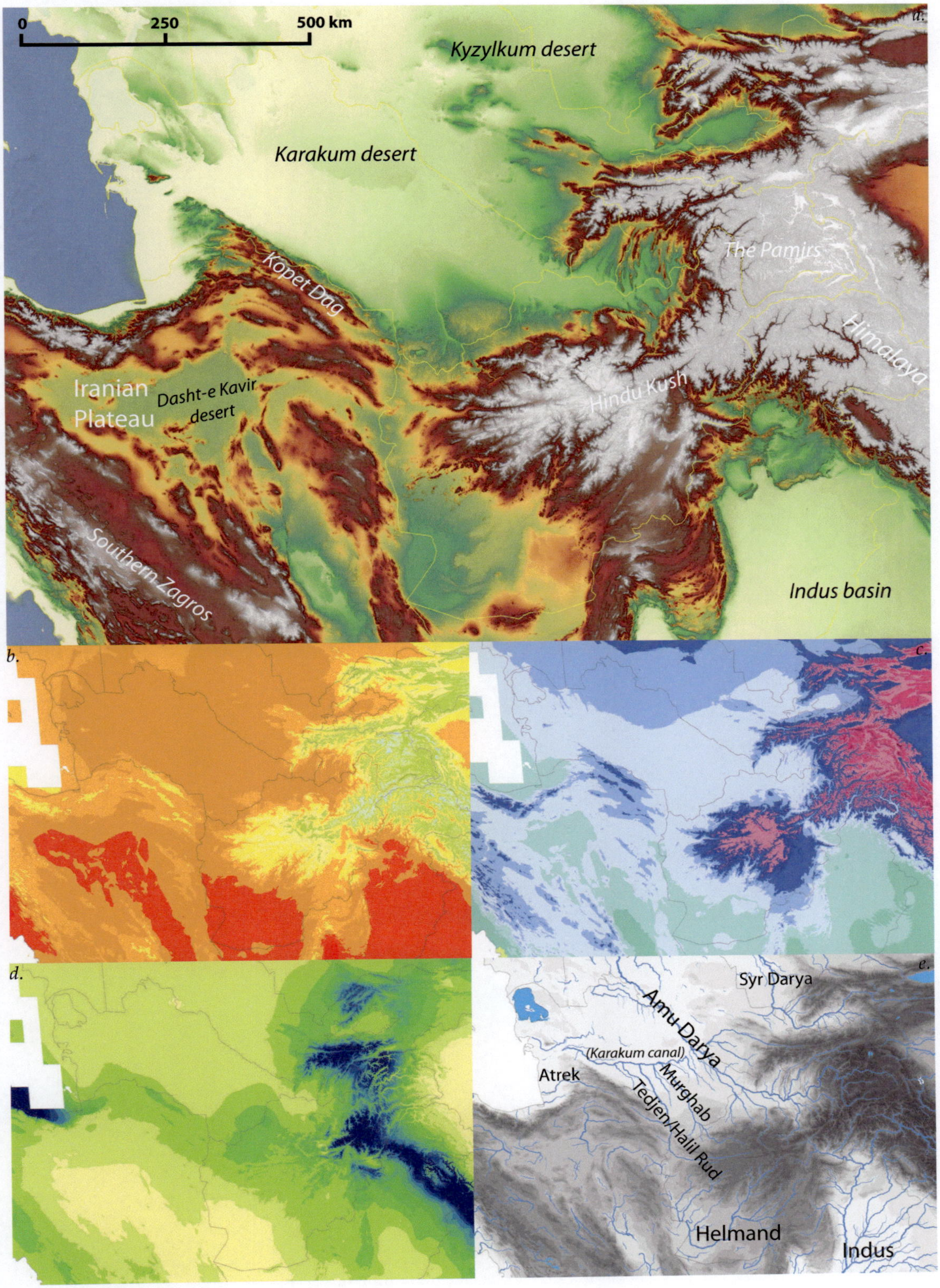

Figure 1.5. The geographical and modern political landscape of western Central Asia. (a) Elevation and topography; (b) Annual high temperature; (c) Annual low temperature; (d) Annual precipitation; (e) Major rivers. (Based on data from: a. NASA Shuttle Radar Topography Mission; b. c. d. WorldClim.org bioenvironmental database; e. HydroSHEDS).

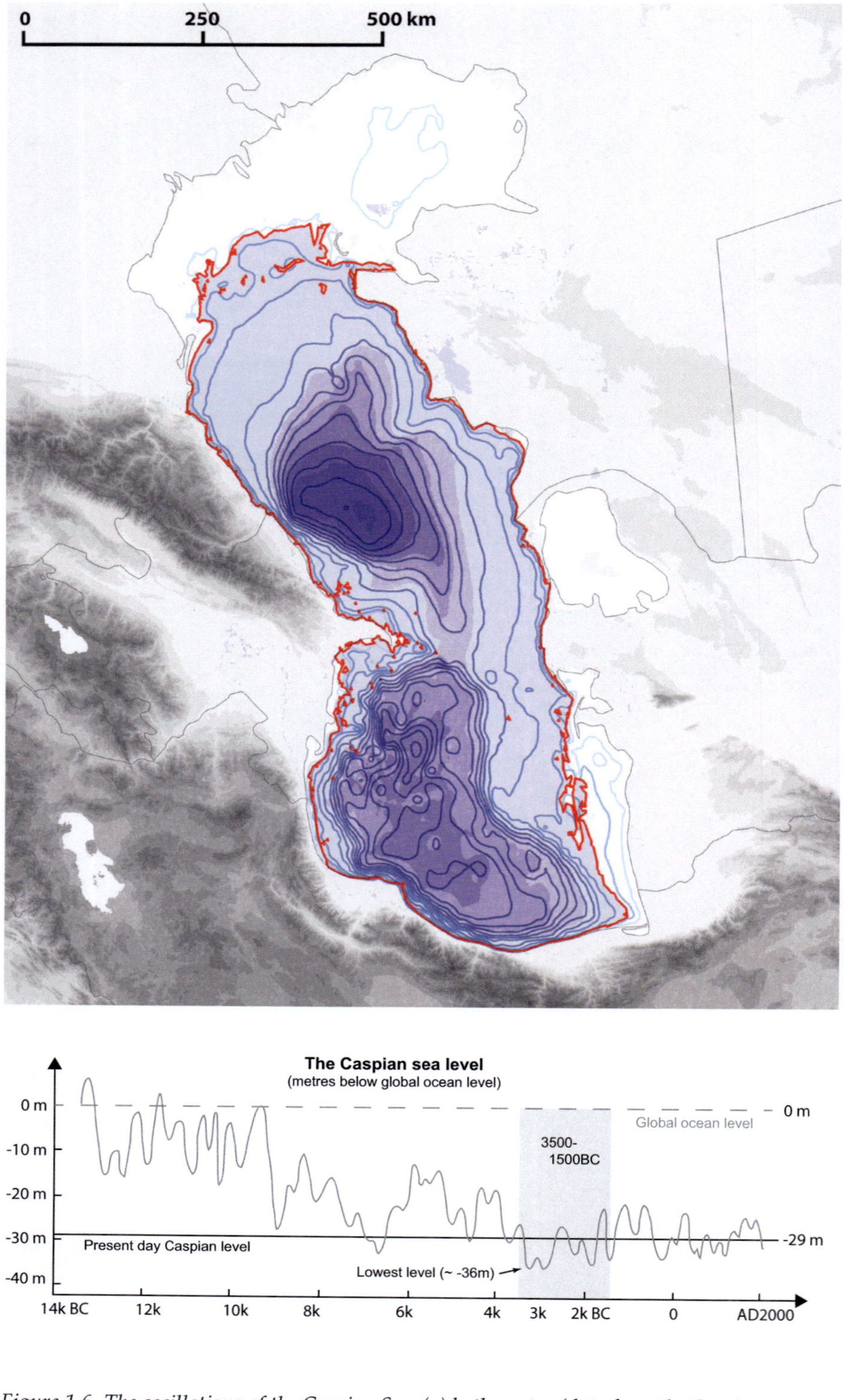

Figure 1.6. The oscillations of the Caspian Sea: (a) bathymetry (data from the Caspian Environment Programme); the red line shows reconstructed coastline of the Caspian at its lowest level (34m) below global sea level c. 3300-2900BC based on this modern bathymetric data; (b) a graph showing estimated sea-level changes through time (based on Kosarev and Yablonskaya 1987).

1.3 Archaeological frameworks: chronologies, sites and assemblages

The chronological framework of the research presented here focuses primarily on the period between 3000 and 1500BC although evidence from before and after this range is referred to where relevant to the arguments and data discussed. To an extent the precise chronological boundaries are of course arbitrary: like the geographical case-studies, they have been selected to filter down the amount of data which needs to be processed to a manageable level. It could be just as interesting to use the same methods applied here to both earlier and later periods. Further, a flexible approach to chronology is in any case necessary, since there is limited agreement on the absolute dating of many relevant sequences. In most instances, especially for the 3rd millennium BC, precision of no more than a century can be given, and anything cited more precisely should be treated cautiously.

The period 3000-1500BC does have some particular historical significance, however. More details will be explored further in the appropriate chapters of the book, but it is worth highlighting, in simple terms, a number of features that have relevance to changes in trade and interaction over the long-term. First, during this period, metallurgy (especially arsenic-bronze, gold, silver and later tin-bronze) becomes widely integrated into both symbolic rituals and everyday lives across the whole of the 'Near East'. Metal objects, principally of copper, were in use earlier of course, but their role appears to have been more restricted. After 3000BC, precious stones, such as lapis lazuli, carnelian and amber, were transported over ever-larger distances. Transportation technology seems to have been revolutionized in a series of steps including the widespread adoption of the wheel, and the use of beasts of burden such as the donkey, the horse and the camel, discussed in more detail below (see Section 1.5). Finally, aside from Egypt, this period is also one in which wool becomes the fabric of choice across a very large area, facilitating the production and consumption of multi-coloured clothing and textiles. In the Marixian critique typical of large-scale historical account, these features have been linked to the emergence of elite groups with particular shared forms of elite practice associated with spreading urbanism and social complexity: namely the promotion and control of a set of resources/materials in order to position certain groups apart from others and justify their power (Lamberg-Karlovsky 1975; 1989). Many of these materials were assembled from distant sources – illustrated most powerfully by the 'rich' and eclectic international content of the funerary assemblages from the famous 'Royal Cemetery' of Ur (Woolley 1934). Whilst this historical caricature involving a 'diffusion' of techniques of social stratification may have some truth, the process does not seem to have applied equally to all regions and we should not assume a monolithic adoption of Mesopotamian practices. The most obvious example is the Harappan (or Indus) civilization which, although apparently well-connected to the west (with many Indus items having been uncovered in Mesopotamia and the Persian Gulf region, see Ratnagar 2006; *cf.* Possehl 2002b), appears to have maintained or developed a very different social system (Possehl 2002a). This raises questions about the effects of 3rd and 2nd millennium interconnectivity: what consequences did it have in different regions? To what extent did ideas and 'systems' as well as people and materials and resources flow along routes of interaction?

Terminological mismatch	Transcaucasia/eastern Anatolia, BC	Western Central Asia, BC
Chalcolithic	5800-3000	?3500-2800
Early Bronze Age	3000-2200	2800-2400
Middle Bronze Age	2200-1500	2400-1950
Late Bronze Age	1500-1000	1950-1450
Final Bronze Age	(Not used)	?1450-?1300
Early Iron Age	1000-800	?1300-?700

Table 1.1. The terminological mismatch between Transcaucasian/eastern Anatolian and western Central Asian 'Bronze Ages' (comparing Sagona 2000 to Hiebert 1994).

1.3.1 Constructing inter-regional chronologies

Since CJ Thomsen first introduced the 'Three Age System' as an organizational scheme for the Danish National Museum, the terms Stone, Bronze and Iron Ages have been adopted nearly globally. On the one hand, these terms identified a broad structural change in the nature of human economies and material exploitation through time on a broad scale. However, as with all deeply embedded terminologies, their metaphorical implications seep unconsciously into ways of thinking: as does the tendency to divide time periods into three. Assumptions are often made when comparing regions whose apparently co-named 'ages' are asynchronous; and the shift from one period to another becomes psychologically significant to the archaeologist in a way that does not reflect the reality of the material record. One of the major dangers of early twentieth-century archaeological interpretations, one which the 'New Archaeology' critiqued very effectively, was an automatic assumption of synchronism between different archaeological phenomena (such as megalithic tomb building). These were based on unidirectional models of cultural change and transfer (evolutionism and *ex oriente lux* diffusionism). Unfortunately the result of the successful critique based on radiocarbon dates (*e.g.* Renfrew 1972) was a rejection of or suspicion towards any long-distance interactionism and a new assumption that change was based on internal systemics in 'processualism'. Macro-scale comparative analysis of interaction was pushed into a 'niche' corner (*e.g.* Lamberg-Karlovsky 1985; Frank 1993; A. Sherratt 1993; 1994). Whilst it is important to treat apparent synchronous or chronological proximal phenomena over large distances with some care in the context of uncertain chronological schemes, there remain many patterns which are worth exploring at this macro-level.

To turn to the two case-study regions in this book, it is perhaps unsurprising that both have relative chronologies for which there is considerable debate on their absolute equivalents. It is important to highlight that the divisions of the 'Bronze Age' from eastern Anatolia and Central Asia almost certainly do not match in absolute dates (see Table 1.1) – though they were originally believed to be more or less contemporary. It would therefore be tempting to discard such labels entirely for the purposes of comparison because the terms have a tendency to frame thought in ways that do not reflect archaeological reality, but sadly it remains necessary to refer to these terms because they are so deeply entrenched in specialist scholarship. Indeed it is only through comparative studies such as this one that this cognitive pre-structuring may be highlighted, and critiqued.

The consequence of uncertainties in comparative chronological frameworks is that in many instances the precise beginning and directionality of 'influence' and 'interaction' is obscured. The expansion of digital processing power (unified structured databases modelled on good archaeological theory) may ultimately

Figure 1.7. Unified chronological framework for Near East and adjacent regions, 3500-1000BC. (Some major sources: Dickinson 2002; Hiebert 1994a; Kohl 1984; Kaniuth 2006; Kavtaradze 1999; Koryakova and Epimakhov 2007; Marro 2000; Mellink 1992; Potts et al. 2009; Sagona 2000; Salvatori and Tosi 2008; Shaw 2000; Sherratt 1985; additional gaps filled by entries from Ehrich 1992) .

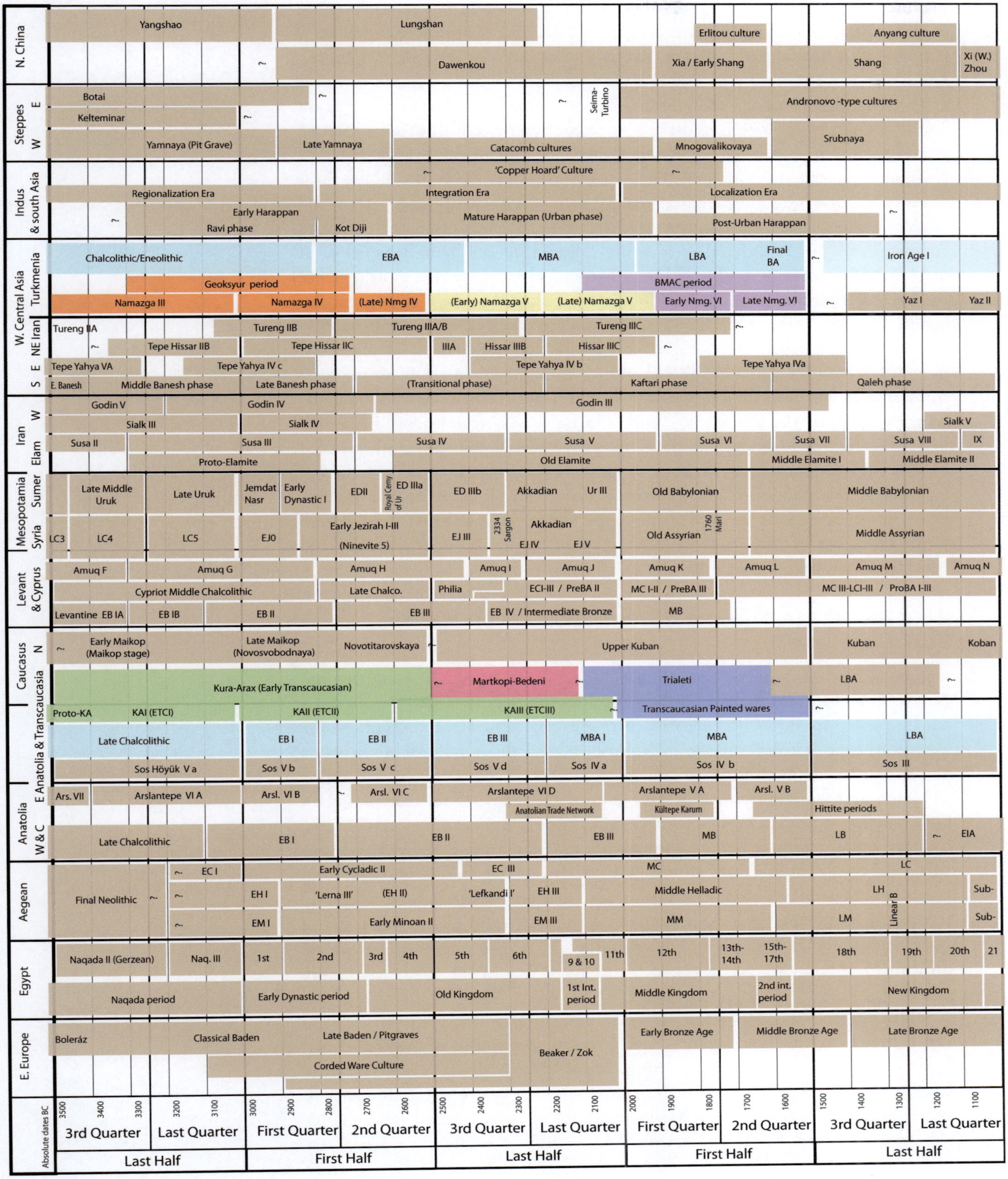
N. China
Yangshao
Lungshan
Erlitou culture
Anyang culture
Dawenkou
Xia / Early Shang
Shang
Xi (W.) Zhou
Steppes E W
Botai
Kelteminar
Seima-Turbino
Andronovo -type cultures
Yamnaya (Pit Grave)
Late Yamnaya
Catacomb cultures
Mnogovalikovaya
Srubnaya
Indus & south Asia
'Copper Hoard' Culture
Regionalization Era
Integration Era
Localization Era
Early Harappan
Ravi phase
Kot Diji
Mature Harappan (Urban phase)
Post-Urban Harappan
W. Central Asia Turkmenia
Chalcolithic/Eneolithic
EBA
MBA
LBA
Final BA
Iron Age I
Geoksyur period
BMAC period
Namazga III
Namazga IV
(Late) Nmg IV
(Early) Namazga V
(Late) Namazga V
Early Nmg. VI
Late Nmg. VI
Yaz I
Yaz II
NE Iran E S
Tureng IIA
Tureng IIB
Tureng IIIA/B
Tureng IIIC
Tepe Hissar IIB
Tepe Hissar IIC
IIIA
Hissar IIIB
Hissar IIIC
Tepe Yahya VA
Tepe Yahya IV c
Tepe Yahya IV b
Tepe Yahya IVa
E. Banesh
Middle Banesh phase
Late Banesh phase
(Transitional phase)
Kaftari phase
Qaleh phase
Iran W Elam
Godin V
Godin IV
Godin III
Sialk III
Sialk IV
Sialk V
Susa II
Susa III
Susa IV
Susa V
Susa VI
Susa VII
Susa VIII
IX
Proto-Elamite
Old Elamite
Middle Elamite I
Middle Elamite II
Mesopotamia Sumer Syria
Late Middle Uruk
Late Uruk
Jemdat Nasr
Early Dynastic I
EDII
Royal Cemy of Ur
ED IIIa
ED IIIb
Akkadian
Ur III
Old Babylonian
Middle Babylonian
LC3
LC4
LC5
EJ0
Early Jezirah I-III
(Ninevite 5)
EJ III
2334 Sargon
Akkadian
EJ IV
EJ V
Old Assyrian
1760 Mari
Middle Assyrian
Levant & Cyprus
Amuq F
Amuq G
Amuq H
Amuq I
Amuq J
Amuq K
Amuq L
Amuq M
Amuq N
Cypriot Middle Chalcolithic
Late Chalco.
Philia
ECI-III / PreBA II
MC I-II / PreBA III
MC III-LCI-III / ProBA I-III
Levantine EB IA
EB IB
EB II
EB III
EB IV / Intermediate Bronze
MB
Caucasus N
Early Maikop (Maikop stage)
Late Maikop (Novosvobodnaya)
Novotitarovskaya
Upper Kuban
Kuban
Koban
Kura-Arax (Early Transcaucasian)
Martkopi-Bedeni
Trialeti
LBA
E Anatolia & Transcaucasia
Proto-KA
KAI (ETCI)
KAII (ETCII)
KAIII (ETCIII)
Transcaucasian Painted wares
Late Chalcolithic
EB I
EB II
EB III
MBA I
MBA
LBA
Sos Höyük V a
Sos V b
Sos V c
Sos V d
Sos IV a
Sos IV b
Sos III
Ars. VII
Arslantepe VI A
Arsl. VI B
Arsl. VI C
Arslantepe VI D
Arslantepe V A
Arsl. V B
Anatolian Trade Network
Kültepe Karum
Hittite periods
Anatolia W & C
Late Chalcolithic
EB I
EB II
EB III
MB
LB
EIA
Aegean
EC I
Early Cycladic II
EC III
MC
LC
Final Neolithic
EH I
'Lerna III'
(EH II)
'Lefkandi I'
EH III
Middle Helladic
LH
Linear B
Sub-
EM I
Early Minoan II
EM III
MM
LM
Sub-
Egypt
Naqada II (Gerzean)
Naq. III
1st
2nd
3rd
4th
5th
6th
9 & 10
11th
12th
13th-14th
15th-17th
18th
19th
20th
21
Naqada period
Early Dynastic period
Old Kingdom
1st Int. period
Middle Kingdom
2nd int. period
New Kingdom
E. Europe
Boleráz
Classical Baden
Late Baden / Pitgraves
Beaker / Zok
Early Bronze Age
Middle Bronze Age
Late Bronze Age
Corded Ware Culture
Absolute dates BC
3500
3400
3300
3200
3100
3000
2900
2800
2700
2600
2500
2400
2300
2200
2100
2000
1900
1800
1700
1600
1500
1400
1300
1200
1100
3rd Quarter
Last Quarter
First Quarter
2nd Quarter
3rd Quarter
Last Quarter
First Quarter
2nd Quarter
3rd Quarter
Last Quarter
Last Half
First Half
Last Half
First Half
Last Half

provide the means to statistical solutions to some problems. In most cases, dating of survey sites, from which much of the current evidence for Bronze Age settlement patterns must be adduced, relies, as in many other parts of the world, upon the identification of pottery types. This is true for both regions in question where extensive landscape surveys still represent the only practical source of information on the general archaeological landscape. Attempts have been made to unify ceramic typologies and hence chronologies across the Near East (such as in the laudable and ambitious ARCANE project). These may bear fruit in the future, but so far it has proven difficult to build links across different regional traditions.

All of these issues are acknowledged in this work, but given that this is a comparative study, which relies to a high degree on synthetic association of material, it is necessary to have at least a working chronology that pulls together the best of the known options. Unlike studies that focus within a particular region, the cross-comparative nature of this work means that broad comparability of dates is essential. The chronological framework used in this book is thus shown in Figure 1.7, and includes broad comparative chronologies not only for the case-study regions but also for Europe, Egypt, the Aegean, Anatolia, Mesopotamia, Iran, the Steppes, the Indus and China. The case-study regions are highlighted in colour on this diagram (the dates shown are based predominantly on: for eastern Anatolia, Conti and Persiani 1993; Kavtaradze 2004; Sagona 2000; and for Central Asia, Kaniuth 2006; Hiebert 1994a, 2002b). Further details on these local chronologies, and the assemblages upon which they are based are discussed in the following sections, as these frame some of the specific historical data and research questions to be discussed in the following chapters.

1.3.2 Assemblages, sites and chronologies in Transcaucasia/eastern Anatolia

Unpicking the chronologies and archaeological material of Transcaucasia and eastern Anatolia is made particularly difficult by the way in which the region is cut by modern political boundaries. There are two 'schools' of chronology, the 'Anatolian' one (created mainly by Turkish, American and European scholars and based predominantly on Anatolian and northern Mesopotamian parallels and sites) and the 'Transcaucasian' one (created by Soviet scholars and based on both internal Caucasian evidence with comparisons to northern/steppe and some Mesopotamian chronologies), with the additional complication that many relevant sites are located within a third political sphere, that of Iran (whose north-western material is nonetheless often linked into the 'Anatolian' school). Only relatively recently have attempts been made to compare and clarify the relationship between (west/central) 'Anatolian' chronologies and those of the Transcaucasus, made difficult by continued uncertainty in both schools (see Edens 1995; Kavtaradze 2004).

Since the 1930s, many attempts have been made to fix Anatolian relative chronologies to the better-understood absolute sequences of Mesopotamia, Egypt or the Aegean (*e.g.* Mellaart 1966; Easton 1976; Mellink 1992). The Early Bronze Age has offered particular difficulties with regards to fixing the dates of archaeological phenomena: this has involved trying to link the diverse relative sequences of only a handful of very widely distributed sites (particularly Troy, Alişar, Tarsus, Kültepe, Beycesultan, Arslantepe, Tell Atchana/the Amuq), all of which lie outside of our focal region. Until the 1970s, very little was known about much of Bronze Age Anatolia, and even today a map of the excavated sites of Turkey shows that large knowledge gaps still remain. In later periods where the 'historical' chronology is relatively undisputed, such as for the Hittite periods, there are still significant problems for archaeological investigations in the Middle and Late Bronze Age sites where there is a great deal of continuity in vessel forms over long time periods, requiring new approaches to dating (*e.g.* Schoop 2006). However, most of these sites lie far to the west of 'eastern Anatolia', and it thus remains difficult to link them to eastern assemblages, such as those of Sos, Arslantepe or Karaz (see Figure 1.8 for site locations).

In Transcaucasia, the long resistance to, or at least uneven application of, the calibration of radiocarbon dates in the Soviet era led to a generally compressed sequence that was difficult to match externally. Whilst settlement sequences from sites such as Shengavit, Arich and Lori Berd (see *e.g.* references in Kushnareva 1997) have provided the bulk of information on changing cultural horizons during the 'Early Bronze Age', including the development of 'Kura-Arax' (or 'Early Transcaucasian') communities, later chronologies are almost entirely dependent on burial assemblages whose relative dating is, unsurprisingly, open to critique and re-interpretation (Edens 1995). Recent revisions of these chronologies based on calibrated dates and a reconsideration of the materials have pushed the likely dating of the earliest kurgan burials (Bedeni/Martkopi) back into the mid 3rd millennium, when previously they were dated to the beginning of the 2nd millennium BC (Kavtaradze 2004). The second millennium remains very poorly understood across the entire region – the paucity of settlement evidence (with a few exceptions such as Kyultepe in Nakhchevan) is mainly responsible for this fact. There is some evidence for the construction of massive fortified settlements in the second half of the 2nd millennium: *e.g.* at Uch Depe in Azerbaijan; and at Metsamor (Kohl 1988) and Tsaghkahovit in Armenia (Lindsay, Smith and Badalyan 2010).

Large areas of eastern Anatolia remain relatively unexplored archaeologically, with only a few key sites having undergone stratigraphic investigation and detailed publication. Much of our information comes from survey work, including those undertaken in advance of dam-building projects, and more recent and intensive projects such as in Bayburt (Sagona and Sagona 2004), Sivas (Ökse 2005, 2007) and Van/Kars regions (Özfırat 2005). However, the wide variations in survey methodologies make it difficult to compare the results. Chronologies of the 3rd millennium have recently benefited from a small number of excavations in eastern Turkey that bridge the geographic gap and add calibrated radiocarbon dates, including those of Arslantepe (Conti and Persiani 1993) and Sos Höyük (*e.g.* Sagona 1999). On the other side, the countries of the Caucasus have been intensively studied by Soviet and local post-Soviet archaeologists (Kushnareva 1970; 1997; Bobokhyan 2008), but have received relatively little attention from 'western' authors until recently. Details about sites and material culture are therefore often hard to obtain in British, German and Turkish libraries. In the last few years, partnerships between local and foreign archaeological teams have begun to make the literature and material of the Caucasus more easily accessible outside the region (Smith 2005; plus see contributions in TÜBA-AR/Türkiye Bilimler Akademisi Arkeoloji Dergisi 2010), through surveys and excavations in Azerbaijan and Nakhichevan (Bakhshaliyev and Marro 2009; Marro, Bakhshaliyev

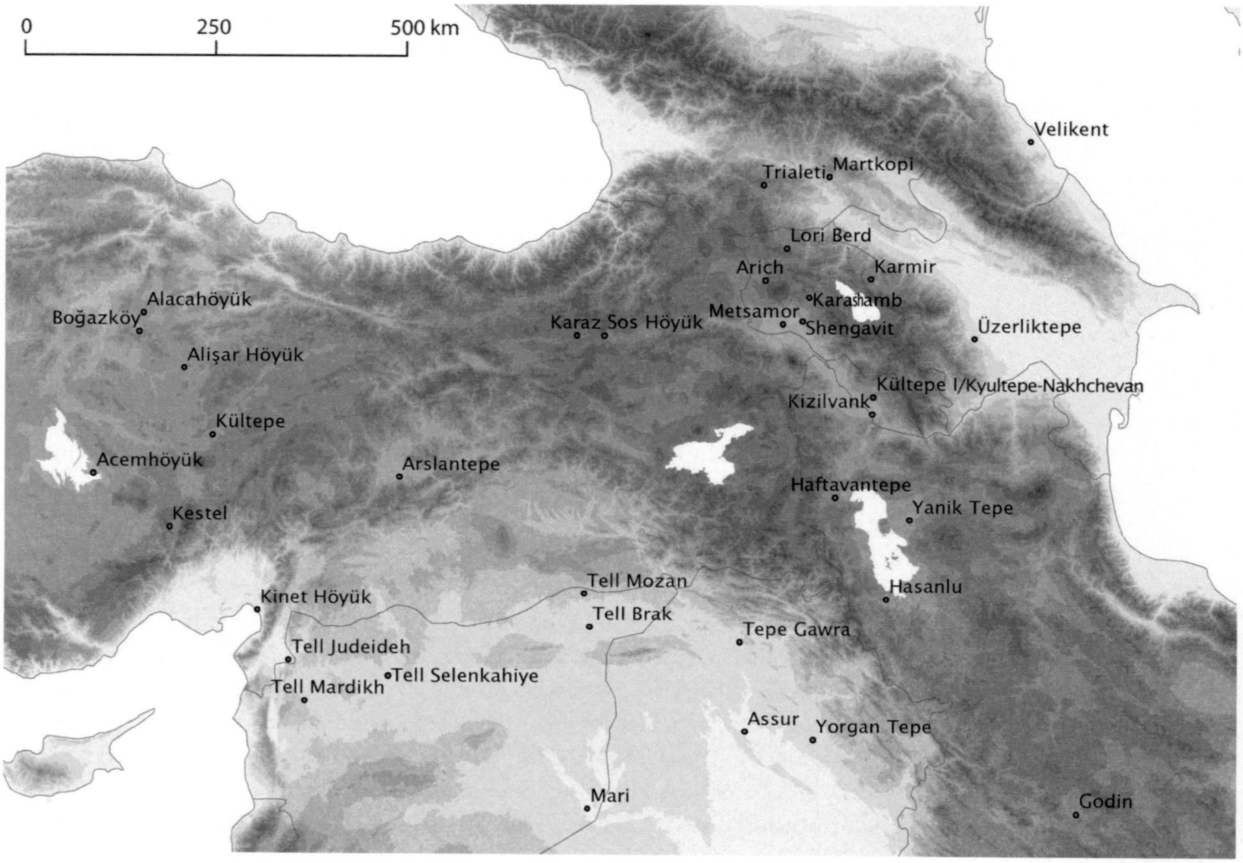

Figure 1.8. Major archaeological sites of the 3rd and 2nd millennium BC in Transcaucasia and eastern Anatolia.

and Sanz 2010; Ristvet, Gopnik *et al.* 2012), Georgia (Sagona, Nikolaishvili, *et al.* 2010; Rova, Puturidze and Makharadze 2010) and Armenia (Smith, Badalian and Avetisyan 2009; K. Wilkinson *et al.* 2012). Many of these projects are in their initial phases of research, and so it may be some time before full results are available. The full publication of other relevant sites, such as Dilkaya Höyük near Van (Çilingiroğlu 1993), and Yanik Tepe in Iranian Azerbaijan (Burney 1961; 1962; 1964; Summer 2004) is also likely to offer further insights into both chronologies and archaeological assemblages for the entire region.

Despite the gaps, the general structure of the archaeological record during the 3rd and 2nd millennia BC can be briefly sketched. Most importantly, the north-east of eastern Anatolia and the adjacent Transcaucasia were dominated to a large degree by 'Kura-Arax' (or Karaz, Early Transcaucasian/ETC) and related assemblages from the late 4th to early 2nd millennium BC. The 'Kura-Arax' complex is defined predominantly by its distinctive pottery, which in its later incarnations is recognizable by its dramatic burnished red-black surfaces (discussed further in Section 5.6.5). The 'culture' is also associated with a distinctive architecture (A. Sagona 1984; 1993), portable and built-in 'andirons' (Kelly-Buccellati 2004; Rahmstorf 2010b) and a few metal types (Kohl 2007: 92-93), including, perhaps, certain spiral-headed pins (Sagona 1981; discussed further in Section 6.7.1). The full distribution of similar red-black Kura-Arax-inspired pottery appears, however, to be much larger than these secondary identifiers, which are generally more restricted in distribution. The 'Red-Black Burnished Ware' and 'Khirbet Kerak'

BC	Western Anatolia	Eastern Anatolia
3000	EBI (3400-2750BC)	KAII (3000-2600BC)
2800	EBII (2750-2200BC)	
2600		KAIII (2600-2000BC)
2400	('Anatolian Trade Network Period')	--Martkopi/Bedeni (?2500-2000BC)
2200	EBIII (2200-1900BC)	MB (2200-1500BC) --Trialeti (?2100-1400BC)
2000	MB (1900-1650BC)	
1800	--Old Assyrian Trading Colonies	
1600	LB (1650-1200BC) --Imperial Hittite	LB (1500-1000BC)

Table 1.2. Broad chronological periods in western versus eastern Anatolia/Transcaucasia (comparing Mellink 1992 to Sagona 2000; Kavtaradze 2004).

pottery assemblages of the northern and southern Levant respectively are, both formally and technologically, strongly interrelated to supposed prototypes from the north-east and this has promoted theories of substantial out-migration from a Transcaucasian 'homeland' (*e.g.* Rothman 2003a; Batiuk 2005; Paz 2009, Kohl 2009b), though so far most explanations and mechanisms for any such migration remain unsatisfying. Alternative theories of 'transhumant' migration have been proposed (see, *e.g.* Frangipane and Palumbi 2007) but the whole subject remains open to further investigations and insights.

Sometime between 2500 and 2000BC the 'core' area of Kura-Arax culture in north-east Anatolia and Transcaucasia becomes dominated by mortuary assemblages: both the large-scale *kurgans* of the Trialeti, Martkopi/Bedeni/Sachkere, and smaller small cemeteries which have yielded a range of painted pottery assemblages known variously as Karmirberd/Kyzilvank, Sevan-Uzerlik, Van-Urmia, 'Aras', or 'Yayla' wares (see Özfırat 2001). In this book I have grouped this general complex of similar wares under the term 'Transcaucasian Painted Wares'/TPW for convenience and they are discussed further with regards to the relationship between their patterns and textiles (Section 6.6.2). Little is known about the settlement or subsistence patterns of the people who made these assemblages. The Trialeti and earlier *kurgans* contain rich metal assemblages, which have been argued to document a newly stratified social system for the region (Edens 1995). Comparisons have been made to northern, southern and western neighbours: the possibly contemporary or slightly earlier royal graves of Ur III and of Alacahöyük (Kavtaradze 2004, 548-549), and the much earlier and impressive Maikop graves (A. Sherratt 1997; Lyonnet 2007; Kohl 2009a) and other diverse kurgan cultures of the Pontic steppes. The relationship seems more likely to be one of structural similarities in economy than any 'genetic' or direct cultural link, however.

1.3.3 Assemblages, sites and chronologies in western Central Asia

Western Central Asia was also split by a major Soviet border, which again has resulted in conflicting chronologies. The chronological debates in Central Asia have focussed particularly on establishing the correct absolute dates for the Middle and Late Bronze Ages and subsequent Iron Ages. As in Transcaucasia, an uneven application of calibration of radiocarbon dates led to a compressed low chronology: reassessment by later scholars pushed the dates for the Middle and Late Bronze Age back by up to 500 years (Kohl 1981; Hiebert 1994a). Relative dates for prehistoric sites in Turkmenistan are based on the sequence from the excavations at Anau and Namazga Depe, with the subsequent 'Iron

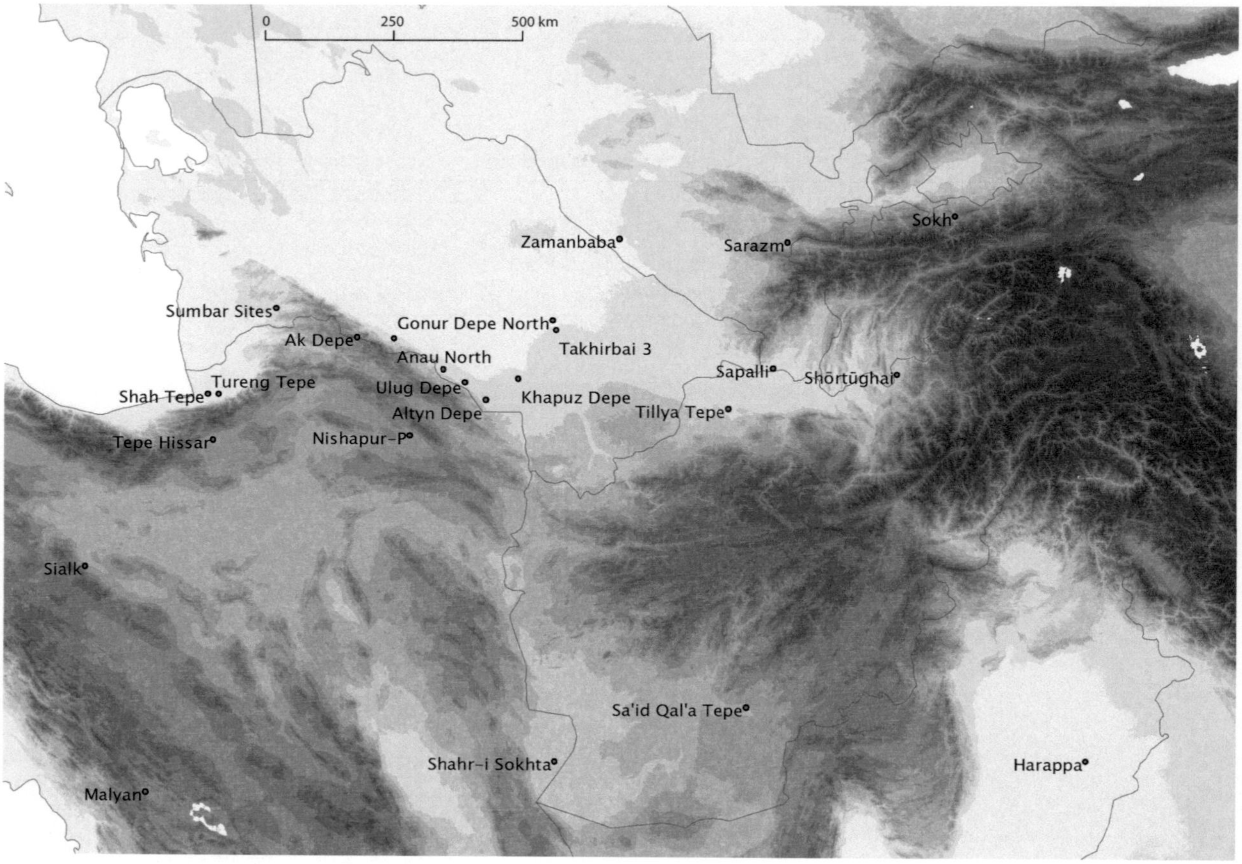

Figure 1.9. Major archaeological sites of the 3rd and 2nd millennium BC in western Central Asia.

Age' periodized according to the sequence from Yaz Depe (see map Figure 1.9). Though arguments between 'Soviet' and 'western' assessment of absolute dates continued for some time, most recent works from both schools view the higher chronology as the more likely (*e.g.* Kaniuth 2006), in part because it appears to agree better with the results from calibrated dates from northern and eastern Iran such as Tepe Hissar, and Shahr-i Sokhta. Meanwhile, however, the status of the relative sequence has also come under some scrutiny: Götzelt (1996), for example, has argued that the internal structure of the 'Namazga V' period in southern Turkmenistan has overly relied on assumptions about the evolutionary development of objects instead of demonstrated stratigraphic relationships. He thus questions many of the chronological assertions made in the past. Whether or not the relative sequence needs a more general review, the lack of stratigraphic excavation and publication at many sites in Central Asia means many nuances of development and local variation may have been missed (see Salvatori 2007 for a full critique of non-stratigraphic practices used in both local and international excavation projects).

Whilst modern archaeological field techniques were first pioneered at Anau, as part of Raphael Pumpelly's excavations from 1904, more recent history of the area has restricted archaeological investigations and interaction between different researchers even more so than in eastern Anatolia/Transcaucasia. Political difficulties prevented many 'western' archaeologists from undertaking work in Afghanistan, Iran and the USSR for large parts of the twentieth century, and the militarized boundaries between modern states have continued to hamper work in

BC	Period	Alternative labels
3000	Namazga III-IV (3500-2700 BC)	Chalcolithic / Eneolithic
2800	Late Namazga IV (2700-2500 BC)	*Early Bronze Age*
2600	Early Namazga V (2500-2200 BC)	
2400		*Middle Bronze Age*
2200	Late Namazga V (2200-1900 BC)	
2000	Namazga VI (1900-1700 BC)	*Late Bronze Age* Bactria-Margiana-
1800	Late Namazga VI (1700-1500 BC)	-Archaeological-Complex Takhirbai period
1600	(Possible hiatus?)	*Final Bronze Age*
1400	Yaz I (?1500 – ?1000 BC)	*Early Iron Age* / Anau IV
1200		

Table 1.3. Broad chronological periods in Central Asia (following Kohl 1984; Hiebert 1994a).

certain areas (for example in the Kopet Dag highlands between Iran and modern Turkmenistan or in the highland passes between central Asian states and China). The transportationally-remote and politically-unstable nature of Afghanistan has prevented scientific investigation in this apparently important region for Bronze Age archaeology, and invasions by both the USSR in the 1980s and NATO after 2001 (*i.e.* after the 'nine-eleven' attacks in New York) and intervening civil war have additionally enabled archaeological plunder and destruction of considerable amounts of cultural heritage (see, for example, Manhart 2001).

These problems notwithstanding, various productive and intensive archaeological research projects were undertaken by Soviet scholars over the course of the twentieth century[6] in Central Asia and in recent years a few international collaborations and an increasing trend toward publishing in English have begun to make the region's literature and material slowly more visible to 'western' scholars. In this book a greater focus is necessarily placed on those areas that have been more thoroughly studied or widely published, including the Murghab delta and prominent piedmont sites such as Altyn Depe. Wider surveys, such as the monumental Italian 'archaeological map' projects have provided a better view of overall settlement structure in the region (Salvatori and Tosi 2008) and innovative new work at the edge of the Murghab delta has revealed details about micro-settlement patterns (Markofsky 2010).

On contrast to Transcaucasia/eastern Anatolia where the third millennium is defined predominantly by a fairly uniform material culture, the material assemblages of western Central Asia can be said to be split between at least three interrelated cultural zones: those defined by the 'Eastern Grey Ware' pottery in north-east Iran (see, *e.g.* Piller 2003); those defined by the 'Namazga'/NMG sequence (particularly Namazga III, IV and V) in and around southern Turkmenistan, typified by sites such as Altyn Depe, Kara Depe, Namazga Depe, Ulug Depe; and those associated with assemblages related to Namazga in the Ferghana and Bactrian regions to the north and east, Sarazm (Isakov 1991; Isakov and Lyonnet 1988), Sapalli Tepe (Askarov 1973; Kaniuth 2006) and Djarkutan (Askarov and Abdullaev 1983). The 'Eastern Grey Ware'/EGW sequence is typified by sites such as Tepe Hissar (Schmidt 1937; Dyson and Howard 1989), Tureng Tepe (*e.g.* Deshayes 1967;

6 Most notably by the Soviet/Russian/Turkmen team, 'IuTAKE', the Southern Turkmenistan Archaeological Complex Expedition (see P'yankova 1994).

1972) and Shah Tepe (Arne 1945). New work at Gohar Tappeh promises to provide more information on this region (Mahfroozi *et al.* 2009).

Overlaps between the EGW and Namazga zones are to be seen in material from sites such as the Parkhai/Sumbar cemeteries, which lie at the northern tip of the Kopet Dag (Khlopin 1981; 1983; 2002). Additionally, at various times the central region of the Namazga zone (the piedmont of the Kopet Dag in modern southern Turkmenistan) shares many material traits with a wide area. The pottery from the Geoksyur phase (c. 3500-2700BC, Namazga III to IV) is characterized by painted patterns (see Section 6.6.3) and vessel shapes whose distribution appears to extend far to the south and north (Sarianidi 1983; Hiebert 1994,168), including to sites in Baluchistan, Afghanistan (Mundigak, Mehrgahr) and eastern Iran (Shahr-i Sohkta). This apparent unity is slowly broken however with increasing regionalization. Towards the middle of the 3rd millennium, the painted wares of Namazga III/early Namazga IV are replaced in the Kopet Dag region by plain wares of the late Namazga IV and V type (discussed in Section 5.6.4); the repertoire of metal and other materials (alabaster vessels) increases; and large settlements – which have been described by some as 'proto-urban' towns (Masson 1968) – grow along streams emerging from the Kopet Dag piedmont region. These towns include Altyn Depe (Masson 1988; Kircho, Korobkova and Masson 2008), Namazga Depe itself (Khlopina 1981) and Khapuz Depe (see Kohl 1981). At some point towards the end of the Namazga V period, in the last centuries of the 3rd millennium BC, these towns contracted considerably in size or are abandoned. At around this time, the delta fringes of the Murghab river appear to have been settled, typified by small sites such as those found in the Kelleli and Egri Bogaz oases. This has inspired various interpretations of migrations from the piedmont toward the north-east (Salvatori 2008), though the mechanisms for de-settlement in one area and re-settlement in another, if that is indeed the case, are far from clear given the chronological uncertainties.

The last century of the 3rd or the early centuries of the 2nd millennium, witness the emergence in the regions of Bactria and Margiana (the Murghab delta) of the so-called 'Bactro-Margiana Archaeological Complex' or 'BMAC', a term first coined by Viktor Sarianidi (Sarianidi 1981; 1990; 2001; 2006). This 'complex' is defined by a set of material features including: a distinctive range of metal artefacts (particularly 'Bactrian' axe-heads, Section 5.3.4); Namazga V/VI-related plain pottery; stone vessels which relate to 'inter-cultural style' types also found in south-east Iran (now often subsumed under the label 'Jiroft'), the northern coast of the Arabian peninsula and parts of Mesopotamia (discussed further in Section 4.4.1); and centralized sites with fortified architecture, such as at Gonur Depe, Adji Kui 9, Togolok 21 or Dashly 3 (see, *e.g.*, P'yankova 1994, 362). As in Transcaucasia, albeit perhaps a little later in date, funerary remains appear to form a much more important part of the record than before. Sadly, much material apparently from BMAC contexts made its way into the art market from the 1960s onwards, presumably illicitly excavated from graves – so considerable data about this period has been irrevocably lost. Only those excavations at settlements and necropoleis such as Gonur Depe have provided more detailed information (Hiebert 1994; Rossi Osmida 2002; Sarianidi 2006), though it is clear that more could be learnt from further work at these and other sites. The end of the BMAC period has long been a subject of considerable uncertainty. It is clear that settlement patterns change fairly dramatically by the following Early Iron Age (or Yaz I) period (Askarov 1992; Boucharlat *et al.* 2005; Bonora and Vidale 2008), but the extent of continuity or disruption in pottery and other assemblages remains an open question. Some hypothesized a hiatus in occupation in former BMAC regions before the emergence of the Yaz communities. More problematically, others identified the transition with the invasions of 'Indo-Europeans', pointing to the apparently simultaneous 'arrival' of new ceramic traditions in southern Turkmenistan, north/eastern Iran and across Afghanistan as far as the Indus and perhaps beyond, and attempting to correlate it with the supposed dating of the Indian Vedic epics, the Avesta (Lal 1992). Leaving aside more general theoretical objections to this model and controversy over the invisible movements of languages, it is telling that exactly which set of assemblages are to be associated with these hypothetical Indo-Europeans varies immensely: sometimes it is the makers of Andronovo pottery, sometimes it is the makers of various forms of Grey Wares (of which Yaz pottery might be included) and sometimes it is the BMAC communities themselves. What is clear from the archaeological data is that through the course of the 2nd millennium, fragmentary remains of 'Andronovo'-type pottery – whose origins appear to lie in the mobile traditions of northern Eurasia (Anthony 1998) – become more common in southern Central Asia (Udemuradov 2002; Hiebert 2002; Cattani 2008). In the latter part of the 2nd millennium, it is possible that such groups became heavily involved in the exploitation of metal resources in the Hissar, Pamir and Tienshan mountains (the significance of tin in this region, in Section 5.2.3). However, exactly how these different communities interacted – emerging, respectively, from the steppe or the sown (Shishlina and Hiebert 1998) – remains ambiguous.

1.4 Frameworks of movement: transportation technologies

During our period of focus, technologies of transportation, *i.e.* the means of movement, also changed through time. Such technological change had the potential to manipulate 'effective' distances (and hence the density of travel) along particular routes either by increasing the *speed* of travel or *amount* of material or people that could be transported. We can safely assume that such changes must have had a significant impact on the socio-economic effects of movement and exchange, even if we cannot be sure of the details. New transportation technologies included wheeled vehicles (wagons, chariots) and sea-craft, and also those animals that were used for their ability to aid transport (ridden to save human energy, attached as traction to pull vehicles, or loaded as 'pack-animals'). The trajectories of domestication and diffusion of these animals are therefore very important. In the following section I therefore try to outline our current knowledge about such transportation technologies to frame the more specific reflections on routes, travel and interaction in later chapters.

1.4.1 The domestication of equids: the donkey and the horse

Undoubtedly the most important transportation technology of the 3rd and 2nd millennia BC was the pack-animal – donkey, horse or camel – which facilitated the transport of much larger quantities of commodities than could be carried by human bearers, and over much more varied terrain than could be transported by wheel-based vehicles. Additionally of course, such animals could also pull wheeled vehicles (as well as ploughs, ards and sledges) more flexibly than oxen or human 'tractors'. As such, the development and distribution of such animals provides a significant indicator of the mobility and intensity of exchange with which ancient peoples were involved. However, it is often difficult to measure the extent of the early exploitation of these animals for transport because such usage tends to leave relatively few faunal remains at settlements and in funerary offerings[7] compared to those animals used for meat, whose copious remains were often discarded in easily accessible archaeological contexts. Clues from iconography and (later) texts must therefore be relied upon to fill in the gaps.

Donkeys[8] (*Equus asinus*) are likely to have been the most important animal given their adaptability to different types of landscapes. They appear to have been initially domesticated over a long period in northern Africa and the southern Levant (Rossel *et al.* 2008), but the earliest representations of donkeys – suggesting their earliest intensive use for freight – are from the Ghassulian period (the 5th millennium BC) in the southern Levant (A. Sherratt 1983). The expansion of their use is difficult to trace precisely, but the early appearance of the cuneiform sign for donkey at Uruk in the late 4th millennium (Oates 2003, 115) gives a hint about the date of its adoption in Syro-Mesopotamian culture. The idea that donkeys played an essential role in the expanding network of contacts of the Uruk period (*i.e.* the later 4th millennium BC), as suggested by A. Sherratt (2003), thus seems plausible – especially in the light of the remains of donkey (alongside horse and onager) in contemporary 'Maikop'-period levels at Galugay 1 (Korenevsky 1995). Mesopotamian texts indicate that by the 3rd millennium donkeys and related hybrids were in widespread use (Oates 2003, 115). The early 2nd millennium Old Assyrian merchants of Kültepe used donkeys to transport metals and textiles from Aššur to central Anatolia (Veenhof 1972, 45). Though they may have been geographically widespread, we have little direct information about how restricted donkeys were socially and culturally: it is possible, for example, that they were more typical of the emergent urban centres of Mesopotamia and Anatolia rather than the non-urbanized highland communities of Transcaucasia and Iran.

Horses appear in archaeological contexts from at least as early as the 5th millennium BC at sites in and adjacent to the Pontic-Caspian steppes where their wild ancestors were presumably tamed or herded first for their meat and/or their ability to dig through snow and hence provide access to fodder for pastoralists' herds (A. Sherratt 1981; Levine 1999; Anthony 2007, 199-204). When exactly people began to exploit horses for transportation remains unclear. Andrew Sherratt (2003), adapting a model proposed by Uerpmann (1990), argued that expansion of the donkey into Caucasus during the Uruk/Maikop period (*i.e.* the mid to late 4th millennium BC) might have provided the stimulus or inspiration for horse domestication here or in the adjacent steppes to the north[9]. Tracing the

---

7 At least until much later, when the horse becomes an elite symbol in burial (see below).

8 Other species of wild equids, such as onagers, may have been used to pull wagons or carry loads on occasions, but were unlikely to have been 'domesticated' (A. Sherratt 1983, 96; Zarins 1978).

9 Anthony and Brown (2011) have recently strongly rejected the influence of Uruk and the model of the donkey as the main driving force of horse domestication *contra* Sherratt – preferring instead to place this process within earlier (late 5th or early 4th millennium) and distinctively local processes relating to hunting and warfare in the steppes of Kazakhstan or Ukraine. The two scenarios are not necessarily as mutually exclusive as presented however, depending on the understanding of the process of domestication. Even if horses were indeed integrated into steppe societies before the Uruk period, it still seems possible that the appearance of donkey caravans bearing exotic goods from the south would have promoted the use of horses as pack-animals in a new secondary or tertiary role.

origin and spread of domesticated horses is complicated by the on-going presence of wild populations of equids in certain regions[10] – and the difficulty we have in differentiating early domesticates from their close wild progenitors, whose female members also appear to have provided pools from which fresh genes may have been 'borrowed' on occasion, according to the evidence from modern DNA (Anthony 2007, 196). The problem may be a terminological one, however, in which we see wild and domesticated as binary oppositions: 'domestication' would be better seen as a process of ongoing changing relationships between humans and animals whose biological impact (or archaeological detectability) may have varied considerably depending on the nature of that relationship. Whilst horses may have spread rapidly across the steppe and forest of eastern Europe, their adoption in the south may have first been in small numbers as a 'tradable commodity' to be interbred with donkeys to produce the hardier mules (donkey-horse hybrids). Horses may have first arrived in the Near East during the late 4th millennium BC (or earlier) through the Caucasus or Anatolia, but they only became widespread in the second half of the 3rd millennium BC (A. Sherratt 1983, 96-97), documented in increasing faunal remains (*e.g.* at Godin Tepe in Iran, Gilbert 1991), records of small numbers of horses in Ur III texts (Oates 2003, 117-119) and depictions which show equids (possibly including horses or horse hybrids) being ridden (Oates 2003, 119-120) or pulling battle-wagons of various kinds (Littauer and Crouwel 1979).

Equids in Transcaucasia and eastern Anatolia

Herds of wild horses were probably present in eastern Anatolia and Transcaucasia before the arrival of domesticates, at least in more open steppe environments. If indeed the extension of donkey usage into the highlands of eastern Anatolia and beyond to the northern Caucasus during the Uruk period had an effect on the exploitation of horses, the faunal remains from this region are critical to understanding this process. Faunal remains from sites such as Korucutepe (A. Sherratt 1983, 96), Sos Höyük (Piro 2009, 405), Godin Tepe (Piro 2009, 284) and others across Transcaucasia[11] indicate that domesticated horses were present in this region relatively early during the late 4th or early 3rd millennium BC. The evidence for donkeys is harder to identify until their mention in Assyrian texts as pack-animals (already noted above) for the trails between Aššur across the western parts of the eastern Anatolian mountains into the Central Anatolian plateau. Kushnareva mentions the presence of donkeys in the 4th or 3rd millennium BC at Elar and Kvatskhela in modern Armenia and Georgia respectively but without providing further references (Kushnareva 1997, 193). Donkeys may have been

10 Which is what is perhaps represented in the 4th millennium horse remains at the sites of Norşuntepe and Geoy Tepe (A. Sherratt 1983, 96), or further south-east at Tal-i Iblis and Choga Mish (Zarins 1978).

11 Chernykh's (1992) unreferenced assertion, presumably from the same source as Kushnareva's similarly unreferenced note (Kushnareva 1997, 193; Kushnareva and Chubinishvili 1970) that there are significant numbers of domesticated horses at many Kura-Arax sites, including Shengavit, Elar, Didube, Kvatskhelebi and Kyul Tepe, needs to be more fully investigated as its consequences are very important for transportation within the Kura-Arax phenomenon – and no doubt for the huge geographical extent within which the Kura-Arax derived pottery is found during the 3rd millennium BC (see Section 5.6.7).

well suited to, or at least sufficiently flexible for, mountain trails across the eastern Anatolian and Transcaucasian highlands, but may have been less useful than horses or donkey-horse hybrids (mules) in winter where passes would be snowed under.

Equids in western Central Asia

Equids appear to have been a relatively late arrival in western Central Asia and the adjacent regions. Anthony asserts that there are "[n]o definite horse remains ... identified in eastern Iran or the Indian subcontinent dated before 2000 BCE" (2007). In contrast, Masson and Sarianidi (Masson and Sarianidi 1972, 20) mention the discovery of clay models of four-wheeled carts with what may represent horses dating to Namazga V contexts (*i.e.* late 3rd millennium), though they do not provide further references or cite the location of these finds. Small numbers of horse-bones have been recovered in Margiana and Bactria in BMAC contexts, dating to the very end of the 3rd or, more likely, the first centuries of the 2nd millennium BC (Sarianidi 2002b, 234-241; Salvatori 2003) and may reflect the increase in contacts between the region and the steppes to the north at this time where horses were abundant[12]. Recent research at Gohar Tappeh, a site along the eastern edge of the southern Caspian littoral, has apparently uncovered horse remains dated to around 3400BC, but before full publication it is difficult to know what to make of this evidence.

1.4.2 The domestication(s) of the camel

Tracing the domestication of the camel is complex because there are in fact two species involved – the two-humped Bactrian camel (*Camelus bactrianus*) and the one-humped 'Arabian' dromedary (*Camelus dromedarius*) – whose trajectories of domestication although separate are complicated by an apparently long history of hybridization (Köhler-Rollefson 1993, 181; Potts 2004a). *C. bactrianus* is adapted to the "high-altitude deserts of Central Asia and remains unfazed by snow and cold temperatures"[13], whilst *C. dromedarius* prefers "hot deserts of northern Africa and western Asia and is sensitive to humidity and low temperatures" (Köhler-Rollefson 1993, 180). The location and chronology of domestication of either species still remains unclear (Kuzmina 2007, 67-69). It was once thought that eastern Iran or southern Turkmenistan provided the most likely arena for the earliest camel domestication, on the evidence from sites like Shahr-i Sokhta in Seistan where "bones, dung and hair" of Bactrian camels were identified in 3rd millennium contexts, and Anau where camelid remains were dated to the 4th millennium BC (Tosi 1974; Compagnoni and Tosi 1978; Sherratt 1983, 97; Potts 2004a, 148-149). There are also parallel distributions of camel bones and representations of dromedaries at Hili and Umm an-Nar in Oman and in Bahrain, probably dating to the later 3rd millennium (A. Sherratt 1983, 98). However, the wild progenitors of the Bactrian camel, *C. ferus*, appear to have been restricted to the eastern steppes (in modern Kazakhstan or beyond), and their appearance in Iran and western Central Asia was a later expansion (Potts 2004a, 144-6). The

---

12 As may the apparent depiction of horses in the form of 'Bactrian axe heads' (*cf.* Section 5.3.4) – though the ones shown by Kuzmina (2007, 162) were unfortunately picked up on the art market and hence remain without context.

13 Indeed winter was apparently the favoured month for Bactrian camel caravans to travel in the Gobi desert (Potts 2004a, 147).

Figure 1.10. A terracotta model of wagon with camel head from Altyn Depe (adapted from Masson 1988, pl. XIX; photo: courtesy of Russian Academy of Sciences.)

clearest early evidence for the human exploitation of camels in transportation comes in the form of camel figurines, such as the terracotta model of a camel head attached to a four-wheeled wagon (Figure 1.10) during the Namazga IV period from Altyn Depe (Kohl 1992, 186; Masson 1968; Masson and Sarianidi 1972, 109). It is possible that, as for other animal domestications, there were two 'stages' in Bactrian domestication: an early management for meat in the eastern steppes, followed by a later intensification of usage for transport in western central Asia and eastern Iran, with subsequent (post 1000BC?) dispersals to other regions.

Camels in western Central Asia

The camel appears to have been an important feature of transportation in southern and western Central Asia by the Namazga III or IV periods, and perhaps formed one of the centres involved with integrating camels more closely with human communities. The earliest camelid remains, already mentioned above, come from Anau (Anau II, c. 3400-2900BC), followed by a much greater density of material from mid to late 3rd millennium BC Namazga V levels (or equivalent) at Ulug Depe, Altyn Depe, Namazga Depe and Shah Depe (Potts 2004a, 149-150), alongside the models of camels with vehicles already mentioned. Bactrian camels subsequently appear frequently as iconographic motifs on metal and soft-stone stamp seals from late 3rd and early 2nd millennium contexts; on metal vessels (Sarianidi 2006, 236-237; Potts 2008a, 172); as "theriomorphic ceramic[s] vessel in the shape of a Bactrian camel" (Potts 2004a, 150); and as inscribed graffiti on gold and silver vessels (Sarianidi 2006, 238) and various ceramic objects.

Camels in Transcaucasua and eastern Anatolia

Despite the association of modern Turkey with camel breeding, no evidence for camels has been identified in eastern Anatolia or Transcaucasia earlier than the more widespread diffusion of camels to western Iran and into Mesopotamia from the Neo-Assyrian era, *i.e.* the very late 2nd or early 1st millennium BC (Potts 2004a, 154-155). If Potts is correct, the Neo-Assyrians appear to have predominantly used dromedaries, with regular but smaller numbers of Bactrian camels to produce stronger hybrids (Potts 2004a).

Figure 1.11. Drawing of a terracotta model of a cart from Arich (Khachatryan 1975, 77 fig. 37, cf. Kohl 2007, 95).

1.4.3 The first vehicles

The first wheeled vehicles in the Old World date to the second half of the 4th millennium BC, and appear to have developed initially out of Sumerian agricultural techniques of ploughing and threshing (A. Sherratt 2003, 242). These animal-powered vehicles had solid wheels and were probably harnessed to cattle or oxen rather than equids. Representations from the later 3rd millennium shows a variety of 'battle-wagons' (Littauer and Crouwel 1979), which were pulled by cattle or later equids (some with four wheels, and four draught animals, others with fewer wheels). How fast and how far such wagons could be transported is unclear – but their extremely heavy construction was clearly a severe limitation to usage over long-distances. As far as we can tell, their main purpose seems to be in warfare, hunting and pageantry rather than for trade or exchange. Sherratt suggests that there is "no evidence from the steppe region that equids were used to pull such devices. The incentive to develop an effective, equid-drawn vehicle came from the military needs of proliferating (and competing) Near Eastern states" (A. Sherratt 2003, 244). Spoked-wheels appear, in contrast, to be a later innovation of the steppes, where the construction "was based on a tradition of woodworking which had its origins, not in solid, sculptural, carpentry but in the bentwood technology used for centuries in making recurved bows" (A. Sherratt 2003, 244) – an innovation that ultimately facilitated the development of the light-weight chariot (see below).

Early wagons in Transcaucasia and eastern Anatolia

The proximity of Transcaucasia and eastern Anatolia to the most likely centre of innovation for early wheeled vehicles, and its intermediate position to the steppes where the spoked-wheel and chariot were subsequently to develop makes this a key region where it is highly likely that wheeled vehicles were in use over a long period (Pogrebova 2003, 403). In central Transcaucasia, a small number of early cart models have been identified apparently dating to the mid-3rd millennium BC. The best known examples are from Badaani (Mirtskhulava 2011) and Arich (Figure 1.11; Kushnareva 1997, 193), but fragments of what may represent wheel models have been found across the region (Sagona 2013: 278-279). Dating such contexts is difficult, and it seems likely that wheeled vehicles may have been in use in Transcaucasia and eastern Anatolia since at least the early 3rd (and perhaps late 4th) millennium but only become archaeologically visible in the later 3rd millennium, particularly during the subsequent kurgan traditions where actual vehicle and wheel remains have been found in rich burials. Support for this probability comes from over 115 burials with wagon wheels and other vehicle

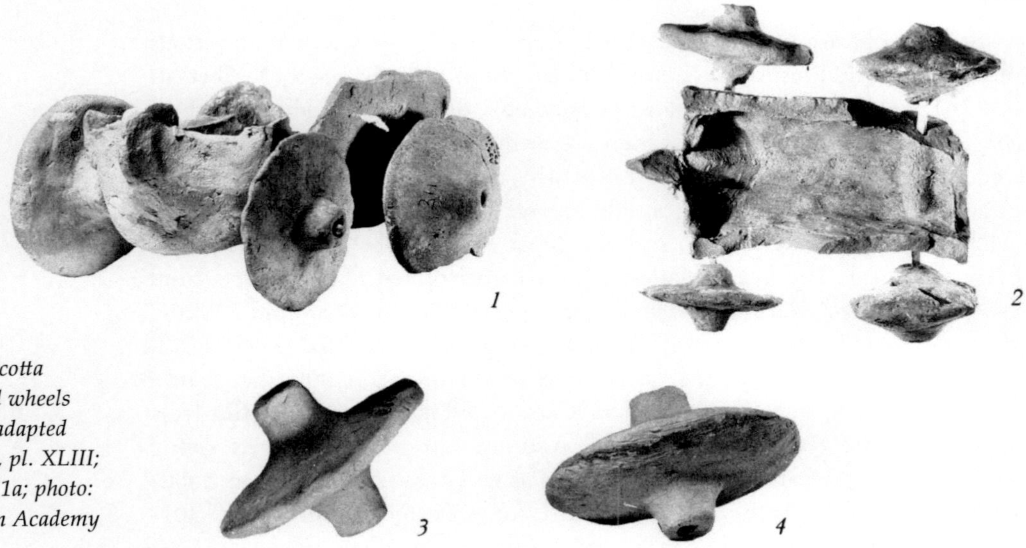

Figure 1.12. Terracotta models of carts and wheels from Altyn Depe (adapted from Masson 1988, pl. XLIII; cf. Kohl 1984, pl. 11a; photo: courtesy of Russian Academy of Sciences).

accoutrements from the Novotitrovskaya culture (early 3rd millennium) of the Kuban steppes, located just to the north-west of the Caucasus (Kohl 2006, 15; citing Gei 2000, 176-177).

Early wagons in western Central Asia

The earliest evidence for wheel transport in Turkmenia is from the Namazga IV period where terracotta wheels – presumably belonging to model carts – have been identified (Masson 1992, 229; Kirtcho 2009). The model of a four-wheeled wagon being pulled by a camel from Altyn Depe has already been mentioned above. This probably dates to the mid-to-late 3rd millennium BC during the 'urbanizing' Namazga V phase (Masson 1992, 240, fig. 6) – associated with a range of ideas including seals and 'ziggurat'-inspired architecture thought by some to have been imported from Mesopotamia (Masson 1992, 237-240). The use of the camel for draught was at one point thought to be exclusive to southern Turkmenistan (according to Kuzmina, cited in Kohl 1984, 114), but petroglyphs from Kazakhstan – although difficult to date – appear to show camels being used for traction (Kuzmina 2007, 178 fig. 34). For what purpose such wagons were used, whether for warfare, hunting, transport or public performance, remains unclear. A wagon with solid wheels was also recovered in a brick-lined mass-grave at Gonur Depe dating to the BMAC period (Sarianidi 2001, 178-183), alongside ten human adults, a dog, a camel and the decapitated body of a horse foal (Anthony 2007, 427). Given the association with a nearby 'royal' burial, this inventory probably represents a sacrifice of prestigious and exotic objects rather than a reflection of everyday use.

1.4.4 Horse riding and the fast chariot

On the basis of various depictions (including on seals), it seems that various domesticates, including donkeys, were ridden from an early stage (A. Sherratt 2003, 241). However, there remains considerable debate as to whether horse-*driving* (*i.e.* the use of horses as traction power for ploughing or pulling wagons) preceded horse-*riding*. There is no reason why our instinctive expectation that the phases of horse domestication should involve a stage of riding before traction or pack-usage should be correct. Indeed, direct evidence for early horse riding is fairly equivocal. The interpretation of certain perforated antler objects from *Trichterbecherkultur*/TRB contexts in northern/central Europe of around 3000BC as cheek-pieces (A. Sherratt 1983), has now been generally rejected (Dietz 1992; A. Sherratt 2003, 244). But the question as to whether 'soft' bits that would not have survived into the archaeological record could have been used effectively to control a horse remains controversial. Anthony has argued for early riding at Dereivka, Ukraine and Botai, Kazakhstan as early as 4200-3700 and 3600-3100BC respectively (Anthony, Brown and George 2006; Anthony 2007, 207-220; Anthony and Brown 2011) on the basis of tooth-wear (that may relate to the use of such soft bits) and mortality profiles of the horse remains.

The earliest unequivocal evidence of horse gear is in fact associated with a group of impressive chariot burials, dating between the 20th and 18th centuries BC by recent radiocarbon analysis (Kuznetsov 2006). These burials, part of the Sintashta-Petrovka complex, are distributed across the steppes of eastern Europe to the Urals. This gear (including bits, bridles and cheek-pieces) could have been used for both pulling chariots and riding. This fast 'war-chariot' seems to have spread quickly across the steppes and south into the Near East, where, by the second half of the 2nd millennium BC, there are numerous depictions of political rulers on chariots (often trampling their enemies) amongst Hittite, Mitanni and Egyptian iconography (Moorey 1986). The extent to which these light chariots were really serious military tools (as opposed to symbols of prestige in which elites would drive or parade around for festivals or at battles) is difficult to judge. Overall it seems most likely that whilst horse-riding was known in the steppes and eastern Europe in the 4th and 3rd millennia BC, it was not a regular or widespread practice, and remained uncommon in the Near East until the later 2nd millennium or even 1st millennium – when riding was transformed for military advantage, associated also with the development of the saddle (Bokovenki 2000). Certainly the first horse-drawn light chariots were unlikely to have been used for carrying goods, even if they may have allowed faster communications. The explanation for this elite-association may be partly cost: relative to donkeys or camels, keeping horses requires immense investment in time or money[14], which means one must either place the horse at the centre of the socio-economic system as happened in the steppes, or else 'pay through the nose' as the elites of Akkadian Mesopotamia, and later elsewhere, were happy to do in order to prove their wealth.

14 A quick online survey of the relative keeping costs for horses or donkeys in the modern day can show this.

Riding and chariots in eastern Anatolia and Transcaucasia

The earliest evidence for chariots in Transcaucasia is from burial assemblages "dated to the beginning of the Late Bronze Age, *i.e.* to the late fifteenth-early fourteenth centuries BC" (Pogrebova 2003, 397). These burials, located near Lchashen and Lori-Berd (both in Armenia), included bronze models of horse-drawn chariots and the remains of wooden vehicles. According to Pogrebova, the designs of the bronze chariots appear to fall into the 'Near Eastern' tradition (as opposed to a Eurasian or steppe tradition) albeit with a greater average number of spokes than is typical for contemporary Syrian or other Near Eastern chariots. She also argues that these are unlikely to be 'battle chariots' since they show serious limitations for manoeuvrability – and concludes that the region is thus unlikely to be the origin of battle chariots more generally. The wooden wagons (upon which the bronze models appear to have been mounted) are of a different design – the 'Y'-shaped yoke perhaps suitable for twisting highland roads according to Pogrebova.

> *"All this evidence enables one to suggest that the population of southern Transcaucasia became familiar with chariots under the direct influence of the Near East, whence it had been borrowed primarily as a marker of a high social position. The mastery of chariots seems to be indicative of the growing power of the local rulers."* (Pogrebova 2003, 404).

In contrast, the first indication of horse-*riding* in Transcaucasia is the appearance of horse burials in tumuli south of the Kura (around the Giandhachai, Gancaçay river) which have been dated to the 13th century BC. Besides the intriguing report of a burial that included horse and mounted horseman together, a series of objects in these burials suggest an association with northern and eastern steppe traditions of horsemanship, including a bone belt plaque and a rein-terret with parallels from Lower Volga (Pogrebova 2003, 405).

Riding and chariots in western Central Asia

Despite western Central Asia appearing closer on the map to the apparent centre of chariot and riding innovation in the central and eastern steppes, evidence for riding and fast chariots in western Central Asia is currently completely lacking, except for a BMAC-style seal "probably looted from a cemetery in Bactria" which "showed a man riding a galloping equid that looks very much like a horse" (Anthony 2007, 427). Given the similarity to an Ur III seal (dated with surprising and unlikely precision to 2050-2040BC), this may possibly reflect familiarity only with the iconography rather than indexing a local tradition of riding *per se*. This suggests either that there was strong cultural resistance to their adoption or that the geographical barriers between southern Central Asia and the steppes, particularly the Karakum desert, were far greater than those on the much longer but less hostile connections via the Caucasus (though the evidence of steppe-type material culture in sites in the Murghab and the eastern fringes of the region which run up to the eastern steppes might be used to argue against this).

1.4.5 Water transport: the boat and the sail

Whilst both eastern Anatolia and western Central Asia as we have defined them are relatively landlocked regions, the development of boat and sailing technology still had consequences for exchange in and between the surrounding regions with major rivers and coasts. Riverine water-based transport along the Euphrates and Tigris undoubtedly played a major role in the development of Mesopotamian trade from at least the 4th millennium BC but perhaps even earlier (Carter 2006), as it did too along the Nile in Egypt and the Indus in South Asia (Bass 2000). Though sails were likely used as early as the Naqada period (mid to late 4th millennium) along the Nile on the basis of depictions on pottery, the development and widespread dispersal of seaworthy sail-craft (from perhaps around the mid 3rd millennium BC) had consequences both for riverine transport and for long-distance sea-crossings in the Mediterranean, the Persian Gulf, the Red Sea and the Black Sea during the 3rd and 2nd millennium BC. Effective long-distance maritime transport transformed the potential 'connectedness' of coastal groups, manipulating the *effective* distance between previously land-orientated coastal zones. Critically, however, we have almost no knowledge about the use of sail –or indeed any form of boats– on the Caspian Sea. It is difficult to imagine that maritime transport was completely unknown here, despite the distance from potential seafaring groups on which to base technological models. This lack of information is most likely to be one of lack of research: exploratory underwater investigation in Azerbaijan undertaken in the Soviet era produced evidence of medieval towns (Kvachidze 2006), which may indicate that relevant information is submerged under the oscillating Caspian sea-level or under complex alluvial deposits (see *e.g.* Hoogendoorn *et al.* 2005, 363; Kosarev and Yablonskaya 1987). Until further research, it is difficult to say what role maritime transport could have played in such an important location between major cultural centres of innovation of the Bronze Age.

1.5 Frameworks of interpretation: indicators for interaction

Besides the means of movement, central to the study of routes and exchange is identifying the kinds of evidence that provide indicators for interaction. Here I want to briefly outline the interpretative approaches to material culture and key threads of analysis that will be followed throughout this book. Many varied sorts of commodities (goods, ideas and living beings) were no doubt circulating over large-distances, but the way in which the archaeological record of prehistory was created[15] means that such flows cannot be directly 'seen' through the extant remains. Various forms of evidence – from direct to indirect, from textual accounts to proxy depictions – must be assembled and interpreted together to form the large picture. As we will see in the following chapters, it is necessary to avoid an overly positivist attitude to the evidence: the majority of the material culture which was exchanged in the past is permanently lost – but this does not mean we should exclude from our analysis of routes and interaction those categories of material which we cannot see directly in the archaeological record. These 'invisible flows', as we could call them, were important structuring features of ancient economies. Similarly, certain otherwise 'visible' materials may have travelled along paths which are difficult to reconstruct because the data is so sparse. On the other hand, objects that did not individually travel long-distances (such as

15 Mostly as discarded debris from settlements (middens or earlier demolished/abandoned settlements under tells); sporadic but sudden accidental preservations (ship wrecks or volcanic eruptions) or deliberate burials which have either long been forgotten (hoards, flat cemeteries) or have remained only as visible monuments (kurgans).

pottery, architectural styles or technologies), can reveal smaller scale networks ('pathways', see Section 3.2.1) through which higher-value materials had to pass, by their greater density of distribution. Thus by bringing together different types of evidence at different scales, and combining multiple overlapping networks of interaction, we can hope to produce a more comprehensive picture of the flow of materials and people during our period of interest.

1.5.1 Visible and invisible flows: materials, people, language

Creating a full list of potential materials that may have flowed along routes of movement between 3000 and 1500BC would be an impossible task[16]. In the archaeological record, stable inorganic materials tend to be most easily identified (stone/mineral, metal, clay-based items), but the contemporary texts we have from Mesopotamia in particular hint at a vast array of organic and volatile substances in use (animal and plant fibres, edible or ingestible plants or animals, dyes and cleaning chemicals), many of which were apparently sourced from distant locations and along extended commodity chains or trade networks (Moorey 1999; Potts 1997). Both 'raw' substances and 'finished' goods appear to have been transported. Whilst the line between 'raw' and 'finished' is a flexible and culturally-specific construct (*cf.* 'the raw and the cooked', Levi-Strauss 1983), there is some worth in differentiating the two since they imply different things about the level of integration between communities. 'Raw' materials (*e.g.* unworked precious stones) may be exchanged more easily across cultural boundaries without, necessarily, a concomitant transfer of meaning or practices – even if the transfer across that boundary is itself important to the material's meaning in the new context (*i.e.* as 'exotic'). 'Finished' objects, which have been transformed significantly (through carving, metalworking or cooking), tend to have been made for a specific cultural context with specific social functions. For example, carved stone vessels (discussed further in Section 4.4.1) were designed, to paraphrase Alfred Gell (1992; 1998), to 'entrap' a specific audience. Thus the implication of a very large number of 'imported' finished goods or the emulation of such goods or styles suggests a deeper degree of cultural integration. Metals are rather unique, in that they permanently lie in an ambiguous state of 'rawness' and 'finishedness'. Old objects can be easily melted down and recycled and this liquidity has had consequences for metal's role in human interaction (something which is discussed further in Chapter 5).

Materials or 'substances' should also be taken to include people, animals and plants of course. Although it will not form part of the analysis of this study, it should be noted that there is considerable scope for mapping the movement of people through the remains of human bodies through isotopic studies and genetic studies of populations with modern and ancient DNA[17]. Though such

---

16 A parallel example from a much later period, namely the Tang era in China, can be found in the form of the vast lists of exotic materials of innumerable type presented to the Tang king and listed by Edward Schafer in his study of the topic, *The Golden Peaches of Samarkand* (Schafer 1963). Whilst a Bronze Age equivalent might have included a smaller range of items from a more limited catchment, the comparison helps to expand the range of commodities and categories of material that we should be prepared to think about being transported.

17 Anthropometric studies that focus on supposed racial and geographical origins of peoples, which are still popular in some archaeological traditions, should, in general, be treated with some scepticism. Where such studies are useful, however, is in the identification of trends in or individual cases of pathological conditions which may offer clues to health, diet and life-cycle of a community's members.

techniques have received wide attention in studies of European prehistory, they are only just beginning to be applied to the Near East and other parts of the Old World. In contrast, whilst we know that demographic and other movements must have caused linguistic change and resulted in substantial linguistic flows, at present there seems little hope that we will ever resolve many of the mysteries surrounding the earliest origins of language groups. In our case-study regions/time-periods, there is very little direct evidence for languages. In Central Asia, written texts do not appear at all in durable materials until the Hellenistic period. Masson's suggestion that certain symbols (found on objects such as terracotta figurines) might indicate a kind of nascent written code, along the Indus analogy (*e.g.* Masson 1968, 182, 184), remains speculative.

In eastern Anatolia/Transcaucasia, the situation is similarly sparse, with the earliest texts dating to the Urartu period (*i.e.* the first half of the 1st millennium BC). In adjacent regions to the south and in overlapping regions to the west, literate groups were active of course: most locally we can include the Old Assyrian tablets from Kültepe, which date to the 19th or 18th centuries BC and the Hittite texts, which date to the 16th to 12th centuries BC (*e.g.* Forlanini 2008; Barjamovic 2011). This select group of fossilized documents from Anatolia and the completely invisible record of Central Asia no doubt mask intricate and constantly transforming linguistic landscapes during the Bronze Age (Sherratt and Sherratt 1997). It seems highly likely that, in these transformations, trade and long-distance exchange played an important part. The traditional models of 'Indo-European' distribution still remain frustratingly unimaginative however, and the persistence of the perception that language and ethnicity are identical (and that ethnic groups can be identified by pottery groups), unhelpful. Whether it will ever be possible to do much more than give posit speculative models is questionable, but since languages are required and come into being precisely for the exchange of ideas and material culture, we should bear in mind the likelihood that some of the material residues of interaction networks also do reflect linguistic *koinai*, and that the communities engaged in such networks were unlikely to be simply passive monolingual carriers of language from one generation to the next, but rather active and motivated polyglots.

1.5.2 Technology and style as indicators of interdependence

With moving people also come moving skills and technologies. Much recent literature in the anthropology of material culture and production has focussed on the (human) body as site of dual processes of transformation of both the physical and structural-symbolic worlds (see, for example Warnier 2007) – as behavioural rituals in craft production. This is something that has emerged from a marriage between the *chaîne opératoire* approach to technological studies (Leroi-Gourhan 1964, 164; *cf.* Soressi and Geneste 2011), and complementary studies on *techniques du corps* or bodily praxis (Mauss 1935). Despite the interesting insights these approaches may have for the history of technology, ethnographic accounts, whilst sometimes comparative, still tend to focus on the local dynamics of technologies within particular cultural contexts. From an archaeological perspective, more interesting is the dynamic flow of technologies and skills in space and time and, potentially, across cultural or geographic boundaries.

Tracing the spread of technological innovations has long been an important thread in archaeological research, although often such studies have focussed on seeking the 'origin' of the techniques rather than on their consequences for social configurations or significance for the level of integration between communities. Philip Kohl has argued that in the third millennium BC, there were few barriers to the transfer of technologies except distance, and that 'innovations' could easily pass from one community to another. He contrasts this with the modern world-system in which various structural and cultural barriers prevent the transfer of technologies from one region to another, and hence create and maintain economic difference (Kohl 1987b; 2011). The extent of technological transferability during the 3rd millennium, remains relatively uninvestigated, however.

If framed in the language of 'praxeology', the spread of many types of technological innovations may be better described as the transfer of bodily techniques, through migration, apprenticeship, mimicry or reverse-engineering (*i.e.* emulation). Given the differential difficulty with which certain innovations can be created in the first place, we should be attentive to the specificity of the techniques in question (see, for example, Rahmstorf 2011). There are differences between different material outcomes (styles, functions) and the amount of technical skill and knowledge required to produce them. There is thus a continuum of intensity of social interaction (emulation, learning, teaching, or migration) required to facilitate the transfer of technology. In a local context, complex skills are passed vertically through apprenticeships and the practice of everyday life, but, for regional interaction, additional mechanisms are required to transmit such knowledge including colleges, travelling craftworkers or gift exchange all of which imply systemic movement of human beings. Such movement need not imply the permanent or unidirectional 'migration' which were typically evoked by early twentieth century cultural historians, but could instead include trickle migration (*e.g.* via the circulation of marriage partners from one village to the next who might bring textile production knowledge), temporary *Gastarbeiter* exchanging skills for labour (*e.g.* itinerant metalsmiths), or structured exchange of 'students' as part of initiatory procedures. The key point is that structured *social institutions* that facilitate interaction and skills transfer are normally required: and this will be reflected in the degree of coherence in the material record. Where objects only look like similar nearby items then the integration and interaction must be less, or at least only focused on the consumptive group in a community; where similar techniques of production are applied but objects are stylistically dissimilar, integration and interaction is likely to be higher at the productive level but not at the consumptive one; and finally where both styles and techniques are similar, interaction across both consumptive and productive spheres of community must be high. Once transferred, however, successful techniques can persist for generations, and be retransferred to other regions that do not have direct connections to the originating communities. It is thus difficult to perceive on-going interactions at a supra-regional scale through technical means alone.

The other aspect of technology is of course, 'style'. Style and technology should not, of course, be considered opposites (*i.e.* decorative vs. functional). In a praxeological sense, 'styles' are also produced by the transfer (learning) of bodily techniques. The effect on the audience should also be considered as part of the 'technical' aspect of the process of production. This is what Alfred Gell (1998) argued when he described art-production as a 'technology of enchantment'.

In this sense, the function of 'stylistic' decoration is to act upon the viewer to enchant, persuade and impress. Like other functional characteristics, stylistic characteristics are only valuable if they address the particular social-technical needs of the consuming community. The distinction between 'function' and 'style' is worth making, however, because it is possible to produce superficially similar looking objects with different bodily techniques. Where it is possible to document it therefore, the transfer of actual bodily techniques suggests more fundamental shifts in social practice, and, of course, in the level of movement, socio-economic integration and exchange between different communities.

Pottery provides a good example to highlight the distinction of 'style'. In the culture-historical model of cultural change, similarly-styled pottery implied the migration of potters. However, as well as rejecting the 'pots as peoples' argument, ever-more-detailed studies of ceramic assemblages have identified differences in production techniques for superficially similar wares that make it very likely that different (and only distantly connected) individuals were involved in production. This forces us to search for different mechanisms for spreading 'styles', perhaps through alternative media. Where techniques are similar, however, it implies a much closer integration. We will return to this again, for example when we discuss the Kura-Arax and related pottery (in Section 5.6.5) and painted pottery wares of the late 3rd and early 2nd millennia BC (in Section 6.6).

1.5.3 Cross-craft interaction: technology, aesthetics and skeuomorphism

Explaining large zones of similar object forms and styles despite apparent local manufacture can often be difficult, but such phenomena must either be a result of frequent contact between technicians across the region, or else they must document the movement of an intermediate medium through which the motifs and surface aesthetics (or 'epiaesthetics') are passed. The nature of disciplinary specialism in archaeology means that different materials are normally studied by different specialists. Whilst this is a necessary division of labour resulting from the need for developed expertise and the sheer time needed to process archaeological finds, a collatoral consequence is that the development of objects in each material category is effectively treated as a "closed-system" (Nakou 2007). The cross-craft and cross-technological links between, for example, metals and ceramics, tend to get obscured, and the explanation for a change in one type rarely sought in the other. In fact, 'stylistic' influence very frequently appears across material boundaries: motifs distinctive to one media (*e.g.* textiles, basketry or netting) may appear as patterns on another (*i.e.* pots); technical features of one media (*e.g.* rivets required to hold together a metal vessel) appear on another (*e.g.* mock rivets in ceramic vessels). The second of these – the mimicry of a feature from another material (*e.g.* shape) that is technically redundant in the new material – is an example of what is strictly meant by 'skeuomorphism'. However, the transfer of techniques and styles between media – 'inter-media flows' – are likely to be far more complicated than we can easily perceive, flowing to-and-fro in multiple directions. Both 'skeuomorphism' and 'migrant motifs' are simply different aspects along a spectrum of 'cross-craft interaction', which should also include other aesthetic mimicry (including colour, sound, taste).

The potential for such cross-craft interaction is particular high when there are overlapping technologies (or rather bodily techniques) between what we tend to identify as separate crafts: the best example of this is of course the overlap of pyrotechnical skills required to work metals or fire clay. Pottery is, of course, one of the best materials in which to identify skeuomorphism or features from other crafts. Clay is a uniquely plastic material, in the sense that it can take on many different forms, colours and 'materialities' (patterns, reflectivity, feel to touch, sound, weight) which mimic other materials. Indeed in the absence of the constellation of materials we call 'plastics' today[18] – which are used in a similar skeuomorphic manner – it is difficult to find a similarly chameleonic material to match clay in the ancient world. Additionally, ceramic objects or fragments survive extremely well in the archaeological record, because they were generally low value, difficult to recycle and not prone to decomposition, and thus were more often deposited in the ground, either as broken waste (on settlements or middens) or as symbolic avatar for its metal, stone or other equivalents (particularly in graves) to be found by archaeologists. Indeed, pottery may often provide the only surviving evidence for similar objects in different media that have not survived, and thus offer clues to patterns of consumption and flows of those materials, which would otherwise remain hidden.

The recognition of skeuomorphism has a long pedigree in the history of archaeology: many of the discipline's earliest practitioners and theorists were concerned with and discussed freely the 'skeuomorphic' aspects and origins of pottery designs (including J. Evans, Childe, A. Evans, Mallowan: see *e.g.* Childe 1915; Schachermeyr 1955). Interest in such inter-craft connections has continued sporadically since (*e.g.* Vickers 1985; McGovern and Notis 1989; Vickers and Gill 1994; Knappett 2002; Nakou 2007), though because of the difficulty of 'proving' skeuomorphism in the absence of prototypes, these approaches are often treated circumspectly or even with downright distrust (see, for example, R. M. Cook 1987). A healthy scepticism towards particular claims for skeuomorphism is always necessary of course, but too often critique arises from an essentialist assumption of material categories. Instead such scepticism should start from an assumption of the 'unboundedness' of materials, and accept the possibility for varying intensities of 'influence' between them, based on synergies of function, form and decoration but also technique and bodily praxis.

An emphasis on cross-craft interaction and on the 'unboundedness' of material categories is an important theme in this book. Cross-craft interaction reminds us of the importance of the social role of materials rather than their abstract archaeological categorization. Pottery in particular appears to offer ways to examine the circulation of otherwise 'invisible flows' through 'inter-media flows': indeed, the distribution of such pottery seems likely to index aspects of the flow of the materials they reference. Because of this, the reader should not be surprised to find pottery discussed in chapters nominally devoted to stone, metal or textile (as skeuomorphs and proxies); metal jewellery discussed under textiles (as adornment); or metal discussed under stones (as weights).

18 The defining feature of this material is often forgotten now it has become so ubiquitous – despite its giveaway name. The complex interaction between plastic jewellery and the precious jewels they tend to mimic could provide a fascinating insight into modern value systems (particularly moral values in clothing, dress and class).

1.5.4 Value, aesthetics and colour 'symbolism'

Related to this focus on cross-craft interaction will be a sustained thread of the analysis devoted to value systems as indicated through material culture. By 'value systems' I refer to the way in which communities promote certain material qualities over others as inter-agent identities are negotiated. This will involve discussion of colour (for example in the form of consumption of coloured stone); the deposition of apparently 'valued' material (such as metals) in deliberate contexts; or the development of patterns (particularly in cloth) to differentiate people and maintain elite status.

Much of this could be argued to return us to the question of the emergence of elites in the 3rd millennium BC (discussed above), and the ways in which they established and maintained their status, but to do so would be to re-inscribe the bias of the archaeological record. We must therefore try to pick out competing and alternative value systems (both elite and 'ordinary') and value systems that can be seen to move geographically in an attempt to understand better the overall ancient economy. To what extent have such value systems remained consistent (and their boundaries remained relatively fixed) through time? And where they have changed or been maintained, what were the mechanisms of such processes?

The context of consumption is of vital importance to unpicking these ideas. For example, metal vessels which are deployed in burial contexts in one region, may be deployed as drinking vessels in another: the first group may also be used in drinking contexts, but the social or political imperatives which drive the final deployment is different, implying a boundary between social transfer of ideas which may contradict the suggestion of free-exchange. One of the challenges of studying interaction is to delineate not just the quantitative nature of interaction, but also the qualitative nature. Is interaction and exchange always peaceable, or can it also be contradictory and oppositional? There is always a tension between desire for difference and similarity in the nature of cross-cultural interaction: the exchange of exotic gifts amongst elite groups for example, relies on the exotic and different nature of the material or object being exchanged. Equally however, the created value for those objects creates a trajectory that pulls desires and styles in the respective regions closer together.

1.6 Research questions: routes and material flows

In this book, three main categories of ancient materials – stone (Chapter 4), metal (Chapter 5) and textiles (Chapter 6)– are examined, and an attempt is made to trace the routes of their exchange or flow through a variety of sources. One of the aims of this work is to attempt to address this question of parallel or synchronic transformations as evidence for the global interconnectivity of peoples in both regions (and, of course, in intermediate areas such as Iran, Mesopotamia and the steppes, or beyond). To do this, it is necessary to compare the flows through different regions of the same sets of materials together in their geographic context, to open possible paths of synergy. Through the medium of these materials, we will also see the transfer of social values, ideas and practices with wider consequences for socio-cultural development. It will soon become clear that direct evidence of these particular materials is often not the most informative source for reconstructing their circulation – and that we must turn to alternative materials (particularly pottery) to provide more detailed and alternative types of information of socio-economic systems. Before turning to the specific data on these materials, however, the first two chapters will deal with the relationship between flows and landscape, through 'routes'. This will involve a review of the concept of 'the route' and our knowledge about prehistoric and historic routes in general (Chapter 2), and then an exploration of modern mapping (or rather 'GIS') techniques which might allow us to visualize and understand the relationship between landscapes and flows of material more clearly, including the proposal of a novel visualization technique based on cost-surface analysis, the outputs of this termed 'archaeotopograms' (Chapter 3).

The main research questions that this book sets out to address include:

- What methods can we use to reconstruct and understand the physical routes of movement and interaction during the 3rd and early 2nd millennium BC in the Near East?
  - Can GIS-based cost-surface analyses help us to visualize routes and the spatiality of exchange better than traditional methodologies, especially where no physical roads have survived?
- How much can we reconstruct about the flows of the three selected material foci (stones, metals and textiles) in this period, using direct and indirect evidence?
  - What were the consequences of the specific 'materialities' of these categories on social values and practice, and what significance did these have for interaction between regions?
  - To what extent can evidence of cross-craft-interaction help us to understand the flow of materials that are poorly represented or otherwise invisible in the archaeological record?
- What effects did the flow of these materials have on the trajectories of communities and cultures in our case-study regions?
  - Can we explain the longevity of the Kura-Arax phenomenon or the emergence of the BMAC, on the basis of information about the flow of materials and techniques, or the lack of it, in and beyond these regions?
- To what extent does the evidence point toward the existence of prehistoric precursors to the famed 'Silk Road'?

Chapter 2

Routes: on the Trail of History and Myth

2.1 Introduction

The traditional method to reconstruct the physical routes of movement and interaction of certain periods for which no extant road systems have been uncovered in the archaeological record (including the 3rd and early 2nd millennium that form the case-studies for this book), relies on two approaches. The first is the examination of the morphology of the landscape itself, which will be dealt with in more detail in Chapter 3. The second, which in practice is often given more weight, is to look at the evidence from later periods from which there is better archaeological or textual data for the location of routes and roads – *i.e.* to create a template of 'prehistoric routes' based on what is known from 'historical' ones: such as Roman roads, known pilgrimage routes (including highly-organized pilgrimages such as the Hajj), 'famous' routes such as the 'Silk road' or 'Khorasan road' or the Achaemenid 'royal road(s)'. This approach relies on the widely-held assumption that routes remain more-or-less stable over very long periods of time and that the connections between regions follow the same paths and roads, over the same mountain passes, and along the same rivers and sea-routes.

This idea could be described as the principle of 'route inertia', and is the founding assumption, to differing extents, of most studies of ancient route or road systems in historical geography. The problem is, however, the principle is invoked – albeit anonymously – more often than it is actually demonstrated to be true empirically, primarily because it is an extremely difficult principle to test. Similarly, because it is normally an unspoken principle, questions about the changing hierarchy of routes or 'density of travel' along them are often ignored in favour of drawing lines of supposed paths on the map. This chapter therefore has two primary aims: the first is to catalogue the evidence for a sample of such 'historical routes' in and beyond our case-study regions, which could, if using the principle of route inertia, be used as comparison for our period of interest. The presentation of this evidence is structured first by the type of evidence (archaeological, textual) and then in reverse chronological order. The second aim of this chapter is, through the aforementioned review, to assess the usefulness of route inertia as a method to analyse dynamic densities of movement and spatiality of ancient interaction, and to consider the source of the principle's academic tenacity (as visual and narrative trope). In this chapter the exact meaning of the term 'route' will be left relatively open, to enable a wide field of data to be assembled.

a

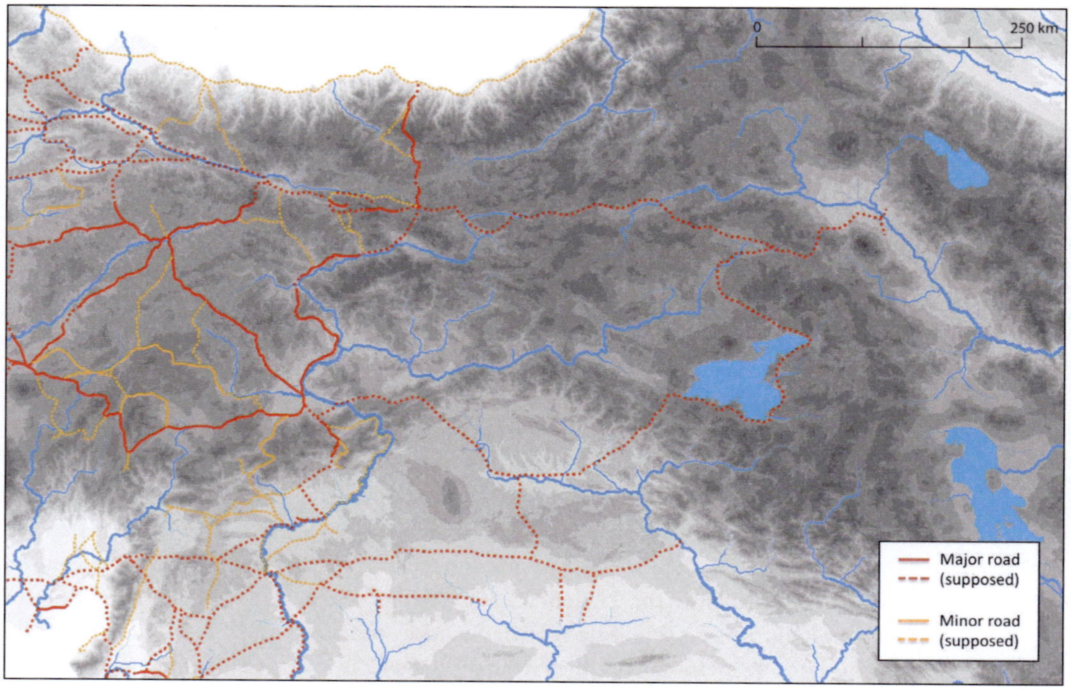

b

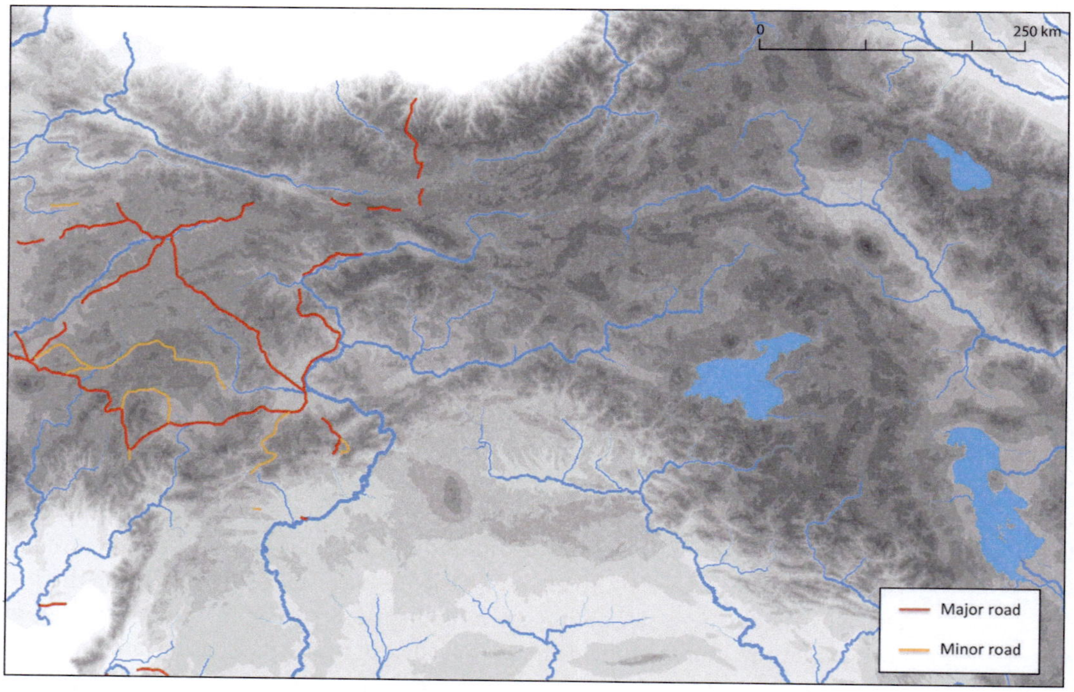

Figure 2.1 Roman roads in eastern Anatolia based on data from the Barrington Atlas: (a) includes those roads which are assumed to have existed but for which no clear archaeological evidence has been uncovered; (b) shows only those roads with clear archaeological evidence.

2.2 Routes and roads from archaeological remains

2.2.1 Roman roads (c. 500BC - AD500)

Roman roads represent, perhaps, the most comprehensively-studied system of roadways and major routes, for a period of around 500-1000 years, though not all Roman roads should be assumed to have remained stable or in continuous use. Since many roads commonly dated to the Roman era are recognisable archaeologically either directly through paving or indirectly from relict lines on the landscape (such as field boundaries or modern roads which follow the same lines) and the location of milestones, we are able to reconstruct a substantial portion of the network. There are also textual itineraries (such as the Antonine Itineraries), which provide distances and place-names for many routes which can be matched to physical remains of towns. Classical archaeologists have long since made the road system and these itineraries a focus of study (see *e.g.* French 1988) with the result that atlases such as the Barrington, which cover the entire Greek and Roman worlds (*e.g.* Talbert 2000)[19], now include detailed location information about roads. Roman roads are potentially useful in that they document both the existence of a route (and often presumably the pre-existence of a route before the construction of a road), but they can also give some idea of a hierarchy of routes through indications from construction (size, quality) and, potentially, use (*e.g.* wear from wheel tracks). That they document earlier roads or non-constructed routes can be deduced by the fact that they are built to connect settlements that we know are much older than the road. Indeed, analysis undertaken by Michele Massa in western Anatolia, comparing Roman roads with the distribution of major Early Bronze Age settlement mounds suggests a degree of correlation which deserves further investigation (see *e.g.* Massa 2011). This represents one of the few solid pieces of evidence that might support the idea that the main routes remained stable for centuries or even millennia – though we should be wary of assuming that the *hierarchies* of routes and the relative *density* of travel along them also remained stable.

Roman roads do have serious limitations for our purposes, however. The first is that of geographic extent: only parts of Transcaucasia/eastern Anatolia were ever part of the empire and Central Asia remained completely outside the reach of Roman engineers, with probable equivalent and contemporary Parthian or Sassanian roads understudied. Even in eastern Anatolia, the number of archaeologically confirmed or physically identified roads (as opposed to supposed roads assumed from texts) are very few (compare Figure 2.1a and 2.1b). We also lack any knowledge of the substrate of more ephemeral local tracks and branch roads, which presumably existed alongside the more durable metalled roads. It is also worth noting that Roman roads were not always built along the most popular directions of travel or trade. For example, where roads were built along *limes*, lines of defence or borders, their route reflects the desire to connect forts and other military stations rather than the most heavily trodden trade route. This seems to be the case, for example, for the north-south road that followed the western side of the Upper Euphrates, the river serving as an eastern frontier of empire for an extended period (Stark 1966). We could thus be easily misled to

19 There is also a project 'Pleiades' to create an online version of the Barrington Atlas, as part of the Ancient World Mapping Center programme (http://pleiades.stoa.org/).

think that certain routes had more 'trade' or 'interactive' significance than they actually had. The extremely straight linearity of many Roman roads (and apparent economical or mathematical rationalism with which they were built) is likely to be unrepresentative of 'routes' over the *longue durée* until the construction of railways in the 19th century AD.

2.2.2 Caravanserais and caravan routes of the Islamic era

During the Islamic era, travel across western and central Asia was aided considerably by the existence of many 'caravanserais' distributed between and within major cities along major highways. The surviving buildings or their archaeological traces are of particular interest for the reconstruction of routes both because their distribution offers indicators of the location of roads between major cities and because their size and numbers appear to indicate the importance of the routes in question, revealing clues to the hierarchy of roads. The word 'caravanserai' itself appears to be a European invention of uncertain coinage (and subsequently re-imported to Middle Eastern languages in tourist brochures), based on Persian-Turkish words *kervan* (a caravan in the sense of a travelling group) and *saray* (a palace or building with courtyard). The native Persian and Turkish equivalent is *han* or *khan*, which was and is still used to describe the countryside or roadside inns we normally associate with caravanserais with lodgings for travellers, animals and cargo. It is also applied to analogous buildings and institutions devoted to trade or commerce at the edges of or within cities, some of which had lodgings although others were only used for commerce or storage of goods. Other words for the same buildings in different regions include: *funduk*, *wakala*, *kaysariya* and *rabat*. The distribution of such buildings is extremely wide: from eastern Europe

Figure 2.3. Caravanserais in eastern Anatolia and environs (13th to 19th century AD). Overlain with Delaunay triangulation to indicate 'natural neighbours'. Those vertices with a length of between 10 and 32km are highlighted in red (to highlight caravanserais spaced less than an average day's walk distance apart, when topography is not taken into account).

Figure 2.4. Caravanserais in western Central Asia and environs (13th to 19th century AD). Overlain with Delaunay triangulation to indicate 'natural neighbours'.

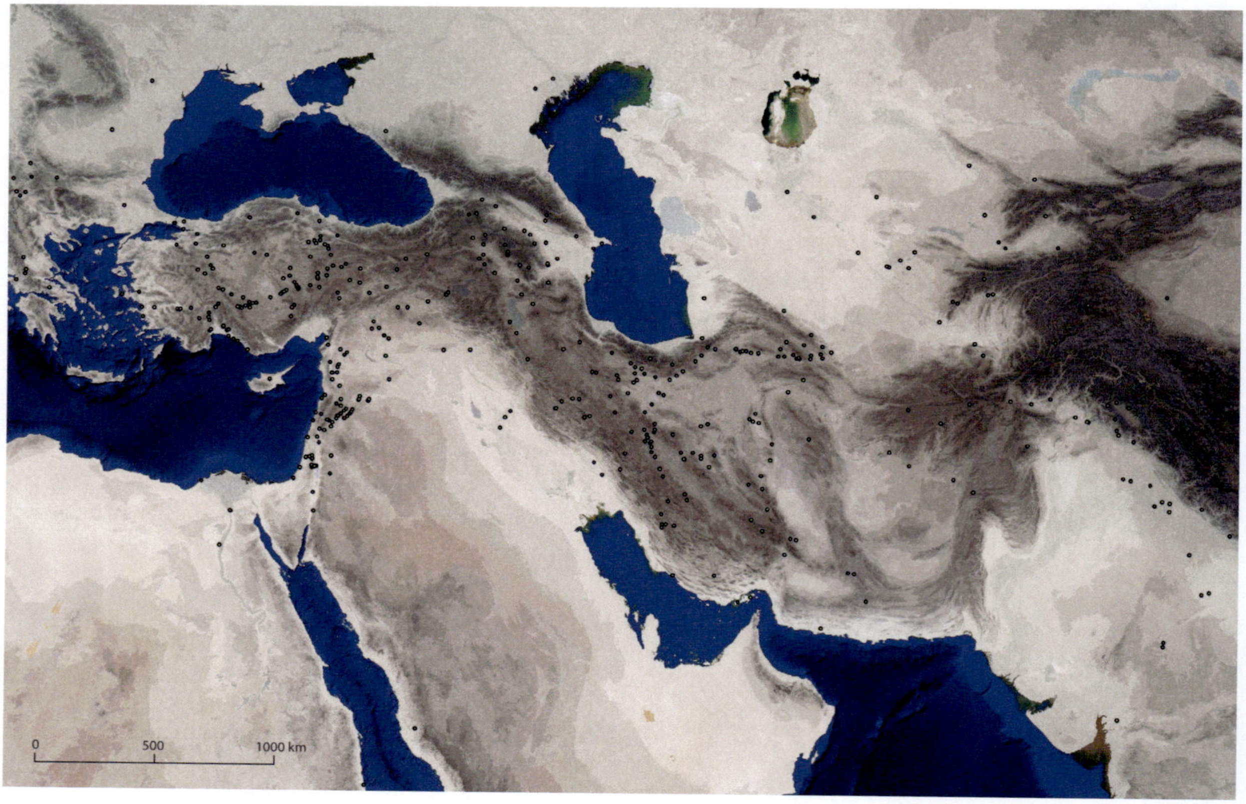

Figure 2.2. Distribution of caravanserais/(k)hans across the Near East (13th to 19th century AD) based on the database of T. M. Ciolek (see http://www.ciolek.com/owtrad.html).

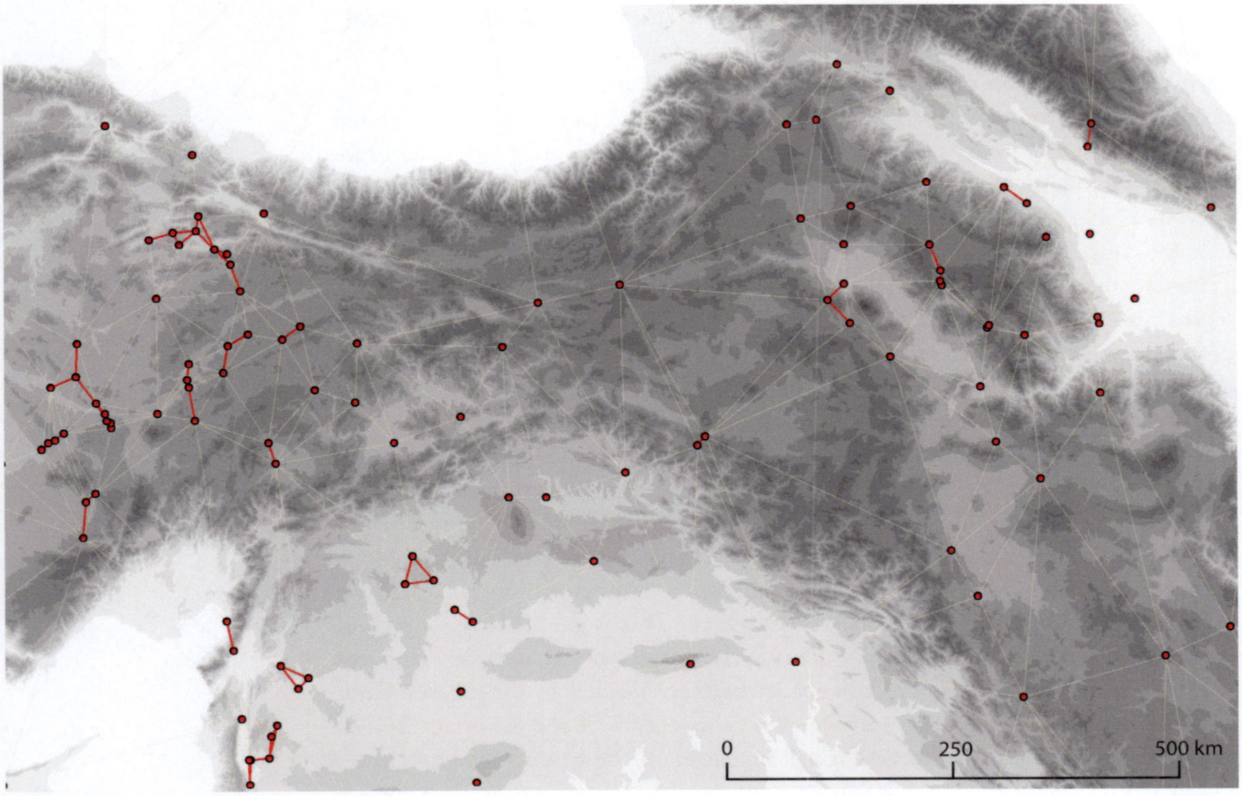
0
250
500 km

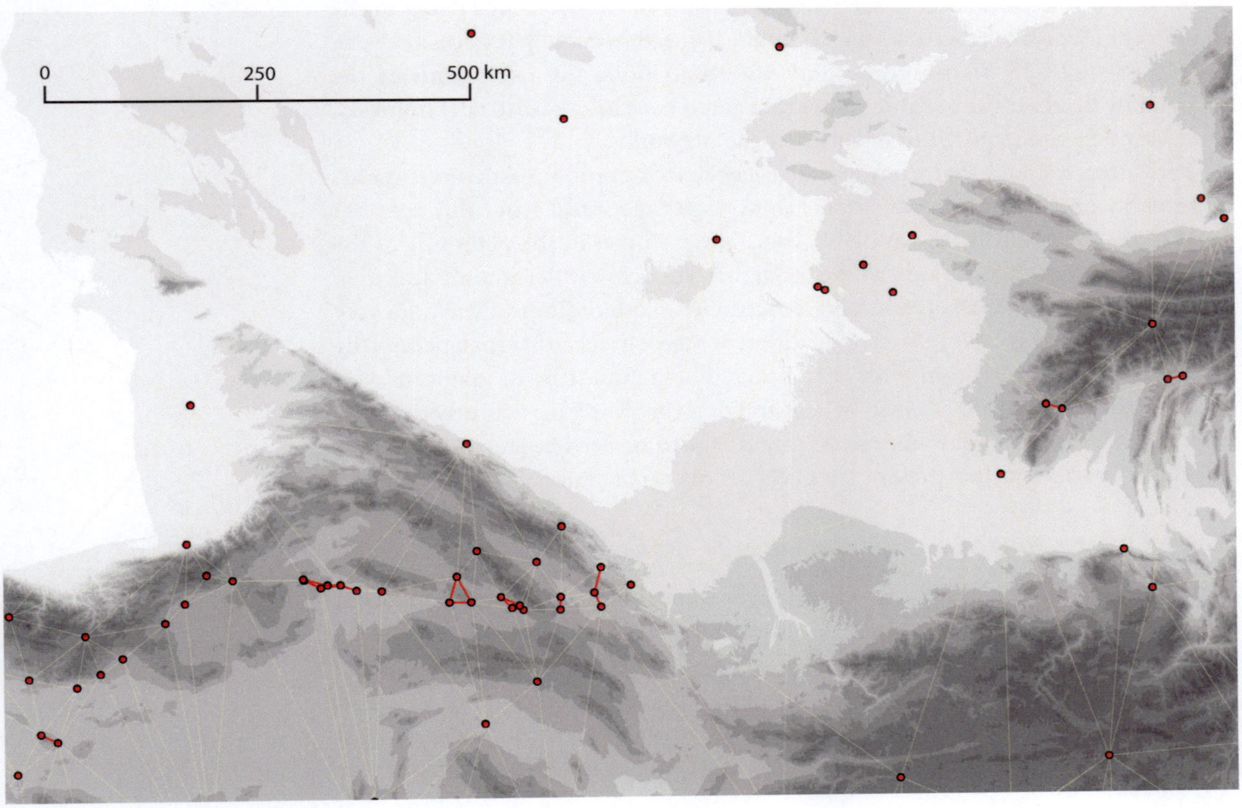
0
250
500 km

to northern India, with considerable architectural, structural and thus presumably functional variations through time and space. A number of research projects have been undertaken on the architecture and distribution of caravanserais (*e.g.* Erdmann and Erdmann 1961; Le Bigre 1996), including a UNESCO-funded database of all known caravanserais which was started soon after 1998 (Le Bigre 2004)[20]. Unfortunately, whilst there have been some very interesting spin-offs (Tavernari 2009; Del and Tavernari 2009), publication of the UNESCO project remains, until now, limited. Seperately, Tom Ciolek's OWTRAD (Old World Traders) project includes a database with over 700 georeferenced caravanserais, represented here as Figure 2.2. Though this database also represents a work-in-progress, its size makes it a useful tool, *e.g.* for the identification of caravanserai clusters which can be taken as indications of high traffic.

The available data shows that both Transcaucasua/eastern Anatolia and western Central Asia contain a considerable number of caravanserais (Figures 2.3 and 2.4). In order to reconstruct a road or route, however, the distribution of caravanserais must meet certain conditions. Primarily there must be a sufficient number of such buildings arranged in a linear manner: too few caravanserais or collections of caravanserais and pattern recognition becomes impossible. In this case we cannot say which 'node' was connected to which. This is particularly a problem in relatively flat regions such as northern Syria, Mesopotamia and the deserts of Turkmenistan. Supplementary evidence is therefore needed in these cases: from both the accounts of medieval travellers (see above; *cf.* Del and Tavernari 2009), and alternative classes of archaeological evidence. In the case of Turkmenistan, recent fieldwork in the Karakum desert has revealed many categories of station below that of the caravanserai, including wells and small rest camps dated to the Islamic era (Wordsworth 2010) which might be used to highlight paths between larger rest-stops. In Transcaucasia and eastern Anatolia the restrictions of the topography decrease the possible routes that could have been realistically followed, so fewer nodes are required to reconstruct the network.

It is often claimed that caravanserais were spaced around 30km apart (based presumably on a common estimate for an average day's walk) – but this does not seem to be borne out by the available data. For example, in the plain just to the west of Lake Van, 3 *hans* are grouped together very closely – certainly less than 30km. The date of construction and structure of the buildings of each are very different so they may have been serving different sets of users rather than necessarily having been in competition with each other. This reminds us to take care with undifferentiated 'dots on the map' when studying caravanserais: it is easy to forget that such institutions were not necessarily designed or used contemporaneously, nor did they necessarily represent channels for the same materials.

2.2.3 Other road fragments: the Pasinler road

In the Pasinler valley, near Erzurum (Figure 2.5), some short segments of constructed flagstone road have been reported (C. Sagona 1999). The investigator, Claudia Sagona, argues that the lack of Roman finds in the vicinity of the road and the date of the sites which it appears to connect mean that it is unlikely to be a Roman road; instead she argues that it is most likely to be a Median or Achaemenid road to the satrapy of Western Armenia, assumed to have been based at Altıntepe, near

20 http://www.unesco.org/culture/dialogue/eastwest/caravan/ Last accessed 7th July 2011.

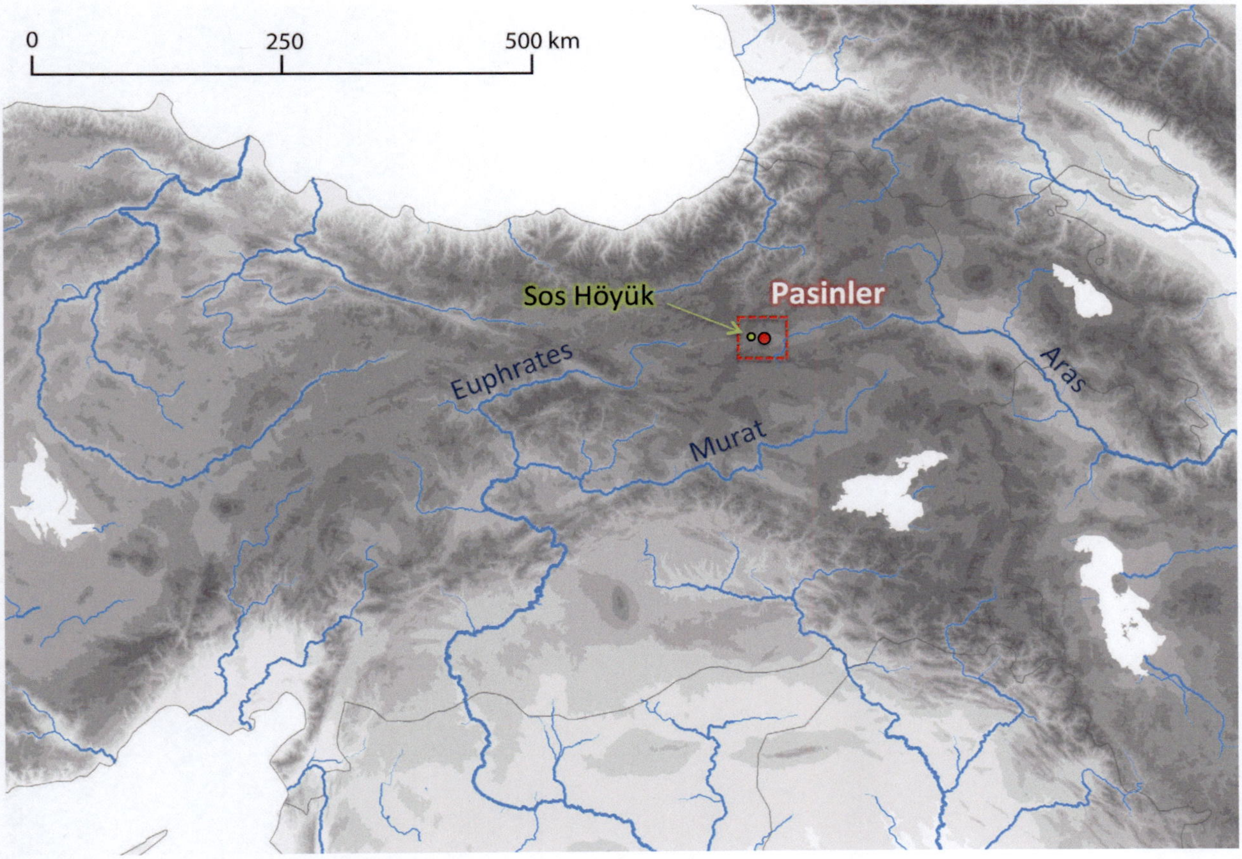

Figure 2.5. The location of the Pasinler valley near Erzurum, relative to 'eastern Anatolia'.

modern Erzıncan, and that the road was one that was followed by Xenophon as reported in *Anabasis* (C. Sagona 2004, 309-310). Though it is difficult to test this hypothesis, it does remind us that we should not assume that all paved roads in the archaeological record are Roman in date. If the road is indeed Achaemenid, it supports the idea that this valley might have had some value as a major route of movement before 500BC – but the short length and fragmentary nature of this information makes it difficult to say much more.

2.2.4 Linear hollow ways in northern Mesopotamia

In certain cases, organically emergent pathways can be inscribed into the landscape and hence show, very specifically, the local routes of movement. This appears to be the case for the so-called linear 'hollow ways' identified around tell sites in northern Syria. Remarkably, ancient non-paved pathways have been preserved as eroded depressions or features which retain water differently to surrounding areas (T. J. Wilkinson 1993; Ur 2009), which show up very well as linear features on aerial and satellite photographs (Ur 2003). Many of these tracks appear to represent radial paths from settlements to pastures (*cf.* the 'pathways' discussed below in Section 3.2.1), though a few must have formed inter-site and inter-regional highways. Dating these hollow ways remains difficult, however. On the basis of which sites they link, and scatters of material found during excavation of a small number of these features, it seems many belong the local Early Bronze Age. If so, they offer a unique insight into 'landscapes of movement' during

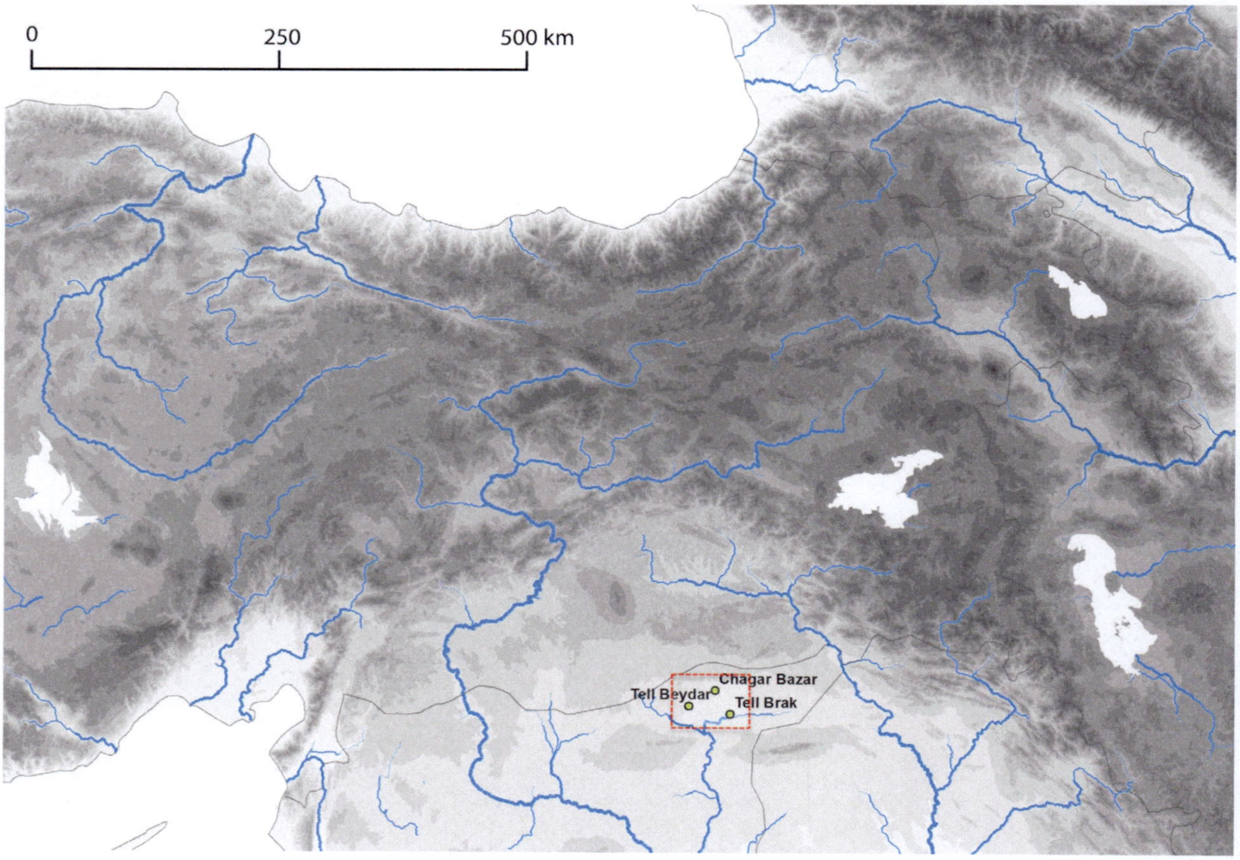
0
250
500 km
Chagar Bazar
Tell Beydar
Tell Brak

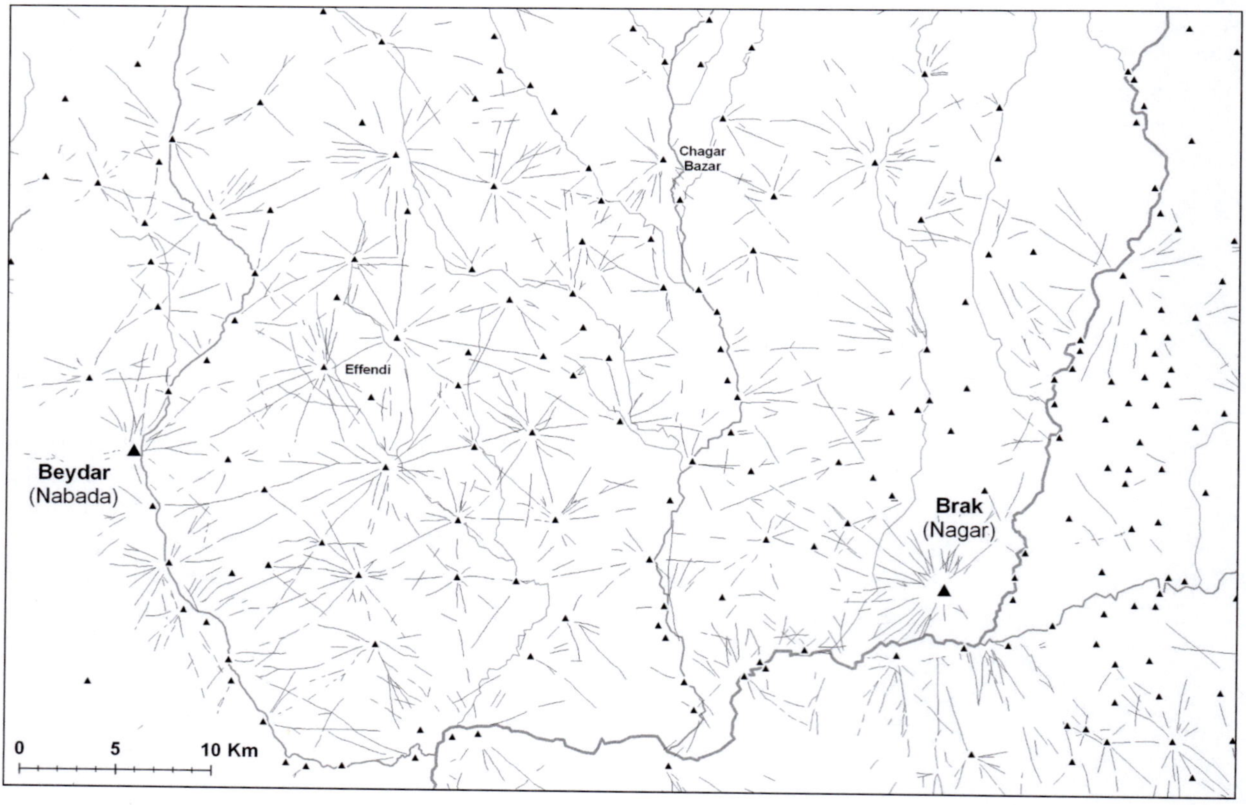
Chagar
Bazar
Effendi
Beydar
(Nabada)
Brak
(Nagar)
0
5
10 Km

the 3rd millennium and allow for detailed reconstruction of human settlement trends (Menze and Ur 2012). Hollow ways have been identified over a very large area of northern Mesopotamia (see Figure 2.6), but the landscape in which they have been identified is rather unique, and atypical of much of the Near East. Whether or not equivalent pathways existed in the past in other areas, for geomorphological reasons, it is unlikely that well-preserved networks will ever be found in the mountainous regions of Transcaucasia and eastern Anatolia and so far nothing similar has been identified in western Central Asia. As such the hollow ways of northern Mesopotamia offer a tantalizing glimpse of the mobility patterns of urban sites in a specific region, but it is difficult to judge how extensive these sorts of networks were.

2.3 Routes and roads in modern or recent reports

2.3.1 Modern road maps

Many modern roads have often been assumed to lie on ancient routes and therefore provide a model for back projection. There are some confirmed examples of this – for example where modern roads have been shown archaeologically (or geographically) to follow the line of Roman roads. However, it is far more difficult to demonstrate this fact for most periods of history since recognizable constructed roads were apparently exceptional in antiquity, and organic braided pathways were more likely to have been typical. It seems at first glance reasonable to assume that, even if the particular paths are not the same, modern highways still give clues to broad 'corridors' of movement between regions and, given the extreme intensity of modern transportation building programmes, it would be surprising if there were no overlap between the modern roads and ancient paths in many locations. But, crucially, the modern map has the potential to be very misleading. First, given the very different requirements of modern and ancient transportation technology, it is unlikely that the route 'hierarchies' of modern and ancient landscapes are identical. For example, long straight roads through empty desert are quite feasible with an automobile (such as those crossing the Karakum in modern Turkmenistan) but are not likely to have been used by travellers on foot or with pack-animals because of the need to find water sources. Water is one aspect of the 'tyranny of distance' that was faced by ancient travellers, and which is hardly felt by modern passengers. Additionally modern administrative divisions or national borders have, in the last centuries, militated against the construction of roads in locations which otherwise may have carried substantial traffic in the past. The borders between Turkey, Iran, Armenia and Azerbaijan in eastern Anatolia, and the border between Iran and Turkmenistan illustrate this well: here there are very few roads crossing these major boundaries, and there are only a small number of official border posts, despite the existence of many natural passes through the hills. Before the modern era, and especially before the invention of the modern state (and later isolation of the USSR), the lack of border restrictions would have made both frontier areas much more porous than they currently appear on the map[21].

Figure 2.6. The location of recognized linear hollow ways of the Near East, relative to 'eastern Anatolia' (below: hollow way map from Ur 2009, 188).

21 Indeed, it is worth noting that the Iranian-Turkish border remains porous to the extent that many recent refugees from Afghanistan and Iran cross here through 'unofficial' routes, if news reports following the Iranian presidential elections of 2010 are correct.

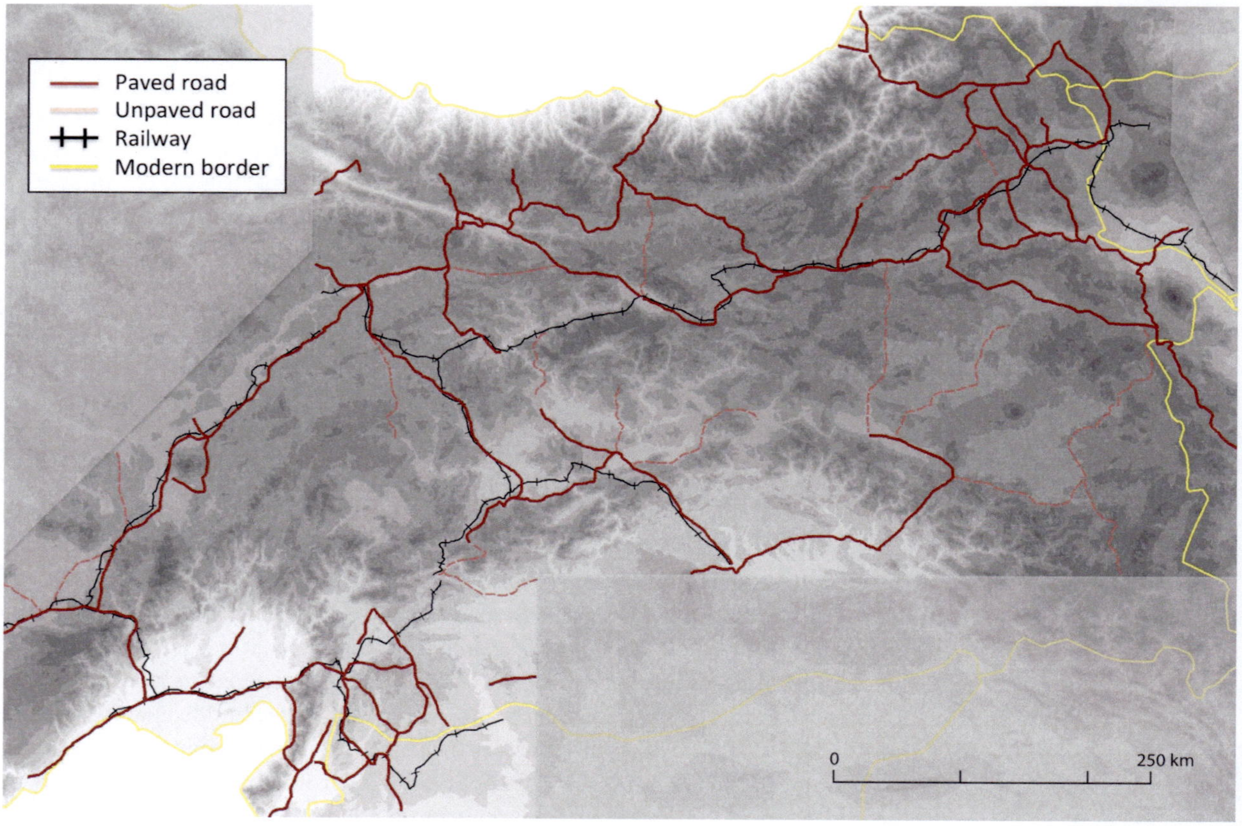
Paved road
Unpaved road
Railway
Modern border
0
250 km

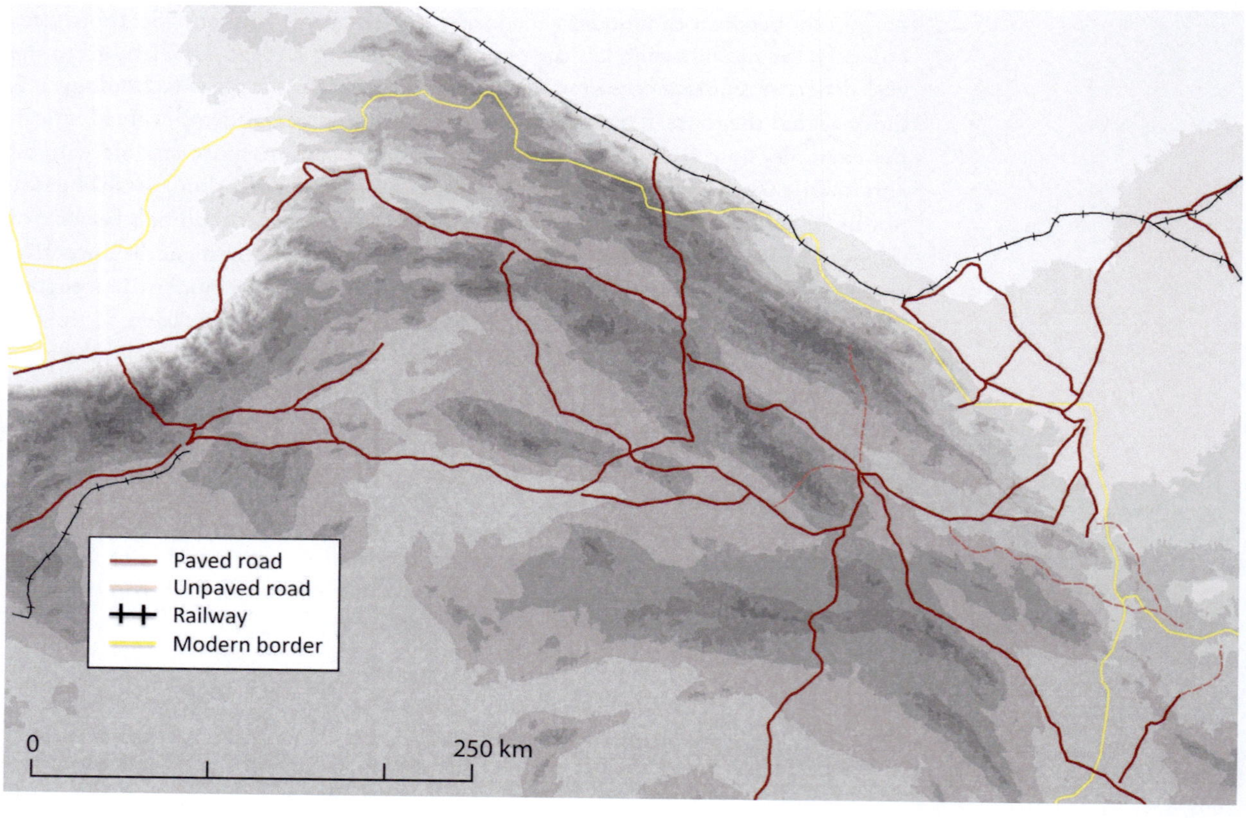
Paved road
Unpaved road
Railway
Modern border
0
250 km

2.3.2 Pre-automobile maps: British Naval Intelligence reports

Information about the state of road-networks in the late 19th and early 20th centuries just prior to mass motorization provides a relatively comprehensive 'pedestrian' map of pre-modern travel in our regions of interest. One set of very useful sources on roads, routes and mountain passes are the originally secret but now long-since declassified handbooks compiled firstly by the Geographical Section of the Naval Intelligence Division of the British Admiralty between 1917-1922 and then by British Naval Intelligence Division between 1941-1946. These reports, published by region or by country, were produced as extremely detailed geographical surveys, apparently designed to support real or contigent planning of military maneouvres within the regions in question. As well as roads, water connections, passes, bridges and methods of transport, later editions also included details of local industries, ethnic and cultural geographies and political and historical background. Different parts of Transcaucasia/eastern Anatolia are covered by a series of editions including *Asia Minor* (4 volumes, 1919), *Mesopotamia* (4 volumes, 1918) and *Turkey* (2 volumes, 1942, 1943), while some parts of south-west Central Asia are covered by *Persia* (2 volumes, 1945), and a few historical and ethnographic notes in the more politically-oriented *Turanians and Pan-Turanianism* (1 volume, 1920). The older volumes include detailed 'marching' maps and textual route gazetteers for the Ottoman empire – which facilitate a 'pedestrian' reconstruction of routes. However, they do not stretch far into the north-east of modern Turkey and certainly not into areas of, by then, Soviet Caucasus. The 1940s editions include some useful large-scale maps (*Turkey*, figs. 34, 41; *Persia*, figs. 12, 60), which, although they post-date the introduction of railways and the beginning of major segmentation by national boundaries, have the merit of highlighting major communication routes before the motor-car had become ubiquitous, Figures 2.7 and 2.8.

Figure 2.7. Major roads in eastern Anatolia according to British Naval Intelligence in 1942 (GBNID 1942).

2.3.3 Ethnographies: nomads, transhumance and nomadic crafts

Transhumance routes can follow both major inter-urban roads and minor pathways between mountains and plains and, though difficult to access, it seems likely that similar paths would have played an important role in the ancient past: the difficulty is in finding suitable data about such routes. Ethnographic accounts of nomadic movement in both eastern Anatolia (such as the *Yörük*, *Türkmen* and 'gypsy' *Çingene* in Anatolia) and central Asia (including variously named *Türkmen, Kazakh, Kyrgyz etc.*) are sadly few – at least in part because of political sensitivities in both regions. This is particularly unfortunate now that many nomadic groups have been settled either forcibly by the respective states or as a result of changing economic and administrative contexts over the last hundred years, so that the knowledge of their routes is being, or has already been, lost. Some attempts to record routes of transhumance have clearly been made (see, for example, the unsourced diagram from the Naval Intelligence Division's report on *Turkey*, fig. 79; adapted here as Figure 2.9). In the absence of direct reports, other sources of information could be exploited. For example, the distribution and transfer of textile motifs between nomadic groups and individual weavers could be used to reveal 'route-based' connections if combined with information about transhumance patterns.

Figure 2.8. Major roads in eastern Iran according to British Naval Intelligence in 1945 (GBNID 1945).

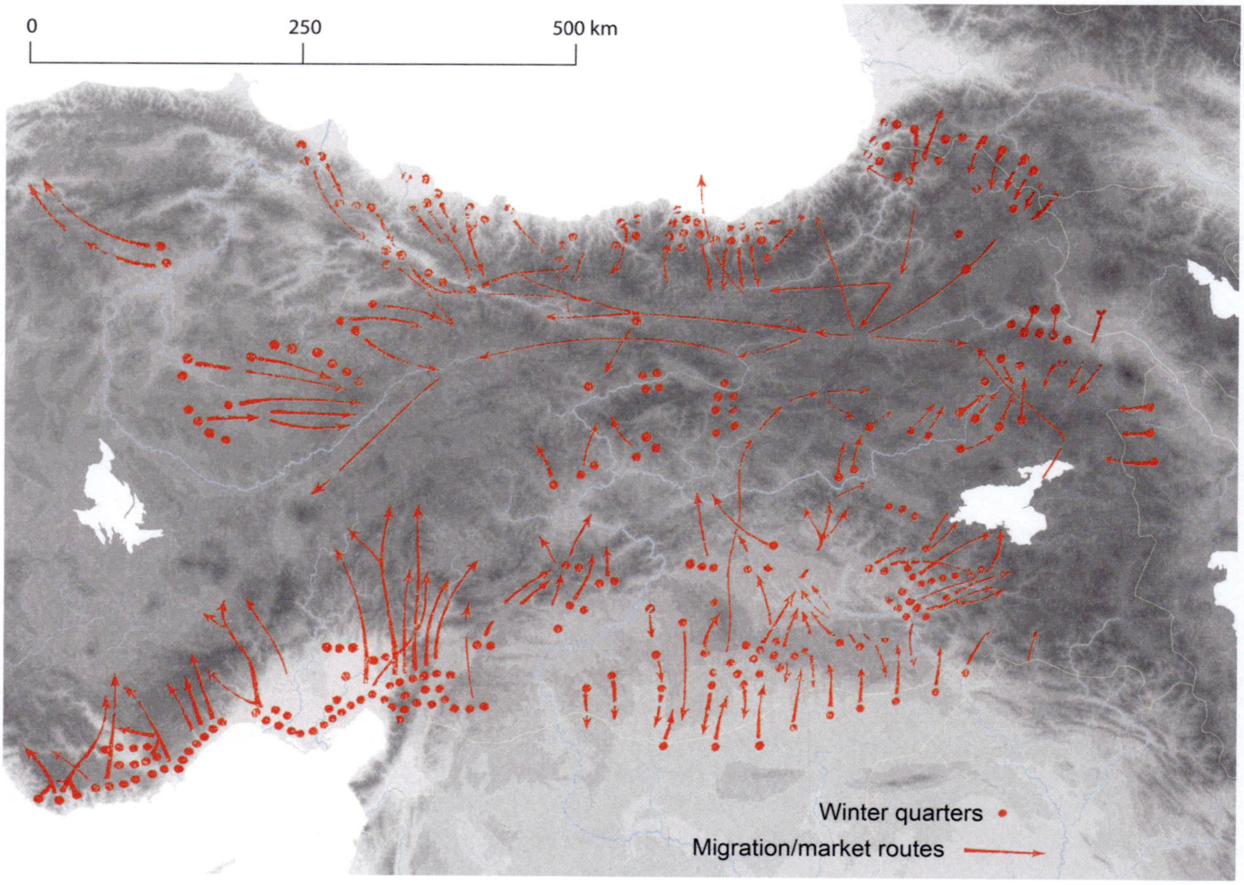

Figure 2.9. Routes of seasonal migration for pasture according to British Naval Intelligence in 1942 (GBNID 1942, 79).

2.4 Routes and roads in textual sources

2.4.1 Geographers, historians and itineraries

Textual geographies or histories provide one of the most obvious sources of information about routes during the classical, medieval and early modern periods. The nature of the texts as historical documents of course has consequences for their interpretation, especially where such geographies draw on older sources themselves. For example, the majority of medieval geographies – both 'Western' and 'Eastern' – were written in the context of an epistemology that valued and prioritized existing written authorities over collecting new empirical observations and reports. This is not to say that new observations or critical commentaries were not incorporated into new geographies at some periods (especially where based on recent travellers' reports), but rather that the traditions of what should and should not be included in such texts remained very conservative. Theoretically, these sources allow us to reconstruct ancient routes where sufficient description of cities and landscapes (or distances between cities) can be fixed to the modern map. Some geographers also started to visualize their geographical knowledge in various cartographic or pseudo-cartographic forms. Visual representations of routes can be found in documents as early as the Peutinger Table, dated to the late Antique period (though presumably based on earlier sources). Most early pre-Cartesian geographical diagrams do not include the sort of information about landscapes or 'objective' projection of the earth that would allow us to reconstruct shifting routes of movement. Where roads between cities are shown on early maps, they are normally symbolic or abstract straight-lines rather than 'accurate' scale representations of distances[22]. The invention of 'objective' scopic regimes that could map landscapes, cities and roads onto a grid, is a relatively recent one, starting during the European Renaissance. There are various documents with more or less abstract representations of connections (both in European and Oriental contexts), which have not been sufficiently mined by historians for the purposes of understanding route systems (see examples in Harley and Woodward 1987; 1992).

Herodotus and others on the Persian 'Royal Road(s)' (600-350BC)

The expansion of the Persian empire as far as the Aegean in the mid-1st millennium represented the largest politically-united geographic entity of its period. Most of our traditional views of the Achaemenid political and cultural structure still depends on the writings of contemporary Greek authors, though examinations of Persepolis tablets is beginning to offer new perspectives on both (Briant, Henkelman and Stolper 2008). The archaeological footprint of this political entity (outside of the imperial capital) is also, slowly, becoming better understood (*e.g.* Boucharlat 2002; Magee *et al.* 2005). Importantly, this power appears to have depended on excellent communication networks, with events monitored by the centre through an extensive and highly organized postal system, supported by an extensive road infrastructure. The admiring description by Herodotus (Hdt. 5.52) of 'the Royal Road' from Sardis (in western Asia Minor) to Susa, provides perhaps our most detailed geographical account of one part of this system. Herodotus indicates the existence of posting stations along the route with fresh horses and riders who could carry messages very quickly. This trans-Anatolian 'Royal Road' appears to have been only one of a number of such trunk roads, as hinted by both texts in the Achaemenid archives and a series of archaeologically identified sites which appear to represent 'posting stations' in Iran and Anatolia (Potts 2008b). The full system seems likely to have stretched far into western central Asia but so far we have few details about the actual location of such roads.

Several researchers have attempted to reconstruct the line of the trans-Anatolian 'Royal Road' on the basis of Herodotus' account, which is said to have been written around 425 BC (see Figure 2.10). Besides a few indisputable place-names, the task of reconstruction relies on fitting in the distances given in the ancient texts with prior knowledge of the landscape. The former orthodoxy about the likely route of the road was that Herodotus had made mistakes in the distances he provides and that the most likely route was one which swung round to the north (probably via Gordion or Angora/Ankara). A recent alternative interpretation, published by David French (the thick red line in Figure 2.10), argues that the account of Xerxes route into Asia, described in another part of Herodotus (Hdt. 7.26-31), appears to represent the same route, and that, by reinterpreting the translation of certain phrases (particularly with regards to 'crossing' the Halys river), an alternative

22 This also reflects the rather low-importance of Cartesian accuracy to the way that most people travel (see Ingold 2000). The London Underground map is an example of a non-geographical cartogram that is far more useful and successful than a true-Cartesian equivalent because it simplifies or removes the confusing surplus information about *space* in favour of the more important focus on *places* and their connections.

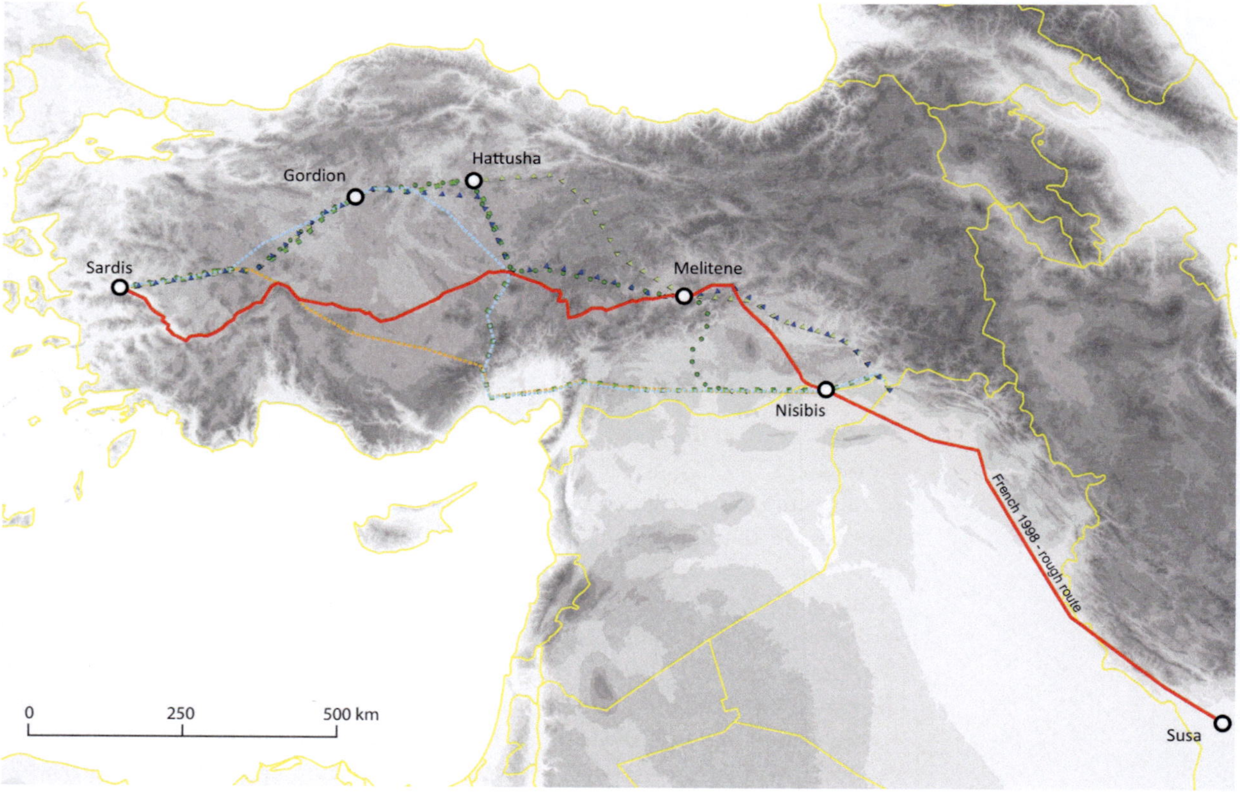

route may be proposed in which Herodotus' distances are accurate (French 1998, 15-16). He also argues that the march of Cyrus to the east described by Xenophon (Anabasis 1.2) written around 370BC, can be shown to more-or-less follow the same routes, as can the 'Common Road' described by Artemidorus and quoted by Strabo (14.2.29, C 663) – whose writings date to the mid-1st century BC to the mid-1st century AD – albeit with an alternative starting point and initial road from Ephesus (French 1998, 19-22). French's account is far more detailed in its understanding of the landscape than the earlier accounts, based on much more extensive fieldwork and the most up-to-date archaeological data on contemporary settlements and the routes of later Roman roads. If we assume that French's alternative reconstruction is correct rather than the orthodox one, the question remains how useful this information is to understanding contemporary interaction and exchange or whether it provides any model for still older route networks?

Figure 2.10. The Persian Royal Road: alternative reconstructions shown as: light green dots (Kiepert 1857), turquoise dots (Ramsey 1890), dark green dots (Macan/Hogarth c. 1895), orange line (Calder 1925), turquoise line (Müller 1994), dark blue dots (TAVO), solid red line (French 1998). For further details and references see Calder (1925), French (1998).

It seems likely that the 'Royal Road' established by the Persians made use of an already extant route of movement across Anatolia. We might also assume then that it represented the most heavily trodden corridor of the time (*i.e.* the 6th to 5th centuries BC). If so, then this route represents a kind of 'fossilization' of a major corridor of movement between the Aegean, Syria and the east. But given that it is only one (land-based) route (and we have no information about any trade conducted along it), we are still left with little idea of route hierarchies. The 'Royal Road' had a very specific purpose: to facilitate the ceremonial but rapid travel of the Persian king (and armies) between capitals and to transmit messages fast. It is possible that the Persians adopted a different route for their postal/ceremonial system from the one that was previously taken by traders, migrants or pilgrims. Given this uncertainty, projecting a particular *reconstructed* path retrospectively in time sounds dangerously anachronistic: it offers a selective fossilized snapshot of what may or may not have been the line of east-west transmission of goods, people or ideas at one particular moment. We can recognize the broad orientation of a corridor of east-west movement across Anatolia: in this sense understanding the Persian Royal Road increases our understanding of the *potential* of the landscape but not necessarily of the particulars of other periods.

2.4.2 Military campaigns

The routes taken by military campaigns represent a special form of textually derived itineraries with particular dynamics. Such campaigns depend on the feasibility of transporting large numbers of soldiers and their equipment down certain routes – in terms of topography, climate and the logistics of feeding an army. But whilst armies may follow the routes of major roads where such roads can expedite forward movement, it is equally possible that less-trodden routes are chosen precisely to provide an element of surprise to the army's opponents – which means we must take care in assuming the routes had wider non-military significance. The precision with which we can reconstruct the routes of particular campaigns also varies enormously. For the campaigns of Genghis Khan that had such a profound effect on the human geography of central Asia, for example, we only know which major cities he visited or destroyed. On the other hand, the routes taken by the Ottoman sultans Selim I in 1514 and Süleyman Kanuni (the Magnificent) in 1532 and 1548 respectively on their campaigns to Iran through eastern Anatolia, Azerbaijan and Armenia were recorded in day-by-day detail by scribes who travelled with the army. Both place-names and dates have survived, which allows us to reconstruct both the routes with a fair degree of precision (see Figure 2.11) and the speed with which the army travelled (typically around 15-20km per day).

Alexander's route into Asia (334-323BC)

The campaigns of Alexander lie somewhere in the middle of this continuum: none of the contemporary written accounts of Alexander's life or campaigns has survived to the present day, though several later commentaries that depended heavily on contemporary sources have allowed classicists to construct an overall narrative of the campaign. Alexander's expedition into Asia is of particular interest for a number of reasons: the relatively early date places it closer to our period of interest, and most importantly Alexander's military campaigns had a very long-term cultural impact over a wide area. Whilst his empire only lived for a short period, the consequences of its creation were much longer: it resulted in considerable eastward cultural borrowing (political, ideological and material) through the creation of 'Greek' cities and city-states as far away as Bactria and the Punjab. Of course, given the involvement of Persian power in the Aegean for at least two centuries before, there must have already been a considerable amount of travel (*i.e.* trade, migration and cultural interaction) in both directions[23] –

---

23 By way of example, we are told in textual sources from the Hellenistic era that following the Ionian revolt in 499-493BC, the family of priests from the temple of Didyma (on the east coast of the Aegean) migrated or were moved to Sogdia (in Central Asia). The full effects of this kind of long-distance 'linking up' by the Achaemenid system remains yet to be fully explored by researchers working on the 1st millennium BC.

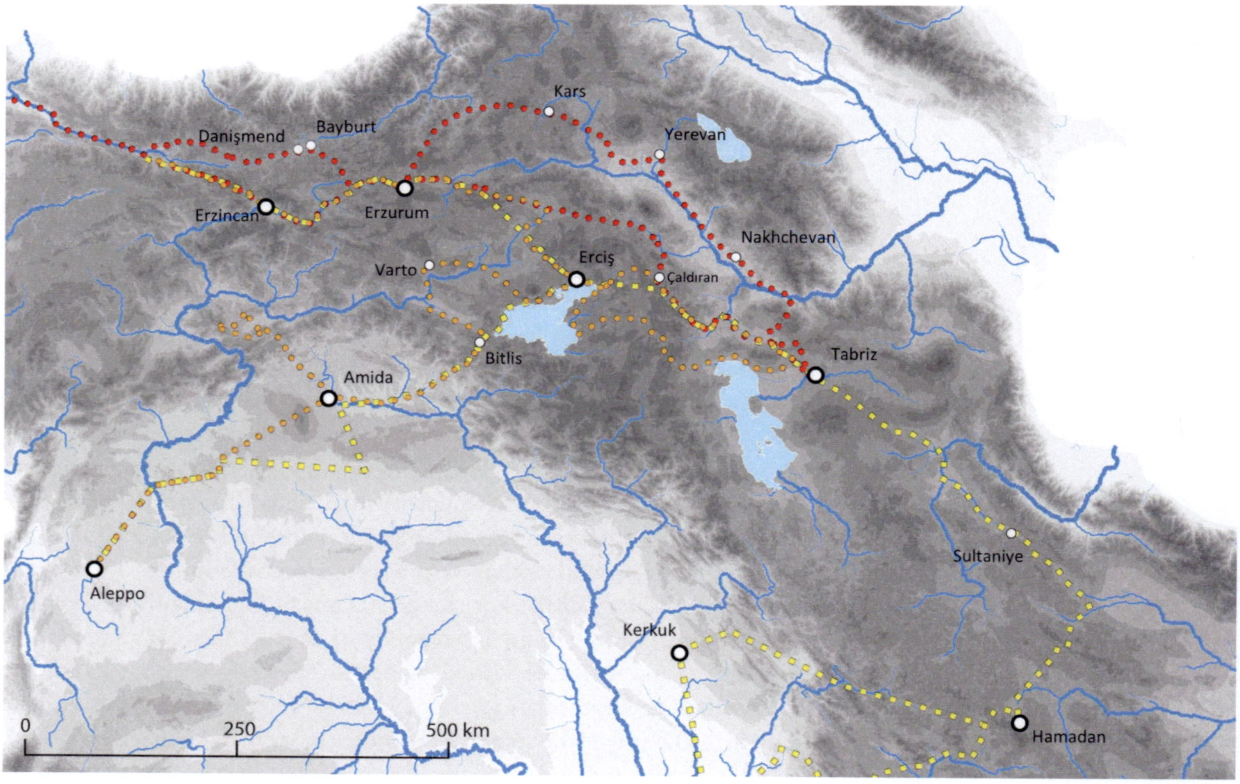
Kars
Danişmend
Bayburt
Yerevan
Erzincan
Erzurum
Nakhchevan
Erciş
Varto
Çaldıran
Bitlis
Tabriz
Amida
Sultaniye
Aleppo
Kerkuk
Hamadan
0
250
500 km

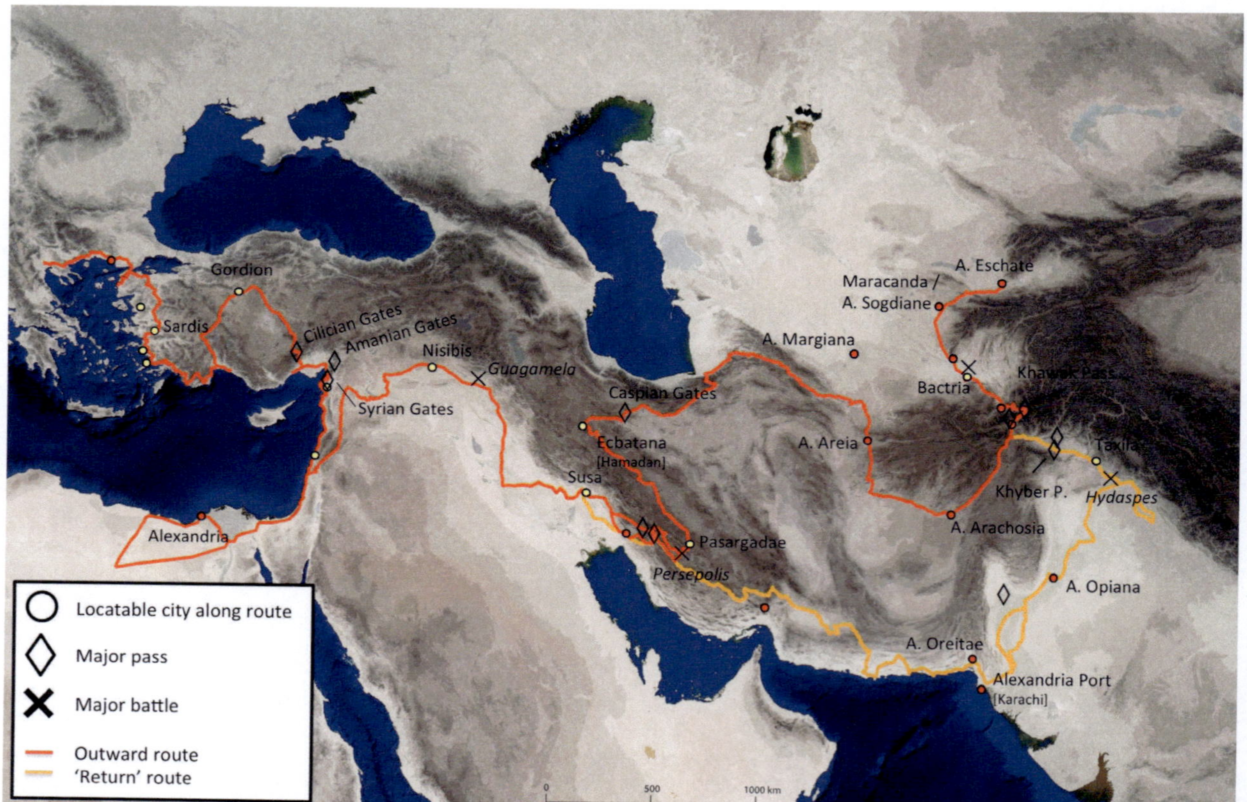
Gordion
Sardis
Cilician Gates
Amanian Gates
Nisibis
Guagamela
Syrian Gates
Alexandria
Caspian Gates
Ecbatana
[Hamadan]
Susa
Pasargadae
Persepolis
A. Margiana
Maracanda /
A. Sogdiane
A. Eschate
Bactria
A. Areia
Taxila
Khyber P.
Hydaspes
A. Arachosia
A. Opiana
A. Oreitae
Alexandria Port
[Karachi]
Locatable city along route
Major pass
Major battle
Outward route
'Return' route
0
500
1000 km

so we should not necessarily assume that the cultural exchange was a result of Alexander's campaigns alone; rather it was part of an organically emerging system of interconnections.

The topic of Alexander's routes has clearly been an attractive one to both classical researchers and a more general audience for travel reports of Asia (a brief search reveals many such publications, for example: Macdonald Kinneir 1818; Williams 1829; Stark 1958; Engels 1978; Stein 1929; MacDermot and Schippman 1999) – originally related to the wider 19th century colonial fascination with routes. The most widely accepted reconstruction of the route is shown in Figure 2.12. Notable is Alexander's reliance on land routes throughout his campaign: *e.g.* travelling along the coasts of Asia Minor and the Levant rather than organization any kind of maritime invasion. He seems to have mainly avoided the trans-Anatolian Persian 'Royal Road', if we assume David French's reconstruction to be correct (see above Section 2.4.1), although his army perhaps followed its continuation from Nisibis to Susa in modern Syria, Iraq and south0west Iran. Whether he was following other well-established routes is difficult to know of course, but it reminds us that the particular paths chosen were militarily strategic – and not always the fastest or busiest roads. His progress through Iran also reflects this: the primary task was first to subdue Persepolis (as capital of the Achaemenid empire) and then track down the fleeing Darius: hence the zigzag route through Persia and into Central Asia. Indeed, unlike the Persian 'Royal Road' as described by Herodotus, Alexander's campaigns did not represent an established system of communication, but rather a single instance of travel along a route predetermined by the promise of acclaim and riches by the defeat of certain enemies. Whilst this should caution us about reading too much into the particular route taken, there is a strong likelihood – albeit speculative – that the routes were already well known for Alexander to have taken them, and that they had some basis in the economic and political interconnections of the period to make them worth following.

Figure 2.11. Ottoman military campaigns in Eastern Anatolia during the 16th century: (red) Selim I, 1514; (orange) Süleyman, 1548; (yellow) Süleyman, 1532 (based on data from Yerasimos 1991).

2.4.3 Travellers' accounts

Travellers' accounts provide the most obvious historical records of late antique, medieval and early modern routes, since they describe actual routes taken by the travellers – with comments on the places visited and about other travellers. Amongst the best known travellers relevant to our regions of interest are Ibn Battuta (Figure 2.13), Marco Polo (Figure 2.14) and Evliya Çelebi[24]. Although these accounts were often based on real journeys and first-hand knowledge, it should be remembered that the texts from which we learn of the travels are not simply factual reports, but active constituents of a literary genre with its own traditions, stylistic demands and developments. This genre crossed the cultural boundaries between western 'Christendom' and eastern 'Islam', at least during the medieval period when Iberian scholars translated texts between Latin, Greek, Persian and Arabic. The texts that we have access to were often not written by the travellers themselves: for example, an original has not survived (and we must reconstruct the journey from a variety of later sources or versions, as is the case for Marco Polo's travels) and/or the journey was written or reported by a ghost writer or at least someone who interviewed the traveller.

Figure 2.12. Alexander the Great's campaign from the Aegean to central Asia.

24 Maps for whose *Seyahatname* have been produced but sadly never published.

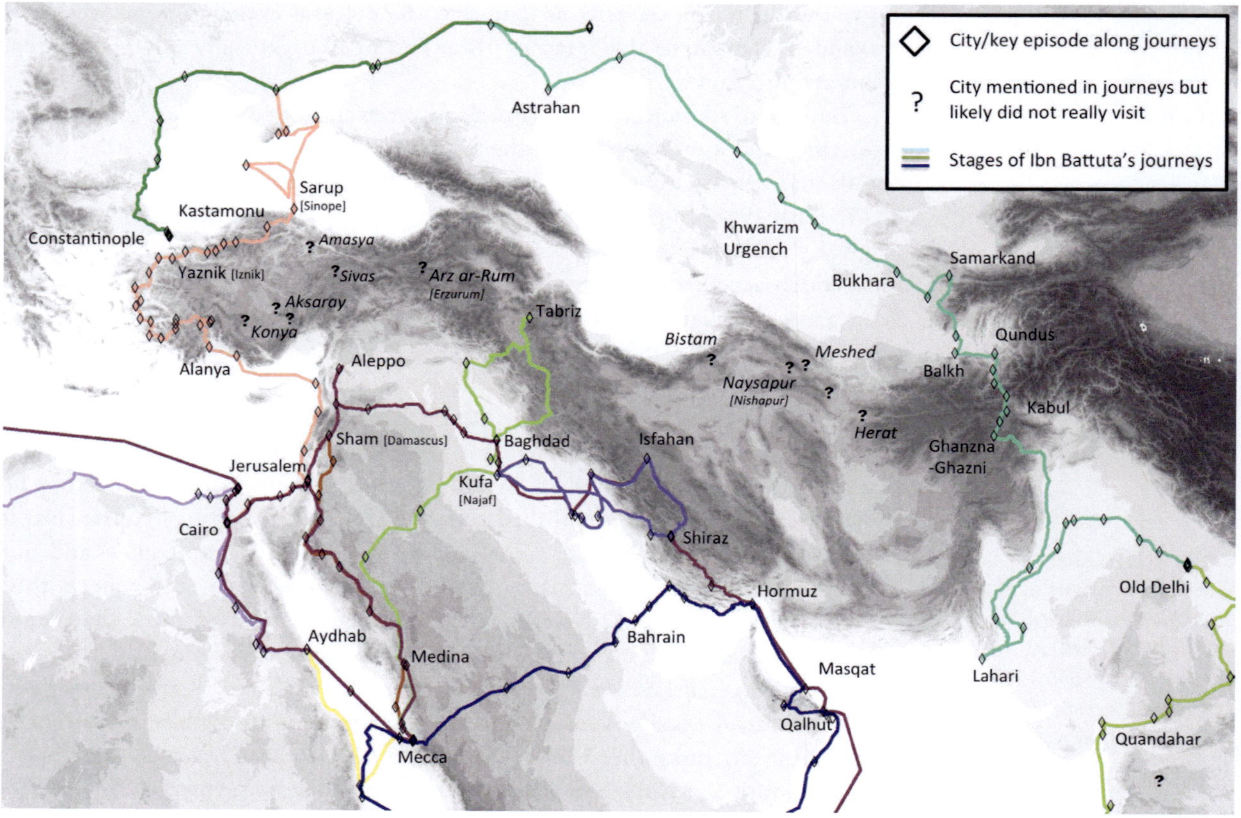
City/key episode along journeys
City mentioned in journeys but likely did not really visit
Stages of Ibn Battuta's journeys
Astrahan
Sarup [Sinope]
Kastamonu
Constantinople
Amasya
Yaznik [Iznik]
Sivas
Arz ar-Rum [Erzurum]
Aksaray
Konya
Alanya
Aleppo
Tabriz
Khwarizm Urgench
Bukhara
Samarkand
Bistam
Meshed
Naysapur [Nishapur]
Qundus
Balkh
Kabul
Herat
Ghanzna -Ghazni
Sham [Damascus]
Baghdad
Isfahan
Jerusalem
Kufa [Najaf]
Cairo
Shiraz
Hormuz
Old Delhi
Aydhab
Bahrain
Medina
Masqat
Lahari
Qalhut
Mecca
Quandahar

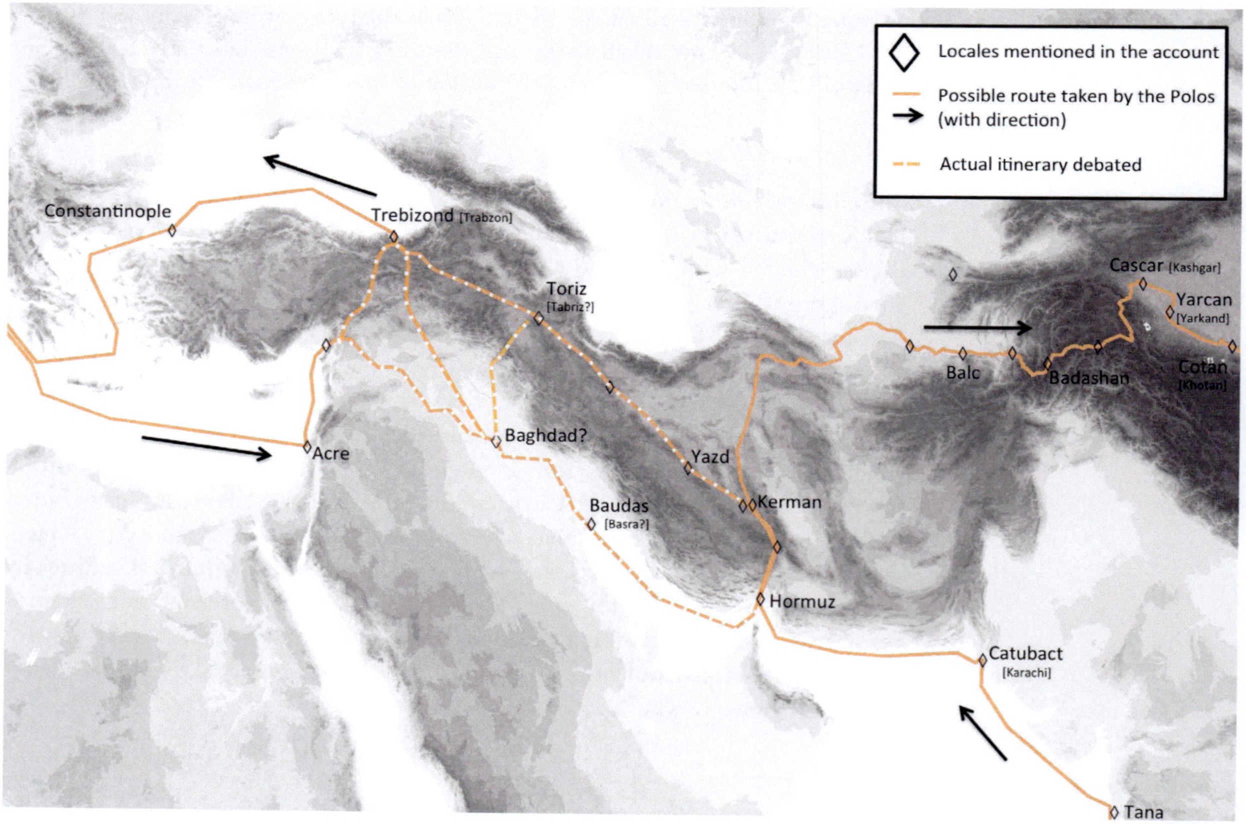
Locales mentioned in the account
Possible route taken by the Polos (with direction)
Actual itinerary debated
Constantinople
Trebizond [Trabzon]
Toriz [Tabriz?]
Cascar [Kashgar]
Yarcan [Yarkand]
Balc
Badashan
Cotan [Khotan]
Acre
Baghdad?
Yazd
Baudas [Basra?]
Kerman
Hormuz
Catubact [Karachi]
Tana

Whilst the conformance of such texts to an overall genre may have been partly a result of editing and selection by curators and copyists (who presumably preserved only what was deemed culturally valuable), the effects on the style and content of such accounts is still significant for the interpretation of their historicity and usefulness as guides to contemporary routes. In some cases where the chronology and geography of a journey do not make sense and where the descriptions are derivative or vague, we may suspect that the traveller did not in fact visit all the cities which they claimed to have done, and instead that the author included the places as a literary device – to increase the comprehensiveness, authenticity or conformance to the genre.

A good example of the effect of writing traditions is to examine what is included and what is excluded from the accounts[25]. Usually, much detail is given about the significant cities at which the travellers arrived, the events and interactions with local leaders (it is normally elites and not commoners who are described), or the unusual customs and objects of the peoples in these cities. Religious affiliations, religious observance and moral codes in general also form an important aspect of most early accounts: whether it is in Marco Polo's mission to represent the Pope and the Catholic church to Mongol authorities, or Ibn Battuta's shock at the different gender values and status in the places he visited. The result is that we often only know the major cities through which these travellers passed, some of whose names we cannot easily match to geographical locations. Famous mountains, rivers and passes are sometimes mentioned, especially in accounts like Evliya Çelebi's, which certainly increase the potential resolution of our reconstructions. But this is exceptional and the normal telegraphic nature of most early travellers' accounts means that we are forced to speculate on the interstitial lines by which travellers themselves and their contemporaries moved between cities. Such speculation can only be based on our prior knowledge of routes and landscapes from other sources.

Figure 2.13. The travels of Ibn Battuta.

Of course, a single account of a journey has rather little to say about the development of route systems in general through time: it is simply proves that a particular route was available to travel, but unless additional details are provided about the relative numbers of other travellers, we cannot know how integrated the route was into the overall economy of the time. Indeed, many of the most famous accounts arguably involve the routes of highest resistance, where the travellers take the role of pioneers along paths that are used by locals but not yet widely-enough known to be seen as a central or normal route for cultural or economic integration. At best, the appearance of a route in such an 'exploratory' account may indicate its recent emergence as economically important. Nonetheless, if we assume that at least some of the journeys include routes that were economically important, the cumulative combination of many different travellers' accounts can support arguments for more general usage of particular routes.

Figure 2.14. The supposed travels of Marco Polo.

25 The interest in comprehensively describing topography and environment seems to have been a recent development in western European art and travel literature, related perhaps to the dual emergence of the 'objective' Cartesian map and the concept of *landschap* in modes of painting alongside enlightenment or colonial classificatory literature as a way of 'enframing' the world 'as a picture' (see *e.g.* Mitchell 1988).

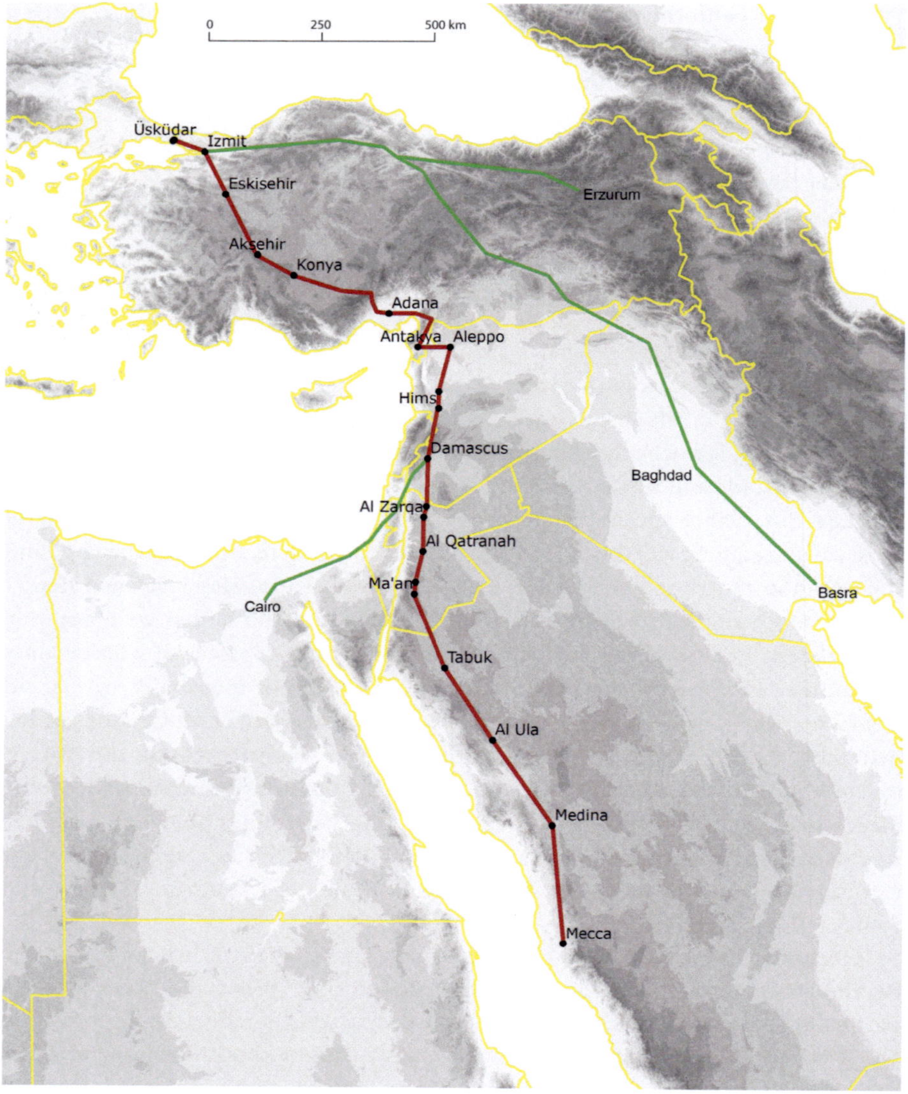

Figure 2.15. Pilgrimage to Mecca: (red) the route of the late Ottoman hajj from Constantinople; alongside (green) contemporary Ottoman postal roads (to Cairo, Basra and Erzurum).

2.4.4 Pilgrimage routes

Religious pilgrimages were a significant phenomenon during the medieval and early modern periods, resulting in considerable cultural interaction and the movement of thousands of people in the Christian, Islamic and Buddhist worlds (and indeed beyond). Whilst one might expect the routes of such pilgrimages to be well known, there have been fairly limited attempts to synthesize data about these routes. Additionally, in contrast to the way we tend to understand modern pilgrimages as 'canonically' fixed, it seems that the exact routes taken by pilgrims did not remain stable, and were heavily affected by political and economic factors. This is unsurprising given the considerable political and religious upheaval in the regions in which we are interested. The majority of our most direct knowledge of pilgrimages comes from the descriptions by travellers like those mentioned above in Section 2.4.3, who reported the stories of others and/or were participants in pilgrimages themselves.

The Hajj to Mecca and Medina (AD700-1900)

An important religious duty in the Islamic world was, and still remains, the performance of the sacred 'Hajj' – the great pilgrimage to Mecca. Whilst the routes of the hajj itself lie mostly outside our case-study regions, a brief excursis into the subject is instructive in in the details of the nature of travels. The most detailed information about the pre-modern hajj comes from the Ottoman period for which we have records of well-organized pilgrimage 'caravans' from major centres, often funded by administrators, aristocracy or the Ottoman sultans themselves (Faroqhi 1994). These caravans included administrative structures of their own and a whole host of people with responsibility for the practical survival of the train: caravan 'commanders', commanders' secretaries or scribes, 'desert pilots' (who were local guides or Bedouin familiar with the terrain), soldiers or private security, one or more *kadi* (judges charged with dealing with various disputes or the property of pilgrims who died *en route*), cooks and water carriers, *etc.* plus the 'ordinary' pilgrims (Faroqhi 1994, 34-35). The typical full size of pilgrimage caravans is difficult to ascertain: estimates of 40,000, 15,000 and 50,000 people or 11,000, 5000 or 16,000 camels are provided by various contemporary observers (Faroqhi 1994, 34-35): but clearly there was great variety in the size of such groups, and it is likely that only the largest examples were recorded by these observers (if we assume they are not exaggerations) since they were the most impressive. The military presence seems to have been necessary because the caravans carried large amounts of cash, both for donations to Mecca and for the practical undertaking of the expedition. Practical needs were central to the organization of pilgrimages: "soldiers and officials had people who handled their food requirements; though ordinary pilgrims had to provide for themselves" (Faroqhi 1994, 36). Various charitable foundations, so-called *waqf,* were thus set up to help ordinary pilgrims. Worries about sufficient water supply were understandably considerable: we are told that in normal years caravans set out from Damascus with 50 camels to carry water (Faroqhi 1994, 43-44). Clearly the exact route of an individual's hajj obviously depended on exactly where the pilgrim started. By way of example, during the high Ottoman period, the hajj from the Ottoman imperial capital started from the Asian town of Üsküdar, on the opposite shore of the Bosphorus (Faroqhi 1994, 37): the journey appears to have been made through central Anatolia to Antakya before again following the inland route through Syria to Damascus and south to Medina and Mecca (see Figure 2.15). The choice of particular routes appears to have been partly practical, based on available roads or as suggested by guides, and on the restrictions in place according to political dynamics, but also religious, based on the particular affiliations of the pilgrim. For example Persian pilgrims were forced by Ottoman restrictions to travel longer routes via Cairo, Damascus, Yemen and not via nearby al-Basra. But Shi'a Persians were also said to have chosen routes that would take them to Shi'a shrines (for example at Karbala or al-Nadjaf) (Faroqhi 1994, 135-137). Obviously the changing political and religious landscape of the Near East and surrounding regions had a major effect on the popularity of particular routes, not least because of the changing numbers of adherents to Islam in each area and therefore the possible pool of pilgrims.

Such descriptions of the organization and execution of the hajj caravans provides a graphic insight into the difficulties of travel before the invention of motorized transport, which we otherwise tend to forget. It is difficult to know whether such routes paralleled those of other forms of interaction and movement

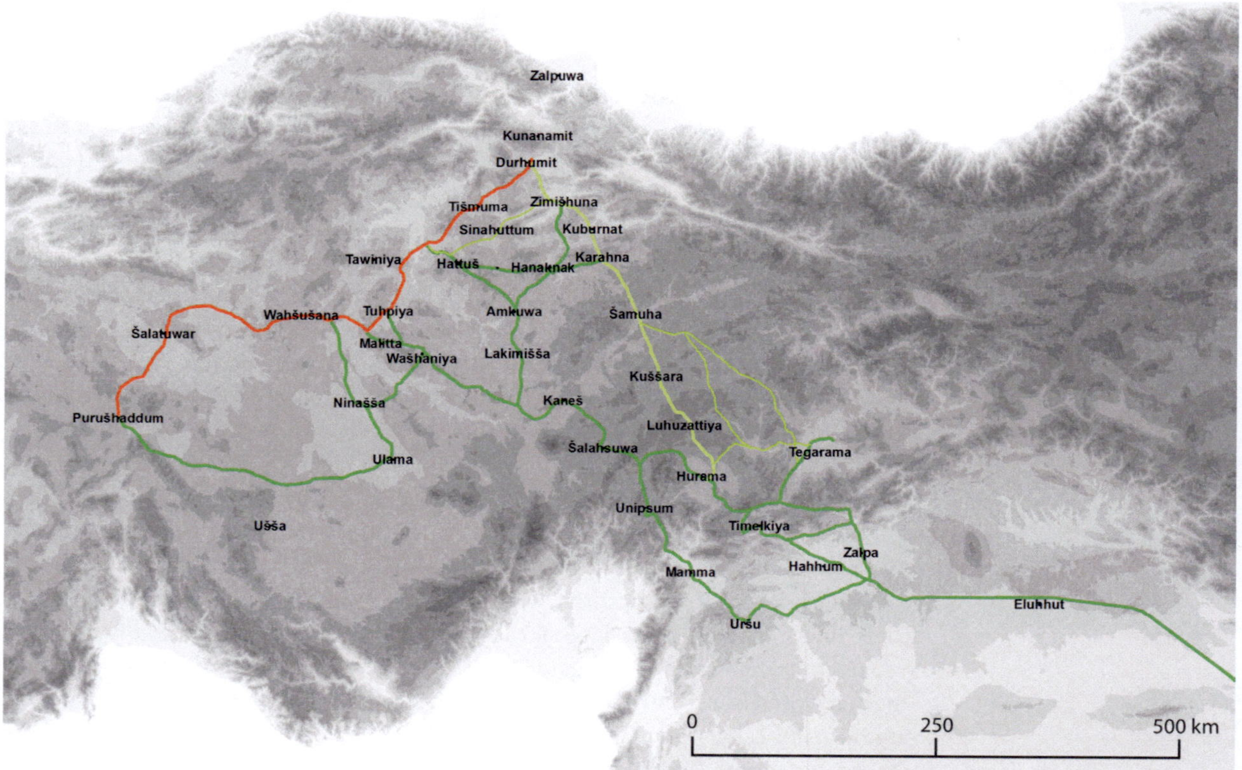

Figure 2.16. Trade routes and settlements of the Old Assyrian Period in Central Anatolia (according to Barjamovic 2011).

such as trade and the exchange of ideas within the medieval and early modern world: but it seems that the pilgrimages involved the translocation of considerable number of goods as well as people, and one of the motivations (or justifications) behind the travels of many Islamic authors was as part of pilgrimage. The use of Bedouin guides also reminds us that there were many groups of people who were habitually migrant: nomads of many different sorts were characteristic of this period. However, it is difficult to take the particular routes of the pilgrimage as a basis to reconstruct the *hierarchies* of earlier routes because the primary engine behind these pilgrimages was so culturally specific. This is not to say that the same corridors were not used in earlier times, rather that the possibility of travel along them gives us little idea about the intensity of the earlier interactions.

2.4.5 Legal and economic accounts

Much more detail about the intensity of interaction between regions along land and sea routes could be garnered from very specific economic accounts or ledgers of medieval and early modern merchants – such as those of Francesco Balducci Pegolotti (Pegolotti 1936) or Hovhannes Joughayetsi (Khachikian 1966) – to look at the origins, destinations and amounts of products, and the usefulness, openness or cost of particular trade routes. Large numbers of these kinds of documents tend not to have survived to the present day, because they were not considered important enough to copy (and therefore preserve), but, where they do survive, then they have the potential to provide greater economic depth to our understanding of routes than is normally given by the standard geographies.

Old Assyrian contracts and caravan routes (1950-1750BC)

The Old Assyrian texts from Cappadocia, particularly Kültepe, represent our earliest written record of a trade network (Larsen 1967; Dercksen 1996; Michel 1992) based on economic and legal accounts. The archives date to within the 19th and 18th centuries BC and thus to within the latter part of the temporal bounds with which this book is primarily concerned. We might, therefore, reasonably hope that they provide the most relevant basis for the reconstruction of 3rd and 2nd millennium routes in central Anatolia – and offer a model for contemporary exchange networks in eastern Anatolia and western Central Asia. The relevant texts, which are in the form of contracts, reports or letters between traders at home in Assyria and in the trading colony (*karum*) in Anatolia, document the kinds of goods which were being exchanged by whom and at what prices. Most famously, tin and textiles were being imported to Anatolia, at a profit of around 100% and 200% respectively, in exchange for silver and gold which were shipped back to Aššur. Some of these texts also record a list of places through which the traders travelled by way of a report of the accounts, noting taxes paid to cross bridges, fees for supplies and sometimes security problems on the road – although the descriptions are not always very specific (*e.g.* 'the river' rather than a named river). In the following examples, *Šinahutum*, *Wahšušana*, *Šalatuwar*, *Purušhaddum* and *Kaneš* are all place names (for identifications of the tablets, 'AKT', 'Kt', see Michel 2003; 2005):

> *I paid two minas of refined copper in Šinahutum … I paid one mina and 16 shekels of copper for three jars at the bank of the river. I paid 32 shekels of copper for our lodgings and I also gave some leather straps.* (Kt c/k 594, translation Barjamovic 2011, 288).

> *From Wahšušana to Šalatuwar 1 5/6 minas 5 shekels came on as fees for donkey fodder and lodgings for each of us. In addition, they deducted 1/3 mina for each donkey at the bridge. 2 minas of copper: fodder for the donkeys in Šalatuwar. 2 ½ mina came on (as fees) for each of us until Purušhaddum. They deducted a fee of 15 shekels for each donkey at the bridge. 1 ½ mina for fodder for the donkeys and our servants in Purušhaddum.* (AKT 3, 34, lines 1-21, translation Barjamovic 2011, 352).

> *The 30 minas of silver –its nishatu-tax added, its saddu'utu-tax paid for– which Dadja entrusted to Kukkulanum, son of Kutaja, and which he carried to the city for buyings – (that) silver belongs to Enlil-bani. From here it will cross the country in the name of Enlil-bani. Goods will leave the city and cross the country in the name of Enlil-bani again. The goods will arrive at Kaneš and Enlil-bani will receive them. Witness by Bazija, son of Ili-kurab, witnessed by Asutaja, son of Ememe, witnessed by Assur-idi, son of Kurub-Istar.* (VAT 13.519, published in Larsen 1967, 8).

The problem with this particular data, of course, is that aside from a few exceptions, the locations of the places named must be reconstructed through association with cognate names from later periods (including those in the Hittite texts – which in turn require reconstruction) or by looking at the inter-urban distances and geographical clues (river crossings, passes) that may be linked to our knowledge of the modern landscape. Some considerable research effort has thus been spent on reconstructing the historical geography of central and south-eastern Anatolia in the Old Assyrian period based on these texts and information from later Hittite sources (Larson 1967; Michel 2001; Forlanini 2008 inter alia; Barjamovic 2011), but, as evidence for the difficulty of this task, some very different reconstructions have been published. The overall picture fluctuates between advocates of a 'minimalist' versus a 'maximalist' model of the geographical extent of the trade described in the texts (Barjamovic 2011, 409-410) – an example of a convincing 'maximalist' version from Gojko Barjamovic is shown in Figure 2.16. One thing that is common to both models is the broad north-west to south-east direction of the routes: those running perpendicular to this into north-east Anatolia or south-west to Cilicia are few or non-existent; and of course the central place of Kaneš (modern Kültepe) as a regional administrative centre for the traders – but also unsurprising given the provenance of most of these tablets. These trends might give the impression that there were no further routes into eastern, western or southern Anatolia – but this is likely to merely indicate the extent of Assyrian contacts and not of wider trade.

The assignment of particular Old Assyrian place-names to their likely geographical location often depends on comparing multiple itinerary lists between two known places using *prior knowledge* of the landscape and the potential routes available, and occasionally trying to fit this with archaeological knowledge of Middle Bronze Age sites. Projecting these reconstructed route systems backward in time thus runs the risk of circular arguments. This is not to say that the reconstructed routes are valueless – they may be very accurate for the period in question – just that we should remember that it would be very difficult to detect situations in which Old Assyrian traders were travelling along routes *which we would not otherwise expect from our prior knowledge of routes from other sources* whether modern and historical. This in turn makes it extremely difficult to judge if the resulting network maps are good models for the wider 3rd and 2nd millennium BC.

2.5 Critique: evaluating 'route inertia'

2.5.1 'Historical routes' as models for earlier ancient route systems

Combining the information from the above sources for eastern Anatolia (Figure 2.17) and western Central Asia (Figure 2.18) onto two maps starkly illustrates the limits of this summary of routes, in particular its differential coverage: the level of detail on the map of western Central Asia is considerably less than that of eastern Anatolia. No doubt further research would produce more comprehensive results, but it is tempting to see these differences as a fair indication of the relative development of and extent of knowledge about historical geography more generally in each region. The overlay of all the 3rd and 2nd millennium archaeological sites in the database created in the course of researching this book also reveals the limitations of the archaeological research in the earlier period. There are some areas (such the Upper Euphrates in Turkey or the Kopet Dag piedmont in Turkmenistan) whose high density of sites may be as much to do with the intensity of archaeological investigation as with the intensity of ancient settlement; conversely there are numerous blank areas with what turned out to be important corridors in later periods. The chief challenge then of any large-scale synthesis such as this is taking account of this uneven knowledge in our analysis.

The above review has emphasized the fact that we cannot simply project the routes of later periods backward and assume that they functioned identically in the past. Far from straightforward continuity, these various sources of information – although incomplete – hint at changing route hierarchies and densities of travel. Though the possibility of travelling along certain corridors in the landscape was available topographically throughout this time, different cultural and political boundaries and the evolution of new social imperatives and economic motivations framed and constrained movement within those landscapes. The development (or disappearance) of new transportation technologies (including road-building, bridges, the wheel, the horse, the donkey, the camel) had considerable effects on both the types of paths selected and on the time taken to travel certain distances, and simultaneously and consequently on the 'reach' of economic systems. The smaller 'reach' of economic (and political) systems is what Gil Stein's 'distance-parity' model was designed to address (Stein 1999). The review has also highlighted the fact that synthetic knowledge of the routes of most historical periods is – in any case – actually rather limited, particularly with regards to the intensity of travel along certain routes. With the exception, perhaps, of Old Assyrian trading accounts, much greater attention has been focussed on individual travellers and one-off itineraries than on the repetitive movements made by unnamed traders, migrants and other 'everyday travellers'. Besides the analysis of roads, studies of material culture that might give clues about these everyday encounters and interaction are often lacking. Additionally, the manner in which the reconstruction of ancient route systems has normally been undertaken is problematic and circular: reconstructions must be based on the reconstructions of routes from later periods. For example, reconstructions of the route of the 'Persian Royal Road' through Anatolia, have relied to a large extent on projecting the knowledge of Roman roads backwards. This is an understandable and perhaps necessary approach with some value, but it should be remembered that whilst there is certainly a connection – earlier routes begat later routes through their pre-existence (hence 'inertia') – *independent* evidence is required to confirm this ancestry. Perhaps more serious is the fact that routes have often been treated uncritically, and the concept assumed to be unproblematic. 'Roads', 'tracks', 'paths', 'itineraries' and 'routes' have sometimes been treated synonymously when it may be better to make distinctions. The result is that routes are almost always inevitably represented visually as emulated modern road maps ('pseudo-road-maps'), which frame our understanding of routes in general. This may be appropriate for some periods, such as the Roman era when roads presumably functioned in a similar way to modern ones – but not for many others in which pathways grew more organically.

2.5.2 Routes in the archaeological imagination

Many of the earliest archaeologists working in the Near East were concerned with routes, including W. Ramsay (1890) and Aurel Stein (1929). The origin of this interest may, in part, be related to the Victorian conception of landscape that had a strong military or colonial components. Essential to the colonial enterprise was detailed understanding of mountain passes, water sources, river

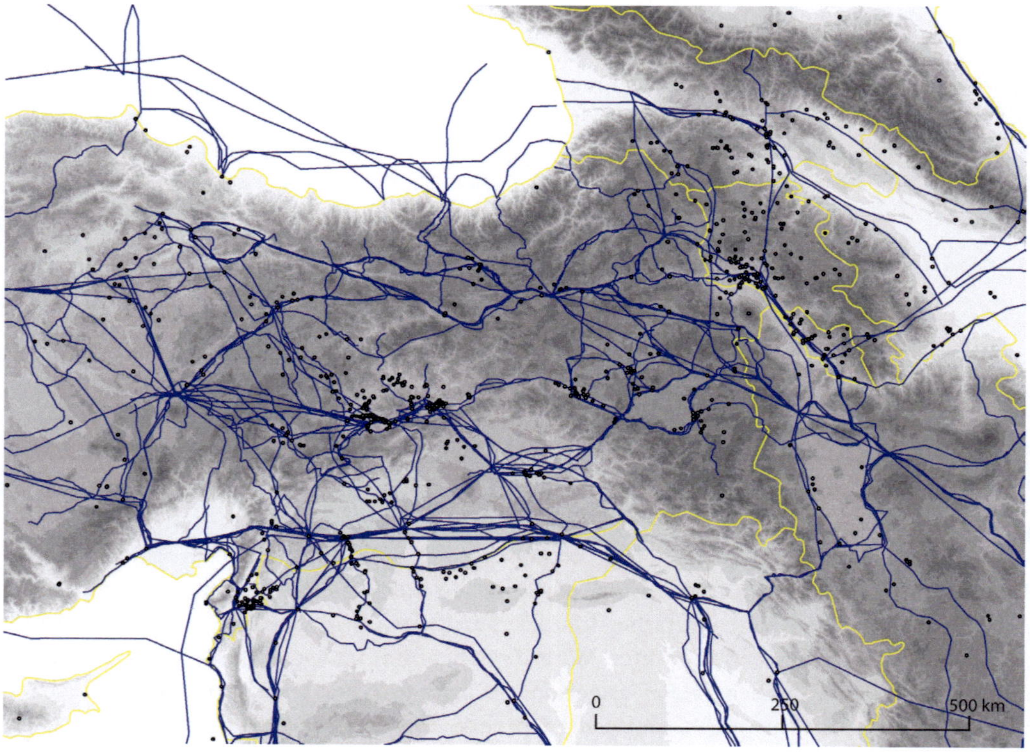

Figure 2.17. Combination of all reconstructed routes and itineraries in eastern Anatolia (blue lines). Black dots represent archaeological sites in database (see Appendix B) dating to 3000-1500BC.

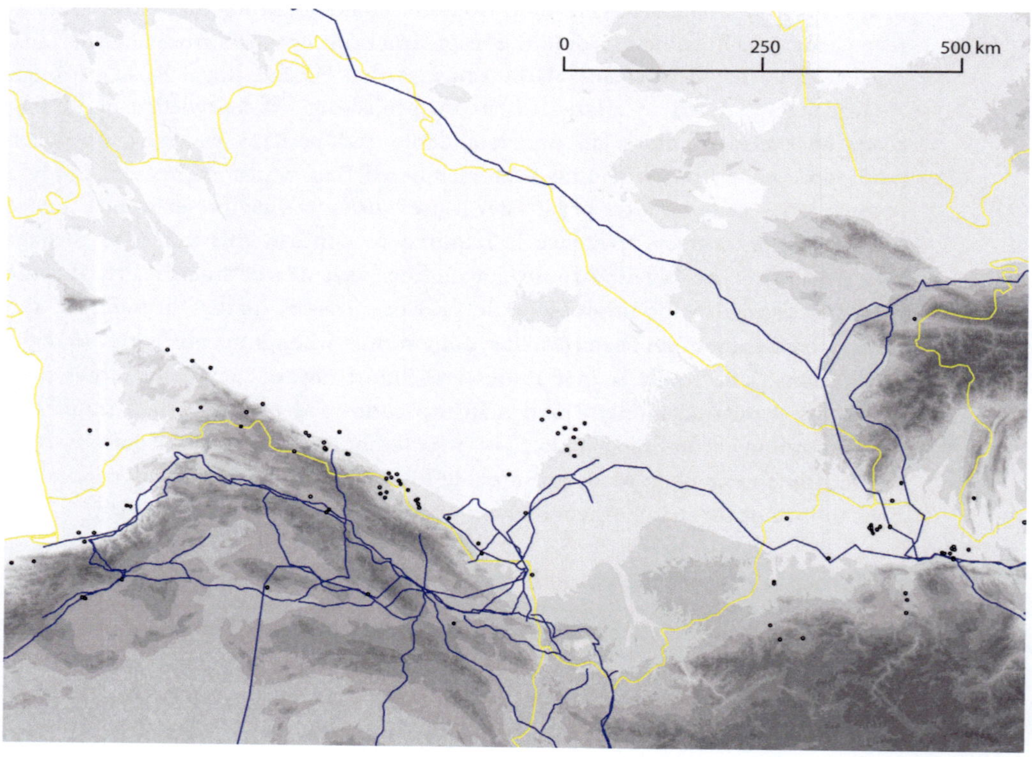

Figure 2.18. Combination of all reconstructed routes and itineraries in western Central Asia (blue lines). Black dots represent archaeological sites in database (see Appendix B) dating to 3000-1500BC.

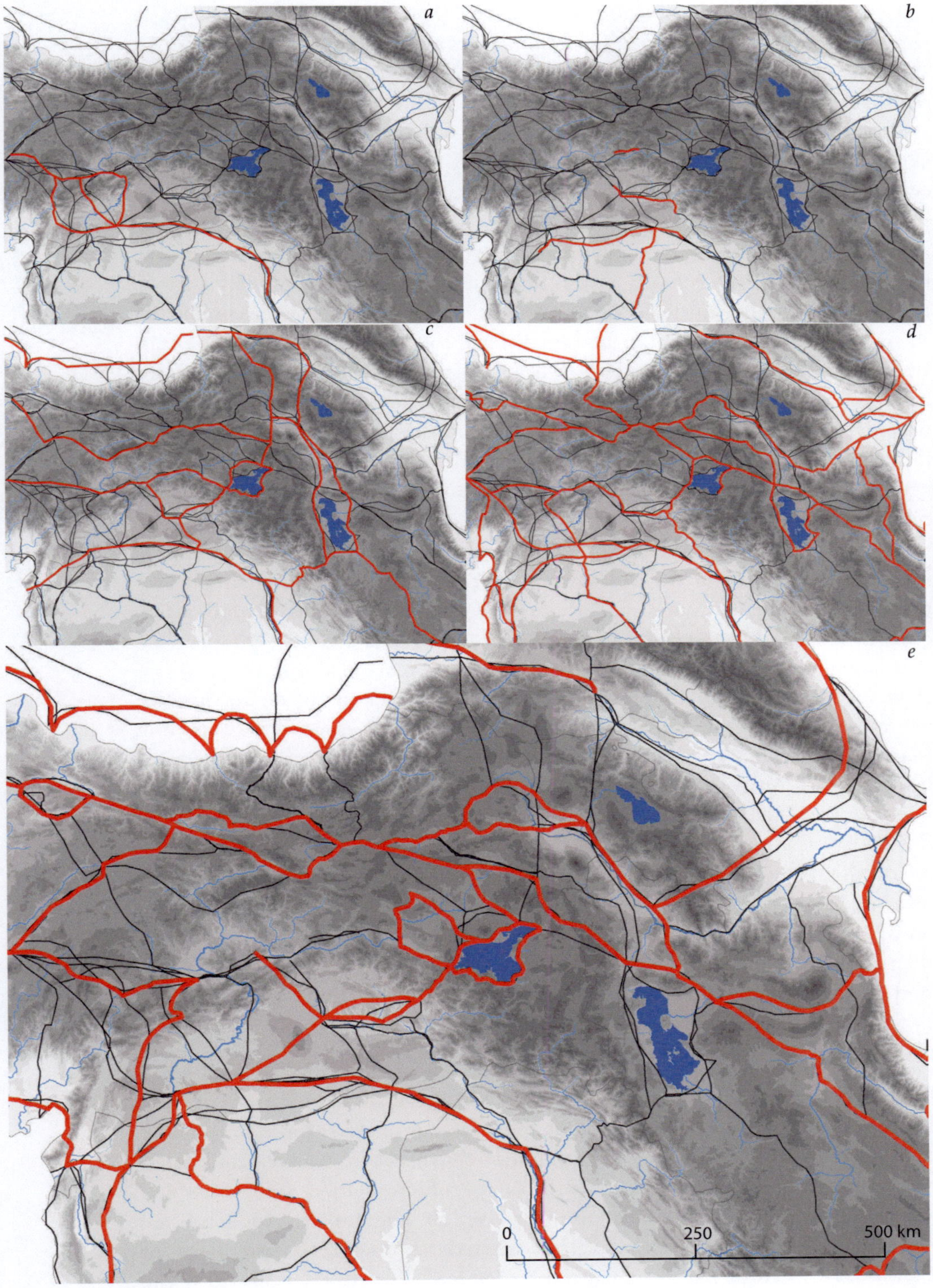

Figure 2.19. The evolution of route systems in eastern Anatolia (according to Marro 2004): (a) Old Assyrian. (b) Urartu and Neo-Assyrian. (c) Hellenistic. (d) Medieval. (e) Early Modern.

valleys and distances in terms of marching days (Sprenger 1864), particularly as part of the 'Great Game' across the Pamirs (see, for example, Meyer and Brysac 1999; Hopkirk 1992). This is starkly illustrated in the British Naval Intelligence reports mentioned earlier in this chapter. The 'road' or route also had romantic connotations – illustrated perhaps by Kipling's sublime description of the Great Trunk Road of India or the passes of the Pamirs in *Kim* – that fed this desire to consume accounts of travels and routes. These same approaches to mapping also played out in the cultural sphere as an interest in reconstructing the route of Alexander's marches into Iran and Central Asia, and identifying the likely passes by which migrant tribes (such as the 'Indo-Europeans') had supposedly entered new regions. The retreat from these forms of historical geography accompanied the retreat of European military and intellectual colonialism. In Anglo-American archaeology, an interest in routes was lost as part of the paradigmatic shift from the 1960s onward to systemic or processual approaches and away from historicism (already mentioned in Chapter 1), in which emphasis was placed instead on (1) indigenous or local developments over interaction and movement; and (2) on generalized systematic processes and patterns over particular historical events. But routes have aspects of both events and processes: the accumulation of particular trips taken repeatedly is a process that constructs what we identify as a route. It is also in the context of the widespread adoption of air travel by archaeologists that the question of 'routes' and the 'tyranny of distance' has become a less obvious subject of research. It is only in a few exceptional accounts lying outside of dominant theoretical discourses, that routes have remained or re-emerged as a topic for discussion: for example, in Tuba Ökse's examination of the topography and routes between the Kızılırmak and Euphrates rivers (Ökse 2005), Catherine Marro's (Marro 2004) attempt to summarize the 'evolution' of route systems across Eastern Anatolia through different periods (her sequence redrawn here as Figure 2.19), Michael C. Astour's (Astour 2000) attempt to look at wider Near Eastern 'road connections' through time, or, on an even larger scale, Andrew Sherratt's graphic reconstructions of expanding route networks across Eurasia (A. Sherratt 2004). In many cases, it is difficult, however, to perceive exactly how (*i.e.* on what specific quantitative or qualitative evidential basis) each author came to the particular characterizations of routes (or roads) illustrated – the reader must simply decide whether to trust the author's judgement.

2.5.3 Mythical routes? The 'Silk Road' as literary trope

One 'historical route' was intentionally left out of the above analysis. The 'Silk Road' (or 'Silk Routes') is certainly the most famous of trade routes in popular culture – and provided the original inspiration for this book. Closer examination of the term, its constituent routes and the sources of information about its nature, soon leads to confusion and uncertainty, however. By common definition, it is the road, or rather set of land routes, which connected the western and eastern ends of Eurasia across Central Asia from the Han period in China (the last centuries of the 1st millennium BC), until sometime after AD 1500-1600 when sea routes across the Indian Ocean started to replace these land routes. The 'high point' of the Silk Road in China is considered to be the Tang era (AD 618-907) when Chinese control over the northwest reached its furthest extent and we have considerable written evidence of commercial and cultural exchange across the Tarim basin (what the Chinese today call Xinjiang, the 'new territories') into Central Asia and beyond into Persia and the Near East. The term 'Silk Road' itself of course references only one of the many valuable commodities to have travelled along these trade routes (see *e.g.* Schafer 1963). The Silk Road is often described in association with Chinese history, such that one would be forgiven for thinking that the whole thing was a purely Chinese enterprise. The reason for this is probably the origin of the phrase. Von Richtenhof's coinage of the term and his student Sven Hedin's popularization were both performed in the context of research into western China (Waugh 2007). However, there were many other groups involved in Central Asia with equal or greater importance: particularly local peoples and polities of the Tarim basin, Sogdia and Bactria who had connections in all compass directions, which the written material uncovered at Dunhuang shows us in vivid detail (see, for example, the selections translated in Whitfield 2001). It should further be remembered that, at least by the Tang period, there was no Chinese monopoly on the manufacture of silk: there were local productions of silk in Sogdia and even Byzantium. Again, by common definition, the Silk Road(s) is/are said to have passed through both our special regions of interest, western Central Asia and eastern Anatolia. Western Central Asia in particular is understood as a 'pivotal' region for this entity, taking in the major cities of Afrasiab, Balkh and Merv, for example. The area of Transcaucasia and eastern Anatolia (especially the cities of Erzurum, Baku, Djulfa, Tebriz) is sometimes represented as an optional alternative to the main 'trunk' of the Silk Road (which is seen as passing south through Iran to Baghdad before climbing up the Euphrates to Anatolia, or via Aleppo and Antioch to the Mediterranean).

The way in which the term 'Silk Road' has evolved is extremely problematic, however. First, as many researchers have stressed and the eastern Anatolian example reminds us, there was no one single road or route involved in the east-west trade of the constituent periods. Many different routes were in use contemporaneously. Further, if one compares the very many representations of the 'Silk Roads' in map form, it is clear that there is little agreement or direct knowledge on the location of the constituent routes. Moreover, there was certainly no contemporary idea of 'the Silk Road' or even a set of 'Silk Roads' for which we could identify the 'official' version from historical texts or archaeological data. What there is instead is a modern attempt to create a fixed identity for a very vague idea about trade across Eurasia in the pre-modern age. In this manner, the term has become useful to academics and other cultural creators alike, since it serves as a shorthand for certain themes (trade, interaction, hybridity, ancient history, 'lost cities'), certain objects or architectures (silk, caravanserais), certain regions or places (China, East-West, Central Asia, the Middle East) and certain times (particularly Tang and Islamic periods). It is sometimes used as a shorthand or lazy way to add time-depth and allusions to more information than we really have. The term's semantic slipperiness has also allowed it to be co-opted by states, NGOs and companies keen to draw on the aura of these same values: in UNESCO World Heritage applications, in the marketing blurb of energy companies, or simply in tourist brochures. This widespread usage generates a counter-flow desire to 'fix' the authentic routes especially on the UNESCO project – a project that unless treated carefully may be doomed to anachronism and false historicity, if it is not properly recognized that the concept is a modern invention.

All of this shows the 'Silk Road(s)' to be a modern myth: a powerful story that we have told ourselves and which can be used for legitimation of various moral and political messages inside and outside of academia (Ball 1998). This lack of analytical value means that whilst we may use the term 'silk routes' loosely to encapsulate the range of east-west routes in use during the 1st and 2nd millennia AD, we should always acknowledge this vagueness and instead try to use more concrete data (specific textual sources, specific classes of archaeological evidence) when we try to compare the route systems of other periods.

2.6 Summary: route inertia and re-visioning routes

In this chapter, we have reviewed the available textual and material sources on 'historical routes' in and around our case-study regions. This has revealed a number of limitations on both an empirical and theoretical front. To begin with, the direct evidence for most periods is rather thin or already relies a great deal on the application of route inertia (*e.g.* the line of the Persian royal road in Anatolia being based on back-projection from Roman roads). We also actually have rather poor understanding over the 'total' systems of movement even for supposedly well-researched periods (such as the Roman) or those with clear physical evidence of lines of communication (where there are fragments of roads as in Pasinler). Moreover, supposedly 'well-known' routes, such as the 'Silk Road' turn out, in many cases, to be more akin to mythical or literary tropes than analytical structures. There are a few areas in which more comprehensive route networks have been identified (such as the hollow ways of northern Mesopotamia), but though they cover a large area, they are restricted in distribution to the rather unique landscape that has preserved them and their date is not completely certain. All of this makes it difficult to consider the principle of route inertia (based on back-projection of supposedly better-known periods, at least) as a universally reliable or applicable guide to the route systems of prehistoric periods. Whilst individual reconstructions based on such a principle are not necessarily unrealistic or incorrect, more effort needs to be devoted to establish particular cases of inertia (or, by contrast, of route dynamism).

Indeed, it seems likely that the hold that the concept of route inertia has over our academic imagination may be hiding all sorts of dynamism in the spatiality of interaction. Critically, though we are able to construct static 'road maps' based on specific itineraries in some cases, the resultant picture gives little idea of the dynamic changes in the amount and nature of movement along these paths (the 'density of travel' or hierarchy of routes). Whilst a number of researchers have attempted to reconstruct changes in route systems (such as Marro's or Ökse's differently-scaled visual sequences), it has clearly proven much more difficult to integrate the movements (of people and materials) that defined and created such routes. Relying on route inertia alone, we are left to speculate on the 'types' and 'amount' of travel taking place in the 3rd and 2nd millennium BC. The following chapter will therefore attempt to address some of the limitations outlined here by refocusing on the nature of the route, on the physical basis of routes (*i.e.* landscape) and by considering novel methods for visualizing and analysing 'routes' without making assumptions about our ability to project back from uncertain 'historical routes'.

Chapter 3

Landscape and Non-linear Networks: Finding Methods to Visualize Ancient Flow of Materials

3.1 Introduction

The previous chapter focused on the principle of route inertia, and in particular its application with evidence from supposedly better-documented periods to reconstruct the route network of those periods without textual indicators or clear archaeological remains. This chapter shifts the analysis of 'routes' onto landscape. By starting from a different perspective, namely the primacy of physical landscape in the formation of 'natural routes', it attempts to develop a set of methodologies that are period-independent – in other words they treat route inertia as an effect of the environment but do not assume that routes always remain the same. It will be necessary to re-examine the definition of 'routes' more closely to free the spatiality of ancient interaction from linear modelling. In order to move beyond the characterization and visualization of routes in archaeological research as static 'pseudo-road-maps', we will also look at how such methodologies might be used to integrate a central but oft-forgotten aspect of 'routes' – namely the flow of objects along them. The representation of landscapes and of 'routes' as networks of interaction is therefore central to this chapter. Techniques of digital mapping and modelling ('GIS' or Geographical Information Systems) offer the potential to create new forms of such visualization and analysis, though there are many challenges to its practical application in this field.

The main assumption of this chapter is that whilst particular physical 'natural routes' (passes, valley corridors *etc.*) may be open to movement over long periods of time, this does not mean that they were actually used in the same way and to the same extent in different periods. Many site reports contain an introductory note that the settlement in question developed at its particular location because of its command of local routes of movement. While this may be true to some degree, in most cases little else about these routes is mentioned, and the idea remains frustratingly incomplete – essentially as an archaeological trope or myth. Of course, this 'travel density' of routes cannot be known from the evidence of one site alone. Routes are a factor of a much larger interaction sphere of landscapes, settlements, socio-political control and material inequalities, and thus must be analysed at regional or even super-regional scales. To move beyond this 'mythical' status of routes, we must be able to find ways to precisely characterize landscape and the distribution of archaeological indicators of interaction together.

3.2 What is a 'route' anyway?

The term 'route' is used to indicate a number of related but distinct entities.

First it is used as a synonym of (1) ***itinerary*** (as in, 'the route taken by Alexander through the mountains') – which is a *specific sequence of movements* from place to place and can be represented textually (as in, 'from Adana to Aleppo', or as a sequence of GPS co-ordinates) or visually (as a line on a map). Itineraries can be repeated and re-followed of course and can be represented in very different levels of detail, but they are always specific.

Such itineraries tend to follow pre-existing (2) ***roads***, *paths* or ***tracks***, with which the word 'route' is also often equated (*cf. la route* as the French word for road). In the modern context, the distinction between road and path is normally between the mode of transport (by wheel or by foot respectively), but over the *longue durée* a potentially more important distinction is between essentially architectural intentionally-built structures (roads, streets, railways) and organically-emergent braided *tracks*. These are closely interrelated of course – one may form the basis of the other when a former track is transformed into a road – but the key point is that roads, paths and tracks are in *specific locations*.

Related to this sort of 'route' is the (3) ***mountain pass***, a specific zone at a lower altitude to the surrounding mountains that hence allows people to pass from one side to the other more easily than elsewhere, for which the term route is sometimes synonymous.

This localized specificity contrasts with a vaguer but often more historically useful definition of a route as (4) a ***landscape corridor of movement***. This refers to the fact that while the specific location and trajectory of a road or path may change, the broad direction of movement remains stable over the long-term within certain bounds: *i.e.* there will always be a road or track between neighbouring places, A and B, (a corridor) but that the specific location of the constituent road(s) or track(s) may change. 'Sea-routes' also illustrate this category of definition, since whilst it is possible to follow an itinerary of ports-of-call, it is rarely possible and indeed irrelevant to attempt to follow exact paths across the sea.

Of course it would be futile to entirely de-link the word 'route' from its various current meanings, but it is still worth differentiating them in the archaeological context because the identification of routes in the ancient world should not be, as sometimes assumed, the same as the search for 'roads' or 'tracks'. Most 'tracks', which probably formed the bulk of those used throughout human-history, are archaeologically ephemeral and will never be identified. On the other hand, route 'corridors' should be identifiable archaeologically by searching for material markers: 'traded' goods and more general similarities between the archaeological remains of adjacent (and distant) sites.

3.2.1 Types of routes: pathways and highways[26]

Approaching routes from another direction, we can also make a distinction between two *types* of routes: what I would call *pathways* and *highways*. Besides the obvious hierarchical implications between these two terms, we can identify qualitative differences that have consequences for their function in their original contexts and their reconstruction. Traditionally in both archaeological and historical research, the focus has been on *highways*. These types of routes are normally reconstructed from historical accounts like those mentioned in Section 2.4. They might

26 An earlier draft of this section formed the first part of an online 'visual essay' on *ArchAtlas*, prepared by this author, which also included explanatory diagrams (*cf.* T. C. Wilkinson 2009).

include: official itineraries; stories or records of military success or failure; pilgrim accounts; or fabulous stories. The common feature of such routes is the narrative-like structure of the journeys described. Accounts of highways are telegraphic, partial and linear. They involve a series of landmarks, or key characters – cities, fords, passes, bridges, or peoples – between origin and destination. They often involve significant journeying over long distances by individuals. Often, though not always, highways are the dominant or 'official' ways by which urban dwellers may pass from one major urban centre to another. Indeed highways are the means by which the elite travel, or send goods, to parallel elites in other centres. They may be reified physically through the construction of metalled roads, official posting stations, or protected by special guards. All these may come at a cost, as tolls and taxes must be paid to access the advantages of the highway. Highways also may cross cultural or political boundaries, as major interregional 'trade routes'. In accounts of journeys along these routes, the differences encountered add to the dramatic effect of their significance. In this sense, highways are linear or bipolar: having clear start and end poles, each of which potentially occupies very different social spaces.

In contrast, the second type of route, a *pathway*, is related to everyday movements between home, work and other settings for action. These rarely get inscribed permanently by being written down. They include the footpaths, animal or cart tracks between villages, houses, fields or, in the case of transhumant societies, between pastures and other bases. Unlike highways, pathways are radial rather than bipolar, in the sense that they involve repetitive travel back and forth between bases. Cumulative unresolved movements may result in gradual drift over the long term, of course, for example in situations where marriage patterns dictate relocation, or where labour or environmental advantages pull and push segments of population. Pathways are, as the name suggests, smaller than highways, and may sometimes involve subversive or unofficial (though probably still socially constrained) movements, such as smuggling. The motivations behind individual selection of pathways are contingent at a local level on ritual, mythical, and personal needs, though cumulatively they are affected by the structure of the landscape, both natural and human.

Whilst the focus of historical and archaeological research has been on establishing the routes of particular ancient highways, archaeological evidence is probably better suited to identifying the remains of pathways. If repetitive small-scale movements create the opportunity for exchange of customs, ideas and gifts, the result in the archaeological record will be areas displaying similar or interrelated cultural practices or material. In other words, pathways are what facilitate the creation of entities that have traditionally been modelled as 'culture areas'. If we treat cultures as the result of dense dynamic networks of pathways – *i.e.* flows of material, ideas and obligations – rather than as reified blocs or essentialist ethnic entities, we can begin to perceive how the cumulative effect of such movements has considerable implications for social change or continuity. Ancient pathways will manifest themselves archaeologically in the distribution of material such as pottery or architecture which requires intense interaction. The archaeological imprint of highways will be shown in materials that involve extensive interaction. Particular classes of high-value commodities, which are distributed sparsely in the material record, provide telegraphic snapshots of the nodes of these routes: 'prestige goods' such as precious stones or high-value textiles are more likely to travel along highways. Simultaneously, chains of pathways might act as interregional highways, carrying materials long distances along apparently small networks. In both instances there are considerable number of items of exchange invisible to us: whether this means religious, social or political ideas on the one hand, or organic materials such as spices or narcotics on the other.

It is important to stress that this distinction between pathways and highways is not *necessarily* based on physical characteristics of the routes as such, but rather on the nature of movement along them, and their role in different types of exchange. Pathways can be characterized as shorter and repetitive; highways are longer and specialized. This might be related to social class: highways are the means by which the elites connect to each other and procure the resources which facilitate their position, whilst pathways connect the masses. In theory, the locations of pathways and highways may coincide, or the same individuals may travel along both pathways and highways depending on the role they are carrying out. The distinctions may be overstated – other lines could be drawn – and the categories should be treated as one potential heuristic device to analyse routes, rather than a universal framework. However, making these distinctions reveals the potential for different types of methodology. Both types of movement facilitate interaction between agents, both people and objects, which may have profound consequences for the trajectories of the societies they connect together. The nature of exchanges that take place in each context may be somewhat different, however, and to be able to trace the evolution of these systems we need to be able to represent such routes –and such exchanges– appropriately.

3.2.2 Modes of representation: road-maps, distribution maps and culture areas

As discussed in Chapter 2, the results of traditional studies of historical routes normally are visualized in one mode of representation, namely a 'pseudo-road-map' in which lines representing the reconstruction of a 'route' or itinerary are shown on an outline map of sufficient scale. This mode of representation reflects a model of routes that is strongly affected by a modern conception of road travel and navigation. Such diagrams are part of an understandable attempt to structure our knowledge of ancient movement in ways that make sense to the modern mind, and they result from subconscious leaching of the highly prescribed and structured nature of modern roads and railways. It is fairly unlikely that ancient travels were structured in the same hierarchical way[27] (except, perhaps, in periods like the Roman one where such road systems were similarly 'fixed'). But we must assume that movement was indeed 'structured' by other features of both natural and human landscapes.

This 'road-map' mode of representation may be useful sometimes, but if we are to find new ways to approach the study of routes, we must move away from our implicit definition of 'routes' as static linear features of the landscape and find new ways to represent movement. Rather than seeing routes as static sets of empty roads available for travel, we should see them as emergent 'corridor-like' entities constituted by processes which balance the opportunities for movement given by the physical (and human) landscape ('natural routes') and the various human

27 And it is unlikely that ancient peoples imagined or visualized their travels and interactions in this way (see thought-provoking comments in Ingold 2000).

motivations for that movement including, of course, the exchange of resources (people, materials and ideas)[28]. Additionally, the development of such routes (what we might alternatively call 'networks of interaction') in turn structures the motivations of each generation by presenting them with pre-existing and endlessly fluctuating social networks. Of course such motivations and networks are invisible in the archaeological record, but the outcomes of many types of movement are inscribed through their material debris: *i.e.* distributions of objects, settlements and people which can be compared to each other and whose material characteristics can reveal inter-relationships (similarities of form, style, diet, common sources) which were the result of different kinds of movement and exchange.

The 'pseudo-road-map' fails to represent two important components of real routes, namely, (1) the materials that travel along them and (2) the density of that travel. The problem then for archaeologists and historians is how to represent routes in a suitably empirical, understandable but chronologically specific way. The archaeologist's solution to representing indications of interaction and long-distance exchange has long been the 'distribution map'. Points marking the find spots of particular categories of object or locations of certain cultural features are used to illustrate the extent of interaction, but it has proven harder to find ways to represent the 'intensity' of such interaction in map form. One solution has been to invent boundaries for so-called 'culture areas' (*e.g.* the outline distribution of certain pottery assemblages) as a way of indicating zones within and between which different types of interaction may have taken place. The problem with both types of distribution map (point-based and area-based) is that they de-contextualize these points and boundaries from their landscape onto a sterile two-dimensional cartographic plane. As with the 'road-map' view of routes, the resultant images are both static and partial. The task, then, is to find new ways to visualize both the archaeological site location information that has been traditionally represented in distribution maps, and the connections between such locations, and to combine this with data on physical landscapes and movement through them.

3.3 Modelling 'natural routes' with GIS

The solution to these representational problems will come, I believe, from the development of new digital mapping technologies – based on current (and as-yet-undeveloped) GIS and database techniques. The processing power of modern computers and the ever-increasing availability of digital data about the Earth's surface (especially those derived from remote sensing satellites and NASA shuttle missions) allow us to create an ever more detailed model of global topography and climate, facilitating the zooming between macro- and micro-scales. They therefore offer the potential, when combined with analytical models, to create new forms of landscape visualization that can take account of multiple dimensions of routes and networks of interaction. Such modelling remains in its infancy, but as part of the research for this book I have experimented with the use of currently available techniques to create what I call 'archaeotopograms' – representations that combine aspects of both landscape and archaeological distributions, discussed further below.

28 This redefinition is akin to Merleau Ponty's phenomenological redefinition of landscape not as abstract space, but as something always in the process of becoming, a view which also inpsired Ingold's approach.

Geographical Information Systems (GIS) may be usefully thought of having two aspects: first the visualization of geographic data, and second the mathematical or statistical functions which may be applied to raw data and subsequently re-visualized. While the first aspect has already revolutionized the speed with which basic archaeological mapping has been undertaken, the second aspect introduces new approaches to archaeological problems, often relying on complex statistics, which require careful interpretation. Amongst the most common of these 'statistical' GIS techniques applied in archaeology are spatial hierarchy analyses, view-shed analyses and cost-surface/least-cost path analyses. Spatial hierarchy analyses have a long history within archaeology as a result of the discipline's encounter with the New Geography in the 1960s and 70s (*e.g.* Chorley and Haggett 1968; Hodder and Orton 1976; Clarke 1977): the drawing of Thiessen polygons (based on the Voronoi tessellation/Delaunay triangulation) is a basic example of this kind of work in which the distance between points (normally archaeological sites) is calculated and this is used to depict a boundary supposedly representing the edge of the 'site-catchment' territory. This has been used to suggest the potential boundaries of control between peer sites – for example megaliths or hillforts (Renfrew 1968; Cunliffe 1971). Of course such analyses are problematic since Thiessen analysis assumes a flat surface – fine if the region in question has no topography – but in most real-world instances the diagrams give a false or misleading impression of potential territories if taken literally (see, for example, critique of the 'central-place' model of hillforts in Sherratt 1996). For this reason a variety of more complex techniques of analysis have been developed which are designed to model space and movement across it in more sophisticated ways.

3.3.1 Impedance along vertices: 'network analysis' and routes

So-called 'Routing' or 'Network analysis' is an analytical technique to calculate efficient routing or paths between locations, which relies on a pre-given model of relationships between nodes. In a network, each relationship (a vertex) between each connected node must be explicitly stated and given one or a number of impedance factors. These impedance factors can be based on geometric distance, time or 'costs' calculated from other spatial analyses (see below). This form of analysis has had a huge application in modern transport navigation tools (such as used by 'SatNav' devices and train timetable calculators), which attempt to solve the 'travelling salesman problem' – *i.e.* how to pass through a set of nodes in the least time or at the least cost. Such techniques might appear to offer a very fertile ground for the analysis of ancient routes (and interaction more generally), but so far their applications have been rather few. The reason for this is that current models cannot handle situations in which all or most of the possible nodes (*i.e.* assumed settlements) and the links between them are not known – and this is, in fact, the most common state of archaeological knowledge.

One recent application of these techniques has been to the Aegean in the 2nd millennium, to model the effects of the Thera eruption on interaction networks by removing nodes and therefore linkages (Knappett, Evans and Rivers 2011). The analysis simulates the consequences of the destruction of settlements by comparing two versions of a network of sites (and the links between them) which are established archaeologically, where the second version does not include those sites destroyed by the eruption or related events. This application is successful partly because the Aegean can be treated as a relatively neutral surface and the impedance costs can be modelled straightforwardly. The knowledge of settlement patterns in the region is also relatively good – so that while the list of nodes may not be entirely complete, it may be close to it. It also addresses a relatively straightforward comparative question of before and after. Similar highly-focussed analysis might be taken to a land-based situation, such as in the northern Syrian plains where a detailed route network can be constructed through the study of hollow-ways and tells (see Section 2.2.4). However, once the analysis is taken to a larger scale, we cannot be sure that we know all of the 'nodes' required for a complete network (*i.e.* there may be many relevant sites that we have not identified), and we are rarely certain about the temporal co-existence of sites in different locations in any case because the resolution of our data is too low.

3.3.2 Friction over a grid: cost-surface analyses

A related but structurally different tecnique, 'Cost-surface analysis' (Conolly and Lake 2006, 214-225, 252-256) aims to create a mathematical or visual model of the relative cost to reach all locations on a surface if travelling from a particular point on that surface using a grid of cells and a model of 'cost' of 'friction'[29]. Cost-surface functions normally require two main inputs: (1) a 'cost-of-passage' or 'friction map' grid (which can be represented as a raster image in many GIS programs), whose contents are based on a model of cost or friction that may include factors such as slope, and (2) the origin cell within the grid from which all accumulative costs are to be measured. The resulting output of this function (an 'accumulated cost surface') is another grid in which each cell contains a value that represents the relative accumulated cost to travel to this cell from the origin. Normally this grid is then represented as a raster image to become a human-readable map of relative or effective distance to a point, once travel cost is taken into account. In the case of an archaeological site, for example, this may give an indication of the relative amount of time/effort it takes to travel from the site to various potential resources.

This is a relatively simple idea in theory, but the usefulness of its application to archaeology depends a great deal on how one is to calculate 'cost' and how well the algorithm used to calculate the accumulated cost can model the real world. There is more than one type of algorithm available to transform a 'cost-of-passage' grid into a 'cost-surface' (Table 3.1). The two main types are 'isotropic' or 'anisotropic'. 'Cost' or 'friction' is most easily understood in terms of time or energy (as opposed to geometric/Euclidean distance). For example, topography can be taken as one friction factor: steep slopes are taken as high 'cost' because they take more energy or more time to cross than flat terrain, and slope can be easily calculated in GIS programs from 'Digital Elevation Models'/'Digital Terrain Models' (DEM/DTM) such as the global coverage 'Shuttle Radar Topography Mission' (SRTM). However, the geometric value of slope is not a good model of the energy or time required for a pedestrian to cross an area, as is obvious to anyone

29 Cost-surface and least-cost-path analysis could in fact be seen as two specialist kinds of network analysis in which 'cells' in a grid rather than 'nodes' in a lattice of relationships are the unit of analysis. In cost-surface analysis, cells can only be directly related to adjacent cells in the grid. Algorithms are used to calculate the 'cost' or 'friction' of movement from one cell to another. This alternative way of modelling 'impedance' is often based on a variety of variables, including but not limited to topography.

GIS suite/program	Function	Algorithm type and Source?	Comments and Further information
ArcGIS (Spatial Analyst)	**CostDistance**	Accumulative cost (isotropic) *Proprietary*	Manual (ArcGIS 9.2 Spatial Analyst): http://webhelp.esri.com/arcgisdesktop/9.2/index.cfm?TopicName=Cost_Distance_algorithm
GRASS ***Also accessible in:*QGIS**	**r.cost**	Accumulative cost (isotropic) ?	Manual (GRASS 6.4): http://grass.osgeo.org/grass64/manuals/html64_user/r.cost.html
SEXTANTE ***Also accessible in:*SAGA / gvSIG / uDig / openJUMP**	Accumulated cost (isotropic)	Accumulative cost (isotropic) ? *Documentation unavailable*	General information: http://forge.osor.eu/projects/sextante/
IDRISI	**COST**	Accumulative cost (isotropic) *Proprietary*	IDRISI 15

Table 3.1. Accumulative cost surface programs (basic/isotropic).

GIS suite/program	Function	Algorithm type and Source?	Comments and Further information
ArcGIS **(Spatial Analyst)**	**PathDistance**	Accumulative cost (anisotropic) *Proprietary and unpublished*	The base algorithm is proprietary and unpublished. Manual (ArcGIS 9.2 Spatial Analyst): http://webhelp.esri.com/arcgisdesktop/9.2/index.cfm?TopicName=Path_Distance:_adding_more_cost_complexity
GRASS ***Also accessible in:*QGIS**	**r.walk**	Accumulative cost (anisotropic) *{Aitken1977wa; Langmuir1984ml}*	Manual (GRASS 6.4): http://grass.osgeo.org/grass64/manuals/html64_user/r.walk.html
SEXTANTE ***Also accessible in:*SAGA / gvSIG / uDig / openJUMP**	Accumulated cost (anisotropic) Accumulated cost (combined)	Accumulative cost (anisotropic) Accumulative cost (combined) *?Documentation unavailable*	Other modules:- Cost in predefined routes. Cost in predefined routes (anisotropic). Generate alternative routes. General information: http://forge.osor.eu/projects/sextante/
IDRISI	**VARCOST**	Accumulative cost (anisotropic) *Proprietary and unpublished*	IDRISI 15

Table 3.2. Accumulative cost surface programs (complex/anisotropic).

GIS suite/program	Function	Algorithm type and Source?	Comments and Further information
ArcGIS **(Spatial Analyst)**	**CostPath** **CostPathAs-Polyline**	Least Cost Path *Proprietary and unpublished*	Manual (ArcGIS 9.2 Spatial Analyst): http://webhelp.esri.com/arcgisdesktop/9.2/index.cfm?TopicName=Cost_Path http://webhelp.esri.com/arcgisdesktop/9.2/index. cfm?TopicName=Least_cost_path_and_least_cost_corridor
GRASS ***Also accessible in:*QGIS**	**r.drain**	Drainage Path	This module can be used to create general Least Cost Paths but is technically a hydrological flow diagram and can be dysfunctional for the purposes of creating a standard least-cost-path because it iteratively chooses the lowest value each cell and terminates when the lowest value is reached like water. Manual (GRASS 6.4) http://grass.osgeo.org/grass64/manuals/r.drain.html
SEXTANTE ***Also accessible in:*** **SAGA / gvSIG / uDig / openJUMP**	**Least cost path**	Least Cost Path *?Documentation unavailable*	
IDRISI	**pathway**	Least Cost Path *Proprietary and unpublished*	

Table 3.3. Least-cost-path programs.

GIS suite/program	Function	Algorithm type and Source?	Comments and Further information
ArcGIS **(Spatial Analyst)**	**Corridor**	Corridor	Manual (ArcGIS 9.2): http://webhelp.esri.com/arcgisdesktop/9.2/index. cfm?TopicName=Corridor

Table 3.4. Least-cost corridors programs.

who has had to climb and descend a mountain. The 'braking' energy required to avoid falling when descending can sometimes be more than that required for climbing at lower gradients. Thus alternative algorithms have been written which are supposed to take account of this 'anisotropic' cost (Table 3.2), some of which have been based on publications of laboratory measurements of the energy cost of walking at different slope (for a summary, see Conolly and Lake 2006, 217-221). The problem is that many of the cost-surface algorithms commonly available for the average archaeologist to use are not well documented, so that it is sometimes difficult to compare or critique them. Exceptionally, the anisotropic algorithm used by GRASS (r.walk) is well referenced (Table 3.2), from which we learn that the formula used is based on data taken from walking in Scotland. Despite the fact that isotropic algorithms are not really applicable to real-world geographical applications (of which archaeological reconstruction is one), the additional complexity has made uptake of anisotropic algorithms relatively slow.

3.3.3 Finding paths: least-cost-path analyses and corridors

Least-cost-path takes cost-surface and cost-of-passage direction models a step further to calculate a specific path between two points according to cost which should, in theory, represent the path with the lowest possible accumulative cost. Least-cost-path analysis has enjoyed some success in the process of modern road planning, as it helps to suggest the optimum economic route to build a road through a landscape: 'costs' such as land-prices and fuel prices translate easily into quantifiable models. Similarly, related algorithms can be used in hydrological modelling (required for flood protection planning, for example) in order to predict the likely path that water will flow downhill. These tools have been borrowed in archaeology to allow researchers to suggest likely paths between a site and its resources, or to compare known routes with predicted models in order to identify the possible factors behind path location (for example Kantner 2004; Bell, Wilson and Wickham 2002; Pelfer 2007; Newhard, Levine and Rutherford 2008). Indeed the generation of least-cost-paths is normally the main aim of cost-surface analysis within archaeological research. Because of the dependence on cost-surface, the usefulness of the least-cost-path is dependent to a high degree on how the model of cost is designed and on the algorithms' fitness for the task. The most accurate way to calculate least-cost is to calculate the cost of all possible paths and then rank them. However this is a highly computationally intensive procedure (and potentially unrealistic from the point of view of the modelled human actor), whose worst-case computation time increases exponentially with the amount of data. It is therefore far more common for programs to approximate least-cost. For example, in a very basic iterative 'drain'-based approximation, the program will search around the current start point to find the adjacent cell with the lower (or lowest) cost, and then look for the lower (or lowest) cost cell adjacent to this new cell iteratively *etc.* until a 'destination' cell is reached. More often a 'direction raster' is used to better calculate the next cell. Different approximation procedures can give rather different results, as was highlighted by Gietl *et al.*'s comparison between GRASS, ArcGIS and IDRISI on the same data (Gietl, Doneus and Fera 2008). Again the base algorithms are not well documented so that it is difficult to compare or critique (Table 3.3). At the small-scale, most of the currently available procedures produce straight lines over steep hills and cannot reproduce the more realistic spiral/zigzag climb which many roads and tracks adopt – an algorithm for which was put forward by Colischonn and Pilar (Collischonn and Pilar 2000).

An alternative type of least-cost is a 'least-cost-corridor'. At its simplest this involves the sum (or average) of two cost-surfaces from two different points; a 'corridor' between the two points can then be established by filtering out costs above a certain (but arbitrary) value. This procedure can be achieved easily in most GIS programs, and ArcGIS even includes a dedicated tool to create it (Table 3.4). This technique has had wide application in ecology, for example in the establishment of natural reserves for deer. So far I am not aware of any previous archaeological applications – but the potential relationship between these mathematical corridors and the restricted 'route corridors' within which roads and trackways shifted suggested possible applications for this study.

3.4 Developing a GIS approach to routes

All of the above GIS techniques were designed to address spatial questions that are relevant to the study of routes and interaction along them. 'Network analyses' involve the explicit definition of relationships between nodes (sites), so they must be firmly linked into particular temporal contexts in which both nodes (sites) and linkages (routes) are already known; only then can the 'connectivity' and accumulation of movement be simulated. In contrast, 'cost-surface analyses' start from a generic geographic perspective (although the landscape model is normally static and so taking account of environment change is different), which is independent of time-period. Subsequently, the accumulation of potential movement can be simulated by performing calculations based on the geographical location(s) of archaeological phenomena/sites. Both sets of analysis may allow the development of complex visualizations of interaction networks, but neither of them offers a straightforward way to 'reconstruct' past route networks that look like the traditional 'road-map'. 'Least-cost-path' analyses, which are often seen as the ultimate aim of cost-surface procedures in archaeological application, do indeed create linear 'road-like' maps, but are very problematic as a predictive tool. It is clear from various studies that a path generated by 'least-cost-path' is fairly arbitrary (Gietl, Doneus and Fera 2008) – it represents one hypothetical path amongst many. This suggests that, at least at the small-scale, 'least-cost-path' should normally only be used for comparison with pre-known paths rather than for reconstructing or predicting unidentified ones.

3.4.1 'Cost-of-passage', 'cost-surface' and 'cost-corridor' approaches

Instead, less-restrictive forms of cost-surface techniques may be the most productive way to access the potential of the landscape in order to create a model of 'natural routes'. Thus 'cost-corridor' may be seen as a more flexible and realistic version of 'least-cost-path' that gives less arbitrary results because it 'brackets' an area rather than suggesting a single path – this certainly fits better with the definition of a route as corridor (see Section 3.2); similarly resistivity models based on the flow of electrons across a field of variable conductance may offer more realistic cumulative maps (for an application from within animal ecology, see Shah and McRae 2008). But there are various potential objections to the application of 'cost-surface' to human behaviour. One of the most serious is the suggestion that the underlying model of movement is defective: *i.e.* that the idea of cost and its assumption of economic rationality produces unrealistic results and cannot make useful predictions. As the anthropological literature on ritual can attest, human agents are interested in many different factors beyond resource-maximization, and paths wind around and past cultural and natural features. Even better walking models (which take anisotropic costs into account or can create more realistic 'spiral' paths up steep hills) remain simplifications of real human motivations (and leave out the other modes of travel and transportation).

There are a number of ways to pre-empt this kind of critique. First (1) I would argue that it is disingenuous to suggest that travel 'cost' did not play a major role in the production of the ancient world and the remains we find: it may be more comfortable for an anthropologically-informed archaeologist to use the word 'friction' in place of 'cost' to disassociate the procedure from monetary economics – though as the Old Assyrian *karum* records attest, monetary and time costs of routes were themselves often of great interest to ancient peoples. Actually, the more serious problem with cost/friction models is to do with the restrictions of the algorithms, which require the user to condense many different factors into a single quantitative value. Real decision-making by individuals (ancient and modern) is much more adaptive, allowing different factors to come to the fore at different times (as we note from the routes taken by armies – for examples, see Section 2.4.2). To some extent this can be overcome by using the more advanced 'anisotropic models' – although the introduction of greater complexity increases the difficulty of creating and comparing models.

Second then, (2) scale should also be considered. Most applications of cost-surface techniques have made use of environmental data, normally topography or resource distribution, to study movement or paths on the *local* scale. The resultant analyses and interpretations often seem abstract and removed from the everyday life of ancient people. Less attention, on the other hand, has been paid to the use of cost-surface techniques on a *macro-* or *supra-regional* scale. By analogy to Braudel's argument that different sets of evidence are appropriate sources to study historical processes for particular temporal scales (Braudel 1996), we may likewise suggest that the larger the geographic and temporal scale of analysis, the more likely that environmental factors (which are the ones most easily modelled with GIS) will determine the structure of human movement. Such GIS approaches are actually far better suited to study macro-scale of inter-site, inter-region and inter-cultural movements than they are to the procurement of resources within a local field because at the macro-scale, localized deviations from straight-line paths (everyday movements along fields, between villages *etc.* – the *pathways* described in Section 3.2.1) will become slowly smoothed out into more general aspects of the landscape (slope, climate). The condensed single variable of 'cost' is thereby less problematic because, at the large-scale, adaptive decisions are subsumed by large-scale processes. Conversely, by providing a baseline model of the landscape, the particular cultural dynamics and route systems of the period can perhaps be brought more starkly into relief.

Finally, (3) we need to dislocate GIS techniques from their traditional association with positivist and hypothesis-testing epistemologies. In a positivist-processualist approach characteristic of much GIS analysis, the analysis of ancient route systems would rely on combining the above factors into a single model or set of rules, which could be programmed into the GIS. The model would then be 'tested' against the evidence to measure its 'goodness of fit'. There are a number of reasons why such an approach is unhelpful, some of which will be briefly summarized below, and not only because of objections derived from phenomenology that mathematical cost-surface analysis itself might be irrelevant to human societies because of its emphasis on non-human values (*cf.* Ingold 2000). Instead we should treat cost-surface and other techniques as tools for visualizing archaeological problems – designed to make historical problems easier to think through. Explicit models allow us to simultaneously foreground our already active but frequently unstated assumptions but apply them rigorously to the data. What this means in practice is that the 'models' created are not abstract ones in the Popper-ian or mathematical sense, but rather visual ones akin to conventionalized drawings of objects or the scale models of ancient architecture or objects. While ultimately based on empirical foundations, these are qualitatively orientated documents that are designed to facilitate understanding or interpretation of the objects in question (in this case landscapes) rather than be manipulated statistically. In my view, rich visualization of interaction across landscapes is far more important to archaeological narratives than the reconstruction of any 'true' road, which is in any case an impossible task especially when the concept of route is so slippery in the first place. This book thus adopts a non-positivist, or rather 'soft' approach to the use of the computer modelling.

The following sections will explore how various factors that contribute to the cost of movement across the landscape can be incorporated into a model of 'cost-of-passage'.

3.4.2 Theory: natural and anthropogenic cost factors

i. Topography – slope

Topography has been the basis for most previous cost-surface analyses, and it is obvious that the shape of the landscape is very important to the ability of armies, traders and migrants to travel. This can be easily gleaned from the weight of importance given to mountain passes in descriptions of military campaigns, real or imagined, from Alexander the Great to the British Navy at the turn of the 20th century[30]. Additionally the topography itself and the environmental conditions such as soil and climate that are formed as a result, may structure the pattern of settlement, in turn structuring movement.

Where bulky, heavy or large volumes of goods are involved, mountainous areas and steep valleys will be avoided at all costs in favour of river valleys and water transport. On the other hand, on certain scales, goods may have high enough value across mountain ranges to be worth transporting in this way. Recent work by Irene Good in the Pamirs of Tajikistan (Good 2006) suggests that mountain valleys may often act as conduits of exchange, an idea which has also been put forward for the Alps. In any case, the slope limits the intensity of movement by reducing the number and 'size' of exchanges possible. If looking at particular kinds of exchange, it might be helpful to categorize slopes into qualitative values ('too steep', 'steep', 'not-steep'), as has been done for some previous analyses. For example, for tracing or predicting Roman roads, whose routes tend to reflect major highways (and thus

30 As shown in the military 'Handbook' to Asia Minor and Mesopotamia researched and published by the British Naval Intelligence Dept. – see Section 2.3.2.

bulk transport routes), some slopes are simply too steep to make sense, and these areas may be discarded in the predictions. On the other hand, if a more general pattern of movement is under analysis, small-scale exchanges between villages and small communities are also important to cultural interaction: thus a fuller degree of nuance can be achieved by using a continuous slope value.

Slope is an obvious measure by which a cost-surface can be obtained, since steep slopes significantly increase the cost of travel. As mentioned earlier, however, its relationship to cost is not straightforward. On a small scale, angle/percentage of the slope cannot be used as a direct value in cost. Instead a model of cost/slope must be applied taking into account relative difficulty of upward and downward slopes. This may differ according to transportation technology, but most archaeological models currently depend on laboratory measurements of energy expenditure or time of walking. Whether this problem also applies to the macro-scale has not yet been theoretically or experimentally investigated. It is possible that at this macro-scale, the differences between transportation types are blurred, and if two-way equal exchange is dominant between all regions, a fairly straightforward equation might be used. Macro-economic theories, especially world-systems, emphasize the inequality of exchange between regions, however. This could militate against such a straightforward equation if inequality is taken to be normal.

ii. Topography – elevation, inequality and directionality

Another related aspect of the topography is relative elevation. Even if the value of exchanges were to be taken as equal, the distribution of resources and their relative bulk or transportability is not. For example, the sources of metals and precious stones may often tend to be found in higher altitude areas, precisely because they are created by the tectonic activity that also creates mountain ranges (such as the Taurus-Zagros, or Pamirs-Himalayas). Transferring bulk goods downhill is significantly easier than transporting them uphill, as is especially the case where water may involved, such as along the river Euphrates[31] – though this is very difficult to model practically. Such directionality in exchange is also important where sea-routes are considered: the rotational currents and winds of the Eastern Mediterranean have long been an important consideration for the study of ancient trade, but it is difficult to see how we could build a comparative model which would include directional aspects of both sea and land together.

iii. Seasonality

Seasonality of travel is a frequently ignored aspect of ancient routes. Archaeologists who tend to travel to their study regions only during favourable summer conditions (normally using motorized transport) rarely see the same landscape under snow, and it is easy to forget the extent to which modern transport and air conditioning technologies considerably lessen the discomfort even in the summer. In the past, regions under considerable amounts of snow may have been relatively or completely impassable for long periods in the year. If data are available, it is relatively easy

31 One example of such 'directionality' in transportation along a river comes from the late 19th century AD: "The Tigris is much utilized by large rafts from Diarbekr [sic] downwards to Baghdad. These consist of about 150 inflated sheepskins tied underneath a light wood framework, and are utilized to carry merchandise or travellers. These rafts cannot ascend the river, but are broken up on arrival in Baghdad and the skins taken back by caravan."(Maunsell 1894, 6).

to impose a modelling 'handicap' on a region on the basis of the number of days of snow cover (or conversely, favour regions which have relatively little), by increasing the cost-calculations appropriately. The only problem here is the extent to which modern data on snow cover can be used to represent past environments: ideally, one would want to take into account periods of climatological cooling and warming – but these data are not easily available over a large geographical area in sufficient detail.

iv. Climate – humidity/aridity and extreme temperatures

Little research has been done into the extent to which general climate, but particularly humidity/aridity in combination with extreme temperatures, can affect travel ability. Anecdotal evidence from regions such as Cilicia shows that high humidity and heat can significantly increase discomfort for travel, though to some extent it may be a matter of acclimatization[32]. Similarly the extreme heat of Taklamakan summers encouraged the tradition of winter caravans across the desert (Potts 2004a, 147). For those travellers engaged in particularly long distance exchange, movement between climatic zones might perhaps represent a cost factor, but it is difficult to place a measure on this. As with other factors, the problem remains a lack of experimental data upon which suggested ratios of 'cost' might incorporate the factor into an equation.

v. Rivers

A related aspect of seasonality is rivers, and their ability to be forded or navigated depending on seasonal variables: this is particularly the case for glacial and rain-fed rivers (though with contrary results) whose levels vary depending on the season. In general navigable rivers facilitate parallel trade, but hinder perpendicular trade. That is to say that it is often easy to travel along the river, on waterborne transport along larger, stable, rivers, or alongside on towpaths, but finding appropriate crossing points may often be difficult, even for smaller streams, and in some cases fraught with danger (flash floods, drowning, and goods/provisions/clothing becoming wet/ruined)[33]. The same is true for crossing irrigation streams: huge detours must sometimes be taken simply to be able to cross safely with vehicles or animals. Rivers also have directionality, related to the slope of the topography, in that downstream is often (though not always) less costly than upstream.

Rivers have been sometimes been included in cost-models by assigning the route of the river a markedly lower cost value than the surrounding land. The problem is how to include the flow direction of the river: whilst theoretically possible to include in an anisotropic model, it would be rather complex. Only certain parts of certain rivers are navigable by boat. As a basic means of modelling the difficulty of crossing the river., an additional parallel 'buffer' of 'high cost' can be added. Where rivers are navigable or fordable, it is usually from particular points, *i.e.* ports, fords and bridges, and not from the entire length of the river.

---

32 An easy way to experience this shift in humidity or temperatures is to travel from the Konya plain across the Taurus into the Göksu valley, or from the Lake Van region into Batman province.

33 E. Newby's *A Short Walk in the Hindu Kush* illustrates the problem of crossing rivers with pack animals in his description of his travels in Nuristan, Afghanistan (Newby 1974, 234-236). Other nineteenth and early twentieth century travel accounts offer similar though less colourful descriptions of the time taken to locate suitable fording or bridging points.

vi. Water sources

Access to clean drinking water sources is a particularly complex problem. The main difficulty with trying to include this factor in route models is in identifying the density of water sources across a landscape: first these have rarely been systematically documented even in modern topographic maps (and maps that might show this are generally harder to obtain); and second there is a strong chance that springs may have changed positions through time depending on water levels, or irrigation/water management schemes. In the hills of the Caucasus and Anatolia, it is rare to be completely without nearby water sources, but, depending on routes taken, the time between sources can prove problematic. By contrast, in the deserts of Turkmenistan, a much better specific local knowledge is required to identify springs or drinkable water – although fountain monuments of the Islamic era have recently been identified along a path between Merv and the Oxus (Wordsworth 2010). Being able to predict availability of clean drinking water has a profound effect on the ability of armies, traders and migrants to move effectively, even if other means of purification (the addition of alcohol, boiling teas) might overcome some of the dangers. Lee Ullman has argued that Hittite monuments are located precisely to take account of the travelling Hittite army's need to camp and replenish near water sources, both bodily and spiritually (Ullmann 2011). Modern springs in Turkey are usually made highly visible by surrounding white-washed brick, stone or concrete constructions, and a water trough for animals – often next to mosques. The common factor in these examples is that of water *monuments* – which are not identifiable in all periods.

Modelling the effect of water sources on travel/exchange routes would be possible if one could establish a comprehensive database of their distributions (converted to a density raster) or else by using proxy indicators such as rainfall or maps of groundwater distribution. The increased 'cost' would be calculated as a factor of the low density of water sources, since greater effort might be spent finding water in 'low density' regions.

vii. Sea

In coastal regions, sea transport can facilitate bulk movement of goods in particular. As mentioned before, sea-transport often has directionality, in that the sea currents and prevailing winds tend to promote movements across water in certain directions. In the case of the Eastern Mediterranean currents and winds favoured an anti-clockwise rotation for sail-based transport, for example. Areas that receive frequent storms and high seas will also resist movement across them. The complicated factor is how one converts this qualitative knowledge into a model which can be directly compared to movement across land. Directionality may be incorporated into the cost-direction model, but the relative weighting between sea movement and land movement is difficult to judge. The conjunction between land and sea routes is also a factor: ships often cannot simply be landed at any point along the coast and transferred into land movements. Long stretches of the Mediterranean and Black Sea coasts of Anatolia are treacherously difficult to land a boat, for example. Harbours and port towns (or suitable locations for them) are therefore important facilitation nodes. Meanwhile, sea changes around the Caspian make it difficult to be sure of the prehistoric boundaries between land and sea. Similar to the effects of rivers above, a 'buffer' between land and sea with a few facilitation points can provide a simple way to include these ideas into a model; however, though the landscape may make some places easier to land than others, the facilitation points (*i.e.* ports) and the suitability of landing places are often historically specific.

viii. Soils, sands, rocks and rain

Anyone who has tried to cross a ploughed field after rain will know that the ability to walk decreases exponentially with the amount of mud picked up on the walker's feet. Movement may be hindered by marsh, sandy soils, or loose rocks, especially under unusual weather conditions. Specific metalled or hardened paths are also thus important not only for the way in which they guide people between places through repeated action, but also because of the practical difficulties of taking alternative routes. These factors are difficult to model easily, however. Given known paths between sites, one might test statistically to see if soil types (or associated vegetation – see below) had an effect on routes chosen, but to guess how this would work *a priori* is problematic. On the other hand, such factors are probably more significant at the small-scale than the macro-scale, and would not need to be included in a macro-scale study.

ix. 'Land-use'/vegetation boundaries: woods and bandits

Whilst 'land-use' is strictly speaking an anthropogenic factor (*i.e.* it is human communities who decide on the use to which particular areas are put, and set the social restrictions which prevent people from crossing certain types of land) and in the ancient world fewer areas of land were managed by humans, vegetation-coverage was no doubt important. Certainly land-use or vegetation-coverage must have played some role in the way in which ancient travellers calculated 'cost'. Woods and forests provide the most obvious example of this. The Mesopotamian *Epic of Gilgamesh*, dating back to at least the 3rd millennium BC, includes a sequence in which Gilgamesh fights alongside his companion Enkesh to defeat an ogre from the Forest of Cedars, of whose trees the ogre is guardian. Woods and forests (and their mythological residents) have a special place in traditional narratives of travel as potentially dangerous, no doubt because of the low visibility they afford the traveller from potential attacking animals or bandits. On the other hand, the wood itself is often an important resource for constructing buildings and ships, and, as with the story of Gilgamesh, the forests have symbolic or mythological meanings (connected to practical ones) that attract pilgrims of different kinds. For these reasons it is difficult to model the significance of this factor to travel.

x. Transportation technology – carts, horses, camels and roads

Transportation technology has already been discussed in Section 1.4. Such technology has a considerable effect on cost, advanced technologies (carts, horses, camels, ships) alternatively increasing the amount that can be carried, the speed with which they can be transported, or, opening the possibility of travel across landscapes which would otherwise be relatively impenetrable. It has been suggested, for example, that the use of camels in Central Asia may have been the catalyst that allowed trade across certain large expanses of desert between oases, which would otherwise have been impossible. On the other hand, pack animals or vehicular transport may be more sensitive to different topography or other conditions than human carriers, as discussed above under slope. Straightforward equations may not therefore be very easy to create, though simple models might give broad indications. Metalled paths can be taken into account in cost models relatively easily, for example, by simply assigning a low cost to their routes (in a similar way to rivers). On the other hand, a particular problem with transportation technology is the difficulty of giving a spatial extent of its usage. While roads may be identified on the ground, it is not always clear where particular pack animals, vehicles or ships are being used, even if archaeological, zoological, iconographic or other evidence hints at their adoption in the region. Unless the aim is specific comparison between modes of transport (*e.g.* comparing possible travel distances with different technologies), cost-based analyses in archaeology still have to rely on the assumption of a pedestrian mode of movement, because it is nearly impossible to measure the relative importance of contemporary transport modes given current models.

xi. Pre-existence of routes and roads

The pre-existence of roads and paths is a key factor in the location of routes of movement. In urbanized societies, towns and villages provide the origin and destinations of most trade and exchange. This factor works as a feedback loop, in that routes of one kind may provide a catalyst for exchange in another field, increasing the importance of the settlement and decreasing the costs of transport, via the economies of scale. Thus routes of wealthy settlements become increasingly 'fossilized' through metalled roads, posting stations or travel inns, and organized transport. The problem is that this particular factor is both circular and historically specific and so cannot be easily built into a model of 'natural routes'.

3.4.3 Praxis: realizing a working model of 'cost-of-passage'

Establishing the theoretical basis of a model is one thing, but actually creating a useable yet realistic quantitative model which can be integrated into a GIS is quite another. The complexity of the task, and the number of unknowns proliferate rapidly. Primarily the concept of 'cost' is very problematic, since there are many different potential interpretations. Ideally a useful archaeological cost-of-passage would represent, say, the sum time taken to travel across an area given the factors defined in the model. In practice, it is difficult to know how to assign real time-quantitative values to these factors, and instead it is necessary to simply give broad relative values.

Besides the specific problems of identifying appropriate data sources (and the sheer amount of processing time required both to create the cost-based models and then use them), one of two of the main problems with going from the theoretical model to practical tool are (1) first, deciding how to transform the source data into a *relative value* of cost; and (2) second, how to decide the *relative weighting* of each of these factors within the final cost model. The simple fact of the matter is that there are no substantial experimental data that could allow us to suggest likely or realistic relative values for anything except for slope, and no information on how other factors could be weighted against slope. A research programme to establish such parameters was outside the remit and time constraints of the current work, but provisional models can still have heuristic value: this is because the data remain fundamentally empirical, and the method by which the data were processed and the way in which each factor is combined is spelled out explicitly. In any case, it is probable that no single formula for calculating supposed 'cost' for all factors, actors and scales could be created, and that in different situations different combinations might be used. In effect, the models should be treated as explicit versions of the normal assumptions about routes that archaeologists work with, but transformed into a form which can be more easily critiqued, revised or rejected.

Three alternative cost-of-passage models of 'natural routes' are shown here in the form of 3 raster grids: Model 1 (Figure 3.2), Model 2 (Figure 3.1) and Model 3 (Figure 3.3). They were created using the principles of cost-surface described above by assigning a value of cost to certain identified 'natural' factors (topography, climate *etc.*) and then combining their relative importance using a set of constants and simple formulas (the source data and description of how each were calculated are described in Appendix A.1). On their own, these graphs of the relative 'ease of travel' should not be taken to equal routes or travel: they represent, instead, a model of travel opportunity which may or may not have been taken by ancient peoples. Only once paired with cultural or material data of some kind (discussed below), do these models begin to say something specific about geographical relationships.

Each model shown here has a slightly different bias: if one takes Model 2 as the 'base model', Model 1 places relatively less emphasis on topography (or rather slope), while Model 3 places greater emphasis on the low availability of water (based on precipitation). These nuances show how pre-defined choices made by the model creator (about the relative importance of the predefined factors) produce differing base models. Numerous details of the resulting maps could be discussed with regards to whether they accurately represent the ease of travel in certain directions. By way of example, the route of the Indus river is depicted as relatively hard in comparison to central India which instinctively seems odd: this is a result of the model's focus on annual precipitation as an imperfect proxy for water availability (and through much of the Indus valley, precipitation is very low). Similarly the foothills of the eastern Taurus and western Zagros seem relatively easier than the adjacent southern plains. Clearly, in the long term, a better model of water availability is required to improve these details. Also, seas and large lakes are excluded from these models completely to avoid complexity. At present, since we have no easy way to control such potential distortions, we must, however, be content with these imperfect models for heuristic purposes.

3.5 Mapping material flows

As set out in the introduction to this book, the aim of the research presented here was to examine the spatial dimensions of interaction between 3000 and 1500BC over a macro-scale in the Near East. For this reason, the establishment of potential 'routes of movement' is an essential first step. However, the aim is not simply to compile a catalogue of potential roads or corridors of movement, but to situate ancient flows of materials in their spatial perspective: in other words to combine natural routes with chronologically and materially-specific distributions in order to better understand ancient interaction. How then can we use GIS approaches, such as the 'cost-surface' techniques discussed in detail above, to 'map' these material flows?

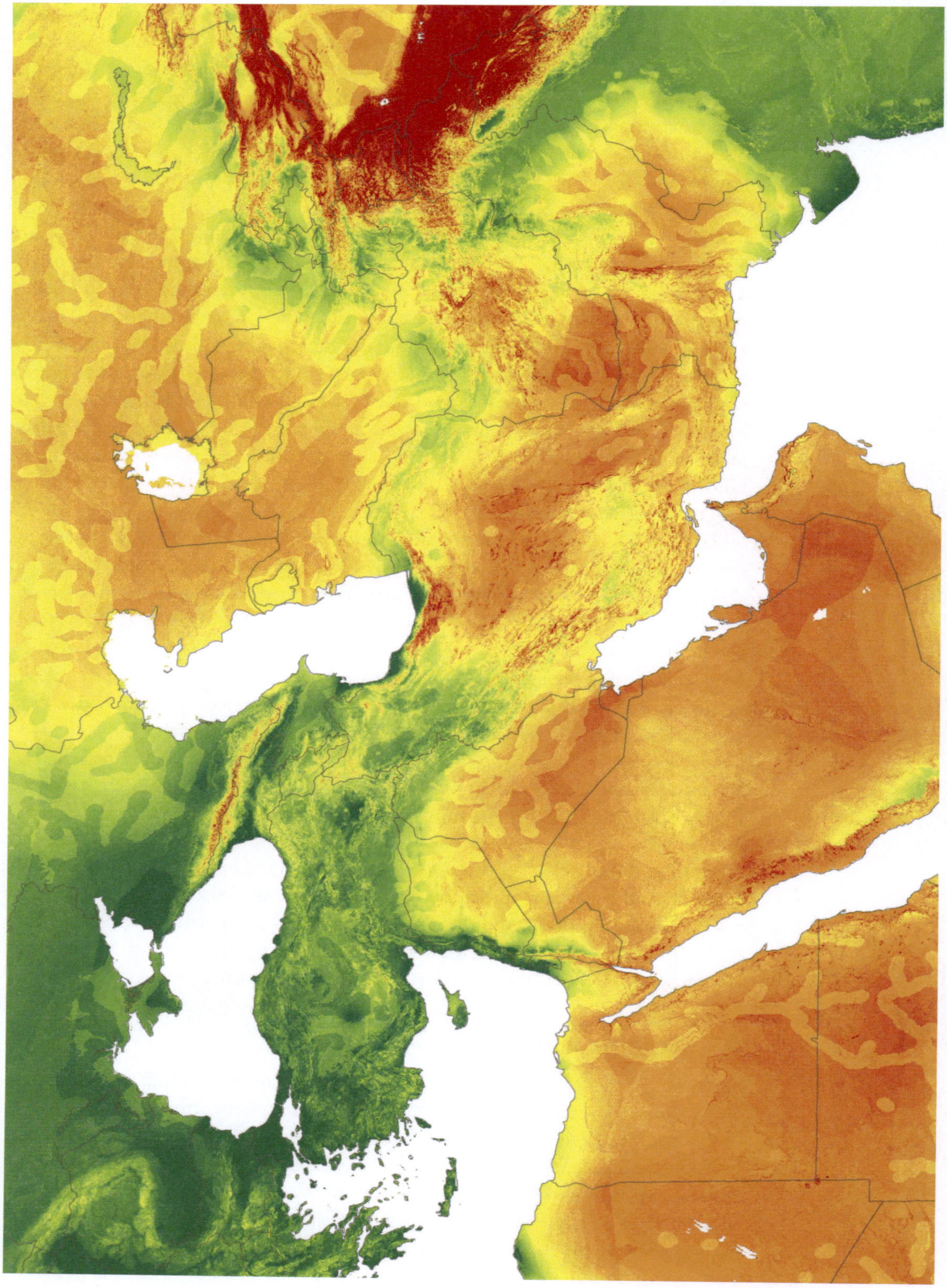

Figure 3.1. 'Cost-of-passage' raster – Model 2. This is the grid model upon which all the cost-based analyses in this book are based. The green areas show the terrain which is the 'least costly' to traverse, and the red shows the 'most costly'. Note that seas are not included in this model.

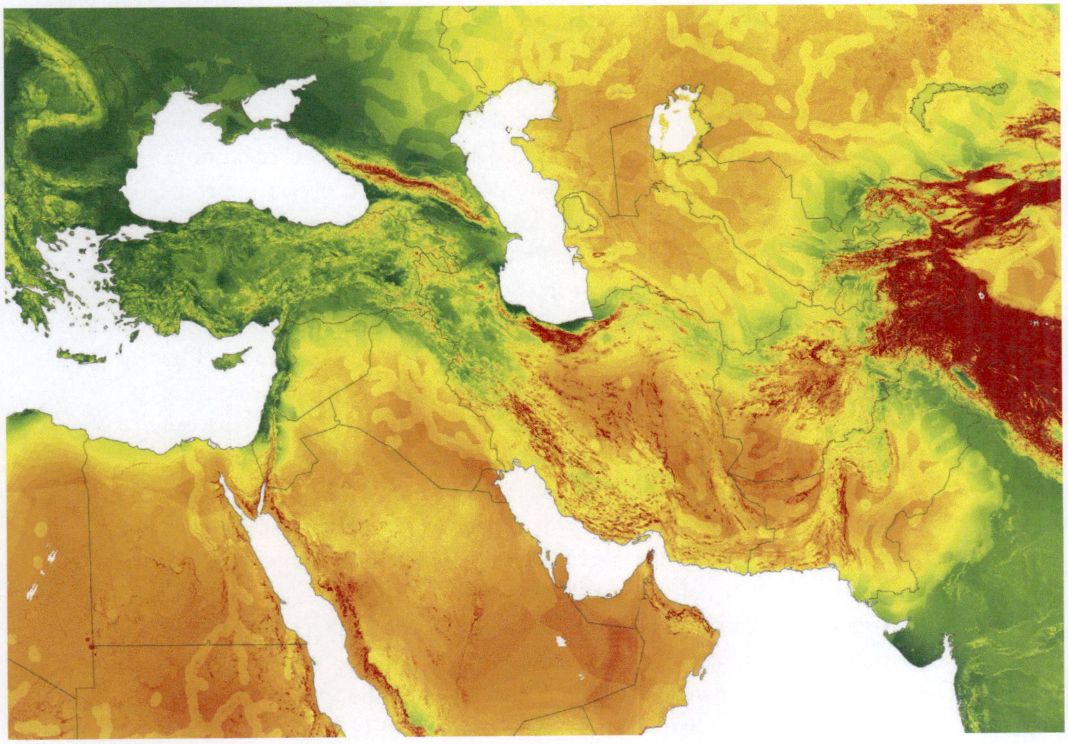

Figure 3.2. 'Cost-of-passage' raster – Model 1. In this model the effects of topography have been lessened. See Figure 3.1 for explanation of colours.

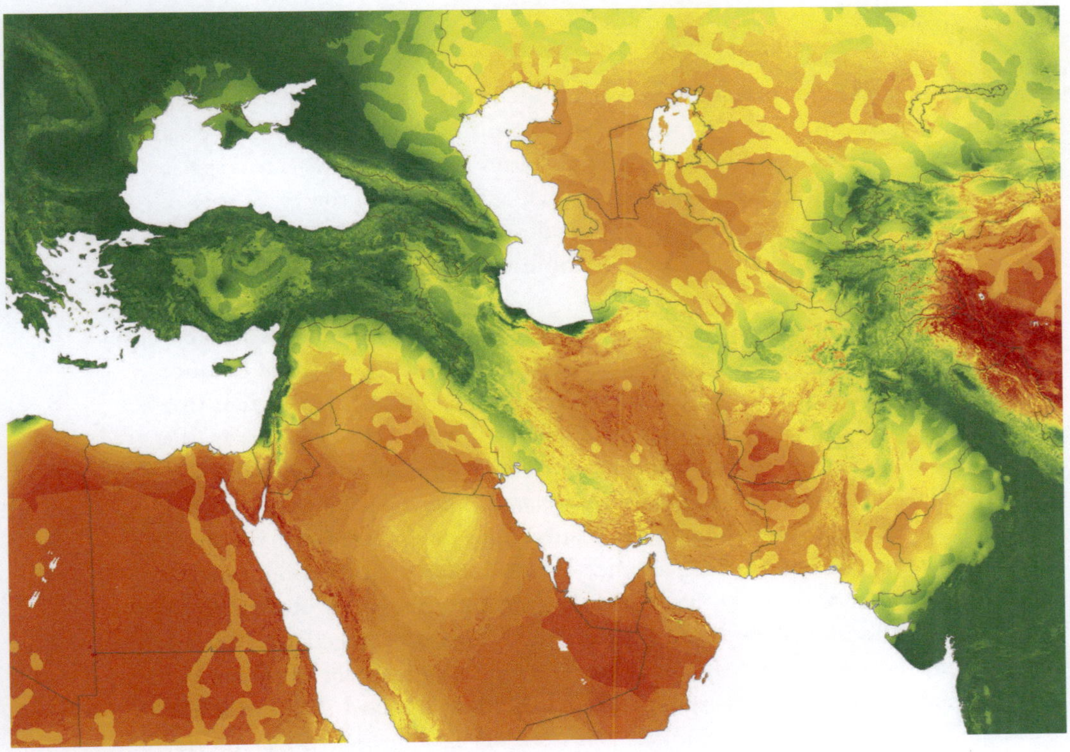

Figure 3.3. 'Cost-of-passage' raster – Model 3. In this model the effects of low-levels of water availability have been heightened. See Figure 3.1 for explanation of colours.

Distribution maps have already been mentioned above as a long trusted, though flawed, tool of the archaeologist (Section 3.2.2). Despite their limitations, we must continue to rely on the founding principle of the distribution map: the spatial location of finds of categorized material. These find spots represent the 'last known address' of the objects in question, and as such index the 'maximum' extent of their circulation. Context remains important: some objects may be 'secondarily' deposited, and others had such long circulation lives that it may be difficult to know how chronologically 'coherent' certain lists of objects are. Various dot distribution maps will be shown in this book. GIS has simultaneously made the production of distribution maps from lists both much easier and much harder: easier because with database technologies, it is possible to make lists cumulative (rather than redraw them from scratch each time); but harder because the epistemology of GIS science demands 'definite' Cartesian locations. It is no longer as easy to hide imprecise information about the origin of objects by the scale of the map because GIS programs require a co-ordinate. Finding accurate and precise co-ordinates for sites and findspots is often extremely difficult, even for relatively well investigated locales (see further discussion in Appendix B), especially given the multiplicity of site spellings.

However, using GIS, distribution maps can be taken further, with more informative basemaps. The 'dots' of the distribution can be overlain onto environmental features, such as topography. This can help the viewer think in terms of landscapes rather than abstract space. More importantly the distributions can be used as factors within the GIS analysis, to create visualizations which can highlight geographical relationships more clearly. This combination of distribution and environmental factors is what I would like to call an '*archaeotopogram*'.

3.5.1 Archaeotopograms: cost-of-passage to cost-surface

Archaeotopograms are graphical tools of archaeological data placed into topographic context to allow comparison between different data sets and to better think through relationships between material culture and their flows across space. They use static models of the environment such as the 'natural routes' or ease of transportation described above (which only indicate the potential for movement), and combine them with archaeological distributions using cost-based analysis to produce meaningful visualizations.

Many different types of archaeotopograms could be designed based on different techniques and different data sets. The most basic example of a simple distance archaeotopogram (what I categorize as type A1 below) would be a diagram that shows the *relative* distance (measured in effort or time) of all locations on the map from a particular site, *when taking account of suitable cost/friction factors such as topography and climate in the cost/friction model.* The resulting map gives an indication of the effective influence that a site might be expected to have on its surroundings (or vice versa) in the absence of, for example, political boundaries. More complex and potentially informative archaeotopograms can be created by combining the results of multiple 'cost-distance' functions together. The key point is that the aim is precisely *not* to create a map of a route in the form of a path or road in the traditional manner, but instead *to represent different aspects of archaeologically-relevant distributions whilst taking into account the landscape.* When compared with other archaeotopograms or other data, these visualizations can be used interpretatively to characterize the nature and shape of 'routes' and the directionality of interaction. As such, these archaeotopograms form merely one tool amongst many that may be deployed to aid the interpretation of the archaeological data.

3.5.2 Some types of 'cost'-based archaeotopograms

The choice of archaeotopogram type will depend on the available data and what needs to be represented. The resultant visualization may not necessarily contain linear patterns that one would normally expect from a 'route' but instead attempt to place objects, sites and collections into their effective landscapes. The following sections suggest some basic archaeotopogram forms and their aim and interpretation. (See also Appendix A.3).

Type A1: Site accessibility archaeotopogram (simple cost-distance)

Essentially this involves a straightforward anisotropic cost-distance analysis from a single node. The resultant diagram effectively shows the relative accessibility of the surrounding landscape to the particular node (*i.e.* site) in question, and hence a general likely sphere of two-way *geographical* influence, though of course the actual cultural sphere of influence at any particular moment in time is also based on technological and socio-political factors. The resultant diagram shows how accessible the location or site is over the long distance, and thus how easy (or how likely) travel to all other locations is. Since this only deals statically with one site, and routes are obviously a dynamic factor of multiple (and changing) locations, this is of limited utility for our project, but it does allow a characterization of the 'distance footprints' of different sites and can be used to help explain cultural and resource links between different sites. Examples are shown here as Figures 3.4-3.8 (where orange indicates relative proximity and green/blue indicates relative distance: the same colour symbology was used in each to make these maps comparable). The sites selected (Ur, Gonur Depe, Shengavit, Arslantepe and Harappa) are distantly located from each other since nearby sites will produce similar results. Each visualization indicates the likely geographical 'trend' of interconnections based on ease of travel: for example, that Gonur Depe is 'relatively' much further away from Ur than Arslantepe, despite a fairly small difference in terms of Euclidean distance. Cultural connections which do not correlate with this relative distance must therefore indicate differences of travel technology or fewer social barriers to travel.

Type A2: Source accessibility archaeotopograms (multiple-source cost-distance)

The background processing is identical in principle to type A1, and is a cost-distance analysis from multiple nodes. Again the resultant diagram shows the relative accessibility of the landscape to these multiple nodes. This is most useful when looking at the relative accessibility of particular natural resources, or (conversely) the cost of travelling to the closest source of this type – no differentiations of the individual characteristics of particular sources are made, however. (For examples, see Figures 4.2a, 4.4, 5.2, 5.3).

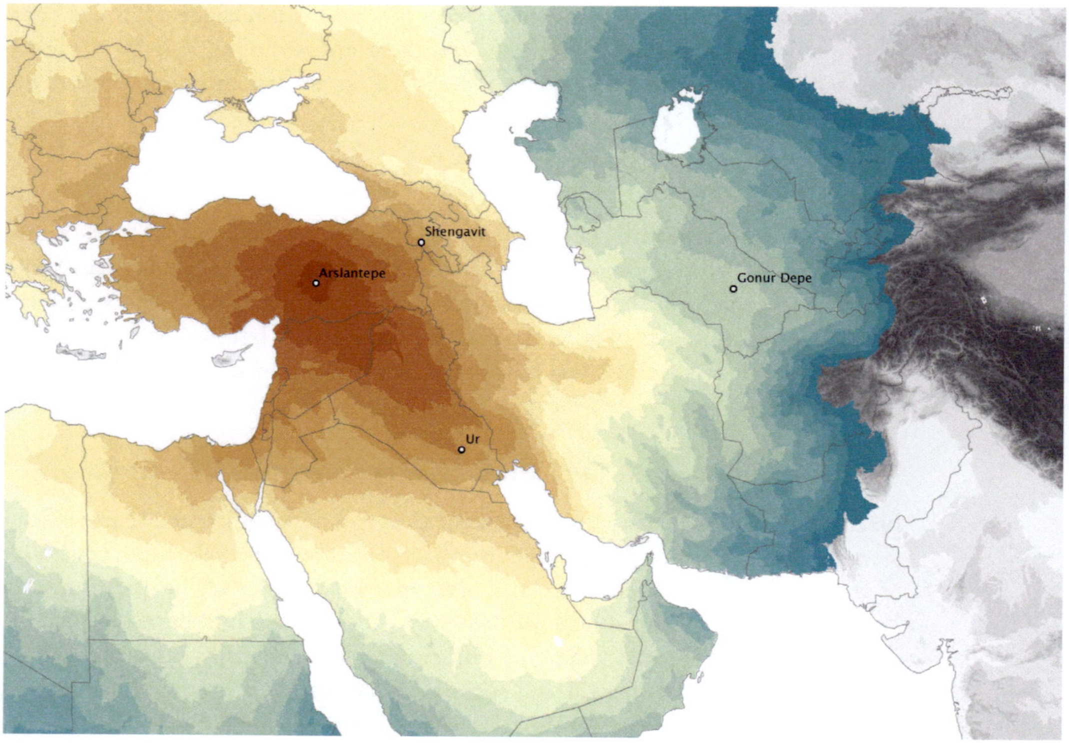

Figure 3.4. Relative distance (site accessibility) from Shengavit (Armenia), based on an archaeotopogram 'type A1'. (Compare Figure 3.8 for key).

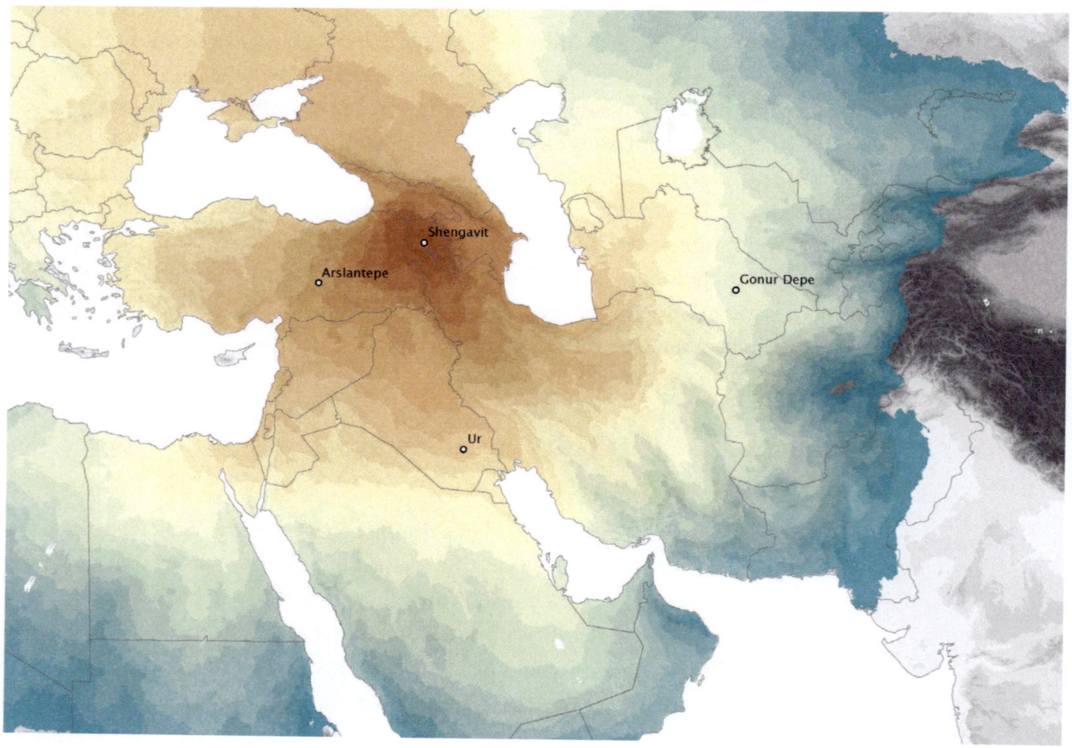

Figure 3.5. Relative distance (site accessibility) from Gonur Depe (Turkmenistan), based on an archaeotopogram 'type A1'. (Compare Figure 3.8 for key).

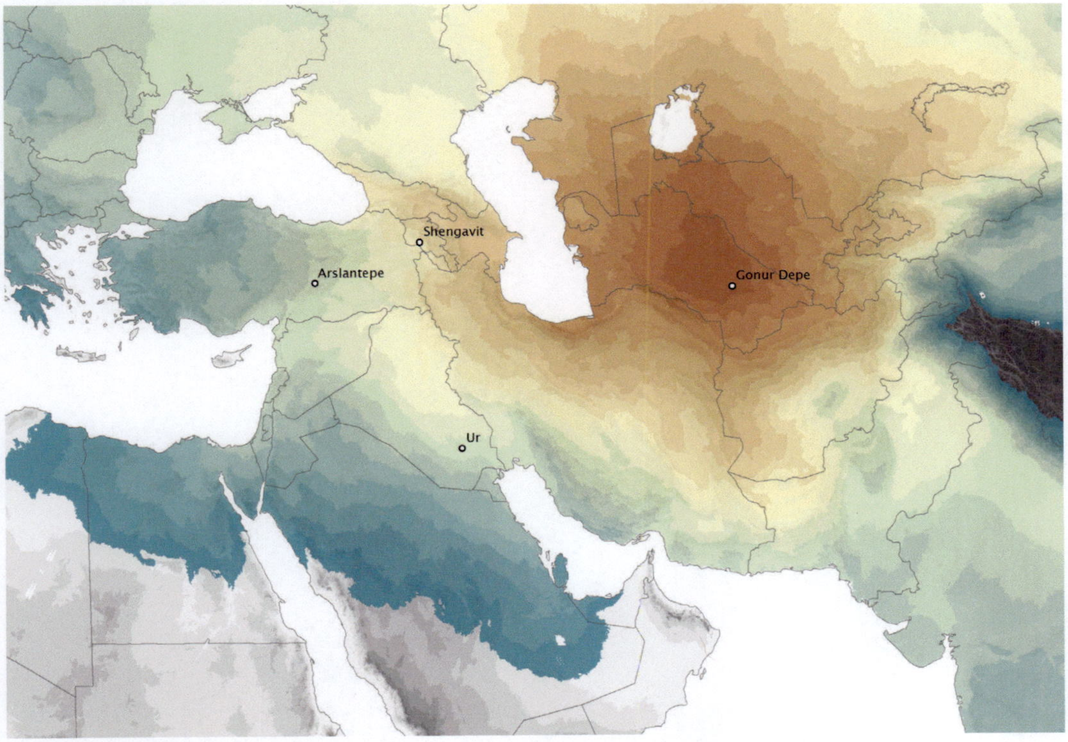

Figure 3.6. Relative distance (site accessibility) from Ur (Iraq), based on an archaeotopogram 'type A1'. (Compare Figure 3.8 for key).

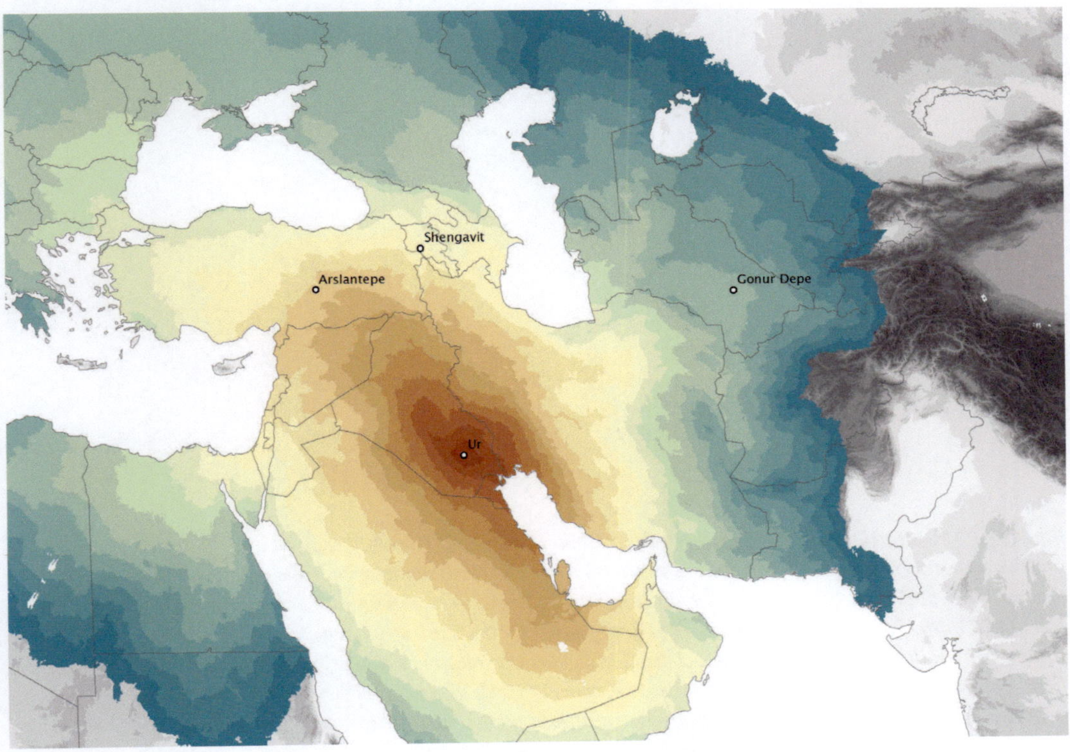

Figure 3.7. Relative distance (site accessibility) from Harappa (Pakistan), based on an archaeotopogram 'type A1'. (Compare Figure 3.8 for key).

Figure 3.8. Relative distance (site accessibility) from Arslantepe (Turkey) based on an archaeotopogram 'type A1'. The coloured contours show the relative distance (time or energy 'cost') required to travel from the source point (in this case Arslantepe) to every space across the map calculated with an anisotropic cost-distance analysis using the 'cost-of-passage' raster shown in Figure 3.1. In the symbology used here, brown indicates low relative distance, and blue relatively high distance. It is interesting to note the south-easterly reach of Arslantepe's position which may help explain the area's early involvement with Syro-Mesopotamian cultures such as the Uruk.

Type B1: Cultural zone archaeotopogram (sum of simple cost-distances)

This type of archaeotopogram uses cost-distance to define a cultural zone geographically by combining the site accessibility of multiple nodes. The idea is that by adding multiple cost-distance results together, this reveals an area in which certain cultural features (*e.g.* a certain object type, or package of objects – however one wants to define the zone) are accessible. This produces something like a 'culture area' map akin to those used by the 'culture history' school of archaeology but with the substantial improvement that boundaries are relative, and rely on the landscape through 'cost'. (For examples, see Figures 5.45, 5.50, 6.40b, where the symbology uses a yellow-red-purple progression from 'proximal' to distant).

Type B2: Resource zone archaeotopogram (sum of multiple-source cost-distances)

This type of archaeotopogram uses cost-distance to define a resource zone geographically by combining two or more 'type A2' archaeotopograms. This reveals the relative accessibility to multiple resources – and thus zones in which the two (or more) resources can most easily be accessed together. (For example, see Figure 5.11).

Type B3: Resource-site zone archaeotopogram (sum of multiple-source cost-distance with simple cost-distance)

This type of archaeotopogram combines one 'type A1' and one 'type A1' or 'A2' archaeotopogram together to highlight the relationship between a site and a set of resources. This reveals a kind of corridor of least cost between the site and its (potential) resources. (For example, see Figure 4.4).

Type C: Corridor archaeotopograms (sum/factor of cost-corridor of multiple natural neighbours)

This type of archaeotopogram is designed to show a set of 'route-like' corridors by combining the cost-corridors of multiple *pairs* of nodes, the pairs selected on the basis of 'natural neighbour' (or other technique). The resultant visualization would show the areas of least-cost between the selected nodes, and hence an overall route map for the particular phenomena being mapped from the nodes. In practical terms, such iterative models are is a very computationally- and time-intensive. (For this reason, no meaningful examples are shown in this book). Resistivity-models described above may, with suitable testing, offer ways to produce equivalent diagrams in a more efficient way in the future.

3.5.3 Applying archaeotopograms

In the following chapters of this book, archaeotopograms are used selectively to help clarify spatial relationships between locations and the location of material flows. In all instances shown, *Model 2* was used as the base cost model for all analysis. This model was combined with an elevation model (*Elevation*) and a Vertical Factor (*Slope*) into an 'Anisotropic Accumulated Cost Surface' procedure (see Appendix A.2 for details), and processed using whichever distribution data was relevant. The final GIS files for this model are available for download from the web address listed in Appendix A.

3.6 Summary: routes, landscape and travel density

This chapter set out to find a period-independent methodology focused on landscape, which might enable us to track dynamism (as well as inertia) in the spatiality of interaction in the past. As has been argued, routes are best modelled as dynamic 'corridors' for movement (in which the exact lines of movement are less important than the cumulative effect of many different journeys), rather than static linear entities like roads. Current GIS techniques relating to travel and movement in the landscape, including network and cost-surface analyses have been reviewed and critiqued. It has been argued that whilst much attention has been recently focussed on network-based analysis, cost-surface techniques offer a clearer way to model the effects of landscape on human movement and in a period-independent manner. This is because the initial model (1) does not rely on the *a priori* establishment of relationships; and (2) does not require a 'complete' record of nodes (or settlements/points of interest) in order to interrogate the model. Using this 'cost-surface' base, a novel mode of visualization is proposed – termed archaeotopogram – that combines the effects of landscape upon travel (the cost surface model) with particular archaeological phenomena (primarily object distributions) in order to produce material-specific visualizations of the spatiality of interaction. It should have been clear from the discussion above that these spatial analysis techniques offer no simple 'panacea' to solve the problems associated with macro-scale analysis of routes. For heuristic purposes we must, for example, ignore environmental changes that might affect the details of the cost model outlined above. The proposed techniques may raise as many questions as they answer. However, the combination of database technologies, easier access to detailed base-maps and these new forms of visualization of the landscape can allow us to handle larger sets of information concurrently.

Chapter 4

Mapping Material Flows: Stone and Stone Objects

4.1 Introduction

The previous two chapters have focussed on traditional and new methodologies of reconstructing the spatial paths – *i.e.* routes – of ancient human interactions. As has been argued, 'routes' should not be envisaged abstractly as static linear connections between points (not simply as spatial 'containers' or tracks for movement), but as shifting emergent phenomena created by many cumulative movements. Even though the actual paths of these individual movements were, aside from certain exceptional examples (*e.g.* hollow ways), too ephemeral to leave any physical trace, the results of these movements are of course documented in the archaeological record by the evidence for 'displaced' objects (or human remains) and technology whose distribution can be mapped and compared with our models of landscape. In the following three chapters, the focus will turn precisely to the material indicators for interaction through varying levels of direct and indirect data. Moreover, attention will be paid to the effects of distinctive characteristics of particular material categories – what we might call their 'materialities' – on social values and practice, which may have had significance for inter-regional interaction. Whilst each chapter is arranged around a major material category – specifically stones, metals and textiles – the reader should not be surprised to find such categories blur to reflect both the 'unboundedness' of such materials to ancient users (as introduced in Chapter 1) and the potential related materials can have as indirect evidence for the interaction of others.

This chapter takes stone (both 'precious' and 'functional') as its central theme. Precious stones form a significant part of the evidence for very long-distance exchange in the ancient world since, by definition, they are precious precisely because of their limited distribution and obtainability. But certain kinds of non-precious stone objects have also been found to indicate the existence of trans-regional connections. This chapter has two simultaneous aims: the first is to look at the evidence relating to a series of important stone materials, and assess whether it is possible to say anything about their routes of exchange during the 3rd and 2nd millennia BC across Eurasia. The second is to broadly situate these materials in the context of their social value, by looking at the differential distribution of sources and objects and the contrasting use of the same stones across different regions. I will thus examine the relevant geological, archaeological and, where appropriate, analytical data on the origin and therefore movement of certain stones (obsidian, steatite, carnelian and lapis), and certain classes of stone objects (for example 'intercultural vessels', stone figurines and weights).

4.2 Stone and its values

There may sometimes be a tendency in archaeology to under-estimate the importance of stone, particularly precious stone, to ancient economies and social systems. This is, in part, because, despite their taken-for-granted 'economic value', the archaeological usefulness of precious stones with regards to dating, is rather low compared to more variable and more commonly found materials like pottery. Simultaneously, however, the aesthetics of precious stones have been somewhat undermined by the ease with which their qualities may be mimicked by man-made materials, first by faience and glass and more recently by plastics. Glass and plastic beads which mimic precious stones like carnelian, diamond and rubies are so ubiquitous that in order for original precious stones to have any impact, their authenticity must be verified by a whole host of material experts who mediate between those who can afford the real thing and those who may never even have seen a real diamond, let alone recognize the difference[34]. We may presume that, despite the probable existence of authenticating 'scientific' tests in the past[35], such tests were limited to a few specialists and so on a day-to-day basis trust systems must have been created which relied on specialist mediators who were able to comment authoritatively on authenticity – as is also the case today. We may speculate as to how important such 'authenticity' may have been to the majority of consumers (*cf.* Sherratt 2008b). However, the relationship between the introduction of materials, such as glass, that mimic precious stones, such as lapis lazuli, and the development of social systems of authentication must have made a significant contribution to the economic trajectories of the Bronze Age when (it is frequently asserted) special materials were increasingly being deployed to differentiate individuals and groups of people, often hierarchically.

The value of stone emerges then from the interplay between the gradual creation of social traditions which value certain material qualities and the stones' 'resistance to being owned' – *i.e.* their scarcity, distant origin or field of control. Certain material qualities are distinctive to stone when compared to other materials: hardness (though each stone type is different); un-malleability (unless predominantly metallic, most stones cannot be reformed through physical force even when heated) and thus must be shaped through reduction (chipping, carving or other abrasive techniques). Another common factor is in fact stone's heterogeneity: different stones have different colour and visual qualities, and of course different levels of rarity, even between particular examples that we would categorize as the same stone geologically or through folk taxonomy. This heterogeneity gives fragments of stone unique identities. This should remind us to be flexible about our terminological categories since ancient folk taxonomies may not have coincided with our own definitions: ancient stone users may have placed emphasis on differences which we cannot perceive or do not consider important, or may have lumped together similar types of stone which we are taught to differentiate.

34 Faking materials seems to have a long history. By way of comparison, a set of slag-centred 'copper' ingots from Oman appear to represent ancient fakes, intentionally-produced (Weisgerber and Yule 2003, esp. 49-52).

35 For example, compare the re-melting of metals by traders in Kültepe-Kaneš to ensure or measure purity (Dercksen 2005, 23).

4.3 'Precious' stones

4.3.1 Lapis lazuli

Lapis lazuli (literally, the 'blue stone') is undoubtedly the best known of the precious stones to have been used during the 3rd and 2nd millennia BC both from archaeological and textual evidence (Moorey 1999, 85ff). 'True' lapis lazuli, as opposed to other stones like azurite with similar blue colouring, is a composite rock with a deep blue ('ultramarine') or green colour with an iridescent golden shine produced by specks of pyrite. It is found geologically in only a few regions of the Old World. Whether every archaeological object described as being of lapis lazuli is indeed from this 'true lapis' is difficult to confirm. The best-known sources of lapis lazuli are the deposits of Sar-i Sang (by the Kokcha river, 35 miles south of Jurm in the Badakhshan province of north-eastern Afghanistan) – which has often been thought to represent the only source to have been exploited in prehistory. There are other confirmed sources, including in the very high Pamirs of Tajikistan and near Lake Baykal (somewhat further to the north-east), though on both geographical and chemical grounds, it seems unlikely that these were major sources of Near Eastern lapis lazuli. While we should not necessarily exclude exploitation on the basis of distance alone, the claims for Near Eastern use of Baykal-sourced lapis lazuli remain tenuous (Zöldföldi and Kasztovsky 2009; Buchanan 1966, 28, no. 133). Various pieces of (unworked) lapis lazuli allegedly collected around Bi-Bi-Dick in the Chagai Hills region (in the Baluchistan province of Pakistan) led to the suggestion that there was another source here with apparently similar kinds of lapis to that of Badakhshan (Delmas and Casanova 1990; Casanova 1992). However, attempts to find an actual geological deposit for these pieces have so-far proven fruitless, and given the general geological map of the hills – which do not consist of rocks that are likely to produce lapis lazuli – it looks increasingly likely, as Randall Law has recently argued, that there is, in fact, no Chagai source and that the samples recovered were sold to its original collectors as 'local' in order to avoid accusations of cross-border smuggling from Afghanistan to Pakistan (Law 2011, 528-543). Whether or not our geological knowledge is comprehensive enough to conclude that Badakhshan therefore formed the only source for lapis lazuli during our period of interest remains difficult to assess. Vague references to Iranian sources, mentioned in historical texts of the Islamic era (von Rosen 1988), have encouraged the idea that we have yet to identify all the potential sources, perhaps because the mines have been exhausted (Moorey 1999, 86-87). Equally however, the tradition of an Iranian origin for some lapis lazuli may simply have arisen in a similar way to the Chagai sources. Besides issues of tax and law, medieval salesmen (many of whom presumably travelled through or were based in Iran) were keen to differentiate their wares from other salesmen and therefore may have claimed alternative origins, or else locals may have claimed nearby sources of such a valuable stone for status reasons. No doubt there was similar uncertainty during the 3rd and 2nd millennium – and it seems likely that many ancient consumers of lapis lazuli had an extremely dim understanding of the geographical origin of the stone. Indeed its ritual or symbolic origins (and material qualities) were likely the more important factors to its authenticity.

Lapis lazuli was transformed into a variety of forms: beads, cylinder seals, amulets, stone axes (apparently 'symbolical'), bowls, and very commonly as an inlay or 'jewel' within composite objects like anthropomorphic, zoomorphic and mythological figurines, musical instruments, jewellery, stone bowls and so on (some examples shown in Figure 4.1). It has been argued that the consumption of lapis lazuli as material was driven by two interrelated factors. First as a material which required considerable investment of time to obtain (its sources being so remote from its centres of consumption) and fashion (the manufacture of lapis beads alone must have taken considerable time), it acted as one of a number of ways to differentiate individuals within the developing hierarchical relations of urbanization Mesopotamian society. Only those able to leverage the labour of others to procure and shape the stone could use it. Second, hints from later textual materials shows that the material also had a range of 'symbolic' associations relating to life, to the gods and to apotropaic power[36] (Casanova 2000, 178-179) – in common with many other sorts of stones in the ancient Near East (Robson 2001). The use of lapis lazuli as inlay in the eyes of figured gods or idols supports this impression of symbolic or magical associations. The deployment of very large amounts of lapis lazuli objects in the Royal Cemetery of Ur illustrates both motivations: the sacrifice of a massive amount of such a costly material seems clearly designed to impress the economic prowess of the ones able to bury so much material wealth, but it may also indicate a desire to protect those buried in any 'afterlife' with a material that had magical associations. It is tempting to suggest that the symbolic factors must have facilitated the later appropriation of the material as index of power, but more likely both aspects were created in tandem.

The full distribution of lapis objects (and lapis debris) found in archaeological contexts is extremely wide during the 3rd and 2nd millennia, stretching from the Pamirs to Egypt and the Aegean (see Figure 4.2). The details of lapis consumption patterns remain somewhat unclear, however, since whilst there have been various regional studies of lapis lazuli assemblages (for example: Crowfoot Payne 1968; Casanova 1993) and small amounts of lapis turn up in many excavation reports, less work has been focussed on detailed patterning of consumption to combine geography and archaeological context. Exceptionally Michèle Casanova's work (including the recent publication of a major monograph on the topic, Casanova 2013) has highlighted that the main geographical concentration of lapis finds recovered is in Syro-Mesopotamia (around 88.7% of 3rd and 2nd millennium finds), and not in areas close to the geological sources such as Iran, Baluchistan and Central Asia (Casanova 2000, 172). Indeed, at intermediate sites in central Iran, relatively few objects of lapis lazuli have been found. The site of Ur, and particularly the Royal Cemetery, has produced the largest number of finds – corresponding to 73.9% of attestations of lapis lazuli in Casanova's database (Casanova 2001, 158). This concentration should obviously be set in its context – the massive sacrifice of material goods in the Royal Cemetery is unique in Mesopotamian archaeology – and we should not assume that this shows that the elites of Ur were the only ones consuming large amounts of lapis lazuli. Equivalent

36 Arguably of course, in many parts of the modern Near East and eastern Mediterranean, we may still see the continuation of this belief in the protective force of talismanic *nazar boncuk* against the 'evil eye' – essentially these objects are blue-stones with a central 'eye', albeit today often made from cheaper glass alternatives rather than stone.

Figure 4.1. Selection of objects dating to the 3rd millennium BC made from lapis lazuli: a. bowl, Puabi's Tomb, Ur (Aruz and Wallenfels 2003, 117, cat. 68; photo: Penn Museum, #B17167); b. lapis beads from necklace, PG453, Ur (Aruz and Wallenfels 2003, 130-131, cat. 79a; image: Penn Museum, #B16799); c. detail from 'Royal Standard of Ur', Ur (photo: Iraq Heritage Program/Greenhalgh 2006); d. wings of portable mythical creature, Mari (Aruz and Wallenfels 2003, 140, cat. 81; National Museum of Damascus); e. lazurite axe-head, Troy (Tolstikow and Trejster 1996, 152; Pushkin Museum). Not to scale.

EDII 'royal' burials have not been identified from other sites, but this does not mean they did not exist. Alternatively, other elites may have chosen not to consign their lapis wealth to the ground in such a spectacular way (*cf.* discussions of metal sacrifice in Section 5.4.2).

According to Casanova, the most frequently found category of lapis object throughout the Chalcolithic and Bronze Age is the lapis lazuli bead (Casanova 2000, 174), albeit with many variations in form. In this sense we might justifiably consider lapis lazuli under the category of personal adornment and relate objects made from lapis to the same issues of 'identity' that are associated with clothing (*cf.* Section 6.7). Less common (in decreasing percentages) are inlay or other 'incrustations', manufacturing debris, seals and other objects. Again, the greatest variety of forms is found in Syro-Mesopotamia (and not the 'source' regions of Central Asia, Afghanistan or Iran), with the peak of variation dating to the 3rd millennium, before a decreasing range in the 2nd millennium. This fact, the Mesopotamian 'style' of objects found in Mesopotamia and the evidence for lapis lazuli debris at various Mesopotamian sites appear to suggest that lapis was usually shipped in raw blocks from source regions (perhaps with limited working to reduce the weight from non-desirable calcite encrustations on the rock) rather than as finished objects (Casanova 2000, 178).

Finds of lapis lazuli in eastern Anatolia and Transcaucasia are very few during the 3rd and 2nd millennia BC. The furthest north and west that lapis objects have been found during the 3rd millennium, according to Rahmstorf (2010b; *cf.* Rahmstorf 2010a, 97, fig. 8.6), is in the form of a few beads in Maikop-type burials (north of the Caucasus) and the exceptional lapis axe (Figure 4.1a) from Treasure L at Troy, item no. 6058 (Easton 1995, 13). We should probably assume that these few finds represent only the visible remains of the full amount of lapis

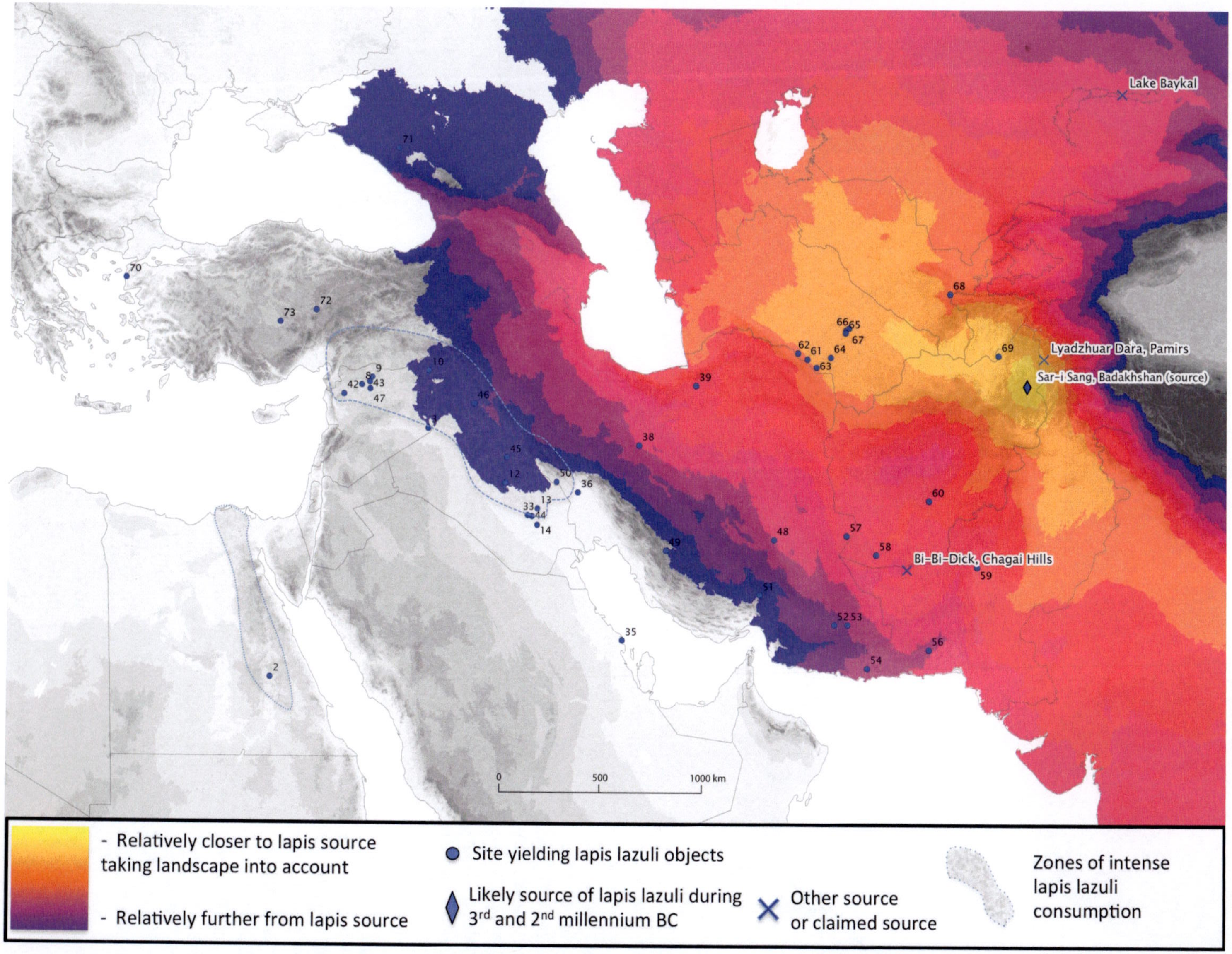

Figure 4.2. Distribution of known lapis lazuli objects and regions of intense consumption in relation to the material's sources. Relative distances from the lapis lazuli source in Badakshan (shown as diamond) are shown by archaeotopogram 'type A2' (yellow = close, purple = far). See Appendix C.1.1 for key to numbers, and Appendix D.1.1 for discussed possible lapis sources.

in circulation, and that the material was simply too valuable to be consigned to the ground in normal situations.

In eastern Anatolia during the 3rd millennium, where Kura-Arax groups were presumably aware of the Mesopotamian predilection for the blue stone, the absence of evidence might be explained by the lack of suitable burial contexts (from which the majority of Mesopotamian finds of lapis lazuli come). Though lapis is also missing from the later richly-endowed Trialeti grave assemblages of the late 3rd or 2nd millennium, beads of blue faience (likely first created as an artificial lapis lazuli) are found plentifully alongside carnelian (see catalogues in Zischow 2004). Alternatively, it is also tempting to consider the hypothesis that lapis did not have the same symbolic power or economic value in this region as in Egypt or Syro-Mesopotamia: certainly the Early Transcaucasian/Kura-Arax communities appear to have been based on a very different social and economic system to the Mesopotamian model (and perhaps different systems of religion), perhaps preventing the entry of Mesopotamian ideas about lapis as a material of the gods or as mediator of social hierarchies.

In western Central Asia during the 3rd millennium, small finds of lapis lazuli are more common: lapis beads are commonly found in graves, for example at Altyn Depe (Masson 1988, 40, 68). The site of Shortugai, in southern Bactria, has been argued to represent an Harappan 'colony' of the late 3rd millennium on the basis of its material assemblages, located north of the mountains to control the sources of lapis lazuli in Afghanistan, with remains of lapis lazuli workshops (or rather chipped lapis debris) as evidence for local working (Francfort 1989). A similar case has been made for the 'middleman' role of Shahr-i Sokhta in the lapis trade (Tosi 1974), this site located in Seistan, between Afghanistan and the route into Iran and west, with considerable evidence for working of lapis.

The question is whether our knowledge of geological deposits, the distribution of archaeological objects and an insight into the motivations behind its consumption allow us to reconstruct the routes of movement of this material. The route by which lapis travelled has been a topic of frequent discussion (Sarianidi 1971; Tosi 1974; Majidzadeh 1982), but most studies represent broad-brush reconstructions of a material 'highway' (as described in Section 3.2.1) with little detail of the changing intensity of trade. The reason for this is simple: the density and resolution of the evidence remain too low and our distribution map is incomplete. Objects in lapis tend to be found only in excavation rather than from survey, and the excavation record is uneven (especially in the relevant regions of Iran and Afghanistan). Similarly, lapis objects also end up more easily on the art market without true context. Leaving aside a few notable exceptions (Shahr-i Sokhta and Shortugai), relatively small amounts lapis lazuli has been found at sites between Mesopotamia and Badakshan, providing no 'fall-off' footprint with which to track waxing and waning exchange networks. Additionally, the context of most finds of lapis is in graves, and the number of objects deposited intentionally does not necessarily reflect the amount of lapis in circulation in life. Arguably this may reflect the real nature of restricted circulation of lapis as a material whose main consumers were the elite of Mesopotamia and Egypt and those who tried to emulate them – but this does not help us reconstruct the route networks. The textual evidence gives us few clues, except to indicate that lapis lazuli was imported (through a variety of mechanisms: trade, gifts, war booty *etc.*) either via 'Aratta', Meluḫḫa, Marhaši or Dilmun (Moorey 1999, 86-87), all regions to the east of Mesopotamia proper,

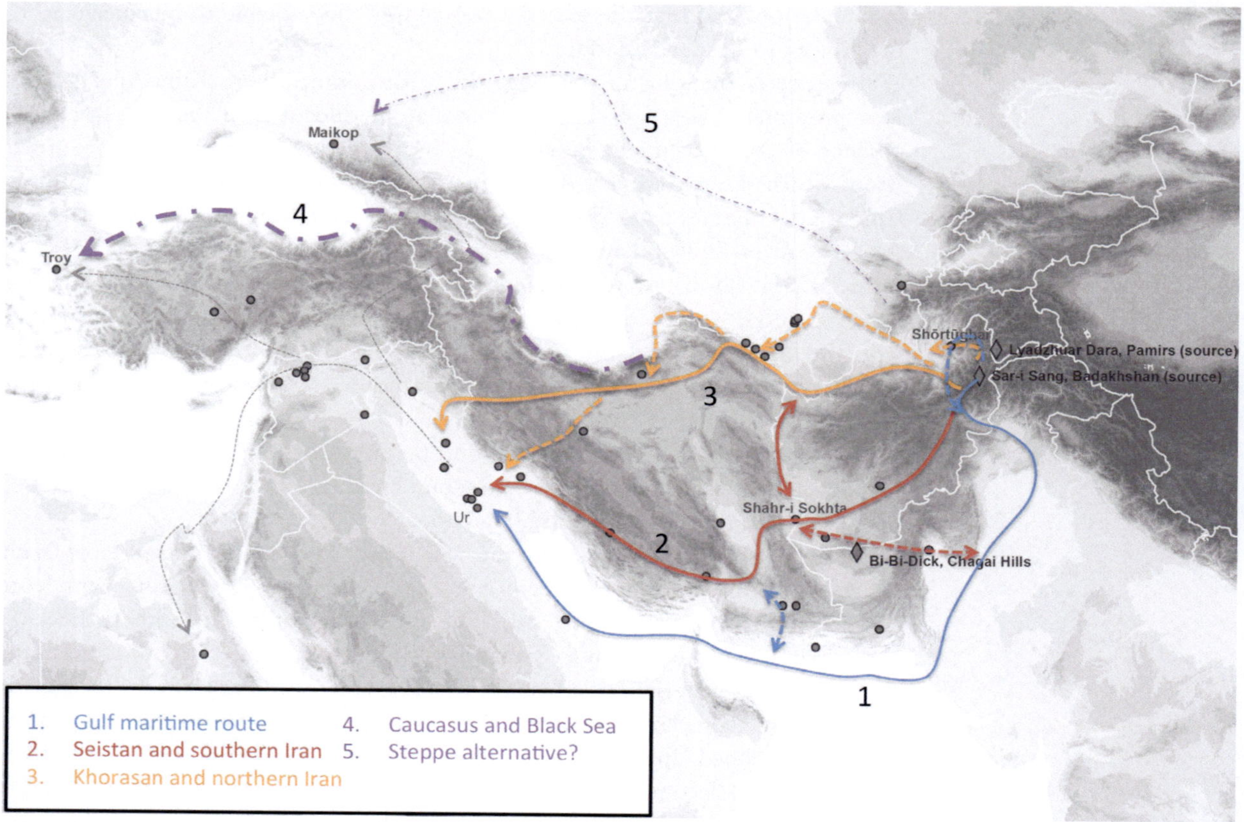

Figure 4.3. Possible alternative distribution routes for lapis lazuli during the 3rd and 2nd millennium BC.

whether these are real geographical terms or simply literary devices (as may be the case for 'Aratta', for discussion see Potts 2004b). Despite the poor evidence, the fact the sources of lapis lazuli are apparently so restricted geographically (*i.e.* only in Badakhshan), four to five corridors of possible transmission have nonetheless been proposed on general geographical grounds (see Figure 4.3):

1. A mostly maritime route: across the passes of the Afghan hills to the Indus and then via coastal or maritime transport up the Persian/Arabian Gulf;
2. The southern route: through Baluchistan and/or Seistan along inland or coastal southern Iran;
3. The former 'Khorasan road': through Bactria and/or Khorasan, probably south of the Elburz across central Iran into north-west Mesopotamia;
4. The Caspian route: again through Bactria, either to south of the Karakum desert, around the Caspian and then southwards, or else westwards into the Black Sea or beyond;
5. The steppe route: an alternative, very long and rarely-considered connection from Bactria north along the Oxus to Transaralia and Transcaspia across the southern Russian steppes to the Caucasus or the Black Sea.

Of these, options 1 and 3 have been the most popular. Georgina Herrmann (Herrmann 1968) suggested some time ago, on the basis of the evidence available in 1968, that the relative number of lapis lazuli finds in the Mesopotamian record might allow us to trace the changing routes by which lapis arrived from the east. On the basis of early concentration of lapis in graves at Tepe Gawra, she suggested the earliest trade (during the Jemdat Nasr period c. 3100-2900BC) was via northern Iran (route '3'), before shifting southwards via an Elamite route in the Uruk IV period (route '2'). After a 'break' in trade during the Early Dynastic I period, she reconstructs a rapid rise in popularity in the Early Dynastic II and III periods (the mid 3rd millennium), as part of cultural convergence between Mesopotamia and Elamite Iran (via the southern route '2'), before again declining in the Akkadian period (late 3rd millennium). In this context it is interesting to plot the 'least-cost-path' between Ur (where such large amounts of lapis have been found) and the apparently sole source of lapis lazuli in Badakhshan (see Figure 4.4). This path curiously travels along a northern route (mostly following hypothetical route 4) around the Caspian Sea. By contrast, if lapis first travelled via the Seistan region (*e.g.* via Shahr-i Sokhta), the least-cost path follows the southern Iranian route (mostly hypothetical route 2). Of course, these paths should not be taken too literally since the exact lines may be an artefact of the model. Additionally, since we know relatively little about the archaeology of the southern Caspian littoral, it is difficult to comment on the realism of the Caspian path at present, though we might not necessarily find much lapis at Caspian sites in anycase. Nonetheless this path is closer to Hermann's idea of an early 'northern' route (since a cost-path to a northerly site like Tepe Gawra would produce a similar path). If the southern route became more important in later periods, this would encourage the idea that the role of intermediate communities – or rather the relationships with and between chains of such intermediate communities – was more important to the transmission of the material than the total or raw geographic 'least-cost'. This might also contradict the suggestion that the expanding Mesopotamian economic system was a purely 'cost'-driven machine, although given that lapis lazuli is only one material out of many commodities being exchanged, we should be wary to read too much into this alone.

However, this 'least-cost' map misses out one key feature, of course, namely the maritime routes and connection to the Indus. Indeed Maurizio Tosi (1974) argued, that the northern route (route '3' or '4') played only a minor role throughout the history of the lapis lazuli trade. Based on the indications for early connections between Uruk Mesopotamia and eastern Iran, he suggested that a southern land route via Seistan and Elam (route '2') formed the primary transport route of lapis lazuli until the mid 3rd millennium BC, at which point maritime movements between Indus and Mesopotamia along the Persian Gulf (route '1') may have taken over. It is certainly tempting to link the circulation of lapis in the mid-to-late 3rd millennium to a wider 'âge des échanges inter-iraniens' (Amiet 1986) or 'Middle Asian interaction sphere' (Possehl 2002, 215-236). During this period there appear to have been intense maritime connections between Mesopotamia and the Indus (and the cultures on both sides of the Persian/Arabian gulf) and equally important sets of interaction across the Iranian plateau from Elam to Central Asia – which are witnessed in shared iconographic styles (see Section 3.4; Amiet 1986) and recorded exchange of goods (Ratnagar 2006). It is notable that there is very little lapis lazuli used in the Indus at this time. This might be a case of differential survival, but the significance of other types of stone suggest instead that lapis lazuli did not have the same associations or status as it developed in Mesopotamia, Egypt and the eastern Mediterranean (see Law 2011).

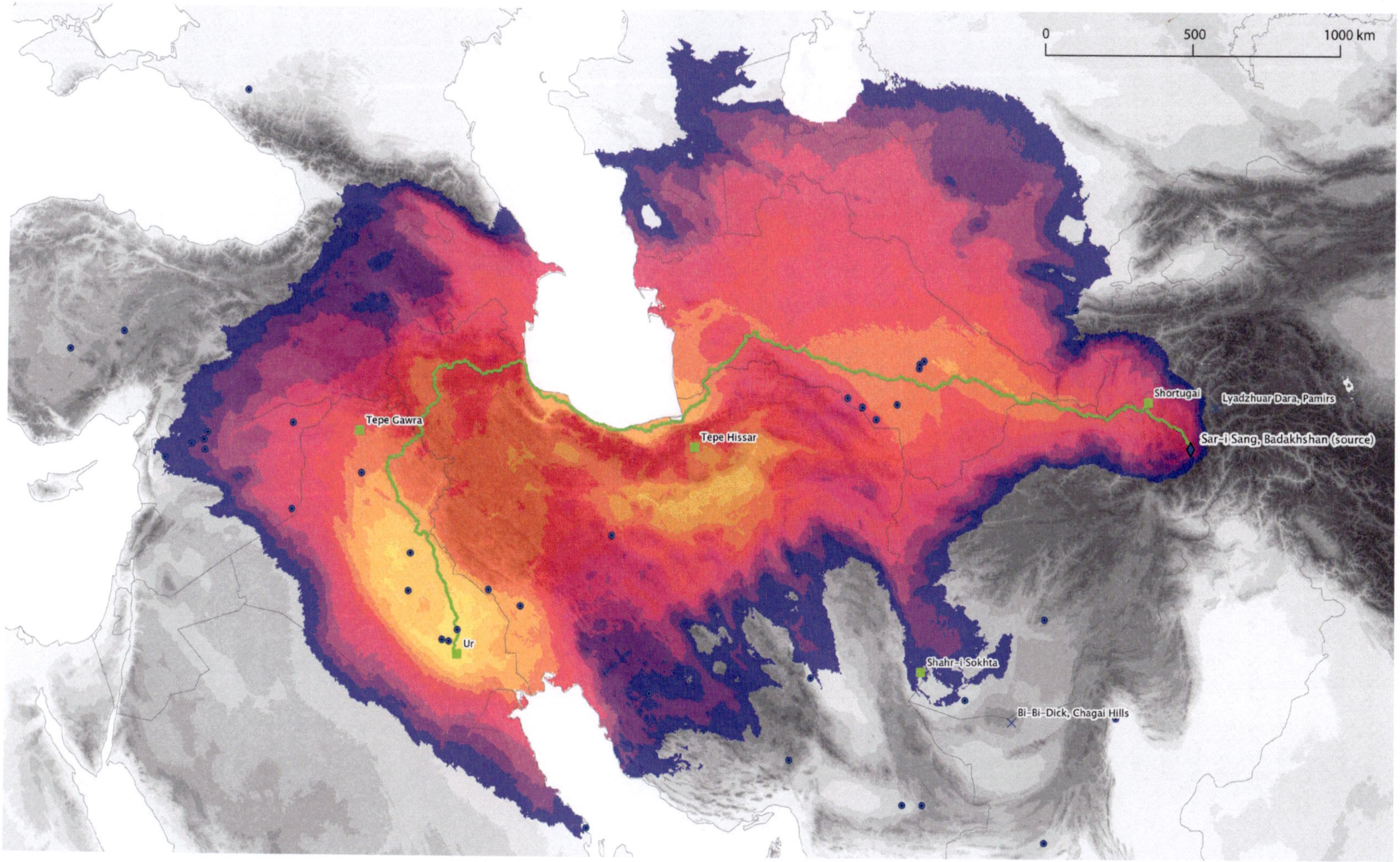

Figure 4.4. The least-cost-path from Ur to the (probable only) source of lapis lazuli at Badakhshan (green line); overlain onto a 'type B3' archaeotopogram (combining a 'type A1' from Ur and 'type A2' from Sar-i Song, Badakshan). Also marked: lapis finds (blue dots) major lapis-working sites of 3rd millennium (including Shahr-i Sokhta, Shortugai, Tepe Hissar).

An association has sometimes been made between the circulation of lapis lazuli and the circulation of tin during the 3rd millennium BC. This idea was partly based on the apparent co-presence of very early tin-bronze and the lapis axe-head at Troy but also on the geographical co-location of geological sources in Afghanistan (*e.g.* Stech and Pigott 1986; Muhly 1999). The emergence of the first tin-bronzes and the height of lapis lazuli consumption in Mesopotamia does appear to coincide, and certainly in the 3rd millennium, both are relatively rare commodities, both associated with the 'east' and both appear to be have been used as status markers. However, the geographical association remains equivocal, because tin may have been sourced more widely than just Afghanistan even during the 3rd millennium (see Section 5.1.2 for summary of the sources of tin and the tiresome 'tin problem'): by the 2nd millennium, any such putative link between the movement of each material would appear to be broken. In any case, there is also an amber bead (presumably sourced from the Baltic) in the same 'treasure' as this lapis axe – making a one-to-one association with Afghan sources of tin even less persuasive.

Ultimately, our ability to be any more geographically (or chronologically) specific about lapis lazuli routes, or test the various hypotheses about such paths is hampered by the nature of the stone's distribution and appearance in the archaeological record. Whilst we can assert the existence of long-distance exchange networks, the relatively low density of lapis finds between the geological source and the main centres of consumption means that tracing the intermediate nodes of travel cannot be achieved from direct evidence of lapis lazuli.

4.3.2 Carnelian and etched carnelian beads

Alongside lapis lazuli, carnelian[37] forms one of the most important precious stones used for making beads and inlay during the 3rd and 2nd millennia BC (Moorey 1994, 97; see Figure 4.5 for examples). A symbolic and economic relationship between lapis lazuli, carnelian and gold appears important to understanding the circulation of each of these materials (Casanova 2000). Like lapis, carnelian had some symbolic associations in Mesopotamian literature, and therefore presumably in wider beliefs: speculatively its orange-reddish hues signalling 'death' against lapis lazuli's blue-green association with 'life' (Casanova 2001, 166-167). Finished objects or pebbles of carnelian appear to have been traded considerable distances. If the routes of its exchange have been less frequently discussed than lapis lazuli, it may be in part because there were a greater number of potential sources from which the stone may have been extracted (see Figure 4.6). Regions with deposits of carnelian are known from eastern Iran and south-west Afghanistan (around the Helmand basin), Bushehr (in south-west Iran), the Elburz mountains (northern Iran), the Deccan plateau and the environs of Ratnapur (in India), parts of the Arabian peninsula, Bactria, Azerbaijan, Anatolia and Egypt (for references, see De Waele and Haerinck 2006, 32; Moorey 1994, 97). But given that the stone may be found widely as pebbles in alluvial deposits, it is difficult to be very specific, especially in the absence of systematic petrographic or other provenance studies. The Indian sources are often assumed to have been the most important for our period of interest, but in fact though detailed study of material from Harappa has confirmed the local importance of Gujarati carnelian (Law 2011, 262-299), so

37 Also spelt 'cornelian', *Karneol, cornile.*

Figure 4.5. Selection of (orange-red) carnelian objects including: a. etched carnelian beads used in necklace (cf. also Figure 3.1c), Ur (Aruz and Wallenfels 2003, 130-131, cat. 79b; photo: courtesy of the Penn Museum, #30-12-673); b. long thin beads from Indus valley, Mohenjo-Daro (Aruz and Wallenfels 2003, 392-393, cat. 279).

far little attempt has been made to quantify the assumption at an international level. Unlike lapis, which seems to have been transported in raw forms during the 3rd millennium, many carnelian objects appear to have been manufactured in Iran, Central Asia or the Indus (Casanova 2000, 178) and then exported to Mesopotamia and elsewhere. But as with lapis lazuli, the Indus valley civilization (and to a lesser extent Central Asia) appears to offer a key link in the chain of exchange leading to the consumption of carnelian in Mesopotamia and the west, at least during the end of the 3rd and early 2nd millennia.

One category of carnelian objects that has received more attention is that of 'etched' carnelian beads (Beck 1933; During Caspers 1972; Reade 1979; Inizan 1995; Kenoyer 1997; 2003; Aruz 2003, 241-243; De Waele and Haerinck 2006; Rahmstorf 2010a, 97). The technique of etching is normally said to have originated in the Indus region given its early distribution there (Reade 1979), but the full distribution of etched beads made from carnelian (or other similar-looking stones) is widespread through the eastern part of the Near East during the 3rd millennium BC (see square points on Figure 4.6). It is not certain that etched beads were exclusively exported from Harappan workshops, such as those at Chanhu-Daro and Lothal (Possehl 1996, 158), but the production and circulation of such etched beads certainly seems to have ceased after the Indus civilization disappeared (De Waele and Haerinck 2006, 32), sometime in the early

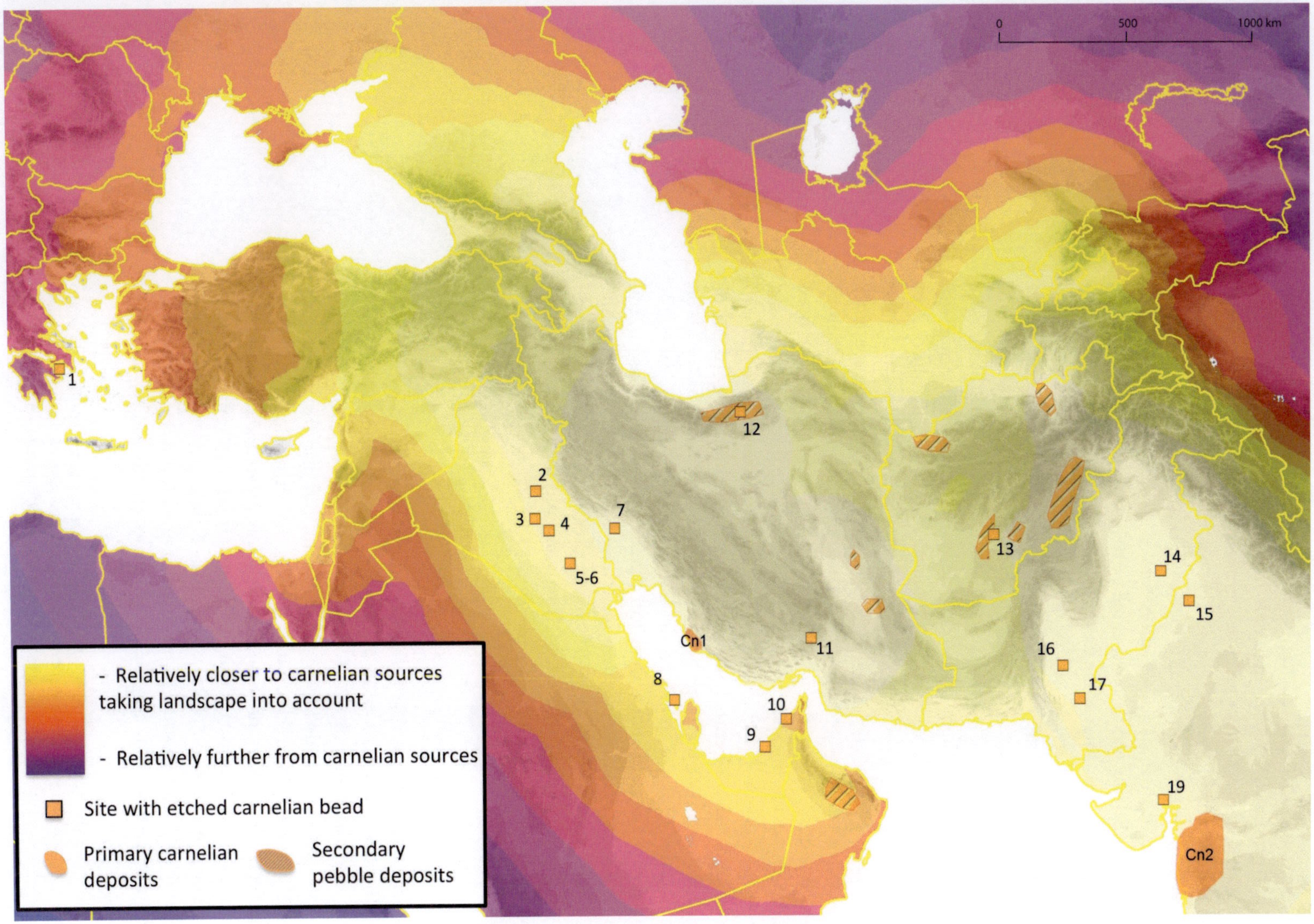

Figure 4.6. Distribution of etched carnelian beads (orange squares) with a representation of the relative distances from carnelian sources (shown as solid orange or hatched orange zones) using a archaeotopogram 'type A2' (yellow = close, purple = far). See Appendix C.1.1 for key to numbers; distribution of carnelian sources based on T. Potts 1994, 196.

2nd millennium BC[38], though other forms of carnelian beads continue to be used in Mesopotamian contexts after this time (Moorey 1994, 98). So far, only one example of an etched carnelian bead has been recorded west of the Euphrates (unlike the other presumed 'eastern' materials like lapis lazuli and tin), which may suggest that these beads were more closely embedded into a distinctive set of 'eastern' cultural imperatives, whose precepts were not exported further west[39].

If we analyse the distribution of all types of carnelian objects in order to reconstruct the route of their trade, we suffer from a similar set of obstacles as those encountered with lapis lazuli. First, since carnelian was almost exclusively used for beads (even more so than is the case for lapis lazuli), the most common context of discovery is mortuary, as part of intentionally deposited grave goods and bodily ornament. Second, the relatively low density of finds and multiplicity of potential sources similarly does not allow detailed reconstruction of routes of exchange. The same 4 or 5 broad corridors as described above for lapis lazuli were available (*cf.* Figure 4.3), but there is currently no way to measure the intensity of trade along them from direct carnelian evidence. Rather the distribution of carnelian objects merely marks the zone in which the material was circulated – or at least the minimum distance that it must have travelled. There were apparently workshops for carnelian in Tepe Malyan, Shahdad and Shahr-i Sokhta (Moorey 1994, 98), which suggest multiple sources and multiple routes could have been in use given the very wide distance between these southern Iranian sites. As a general impression, it seems likely that, as for lapis lazuli, the earliest (pre-3rd millennium) transportation of carnelian was via the land routes of northern or southern Iran. Later maritime exchange between the Indus, Mesopotamia and the Gulf states (particularly for the etched beads) became very important, though it is difficult to know if the land routes were still functioning. The collapse in the circulation of lapis lazuli and carnelian may again have coincided with the decline of the Indus civilization. Moorey (1994) has suggested however, that maritime networks may have only served southern Mesopotamia and that northern (Syro-)Mesopotamia may have been supplied from sources to the north and east (in Anatolia, Azerbaijan or western Iran).

4.3.3 Colour, symbolism and visual arrest

Colour symbolism and dramatic visual effects appear to have played a major role in the desirability and economic value of certain 'precious' stones, and hence their likelihood to be transported over long distances. This is particularly true of the stones just described, lapis lazuli and carnelian, but also for others such as turquoise, agate, amber and rock crystal. During the 3rd millennium BC, it seems that the social desirability of these materials was created and maintained through their use in elite performance in public rituals (*cf.* the 'tournaments of value' described by Appadurai 1986, 21), which placed their users apart from the general population. The distribution of these stones thus indicates not just the extent of a resource network, but also the extent of the circulation of particular

38 As with lapis lazuli, we can speculate whether other materials came to replace carnelian and hence undermine the economics upon which their trade was based. In this case, amber from the Baltic may have provided an alternative yellow-orange stone in the east Mediterranean and Mesopotamian markets during the 2nd millennium BC.

39 Presumably related to other material and social features of the wider 'inter-Iranian' cultural sphere as described by Amiet (1986), which also must have included the Indus.

ideas about and attitudes to the consumption of these stones. Their precise values and roles within the socio-economic systems of each region may not be the same, but where there is other evidence of interaction, the value systems are likely to be systemically interrelated.

4.4 Softstones: steatite, chlorite

A variety of softstones (including steatite, chlorite, serpentine) with similar properties and colours (white, red, and most commonly shades of green) can be grouped under the same label since they are frequently confused or combined in archaeological literature (Moorey 1994, 100) and were apparently considered sufficiently alike in antiquity to be used for similar purposes. Importantly these stones are relatively soft and easily carved, which led to them being used both for beads and seals, but also very commonly for vessels. Given the diversity of stones that fall under this category, it is rather difficult to create a distribution of their geological sources, though of course some of the urban centres of Mesopotamia, Egypt and the Indus which used these stones and the objects manufactured from them needed to procure the material (or finished goods) from distant highland regions.

4.4.1 'Intercultural style' vessels and 'weights'

Steatite and chlorite are important for reconstructing exchange routes not so much for the material itself (though presumably the material characteristics were important), but for a group of distinctively formed and decorated vessels and other objects that have been dubbed 'intercultural style' (Kohl 1978). The term is somewhat unwieldy since it has such a general meaning, but it is the most convenient and commonly understood for the group of objects described. This label encompasses a fairly diverse set of objects dating to the 3rd millennium including bowls, vases and 'weight-like' shapes whose real function remains unclear (see Figure 4.7). Various iconographic themes occur frequently as decoration on these objects: particularly 'hut', 'woven-mat', 'combatant snake', 'bevelled square', 'guilloche', 'date-palm' motifs (to follow Kohl's terminology), alongside a few other anthropomorphic or zoomorphic motifs. Many show inlaid decoration, using stones of other colours (often remaining inlay includes a white version of steatite, but sometimes other stones like lapis lazuli or carnelian). Whilst there are many finds of these sorts of objects in Mesopotamia (where they were first identified), it now seems there is a particularly strong relationship with south-eastern Iran where chlorite/steatite vessel workshops have been identified (for example at Tepe Yahya, see Kohl 1978), and where many hundreds of objects have been uncovered recently around Jiroft/Hari Rud region. Sadly, in the latter case, the majority were from uncontrolled excavations and recovered after the looting of graves (Lawler 2003; Perrot and Madjidzadeh 2006, 2005; *cf.* Muscarella 2001). Their dating is therefore difficult to pin down, but given the better-dated 'intercultural style' vessels from other sites, the second half of the 3rd millennium seems most likely for the Jiroft cemeteries (Muscarella 2001, 178-179; Madjidzadeh 2003; Majidzadeh 2008; Potts 2005).

In general, like etched carnelian beads, these objects represent a distinctively 'eastern' commodity in the sense that their distribution seems to involve an exchange network stretching eastwards from Mesopotamia: *i.e.* the Gulf, Iran, south central

Figure 4.7. A range of 'inter-cultural' style vessels and related objects: a. tall conical vessel, 'Iran' (Shinji Shumeikai, Japan, SS 1498; Aruz and Wallenfels 2003, 338-339, cat. 235); b. shallow cylindrical vessel with figured decoration and cavities for inlay, 'Mesopotamia' (BM 1288887, photo: © Trustees of the British Museum #AN00032381; Aruz and Wallenfels 2003, 330-331, cat. 227); c. handled 'weight' with snakes, Soch, Uzbekistan (State Museum of the History of Uzbekistan, Tashkent; Aruz and Wallenfels 2003, 339, cat. 236); d. 'guilloche'-pattern small vessel/vial, Tarut island (National Museum, Riyadh, 2633; cf. Aruz and Wallenfels 2003, 326, cat. 224f); e. shallow cylindrical vessel with zigzag patterns and 'hut motif', Saar, Bahrain (Bahrain National Museum, Manama 358-2-88; Aruz and Wallenfels 2003, 341, cat. 239); f. miniature vial possibly used for cosmetics, Gonur Depe (Sarianidi 2006, 198, fig. 64).

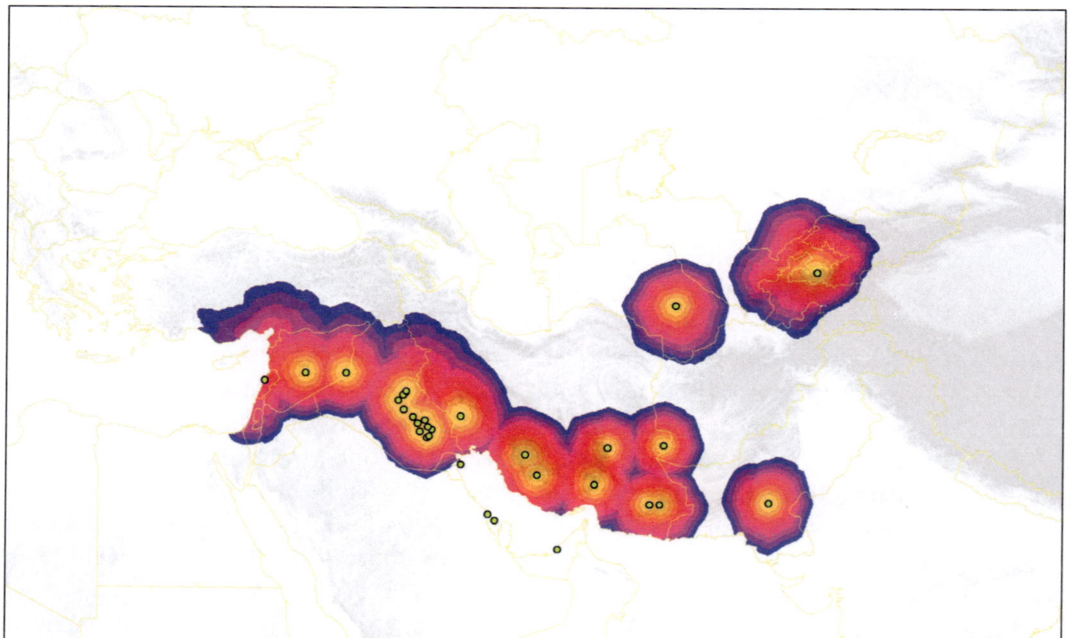

Asia and the Indus (see Figure 4.8). No 'intercultural style' objects are known west of Mari, and for some reason they did not seem to penetrate the Egyptian cultural sphere (unlike lapis lazuli), nor have any examples been so far found in Anatolia, the Caucasus or the Aegean, where lapis lazuli has been found in small quantities. Again this implies they had a particular 'intercultural appeal' (Aruz 2003, 244-245), but one whose power did not extend outside this eastern interaction sphere. It is difficult to assess what kind of exchange items these objects represented in the different regions in which they were circulating. It seems likely that their cultural meaning or value was rather different in each locale: in south-east Iran, where they appear to have been manufactured, a close relationship to particular local cultural and ritual meanings seems likely, compared to a potentially more abstract role as indicator of wealth and power to obtain foreign goods in Mesopotamia. Presumably the distribution of these objects also indexes the movement of people between regions (as all these distributions must do), but the extent and nature of such movement cannot be easily assessed. We do not know if such objects had 'ethnic' associations or were imported by immigrants from the Jiroft region, for example[40]. Perrot has suggested that the intended contents of these vessels may have been the prime motivation behind their distribution – and argues on the basis of the repeated depiction of an unidentified plant[41] that these may well have been medicinal, narcotic, seasoning or aphrodisiac plants which would have been more easily available on the Iranian plateau but not in Mesopotamia (Perrot and Madjidzadeh 2005, 145, 148).

Southern Iranian land routes and the Persian Gulf maritime routes were most important to the flow of these 'intercultural style' objects, presumably mostly in finished form. The footprint shows a particular network of interaction within this eastern sphere which linked southern Mesopotamia and the Gulf region to eastern Iran and parts of Central Asia (Figure 4.8). However, as with lapis lazuli and carnelian, the density of discovery – or at least the way in which they are currently published – does not facilitate the reconstruction of *local* social networks.

4.4.2 Central Asian softstone objects: composite figurines and miniature columns

A small number of 'intercultural style' items find their way into Central Asia during the late 3rd millennium: for example two vessels were found in graves at Gonur Depe and a 'weight' was found in uncontrolled circumstances in the vicinity of Soch, Uzbekistan (Figure 4.7c). These might well have been imported items given their low number. However, it is interesting to note the occurrence in this region of a range of other objects made from softstone (especially steatite, chlorite and alabaster). These objects include: composite figurines; stone vessels, including miniature vials thought to have contained perfumes or other cosmetics (Pottier 1984, 28-9); steatite and alabaster seals (Sarianidi 1981, 178); stone 'pocket books',

Figure 4.8. Distribution of 'inter-cultural' style vessels of known provenance with archaeotopogram 'type A2' from these sites indicating the 'zones of interaction' in which these objects are likely to have circulated. Limitations of the model (i.e. the inability to 'travel' across the sea, mean that the items on the south-coast of the Persian Gulf do not have the expected 'halos'. See Appendix C.1.1 for the database of objects.

40 Whilst avoiding an essentialist view of ethnicity as an immutable property of individuals, groups or their material culture, it is still possible that different individuals and groups could 'buy into' material culture with particular 'ethnic' associations, in this case an 'eastern' one for Mesopotamians, or else migrants from the 'east' could demonstrate their distinctiveness through the use of such material.

41 Perrot likens the status of this unknown plant to that of *silphinum*, whose botanical identity is also unknown but which was an important plant for seasoning of food and as aphrodisiac for Greeks and Romans (Perrot and Madjidzadeh 2005).

which might possibly have been used as loom weights (Hiebert 1994b, 382); and 'miniature columns' (Hiebert 1994a, 154-157), staffs and maceheads whose intended functions remain elusive. The presence of these objects, at least some of them apparently locally manufactured (Hiebert 1994b, 386), even if the material was sourced from the Kopet Dag or Pamir mountains to the south, demonstrates that central Asian communities were well able to manufacture complex objects in stone. It also suggests that they were much less interested in, or less able to obtain, foreign-made 'intercultural style' vessels than is the case for, say, Gulf and Mesopotamian communities[42].

There were at least two types of composite figurines used during the BMAC period: a seated female type which, whilst many of the examples have come from uncontrolled contexts in Bactria (Ligabue and Salvatori 1990, figs. 108-113), is now fairly confidently linked to BMAC contexts (Sarianidi 1990b; Hiebert and Moore 1993), and the perhaps related, though even less clearly contextualized, male type with gashed face (Aruz and Wallenfels 2003, 244-245) found in eastern Iran. The role of these figurines is of course unclear: it has been speculated that the female types represent a princess or goddess (Potts 2008a) whilst the male types might be some kind of god, demon or 'dragon' (Francfort 1994)[43]. Whilst these figurines may represent clues about the circulation of other materials, such as textiles (see Section 6.5.3) or iconographic styles, given the lack of good archaeological contexts it is rather difficult to say anything more directly about the function or circulation of the figurines themselves.

Softstone vessels of alabaster (sources of which are known from the Kopet Dag mountains) were used by earlier Namazga communities near the piedmont during the early and mid 3rd millennium (Hiebert 1994b, 376)[44]. The miniature vials which date to the end of the 3rd millennium or early 2nd millennium BC (*e.g.* Figure 4.7f) appear to be related to very similar steatite vials from south-east Iran and the Persian Gulf, which show different decoration schemes. Scientific analysis of their composition and contents, which could reveal both greater detail about their circulation (*e.g.* whether the vials are locally circulated or exported) and related function (*e.g.* whether they were used to transport or sell the contained liquid, or just temporarily store it), has sadly not been undertaken to date. Hiebert (1994b) suggests that the distribution of the various types of vial may indicate changing contacts or alliances between Central Asia, eastern Iran and the Gulf, but does not go on to suggest why or by what mechanism the vials would have travelled. If they are indeed cosmetic bottles containing perfumes or make-up then their distribution is of course interesting as a proxy for the exchange or expansion of 'cosmetic' fashions or rather of particular organic substances used for the cosmetics (*cf.* also later 3rd millennium 'Syrian bottles', see Section 5.6.3). Whilst we should not assume that 'cosmetic' products were exclusive to women, if these vials could be linked to female burials it could be that the movement of

42 On the other hand, this may be partly a matter of economic scale – central Asian communities were apparently much smaller – hence the smaller 'intercultural style' corpus.

43 Certainly, given the standardized depiction, it seems plausible that both represent important generic characters with 'mythological' associations rather than historical political personages in the Renaissance sense (*i.e.* they are iconographic rather than photographic) but the same can be said for much of the imagery of the 3rd millennium in the Near East.

44 The shapes and colours of which might, contra arguments set out later in Section 5.6.4, have provided a model for the new plain aesthetics of the pottery of the Namazga V period (for examples, see Masson 1988, pl. 7, 35).

these objects was facilitated by migration through elite marriage exchanges and alliances, for example (especially if isotopic analysis of female skeletons showed evidence for immigration). Unfortunately, data to confirm or reject this hypothesis are currently lacking.

4.5 Chipped stone: obsidian

Chipped stone tools become a much less important part of the archaeological study from the 3rd millennium onwards compared to earlier periods, although in fact they continued to be used extensively in many regions, and our poor knowledge about them is in many instances a result of a lack of investigation rather than a lack of evidence. Most chipped stone appears to have been sourced locally, however, so, whilst a detailed study of form types and selected materials might reveal networks at the micro-scale, it is only really obsidian, known to have been transported over much larger distances, which is potentially relevant for the discussion of macro-scale routes of interaction and exchange.

Obsidian is a volcanic glass ranging in colour from opaque black to almost transparent grey, with, depending on the particular source and therefore chemical impurities, slight greenish or reddish tints. Sources of obsidian are fairly restricted in distribution and these sources can be characterized by various scientific techniques (see *e.g.* Gratuze 1999). Major or well-known sources of obsidian include those in Cappadocia (*e.g.* Acıgöl, Nenezi Dağ, Hasan Dağ, Göllü Dağ), eastern Anatolia (Bingöl, Nemrut Dağı[45], Süpheli Dağ), various locales in Transcaucasia (see Badalyan, Chataigner and Kohl 2004; Cauvin *et al.* 1998) and Melos and Giali in the Aegean (see Figure 4.9). The most intensive use of obsidian belongs primarily to earlier periods (especially the Neolithic and Chalcolithic) during which this material travelled considerable distances. The high extent to which obsidian may be characterized and its widespread use offer many insights into long-distance exchange of materials in these early periods (Cann and Renfrew 1964; Williams-Thorpe 1995). It also has provided evidence for the configuration of regional social networks (*e.g.* Healey 2007). While there are certain zones in which obsidian from particular sources tends to dominate the obsidian repertoire (*cf.* Figure 4.9), individual pieces of obsidian from different sources turn up in different zones, suggesting a complex pattern of exchange during the Neolithic (A. Sherratt 2005).

Obsidian continues to be used into the 4th, 3rd and 2nd millennia for blades, microliths, beads and even vessels[46] in certain regions. In Mesopotamia, obsidian was counted amongst the most important precious stones for beads to accompany gold: the others were lapis lazuli, carnelian and calcite/alabaster (Moorey 1999, 70). Unfortunately, the obsidian objects from these later periods have received much less investigation, so our understanding of the circulation of this material is less than clear. It also remains unclear whether obsidian continued to be transported

45 The volcanic Nemrut Dağı that is next to Tatvan, by Lake Van, not the identically named mountain with which it is frequently confused that lies between Malatya and Adıyaman and has the massive artificial tumulus-like funerary for Antiochus I Theos of Commagene.

46 Including, famously, the obsidian-marble 'chalice' from Tell Brak dating to the early 4th millennium (Oates *et al.* 2007); a bowl from the Puabi's grave, from Royal Cemetery of Ur (Figure 4.10) – now in the British Museum (ME 121690); with later examples of obsidian vessels or fragments from Tell Malyan, Iran and including at Acemhöyük and Kültepe in central Anatolia (dating the early second millennium BC) (Moorey 1999, 70).

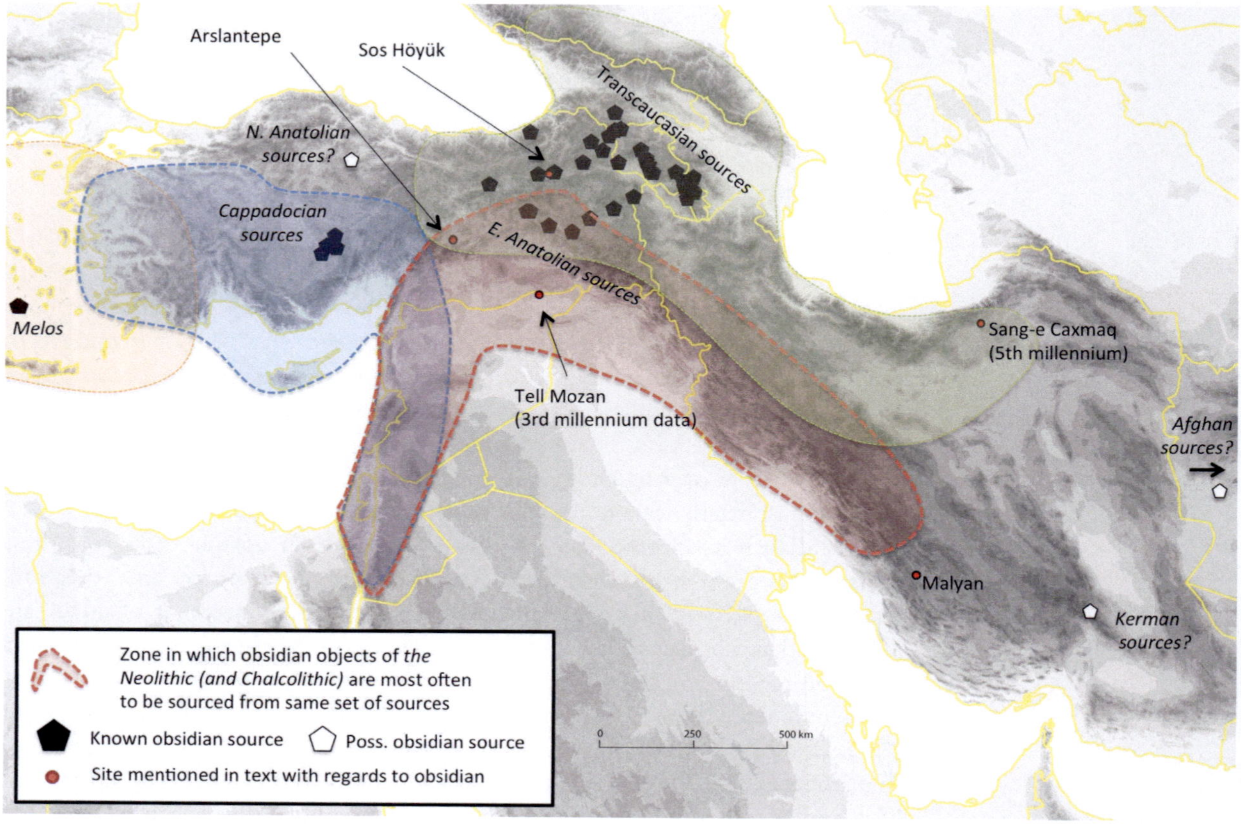

Figure 4.9. Distribution of some major obsidian sources across Near East with broad zones, served predominantly by these sources during the Neolithic and Chalcolithic periods (cf. Frahm 2010). See Appendix D.1.1 for list of obsidian sources. These zones represent very broad generalizations of a much more complex dataset. Sites relevant to obsidian use in the 3rd millennium BC and mentioned in the text are also marked.

over long distances as a high value item, as it appears to have done during previous periods, or whether by the 3rd and 2nd millennia it has simply become one of a number of different local resources (valued simply like other chipped stone), as its technological and social functions were increasingly subsumed by other materials, particularly metals. The apparent rising trend of using obsidian to make vessels (see, for example, Figure 4.10), especially during the 2nd millennium BC hints at a change in the perceived social and functional emphasis of the material through time (Coqueugniot 1998).

4.5.1 Obsidian use in eastern Anatolia and Transcaucasia

In eastern Anatolia and the Caucasus, obsidian represents an important material for stone-based tools into the Bronze Age, despite the presumed emergence of a wide-ranging metal toolkit (see Section 5.3.1). Analysis of samples of obsidian from Late Chalcolithic levels of Arslantepe (91% of which were arrowheads), dating to the late 4th millennium has shown them to be predominantly sourced from a variety of eastern Anatolian sources, though with some from Cappadocian sources (Taddeucci *et al.* 1975). The obsidian blades from Early Bronze Age levels at Sos Höyük (Sagona *et al.* 1996, 38, 47-48, 49) do not seem to have particularly distinctive forms (being characterized by a wide variety of rough shapes and microliths) and there is no obvious functional differentiation (Figure 4.11). Nearby sites like Karaz, Pulur and Güzelova contain similar assemblages. Sagona suggests the obsidian from Sos Höyük comes from alluvial pebbles in the local

Figure 4.10. Obsidian bowl from the grave of Queen Puabi, Royal Cemetery of Ur, Early Dynastic period, about 2600-2400 BC (photo: © Trustees of the British Museum, #AN32556001).

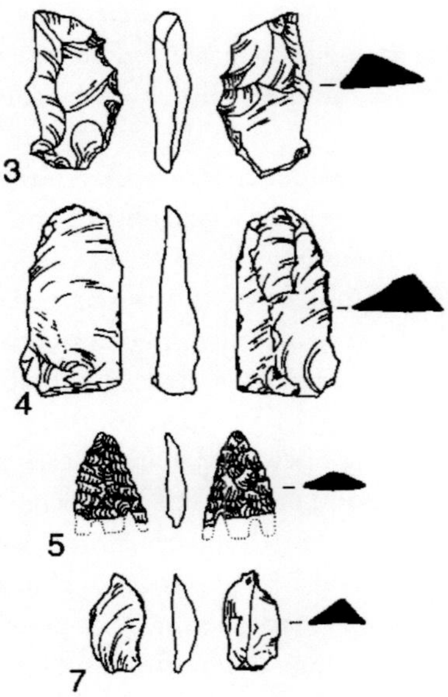

Figure 4.11. Examples of obsidian artefacts from 3rd millennium levels at Sos Höyük (Sagona, Sagona and Özkorucuklu 1995, 209, fig. 12).

Pasinler region, rather than having been procured from distant major obsidian sources. If this is true, it suggests that obsidian is simply a local resource, and perhaps not considered a high-value traded commodity.

The evidence from Tell Mozan (ancient Urkesh) confirms that the use of obsidian is not restricted to the highland regions (Frahm 2010; Frahm and Feinberg 2013a; 2013b). Here a substantial corpus of obsidian blades has been identified in layers dating to the 3rd and 2nd millennia BC. The provenance analysis undertaken by Frahm and Feinberg on this corpus reveals a relatively complex mixture of sources, with the majority of the material apparently from eastern Anatolian sources, dominately particularly by sources at Nemrut Dağ and Bingöl, but small numbers of examples from other sources including 3 samples from Cappadocia (Göllü Dağ) suggesting a wide network (Frahm 2010, 658)[47]. Whilst this work shows obsidian continuing to be used in a North Syrian urban

47 Of course we cannot say for sure how long these distant obsidian pieces were in circulation, or what the biography of their travels really were:- for example, it is possible that obsidian cores or flakes were sourced from Cappadocian centuries or millennia earlier than their final use and deposit at Tell Mozan.

context, it is difficult to know how representative this is of wider Syrian and Mesopotamian contexts, in the absence of other investigations (*cf.* work in the east Mediterranean and Aegean, *e.g.* Carter 2008). It should perhaps be remembered that Tell Mozan has often been claimed to have had a special relationship with the eastern Anatolian highlands, for example in the presence of small amounts of Kura-Arax ware, a presumed role in the metals trade from the north and portable andirons which bear similarities to the Early Transcaucasian types (Kelly-Buccellati 2004; though *cf.* wide distribution of different types of andirons or 'fire stands' in 3rd millennium Rahmstorf 2011, 273-277). The presence of so much obsidian may therefore relate to Tell Mozan's particular location near to highland communities. Whilst this relationship has often been tied up, unhelpfully, with the identity and origins of the Hurrians, who according to contemporary texts become a powerful group in the Mitanni and Hittite cultural spheres during the second half of the 2nd millennium BC, it is clear that northern connections are very important, and it least suggests that obsidian was still valuable enough to be worth sourcing from several hundred kilometres by some members of the Mozan urban community.

One nebulous thread of evidence hints that obsidian may have continued to play a symbolic aesthetic role in Chalcolithic and Early Bronze societies in Transcaucasia and eastern Anatolia. The region is characterized in the late 4th and early 3rd millennia by Kura-Arax or Early Transcaucasian pottery. The later Kura-Arax pottery is defined by red-black burnished aesthetic, with incised decoration, which could be compared to a metallic aesthetic, perhaps indicating metal (copper and/or silver) skeuomorphism (see Section 5.6.5). However, this dual red-black colouring appears to have developed later than the earliest Kura-Arax vessels in the southern Caucasus – the dominance of red-black colour combinations is a later trend, apparently a result of interaction with central or eastern Anatolian pottery traditions (Palumbi 2008b, 205, 311, passim). The earliest vessels are instead mostly black-burnished or very dark brown. It is possible that the aesthetic of a dark or black-burnished colour for this pottery (and indeed for the new material of metal) may have been derived from earlier aesthetic colour schemes, for which obsidian offers the obvious model[48]. Whilst this is difficult to prove, there is one possible piece of evidence that could promote this interpretation further: namely the presence of obsidian as one constituent inclusion in Kura-Arax pottery from at least two Transcaucasian sites (Iserlis *et al.* 2010, 254-255)[49]. Of course, this may simply be related to the local geology in Transcaucasia, where obsidian fragments are a common constituent of sand matrices, rather than intentional addition of crushed volcanic glass; and obsidian was not found in the paste of the equivalent Khirbet Kerak ware ceramics from Bet Yerah[50] (instead basalt was present, also

---

48 Both metals and pottery can, after all, take on a variety of colours depending on their production or treatment. The black-burnished pottery may be a direct reference to obsidian, or filtered through black-coloured metal objects.

49 Iserlis *et al.*, by contrast, compare the intense attention to detail given to surface treatment on Kura-Arax and Khirbet Kerak wares to a 'skin' (see more explicit theoretical discussion in Iserlis 2009, 191-192) – and it is tempting to take their observation literally and speculate whether the Kura-Arax aesthetic may relate to leather (another skin with intense surface treatment) rather than obsidian or metal.

50 The site whose alternative name, Khirbet Kerak, provided the type assemblage for this ware. Note other alternative spellings and names: Beth Yerah, Al-Sinnabra.

black, but not shiny). However, the possibility deserves further investigation with more technological studies following Iserlis *et al.*

4.5.2 The lack of obsidian in central Asia

In contrast to Caucasian early farming sites, obsidian forms either none or only a very small fraction of the stone tool repertoire even during the local Neolithic in western Central Asia (Korobkova 1981). Obsidian blades have been found in certain zones of the site of Tepe Sang-e Caxmaq (Masuda 1974), for example, though the material has not been fully analysed or published. But Djeitun stone tools are characterized only by flint, sandstone, schist and diorite (Berezkin *et al.* 1993, 335). Obsidian is likewise absent at Anau (Hiebert and Kurbansakhatov 2003, 89-93). The exchange networks that supplied a wide area of the western Near East during the Neolithic and Chalcolithic from highland sources in Transcaucasia and Anatolia, thus do not seem to have extended into Central Asia. This is despite the fact that the assemblage of domesticates of the early farming Djeitun culture, appear to come directly from southwest Asia (*i.e.* the 'Fertile Crescent'), rather than via south Asia (Berezkin *et al.* 1993).

This negative evidence is mentioned here to highlight the changing nature of interconnections between Central Asia and the communities to the west and south from the 6th to the 3rd millennia BC. Whilst domesticates must have been imported (perhaps through a slow 'wave-of-advance' process), a favourite material with both practical and symbolic value (*i.e.* obsidian) was not, and local materials were used. This suggests a very low-level of interaction between Central Asia and the west. Instead, sometime in the 4th millennium, stones with social functions of adornment rather than manipulation (*i.e.* lapis lazuli, carnelian and turquoise) began to travel in the opposite direction (from Central Asia and Afghanistan to the west) – but as yet, we do not know for what material they were being exchanged.

In this context it is also interesting to note that the pottery repertoire of Central Asia almost never includes black or dark burnished surfaces, though black painted lines are not uncommon. The early part of the Namazga sequence include mostly reds and whites, with various painted patterns, perhaps related to textiles (see Section 6.6.3), and the plain wares of Namazga V and VI, which appear to have been inspired by metal prototypes (see Section 5.6.4), are light-coloured creams, reds and beiges. Whilst this could simply be a technical aspect of the clay, it could also suggest that the shiny black of obsidian never played a role in the social-colour symbolism/value system of Central Asian communities. This difference between eastern Anatolian (and a much wider area of the western part of the Near East) and Central Asia hints at long-lived traditions of colour preferences which survive through many other social transformations and manifest themselves in different materials.

4.6 Weights: stone as metric mediator

Whilst the concept of a stone 'weight' (often of a more or less standard mass) used for technical purposes, such as for looms, was apparently known well before the 3rd millennium, concrete remains of a standardized weighing *system*, with objects of identifiable multiples and fractions of abstract units, seem to have appeared only sometime in the mid-third millennium BC. In fact several types of weights and weighing systems have been identified, with differing distributions. Three discrete systems have been identified across the Near East: the 'Mesopotamian', variations of the 'Syrian' and the more restricted 'Indus' type (Rahmstorf 2010a). Whilst the particular multiples and material forms of the weights differ slightly in each region, as Rahmstorf has argued, it now seems very likely that, excluding the southern Mesopotamian/Sumerian system, all of the weighing systems of the 3rd millennium either had a common source or were designed to facilitate relatively easy inter-conversion (see Table 4.1 for a table of conversion based on Rahmstorf's arguments)[51]. In the archaeological record, various recognisable material forms are prevalent, the most easily recognized from 3rd millennium contexts are the 'Syrian' haematite types (Figure 4.12a), Aegean spool-shaped types (Figure 4.12b), and Harappan cubic types (Figure 4.12c). The significance of the particular forms is currently not obvious, though we may speculate that there were some cultural prototypes before the introduction of weighing to a region. The majority of weights recognized in the archaeological record were made from stone, such as haematite and marble, though lead weights are known from 2nd millennium contexts in the Aegean (Michailidou 2010, 72-73, fig. 7.3, 7.14). We should bear in mind, however, that weights could easily have been made in less durable or more recyclable materials (such as sand-bags or metals). Arguably, standardized ingots also represent a kind of 'weight' (see Section 5.2.5, 5.6.1). This aside, it is interesting to consider why stone appears to have become such an important medium for precision weight measurement: the material qualities of stone as un-fluid, un-changing and subtractive presumably played a role in this.

Standardized weighing systems are significant for exchange and routes because they index the need or desire to consistently and accurately compare material goods which were sourced over large distances (Figure 4.13a). The inter-calculability of the systems that emerge in the 3rd millennium provides an indication of the common acceptance of the necessity of standardization and also suggests that similar types of goods were being exchanged. It is normally assumed that the emergence of an international metals trade must have been the driving factor for the creation of weighing systems, particularly for economies in which metals were becoming central mediators of value (see Section 5.4.1). Textual evidence from 3rd millennium Mesopotamia and earlier 2nd millennium Anatolia reminds us that, unsurprisingly, metals were not the only item to be weighed with these systems however. Materials and commodities such as wine, beer, oils, barley, wheat, flour, pulses and figs appear to have usually been measured by *volume* (presumably using standardized vessel sizes[52] – something which may explain

51 In fact the 'standard' value given for the southern Mesopotamian shekel (of 8.33g) cannot fit into this system, only by using a higher value of 8.54g (or else 8.616g) can it be made to be easily convertible, suggesting an alternative (and competing?) metric system in use (Rahmstorf 2010a, 102).

52 Compare the interpretation of the Urukian 'bevel-rimmed bowl' as a standardized ration of wheat or bread (Goulder 2010).

Figure 4.12. Selection of objects confirmed as weights in recognizable weighing systems, alongside unconfirmed or possible weights of the 3rd millennium BC: (a) 'Syrian' haematite weights (Rahmstorf 2010a, 93, fig. 8.3); (b) Aegean spool-shaped weights (Rahmstorf 2010a, 89, fig. 8.1); c. 'Harappan' cubic weights from Chanhu-Daro (Aruz and Wallenfels 2003, 401-402, cat. 292a; photo: © 2014 Museum of Fine Arts, Boston, Joint Expedition of the American School of Indic and Iranian Studies and the Museum of Fine Arts, 1935–1936 Season, 36.2274, 36.2297, 36.2299 36.2305, 36.2308, 36.2316, 36.2322, 36.2325, 36.2328); d. possible lead weight, shaped as a bull, unprovenanced (Aruz and Wallenfels 2003, 363, cat. 255); e. 'miniature columns' of unknown function, Togolok 21, Murghab delta (Sarianidi 2006, 269, fig. 122); f. an example of an 'intercultural-style' handbag 'weight' or 'stone pocketbook' from eastern Iran, the Gulf region or Central Asia, unknown provenance (Aruz and Wallenfels 2003, 328-329, cat. 225b; photo: Metropolitan Museum of Art. Gift of Norbert Schimmel Trust, 1989.281.40). Note, however the very different scales.

the increasing use of wheel-made pottery from the 3rd millennium onwards), whereas metals, precious stones, wool and goat's hair, linen, yarn, ropes, alum, wood (occasionally), ivory (as fragments), certain spices, perfumes and dyes, celery, wax and tendons, finished textiles (occasionally, when compared to raw materials), hides (occasionally), and (very rarely) bread, fish and possibly meat are all recorded as being measured by *weight* (Michailidou 2010, 74-75). That textiles and precious stones were also measured by weight suggests a wider interest in ensuring trust and equality of exchange in many different materials as a complex package of new exchange traditions.

"Sumerian"*	"Ugarit"	"Hatti"	"Carchemish"	"Harappan"	Modern metric	Rahmstorf
m (8.54g)	*u (9.4g)*	*h (11.75g)*	*k (7.83g)*	*i (13.71g)*	*grams*	*units*
						10
		2	3		23.5	30
	5	4	6		47	60
			7	4	54.83333333	70
		6	9		70.5	90
11	10	8	12		94	120
			14	8	109.6666667	140
		10	15		117.5	150
	15	12	18		141	180
		14	21	12	164.5	210
22	20	16	24		188	240
		18	27		211.5	270
			28	16	219.3333333	280
	25	20	30		235	300
		22	33		258.5	330
			35	20	274.1666667	350
33	30	24	36		282	360
		26	39		305.5	390
	35	28	42	24	329	420
		30	45		352.5	450
44	40	32	48		376	480
			49	28	383.8333333	490
		34	51		399.5	510
	45	36	54		423	540
			56	32	438.6666667	560
		38	57		446.5	570
55	50	40	60		470	600

Table 4.1. Table showing common multiples allowing inter-conversion of 3rd millennium weighing systems (following Rahmstorf 2010a).

Despite the widespread distribution of weighing systems, there are areas in which weighing systems have definitely not (yet) been identified, and at present this includes both our case-study regions. 'Intercultural-style' weights, also known as 'stone pocketbooks' (Hiebert 1994a, 157) – whose distribution stretches from the Persian Gulf to the Pamirs– have already been mentioned above (Figure 4.7c). Their small number and lack of sufficiently detailed publication have generally prevented the possibility of identifying an indisputable weighing system; and the same is true of the exceptional 'bull weight', thought to have come from eastern Iran or central Asia (Figure 4.12d), and the spool-like 'miniature columns' typical of the Bactria-Margiana Archaeological Complex (Figure 4.12e), whose function remains elusive but which bear a formal resemblance to Aegean spool-weights even if they are much larger. Too often the mass of objects, including those thought to be weights, is not published, making it impossible to confirm or reject the hypothesis that certain objects might be part of a weighing system. Given the extent of interaction between Mesopotamia, the Indus and Central Asia in the second half of the 3rd millennium (seen already in some of the iconographic links between the regions), and the fact that two of the commodities assumed to have been transported from Central Asia to Mesopotamia (precious stones and

Figure 4.13. Distribution of different weighing systems during the (a) 3rd millennium BC; (b) 2nd millennium (following Rahmstorf 2010a, 96).

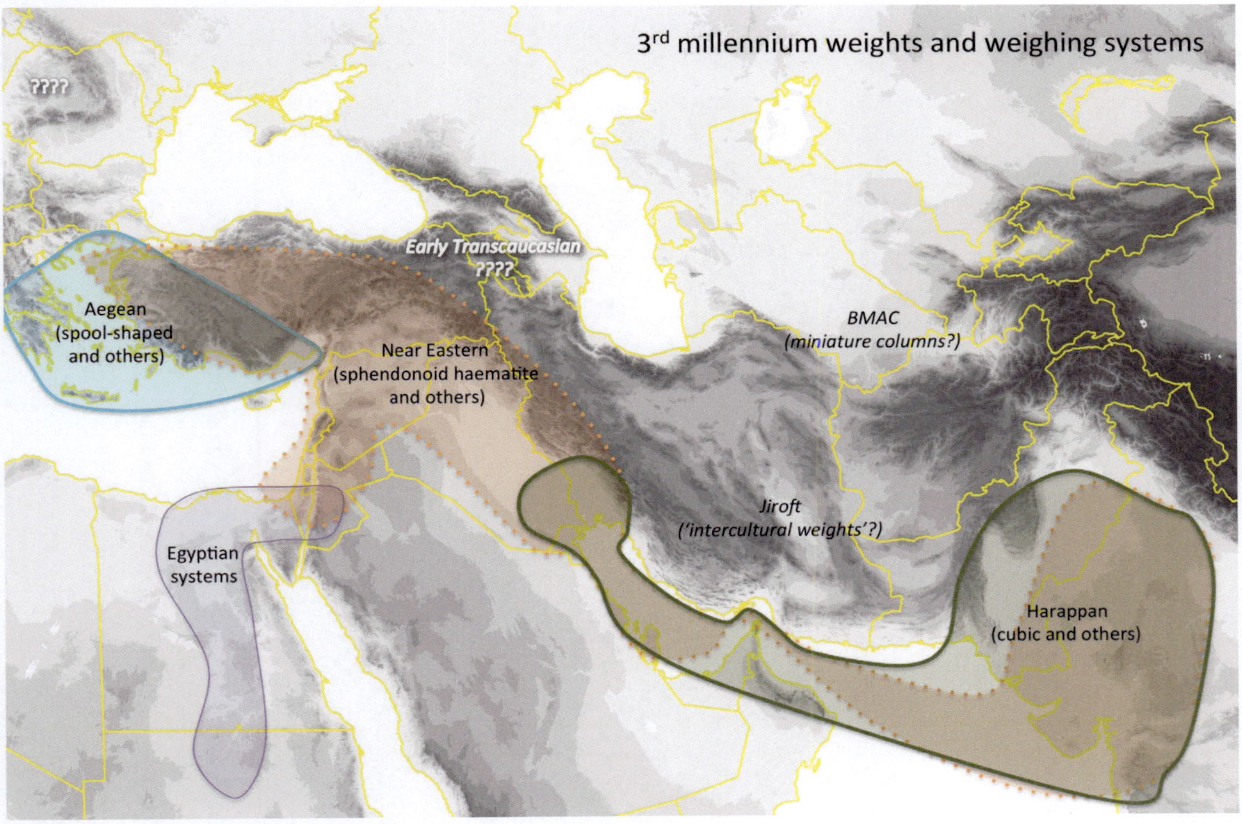
a
3rd millennium weights and weighing systems
????
Early Transcaucasian
????
Aegean
(spool-shaped
and others)
Near Eastern
(sphendonoid haematite
and others)
BMAC
(miniature columns?)
Egyptian
systems
Jiroft
('intercultural weights'?)
Harappan
(cubic and others)

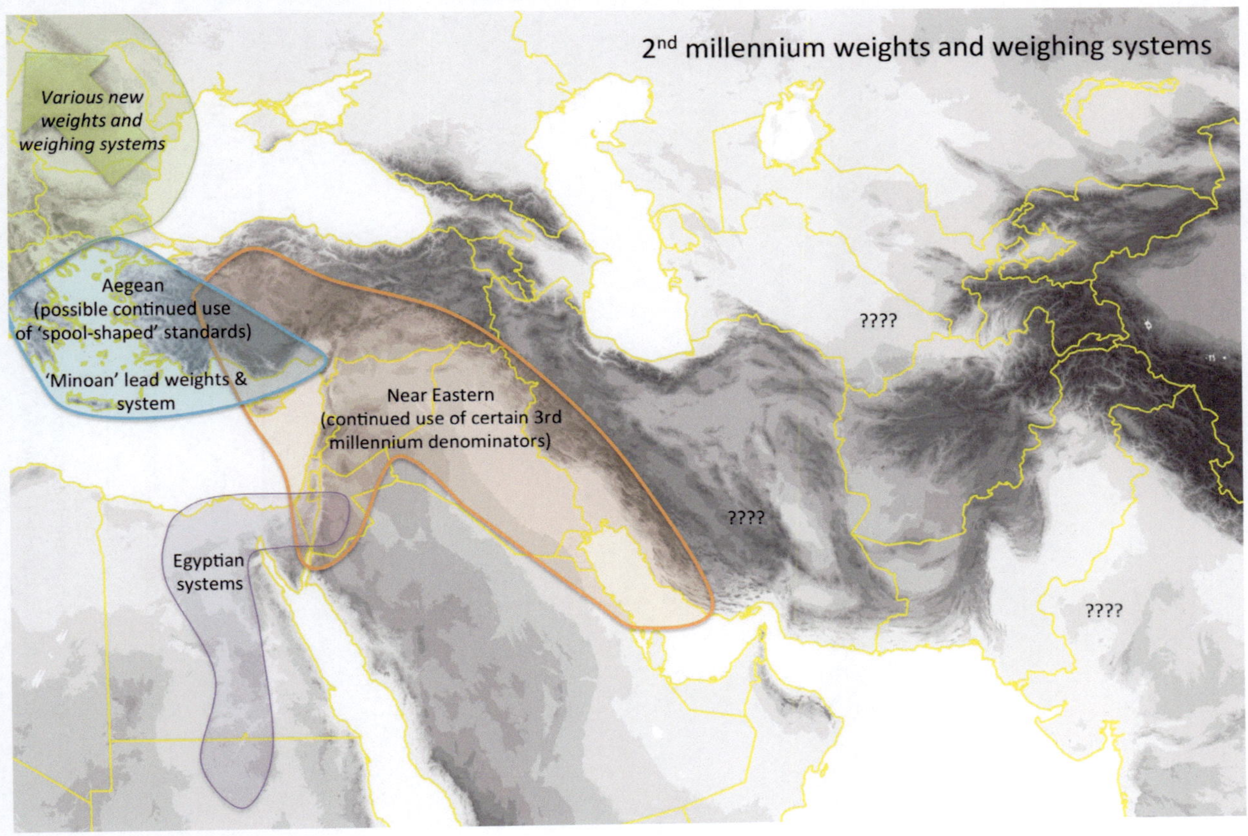
b
2nd millennium weights and weighing systems
Various new
weights and
weighing systems
Aegean
(possible continued use
of 'spool-shaped' standards)
'Minoan' lead weights &
system
Near Eastern
(continued use of certain 3rd
millennium denominators)
????
????
Egyptian
systems
????

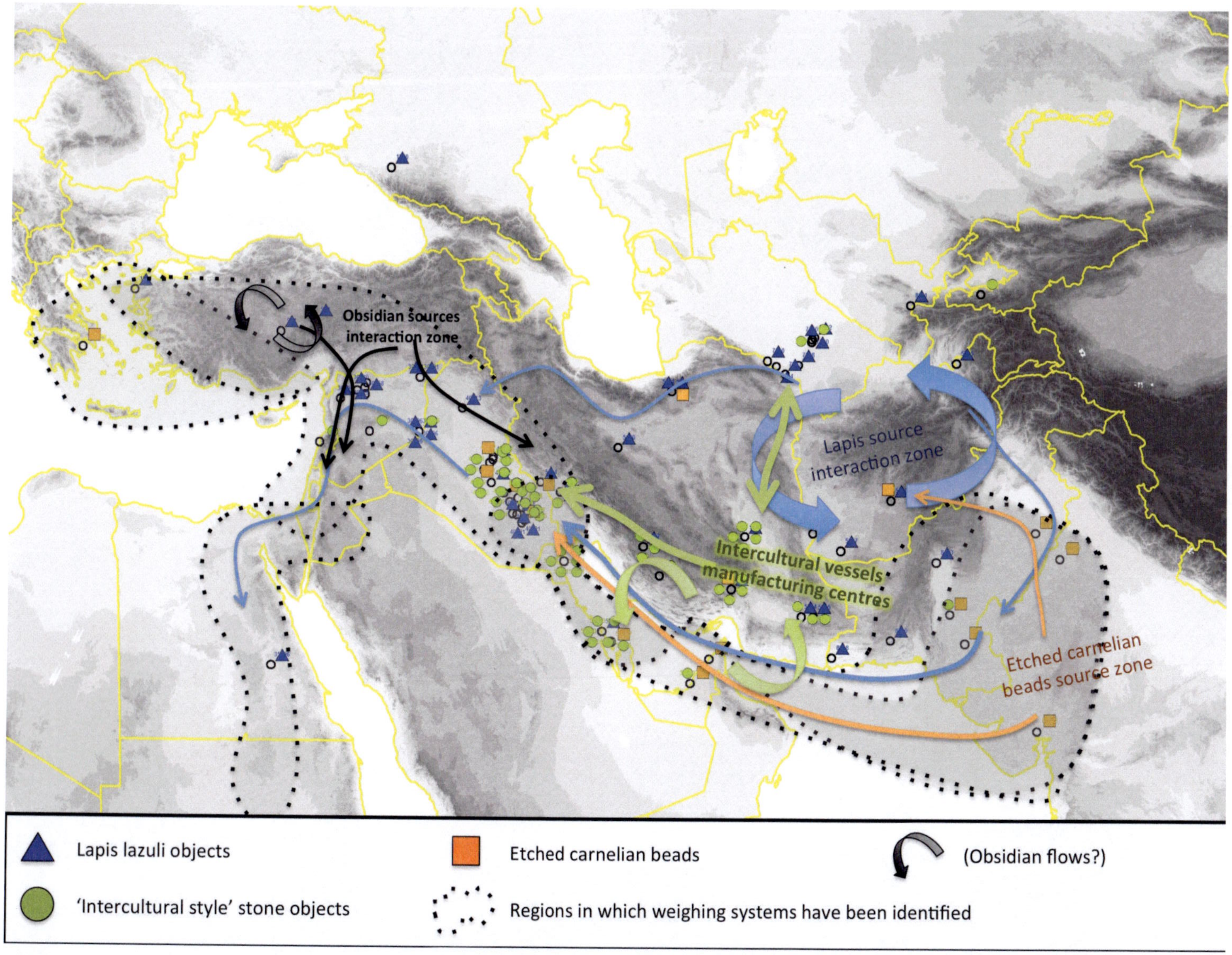

Figure 4.14. Summary of distribution data on lapis lazuli, carnelian, 'intercultural-style' objects and weighing systems for the 3rd millennium BC.

metals) were normally weighed in Mesopotamia, it would be surprising if the communities of Central Asia were not familiar with weighing systems. However, we should also consider the possibility that if the communities of this region had no need of standardized measurements, objects like the 'stone pocketbooks' may represent some kind of pseudo-weights – particularly interesting if the Indus script is indeed a kind of 'pseudo'-writing as has sometimes been argued (Farmer, Sproat and Witzel 2004; Lawler 2004). Indeed the items shown on the right side of Figure 4.12 are much larger than those on the left, supporting alternative perhaps symbolic rather than metrical roles.

If we exclude the urbanizing sites of northern Syria, no objects have been identified which might be identified as weights for measurement have been found in eastern Anatolia. Again this is surprising if we consider the geographic proximity to weight-using cities of Syria and central Anatolia, and in the context of metal extraction in which the eastern Anatolian highlands and Transcaucasia are assumed to have played a role (see Section 5.2.5). What are we to make of this negative evidence for weighing systems? If we exclude the possibility that this is simply a factor of archaeological visibility (*e.g.* that local weights were made in perishable materials), then this absence shows that, despite geographical proximity and deep iconographic and material connections, local communities did not see the need to measure objects using standardized systems. This implies either that local social systems of control were not strong enough to enforce standardization, or indeed that standardization was not necessary for the kinds of interaction and exchange being enacted in these regions. In other words it makes it more likely, perhaps unsurprisingly, that the economies were based on non-commercial non-regulated exchange systems (like gift-exchange and barter) than on market and prices.

4.7 Summary: routes and stones

A range of different categories of stone object have been catalogued in this chapter to examine their consumption, distribution and movement during the 3rd and 2nd millennium BC. This process has shown how difficult it is to disentangle the study of the flow of stones and their role in social value systems (whether as 'symbolic power' for jewellery, as exotic ethnic items or as static mediators for standardized systems of weight). 'Precious' stones like lapis lazuli and carnelian appear to have moved over very large distances, often in raw form (although the etched carnelian beads seem more likely to have been transported as finished items). The types of objects manufactured from these materials are relatively restricted in range despite the geographical distances involved: beads are the most obvious example, which appear to index a socio-aesthetic system in which the public display of such exotic materials played a role in the assertion or negotiation of hierarchies. 'Intercultural style' chlorite or steatite vessels, were probably transported as finished objects and show both the movement of objects themselves and also of iconographic themes and styles. Similarly, it has also been suggested, their distribution seems to index not just the movement of abstract identity, but also perhaps of some other more ephemeral commodities, such as plant-based narcotics, ingested as part of special (imported?) social practices. A very diverse set of deliberately shaped objects, stone weights, can be identified in distantly located sites, whose weight multiples although being locally specific, appear to indicate global inter-calculability. Thus through focusing on the flow of one material category – stone – we actually reveal the flow of visual aesthetics, socio-economic systems of power and exchange and ritualized social practices.

Plotting these different items together (Figure 4.14), it is clear that certain areas are more integrated into these flows during the later 3rd millennium (which is the period for which most of the above material appears to have been circulating) than others:- Mesopotamia, Syria, Egypt, parts of western Anatolia and the Aegean, parts of Iran (particularly the south), certain parts of the Gulf, the Indus and some parts of Central Asia. These differential distributions can be argued to document differing systems of value accorded to these stone materials. Other areas appear to have remained entirely outside this zone. Transcaucasua/eastern Anatolia is perhaps the most interesting example of this. Its lack of substantial lapis lazuli, carnelian, intercultural-style objects or obvious weighing systems despite its relative proximity to the 'urban core' of northern Syria and Mesopotamia is striking in comparison to the much more distant regions like the Aegean (which has clear weighing systems, but small amounts of lapis, carnelian) and Turkmenistan (which has 'intercultural style' objects, plenty of lapis lazuli and carnelian but no unequivocal evidence for weights). The fact that the 4th millennium kurgans of Maikop, just north of the Caucasus, yielded both lapis lazuli and carnelian objects hints that while such material must have circulated or travelled through eastern Anatolia on occasion, the local attitude towards lapis lazuli or carnelian was very different. This does not however mean that eastern Anatolia was somehow isolated from the urban communities to the south, but, as argued in the following chapter, instead appears to have been connected via different material networks, particularly metals.

Chapter 5

Mapping Material Flows: Metals

5.1 Introduction

The previous chapter attempted to map the flows of a sample of different stones and stone objects during the 3rd and 2nd millennia BC. This suggested regional differences in the level of 'integration' into these flows, which hints, in turn, at varying, cross-cutting or alternative systems of value. This approach, teasing out the detail of material flows in order to compare the social systems of different regions will be continued to this next chapter. Here, we will instead focus on metals: their raw sources, objects made from metals and indirect (or proxy) data about their circulation. Metals have a symbolic role for prehistorians, of course, in that they have provided the definition for an era: for most areas of the Near East, the 3rd and 2nd millennia BC form the major part of what has become known as the 'Bronze Age'. Whilst there is often a danger that this symbolic role may distort research agendas, it seems clear that metals nonetheless played an extremely important role across the Near East during this period, functioning both as the flexible basis for toolkits designed primarily to manipulate the environment (axes, adzes for agriculture) or coerce other people (weaponry such as daggers); but perhaps more importantly providing a medium for symbolic value for literally 'flashy' performance of status (as jewellery and elaborate vessels) and, increasingly in many places during the 3rd millennium, as abstract mediator for exchange (as currency in standardized ingots, weights or values given in surviving texts). All of these roles rely on the unique materiality of metals – its liquidity and malleability – and some of them also rely on metals' more-or-less unique 'aesthetic' qualities (although some of these qualities are possible to mimic). This chapter will approach the reconstruction of the flow of metals by considering the geographical location of its raw sources, the constitution of metal assemblages in our case-study regions and the distribution of certain significant metal objects; by analysing the overall consumption of different types of metals and the differential deposition of metal objects (in graves and hoards); and by comparing a set of indirect or proxy evidence for the circulation of metals (including weighing systems and cross-craft indicators or skeuomorphs).

5.2 Materials, geological sources and analyses of metal provenance

The distribution of mineral resources and their likely exploitation during the period of interest provides an obvious starting point for an examination of the flow of metals. Considerable literature and research effort have been devoted to the identification of potential raw material sources (*e.g.* Artioli *et al.* 2005; Alimov *et al.* 1998; Bayburtoğlu and Yıldırım 2008; Berthoud *et al.* 1982; Boroffka *et al.* 2002; Brovender 2009; Charles 1985; Craddock 1980; de Jesus 1978, 1980; Helwing 2005; Momenzadeh and Sadighi 1989; Muhly 1973, 1985, 1993; Stöllner *et al.* 2004; Stöllner 2005; Wagner, Öztunalı and Eibner 1989; Wagner and Öztunalı 2000; Wertime 1973; Yener 1986; Yener and Vandiver 1993a), extensive programmes of research involving chemical and isotopic analysis of metal objects to attempt to identify the source of their constituent materials (*e.g.* Agrawal 2000; Artioli *et al.* 2005; Begemann *et al.* 2008; Begemann and Schmitt-Strecker 2009; Berthoud *et al.* 1982; Chegini *et al.* 2000, 2004; Courcier 2007; Frame 2010; Gale and Stos-Gale 1999; Gillis *et al.* 2003; Hauptmann and Weisgerber 1980; Hauptmann, Rehren and Schmitt-Strecker 2003; Helwing 2005; Kaniuth 2006; Vatandoust, Parzinger and Helwing 2011; Weeks 2003; Webb *et al.* 2006), and more synthetic reconstructions of regional patterns and the routes that such raw materials took to reach metal consumers (*e.g.* Potts 1994, 143-176; Kroll 2002; Muhly 1993, 1999, 2005; Pigott 1999a; Roberts, Thornton and Pigott 2009; Stech and Pigott 1986).

Our discussions of metal sources and their routes of movements remain necessarily partial and open to reinterpretation, however (Palmieri, Sertok and Chernykh 1993). Four main factors make the investigation of the movement of metals through comparison between geological sources and metal objects difficult, outlined clearly by Chernykh (1992). First, whilst there has been considerable investigation into ancient metal ore sources, there are many gaps in our knowledge. Modern geological research and mining activities may reveal the occurrence of regions rich in particular ores, but given the nature of mining as extractive excavation, positive evidence for ancient exploitation of particular resources in particular periods is extremely difficult to find. For example, whilst the distribution of certain metal types and objects appears to favour the Caucasus as a major exporter of copper alloys to the steppes to the north in the Early Bronze Age, there is paradoxically little or no direct evidence of mining or smelting from the region itself in this period (Chernykh 1992, 59, 276). Smaller sources may be especially difficult to identify, since the minerals may have been exhausted in the past. Added to this is the fact that geological investigations tend to be conducted on a country-by-country basis, and it is often difficult to pull the diverse data together. Secondly, the techniques of characterization of metals and ores, which might allow the identification of raw materials for given objects, are complicated by the geological indeterminacy of mineral sources (and our incomplete knowledge of them) and the effects of metallurgical processing. The transformation of ores to workable metals often results in fractionation of the impurities that would otherwise identify a particular ore source. Thirdly, metals were continually recycled in antiquity, as old, stolen or broken metals were melted down and recast into new shapes, often resulting in mixing of metals from diverse sources and obscuring any identifying signatures[53]. Finally, the main reasons for this recycling, metal's high value and universal exchange functions, also have consequences for its archaeological visibility: in comparison to their likely usage and circulation, metal objects were deposited intentionally only under special circumstances, and less likely to be casually lost or thrown out than some other materials (unlike broken pottery, broken metal is re-used). The corpus of metal objects in the archaeological record is thus only a fraction of the amount that must have been circulating in antiquity, and substantially less representative than

53 It is fascinating to speculate as to the percentage of metals in circulation today that have been continuously remelted for 6000 years!

certain other materials. Despite this, very large databases of objects and analyses can yield useful results in different fields (see Section 5.4).

Whilst modern geological surveys can identify areas of high concentrations of certain minerals (or at least mineralogical zones), and hence offer general ideas about potential sources, unequivocal evidence of exploitation of *particular resources* at *particular periods* is extremely difficult to find. Since mining is an extractive process, resources mined in antiquity may have been exhausted, and therefore remain unrecognized. In Anatolia, where the Turkish MTA (the state geological and prospection service) has investigated and published detailed data on the mineralogy of the region and found evidence of 'old workings', it remains difficult to assign a date (*cf.* Wagner, Öztunalı and Eibner 1989; Wagner and Öztunalı 2000). Based on material remains it has been suggested that Classical, Byzantine and Ottoman periods witnessed substantial mineralogical exploitation, often still remembered in place-names[54], but arguably this reflects a general bias in the material record as a whole. In the Caucasus and especially Central Asia, the political fragmentation of territory – one of the outcomes of the European imperialist 'Great Game' (Meyer and Brysac 1999) – has resulted in fragmented geological knowledge, and few cross-border syntheses. In both regions certain 'headline' sources are cited repeatedly in general literature: for example the Ergani copper and silver sources near Diyarbakır; the copper sources on Cyprus[55]; or the much cited but little investigated sources of tin in Afghanistan and Central Asia.

5.2.1 Social and ritual aspects of mining

The broadly positivist outlook of archaeometallurgical research makes it difficult for newly identified potential sources to enter the canon, and it can be difficult to make sense of the polarized accounts offered by different researchers. For example, the suggestion that tin may have been extracted in the western Taurus at Kestel (Yener and Vandiver 1993a; Yener 2000) during the Early Bronze Age has caused considerable controversy (see discussion below, Section 5.2.3).Whether or not one believes that the particular mining and metallurgical evidence from Kestel/Göltepe proves the extraction of tin in Anatolia at this time, the very possibility should open our minds to the idea that multiple small sources might have been exploited at times (perhaps coming in and out of use), and not just the large sources of ore cited so frequently. Deposits which are economically unviable in the modern day may have been otherwise in the past. Similarly, mining is dependent on *particular* (and therefore culturally-bound) knowledge about earth-based resources, as conceptualized and integrated within the various social spheres through which the materials flowed. Ancient and modern commercial exchange of minerals relied, and still relies upon today, pre-defined and sometimes conservative distribution structures. In order to procure metal objects, one has to go to a metal specialist (the smith), who in turn will procure regularly from a set of sources known to this specialist. In a commercial setting the relationship may sometimes be impersonal and choice of source dependent on market prices, but often long-term relationships between dealers (which in small-scale communities, may

54 For example, in Turkish place names: *gümüş* – silver, *maden* – mine or mineral, *altın* – gold, *bakır* – copper, *kalay* – tin.

55 The name of the island providing (or being provided by) the word for copper in ancient Latin.

intersect with kinship, friendship or other obligation networks), or requirements for ritual/symbolic qualities of the material may play a more important role than mere price in the selection of material.

By way of illsitration, it is very likely that colour and visual properties (*e.g.* lustre) played an important role in ancient prospection, as they did in the production of finished objects. Given the various 'symbolic' values that colours may take on under different situations[56], it is interesting to consider whether choices of ores and production of coloured artefacts were not just functional or abstractly 'aesthetic' but also 'ritual-symbolic' (*cf.* Killick 2009). The earliest usage of copper objects in the form of coloured beads (Muhly 1989), contextualizes copper as merely one 'precious' stone amongst many (such as carnelian, lapis, quartz, obsidian, *cf.* Section 4.3), albeit each must have its own characteristics and associations (A. Sherratt 1976). Technical advances along the road to metallurgy and metalworking must have had cultural-aesthetic roots. For example, Pigott (1999a) has argued that the earliest evidence for copper working in Iran may relate to the 'attractive' colour and naturally malleable properties of native copper, and that the subsequent development of smelting and arsenic-bronzes may have been an accidental discovery due to the colour similarity of certain copper-arsenide ores (algodonite and domeykite), which are indistinguishable from native copper.

The way in which certain ores were recovered may also have differed in the past. Whilst we tend to picture mining as the direct extraction of rock from the ground, 'panning' may have been a more common technique to collect many different types of metals and precious stones. Many minerals are 'worked out' of rock-based deposits by the action of water (rain, rivers or glaciers), and it is not difficult to imagine the symbolic and mythological associations that might be created through this association as water feeds both biological and socio-economic life: our modern divisions between organic and inorganic life may be anachronistic to ancient world-views.

If the above paragraphs appear to over-emphasize the ritual and the symbolic, then it is as a corrective against the most common narratives about ancient raw metal sources which tend to situate activities of ore procurement and metallurgy within a field of de-contextualized or de-ritualized technology (*cf.* similar critique in Budd and Taylor 1995; Rowlands and Warnier 1993). In this orthodox perspective, issues of functional advantage and Lamarckian evolutionary progression come to the fore, at the expense of cultural context and the unpredictable messiness of the material record. The use of certain alloys of copper (especially tin-bronze) or iron have been seen as a line along a 'normal' trajectory of technical progress. Though there are definite patterns of technology, this approach is barren in that ultimately it reduces differences of adoption of certain technologies to how 'advanced' a particular culture is, rather than attempting to analyse the social motive behind the invention, adoption or rejection of technical skills and knowledge (*cf.* Rahmstorf 2011). In response, the technico-positivist perspective can argue that a cultural approach sometimes ignores the specificities of technical process and real 'advantages' of some materials and techniques over others. Integrating the technical and symbolic aspects of metallurgy thus remains difficult, but we must

56 As illustrated by 'imperial' purple of Rome, or the Russian linguistic association between the word for red, красный, and beauty, hence Red Square is, in fact, the 'beautiful' square.

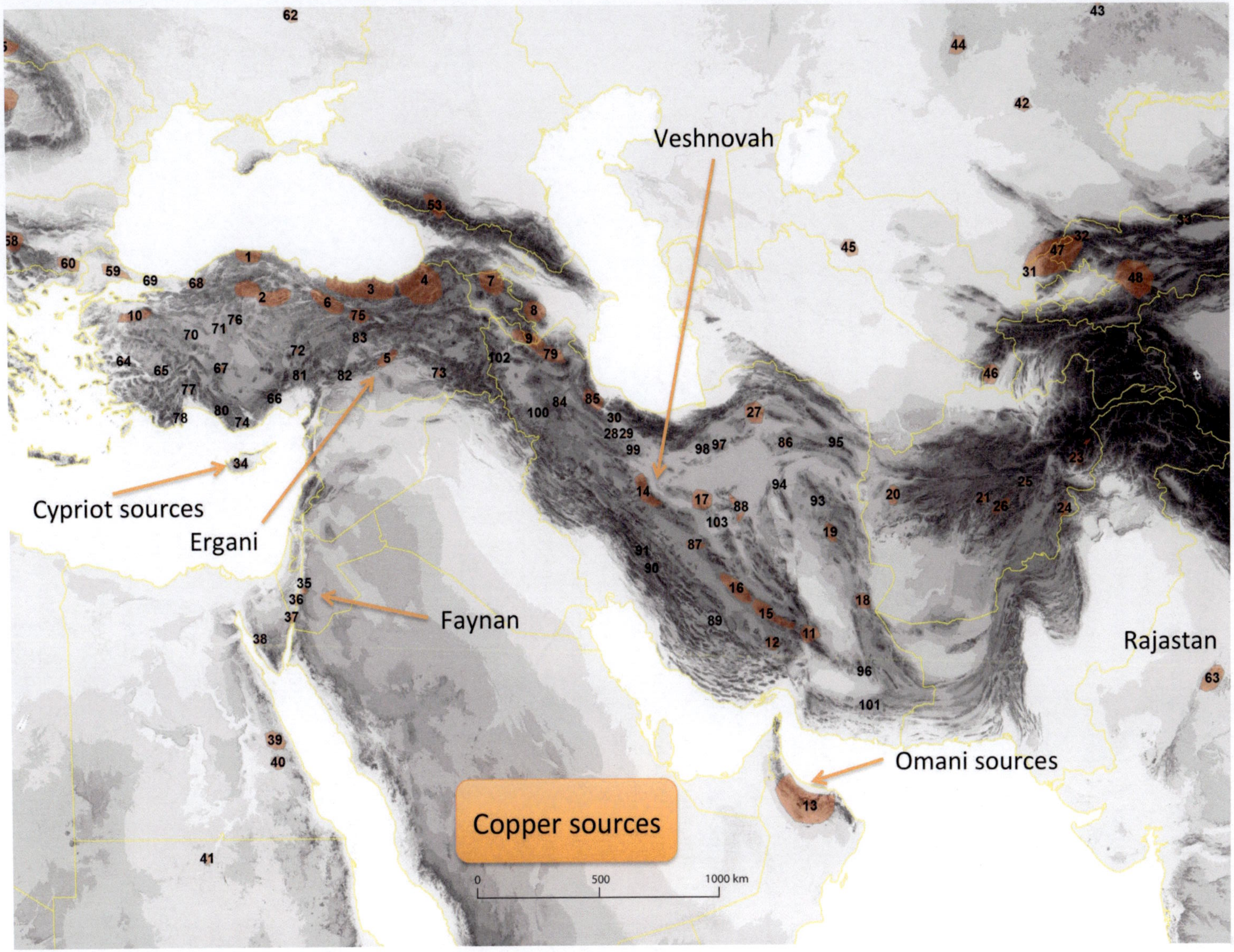

Figure 5.1. Sources of copper ore, including Ergani, Veshnoveh, Omani, Cypriot and Rajasthani regions. See Appendix D.1.2 for key to numbers.

constantly keep them both in mind if we are to have any chance of accurately reconstructing and explaining long-term changes in the flow of metals.

Whilst textual records from Mesopotamia do sometimes provide hints as to the dominant sources in the third and second millennia (Muhly 1973; Moorey 1994, 245-246), more often it is the combination of modern geological knowledge, a very sparse scatter of archaeological clues, and an impressionistic sense of the metallurgical analyses of objects and sources which must be relied upon to hint at the complex dynamics of source exploitation (*cf.* Potts 1994, 145-153; Sherratt 2007).

5.2.2 Sources of copper

Copper is perhaps the most important metal for the study of early metallurgy, because of its widespread use for all types of objects starting from the Chalcolithic period onwards. Modern geological reports suggest that potential sources of copper are and were widely distributed across the Old World (Wagner and Öztunalı 2000, 31), but the actual ore forms are diverse (*e.g.* native copper, malachite, azurite and cuprite), and not all ores may have been recognisable to ancient metallurgists at particular times. Copper ores, though widespread, are by no means evenly spread. A summary map of the distribution of known copper-bearing regions, based on a synthesis of the current archaeological literature on the topic, shows this uneven distribution (Figure 5.1). Certain regions include copper ores with significant impurities of other metals, such as antimony, arsenic, nickel and lead; the impurities were sometimes part of the ore's attraction, even when the additional metal was not known as a separate material in antiquity.

In Anatolia and Transcaucasia, a number of attempts have been made to identify copper sources which were exploited in the prehistoric era by projection back from modern data (*e.g.* see maps in Korfmann 1982, 136-137; Yakar 1985; de Jesus 1978). More systematic and extensive surveys (Wagner and Öztunalı 2000; Palmieri, Sertok and Chernykh 1993), which have attempted to look directly for archaeological evidence, often show large-scale Roman, Byzantine and Ottoman exploitation in many areas, but found it harder to confirm prehistoric exploitation categorically. Hence, while there are known deposits of copper from which it seems highly probable that most ancient copper came (*e.g.* the Ergani sources in eastern Anatolia), it is also possible that smaller deposits outside these regions were known and used periodically. The main deposits include those in the West Anatolian highlands, along the eastern Black Sea, around the 'Hittite' highlands along the Kızılırmak river, in the hills between Malatya and Diyarbakır (especially Ergani) and in areas of modern Georgia, Armenia and Azerbaijan. Problematically, however, direct evidence for copper ore exploitation is rather difficult to find, especially before the Late Bronze Age (Palmieri, Sertok and Chernykh 1993; Chernykh 1992, 276). The Ergani deposits show little positive evidence of exploitation, for example, but the extent of subsequent mining means that any evidence could have been destroyed, so this dearth is not particularly surprising.

Intermediate to our two main regions of interest, the Iranian Plateau contains a large number of copper sources of varying qualities and quantities. The most oft-discussed sources are those of Anarak (Talmessi/Meskani) and Veshnoveh in central Iran (Pigott 1999a; Stöllner *et al.* 2004; Stöllner 2005). Partly because of the extensive evidence for early Chalcolithic metallurgy in these regions and

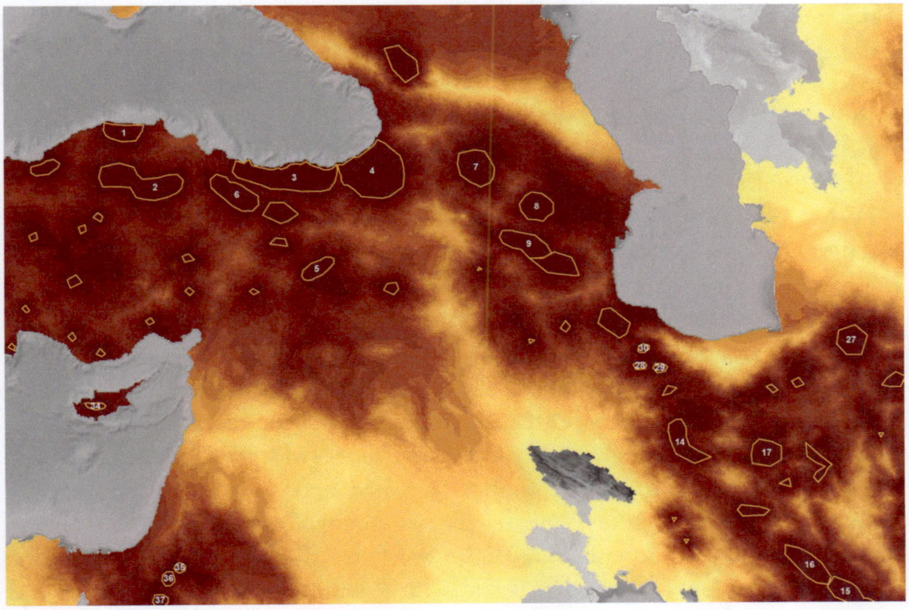

Figure 5.2. Archaeotopogram 'type A2' showing relative distance from copper ore sources around Eastern Anatolia. Darker colour indicates zones which are relatively closer to copper sources.

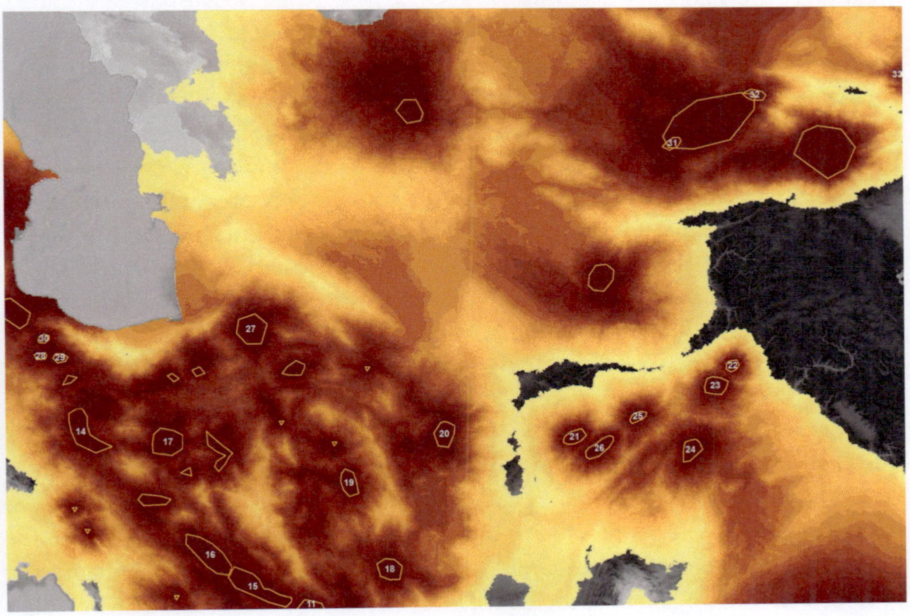

Figure 5.3. Archaeotopogram 'type A2' showing relative distance from copper ore sources around western Central Asia. Darker colour indicates zones which are relatively closer to copper sources.

partly because the arsenic content of some of the deposits accords with the dominant composition of copper-alloys in the Early Bronze Age corpus, many have considered these regions to have formed one of the prime sources for early Mesopotamian copper-working (see Stech and Pigott 1986; Moorey 1994, 247; *cf.* recently Matthews and Fazeli 2004). Again direct datable evidence of exploitation is often elusive, however. A ceramic vessel of Sialk IV type at 'Chale Gahr' has been taken as suggestive evidence of exploitation in the Veshnoveh region datable to the early 4th millennium, for example (Pigott 1999a, 78; Holzer and Momenzadeh 1971, 7), but this is only a single vessel. A more promising confirmation of the region's metal extracting importance comes from the nearby metalworking site of Arisman (Chegini *et al.* 2000, 2004; Pernicka 2004; Helwing 2005; Vatandoust, Parzinger and Helwing 2011). There are numerous other large copper sources across the plateau, and in neighbouring regions of Afghanistan, highlighted by Wertime (Wertime 1973), Berthoud (Berthoud *et al.* 1980, 1982; Cleuziou and Berthoud 1982) and Weisgerber (Weisgerber 1990; Weisgerber *et al.* 1990). As with Anatolia, it is difficult to assess the extent to which different sources played a part, and indeed how comprehensive our knowledge of the material sources is in the first place[57].

Whilst recent work in Iran has considerably updated our knowledge of the metallurgy of certain areas, the most extensive investigations into Afghan sources of ancient copper, namely the French expeditions of Berthoud and team (see Berthoud *et al.* 1982), are now 30 years old. In south-western Central Asia, to the north-east of the Iranian plateau, the landscape is poor in minerals. In the traditionally identified heartland of Namazga cultures, for example, particularly in the areas close to the Kopet Dag, very few copper sources have been recorded on either the Iranian or Turkmen side of the borders. The copper from metal objects found in this region must have come from Iran or Afghanistan, or further to the north-east, in Bactria, the Zerafshan or Ferghana. These regions, fragmented by the modern states Uzbekistan, Tajikistan, Kyrgyzstan, northern Afghanistan, Kazakhstan and the Chinese region of Xinjiang, do contain a wide variety of mineral sources including copper ores (Chernykh 1992, 6, 179; see also Rubinstein and Barsky 2002 for a summary of mineral resources in the former the USSR republics). Some additional more distant regions bearing copper should be mentioned with regards to 3rd and 2nd millennium copper circulation, namely: Egypt, the Levant, Cyprus and Oman (Moorey 1994, 245-248), and also the sources of south Asia. Investigations in Oman appear to have confirmed the importance of this region for the early extraction of copper on a large scale for Mesopotamia (Potts 1994, 149-15; Weeks 2003), and support the identification of the Persian Gulf region with '*Magan*' of Mesopotamian texts. The Levantine and Cypriot sources may not have contributed to eastern Anatolian and western Central Asian consumers during the 3rd millennium (Muhly, Maddin and Karageorghis 1982), but it seems likely that they were supplying to local and various eastern Mediterranean communities. In the Levant, copper sources around Feinan and Wadi Arabah may have been important (Moorey 1994, 247; Levy *et al.* 2002), both locally and 'internationally'. Sources of copper in Egypt (for example in the Sinai and Eastern deserts) may also have supplied Egyptian or more distant consumers. The sources of copper used by the peoples of the Indus civilization and their neighbours in the Indian subcontinent remain unclear – there are extensive deposits in Afghanistan, as has already been mentioned, but also to the east in Rajasthan. These could have been major sources, though the evidence remains moot (Hoffman and Miller 2009; Agrawal and Seshadri 1998; Agrawal 2000; Kenoyer and Miller 1999). Omani sources may offer a more likely alternative source (Weeks 2003). Copper may have circulated in the form of standardized ingots by the mid 3rd millennium – documented in the eastern Mediterranean, for example, by the 'Levantine' ingot from Poros, Crete (Doonan, Day and Dimopoulou-Rethemiotaki 2007, 105-106, fig. 6.2). It seems likely then that ingots (perhaps of different types) were also circulating elsewhere, but whether it was on the bulk scale hinted at for the late

57 This is partly a result of the fragile political situation in the region, but there also has been a paradigmatic shift away from certain types of mineral investigations in archaeology generally, as the scientific techniques of provenance analysis have proven less conclusive than initially hoped (see comments in S. Sherratt 2007).

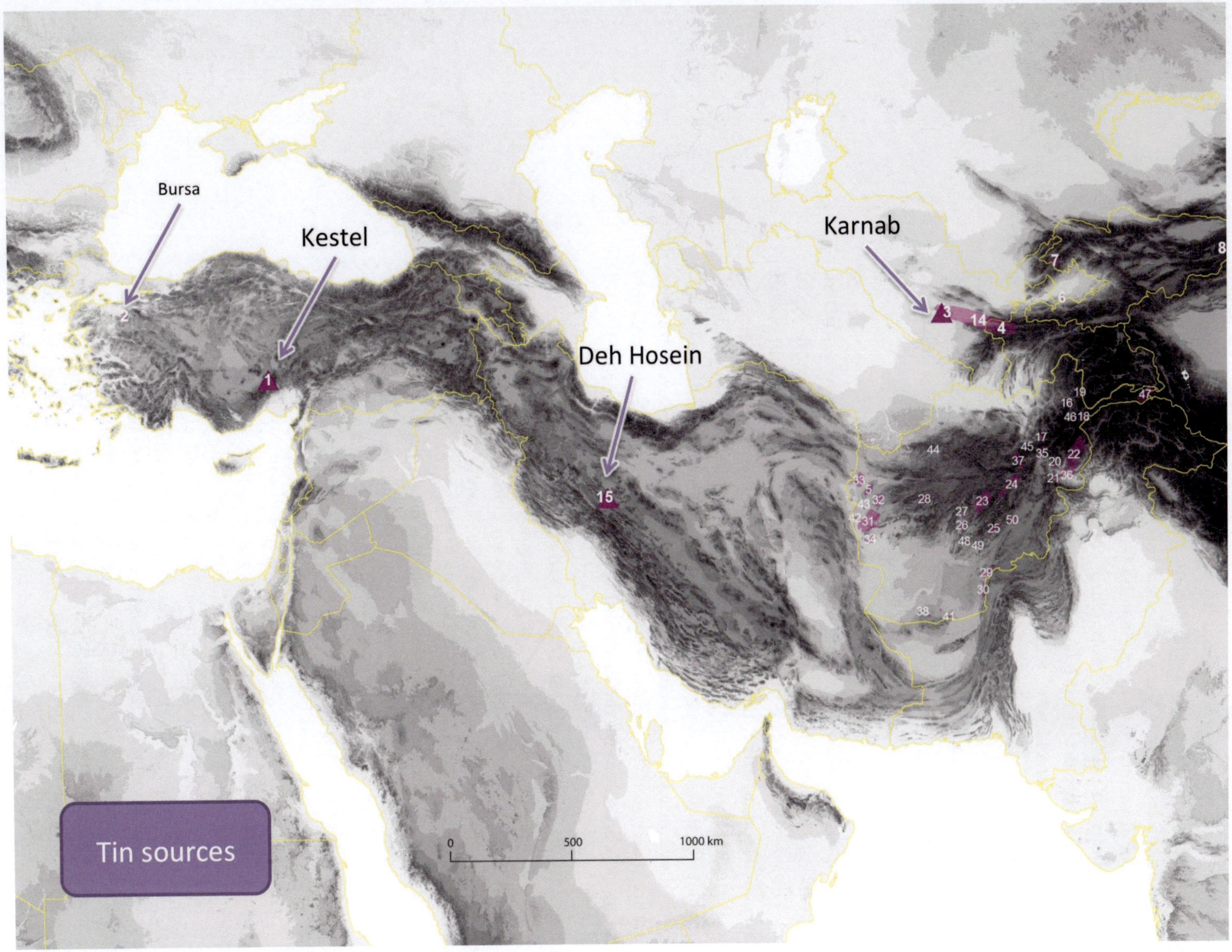

Figure 5.4. Sources of tin, including Deh Hosein, Kestel and Karnab regions. See Appendix D.1.2 for key to numbers.

14th century BC by the ingot-laden wreck at Uluburun (Pulak 1998; Sherratt 2000) is unclear.

The relative accessibility of copper sources to the surrounding regions is shown by the archaeotopograms in Figures 5.2 and 5.3. It is clear that highland regions are (unsurprisingly) disproportionately better resourced in copper than the lowland plains where urban settlements grew exponentially from the 4th millennium onwards. The overall pattern in eastern Anatolia illustrates the relative 'closeness' of Anatolian sources to southern Mesopotamia compared to, say Levantine or Iranian sources. This may explain the faster integration of Assyria into the Uruk cultural sphere since contacts were relatively 'cheaper' than those to the west and east. The map of western Central Asia interestingly shows the Kopet Dag piedmont zone in a kind of accessibility 'shadow' – the comparatively late interest in metals (compared to the metallurgical innovations of 'nearby' north-east Iran, *e.g.* Tepe Hissar), may find part of its explanation in this longer relative distance.

5.2.3 Sources of tin

In contrast to copper, the distribution of tin-bearing ores is much more restricted, and so far it has proven even more difficult to find ways to characterize the tin in copper-alloy objects. Recently, isotopic analysis of tin has been proposed as a possible way to characterize alternative tin sources and objects containing tin (Haustein, Gillis and Pernicka 2010). The distribution of the few earliest tin-bronze objects offers little help for understanding the process of early experimentation with tin-based copper alloys, other than it being more widespread than was earlier thought. The distribution of viable tin sources is no less clear (Figure 5.4). A considerable amount of ink has been spilt on the likely sources of tin exploited during the 3rd and 2nd millennia, and the current state of knowledge still remains confusing and contentious (see Weeks 2003, for detailed and measured consideration of the issues). What follows is a brief summary of the issues.

For some time it has been clear that the textual evidence for the late third and early second millennium strongly supports an 'eastern' origin for the tin used in Mesopotamia during this period (Muhly 1973; *cf.* Moorey 1994, 298). This evidence led researchers to focus their attention on potential sources to the east of Mesopotamia and reconstruct trade routes stretching from Afghanistan to Anatolia or the Aegean (Stech and Pigott 1986). Afghanistan was frequently cited as a major source of tin, following the geological investigations of Berthoud and his colleagues (see *e.g.* Cleuziou and Berthoud 1982). Certainly the region is also rich in alluvial gold and the country contains one of the most likely sources of Near Eastern lapis lazuli – the aura and physical properties of gold and lapis are strongly interrelated in the Mesopotamian imagination, if their frequent combination in statuary and personal ornaments are any kind of indication (see Section 4.3.1). The same exchange networks could have carried gold and tin. All three materials may well have been panned rather than mined[58], hence there may have been fairly little technological difference in their initial prospection and exploitation (Stech and Pigott 1986, 45-46). The association of tin in the form of river-borne cassiterite or stannite pebbles with these other valuable materials may well have been enough to facilitate its early incorporation into alchemical

58 It is easy to imagine the mythological, magical and explanatory narratives which could build up around the appearance of these stones from the water.

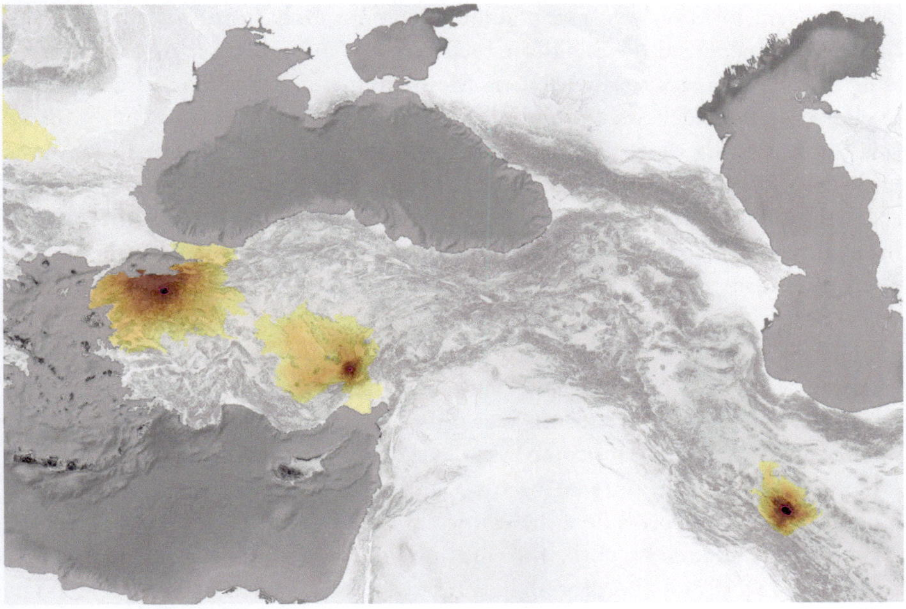

Figure 5.5. Archaeotopogram 'type A2' showing relative distance from tin sources in and around greater Anatolia. Darker colour indicates zones which are relatively closer to tin sources.

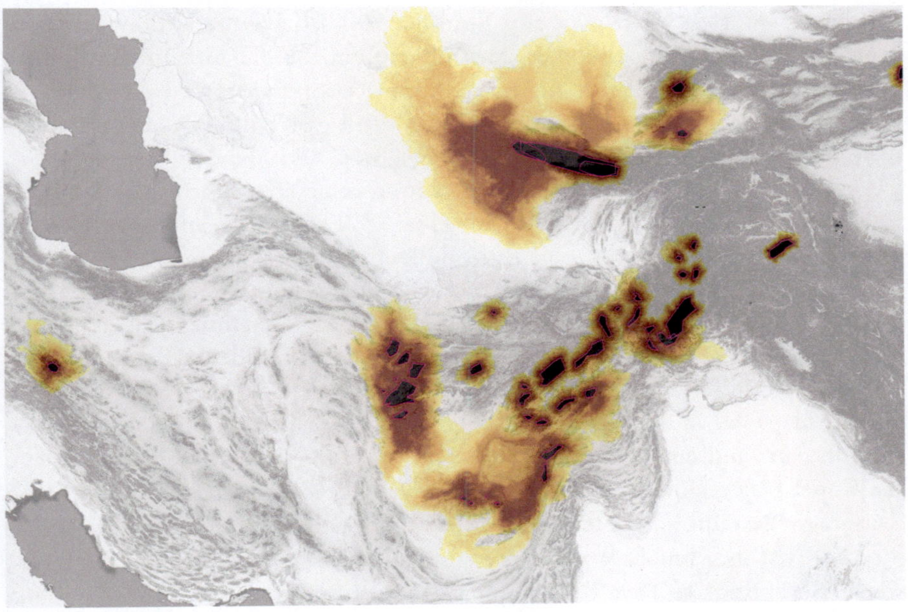

Figure 5.6. Archaeotopogram 'type A2' showing relative distance from tin sources in and around Central Asia. Darker colour indicates zones which are relatively closer to tin sources.

metallurgical experiments. The evidence of early examples of tin-bronze from the mid-4th millennium at the site of Mundigak in southern Afghanistan (Shaffer 1978, 144; Moorey 1982, 99; 1994, 299) suggests either that tin-bearing copper ores were sought, or that tin was already being deliberately added to copper in Afghanistan.

The consensus was broken following the publication of evidence from Kestel/Göltepe. After gold and tin were recognized in several streams near the Bolkardağ hills in the central Taurus, archaeologists identified a tin-bearing mining area with an associated Early Bronze Age metalworking settlement (Yener 2000, 71-76) in which tin was apparently processed in crucibles (Yener and Vandiver 1993a). The suggestion that tin may have been mined from a region nearby to which tin-bronzes seem to have been manufactured from an early stage (*i.e.* Anatolia and the Amuq) challenged the established narratives of long-distance tin exchange, in which Afghan tin was transported via Mesopotamia (*e.g.* Stech and Pigott 1986) or else along the Black Sea and across the Caucasus to Anatolia (*e.g.* Muhly 1999). It is not surprising then, that the Kestel evidence was initially viewed with some scepticism. Whilst some were prepared to accept the idea that tin was being mined at Kestel and then processed at Göltepe, others objected that the interpretation of the evidence seemed implausible, and that it was more likely that other metals (*e.g.* gold) were being procured and processed (Muhly *et al.* 1991; Hall and Steadman 1991).

The main positive result of this controversy was a reinjection of uncertainty, enabling the creation of more imaginative or more complex accounts of ancient tin procurement and distribution. We might now suggest, for example, that many smaller tin sources may have come in and out of use (many of them now unknown, exhausted, or currently unrecognized because of their low 'economic' potential to modern eyes). Known tin deposits whose exploitation has been suggested for the Bronze Age include those of the Bursa region (Yakar 1984, 80-81), Egypt[59] (Rothe and Rapp 1995) and the central Balkans (McGeehan-Liritzis 1987, 291). The recently investigated mines of Karnab, *etc.* in the Zerafshan region in Uzbekistan (Alimov *et al.* 1998; Parzinger and Boroffka 2003), have revealed evidence for exploitation at least by the later 2nd millennium, in the form of Andronovo pottery (Parzinger 2002). Similarly, there now also appears to be evidence of tin-exploitation in western Iran, along the north-east boundary of Luristan, a region somewhat closer to Mesopotamia (Pigott 2004; Fleming *et al.* 2005; *cf.* Thornton 2009, 317). Tin-bronze objects from cemeteries near the Deh Hosein tin deposits appear to match, both in terms of lead isotope and trace element analysis, objects of tin-bronze of third millennium Mesopotamia, Luristan and the Aegean (Nezafati, Pernicka and Momenzadeh 2006). Though suggestive of a possible source near to the earliest evidence for Mesopotamian tin, the dating of the archaeological material is controversial, and the evidence for the start of mining unclear.

The maps shown in Figures 5.5 and 5.6 show the relative distance of tin sources from the surrounding regions, and shows that both highland eastern Anatolia and the Kopet Dag region of western Central Asia had no easy access to tin sources. This might account for the low level of tin adoption until very late in both regions – except that the same lack of access is also true for Mesopotamia and the Indus, where tin appears to have been popular from at least the mid-3rd millennium BC. The sources were probably not all used at the same time, however. A likely scenario – though one which is currently difficult to confirm – is as follows: given the distribution of the earliest tin bronzes, early experimentation with tin was based on mining from various pockets like those at Kestel, Deh Hosein or in Afghanistan. Certain larger sources – perhaps Deh Hosein – may have became the dominant suppliers for Mesopotamia towards the middle of the 3rd millennium. In the later 3rd or early 2nd millennium, other sources like those of the Zerafshan or Afghanistan may have risen to supply increased demand, whilst in the later part of the 2nd millennium (as interaction across the Mediterranean accelerated), distant

59 The late adoption of tin-bronze in Egypt (from the 18th dynasty onwards, in the last centuries of the 2nd millennium) does not necessarily exclude its extraction and export at an earlier period, though there is currently little information that might be used to support such a scenario.

mines like those in the Balkans, Italy, Spain, the central European *Erzgebirge* and Cornwall may have begun to regularly contribute tin to the markets (*cf.* Moorey 1999, 300).

It is tempting to wonder whether too much significance has been given to the 'tin problem', which is only one amongst various other potential questions to be addressed concerning the sourcing of metals[60]. Tin is only one of the materials which was added to copper: there has been no equivalent 'arsenic debate' despite the fact that arsenic was the most common alloyed addition during the 3rd and 2nd millennia BC. This may in part be a consequence of the fact that copper-arsenic is not considered 'true' bronze (and there remains a latent tendency to focus on the technical qualities of bronze within an evolutionary scheme of the Three Ages 'Bronze Age'); but also because of the apparently 'exotic' sourcing of tin and explicit mention in the ancient texts (*e.g.* the Old Assyrian Kültepe tablets) that have added to its importance in the context of archaeological research narratives of long-distance trade and interaction.

5.2.4 Other metal sources: gold, silver, lead and iron

Overall the exploitation of sources of metals other than copper and its alloys has been less intensively investigated by archaeologists and archaeometallurgists. Scientific techniques have been unable to match certain materials to actual ore sources, and destructive techniques are less likely to be approved on precious items than on copper-based ones. In ancient times gold was probably panned from rivers rather than mined directly from the ground (Stech and Pigott 1986). Such alluvial gold is to be found in a number of regions in Anatolia (Bayburtoğlu and Yıldırım 2008), and in large amounts in Afghanistan (Figure 5.7). There are a number of current expeditions in the Caucasus which look set to considerably enhance our knowledge of gold acquisition in the region in the very near future (*e.g.* Kunze *et al.* 2013).

Silver and lead have been most successfully traced using the unique signatures from lead-isotope analysis. Both metals are found rather rarely in comparison to copper objects, however – in the case of silver this is no doubt due in part to its high value and role as currency (see Section 5.5.1), but also because it tends to suffer more severely from destructive taphonomic processes than gold or copper. As well as objects made from either material on its own, both lead and silver are known to have been alloyed with copper on occasions (Hauptmann and Palmieri 2000; Hauptmann, Rehren and Schmitt-Strecker 2003; Pernicka 2004). Silver is found (along with lead) in considerable amounts in different parts of the Aegean (Wagner, Gentner and Gropengiesser 1979; Gale and Stos-Gale 1981) and Anatolia (Yener 1986; Bayburtoğlu and Yıldırım 2008), Iran (Momenzadeh, Hajisoltan and Momenzadeh 2004) and parts of Afghanistan and the highlands of Bactria (Figure 5.8).

60 Stepping back from the details of the debates, one might also consider the effect of geopolitics upon archaeological narratives: many countries are keen to make heightened claims about their past importance. More specifically, archaeometallurgical work has depended on the access by researchers to the countries in question: much of the work in Iran stopped after 1979 (restarting more recently around 1995); and in Afghanistan after 1982 (and has really yet to restart); the minerals of Central Asia were generally difficult to research for western archaeologists until after 1993.

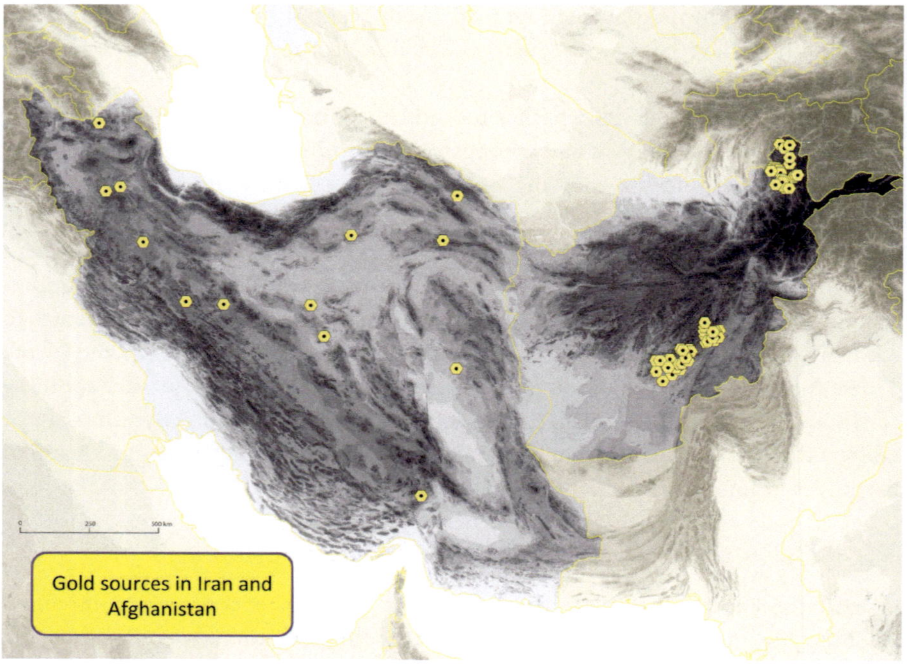

Figure 5.7. Regions with sources of gold in Iran and Afghanistan. (Data on gold sources based on Pigott 1999b, 82, 117; Momenzadeh, Hajisoltan and Momenzadeh 2004, 17).

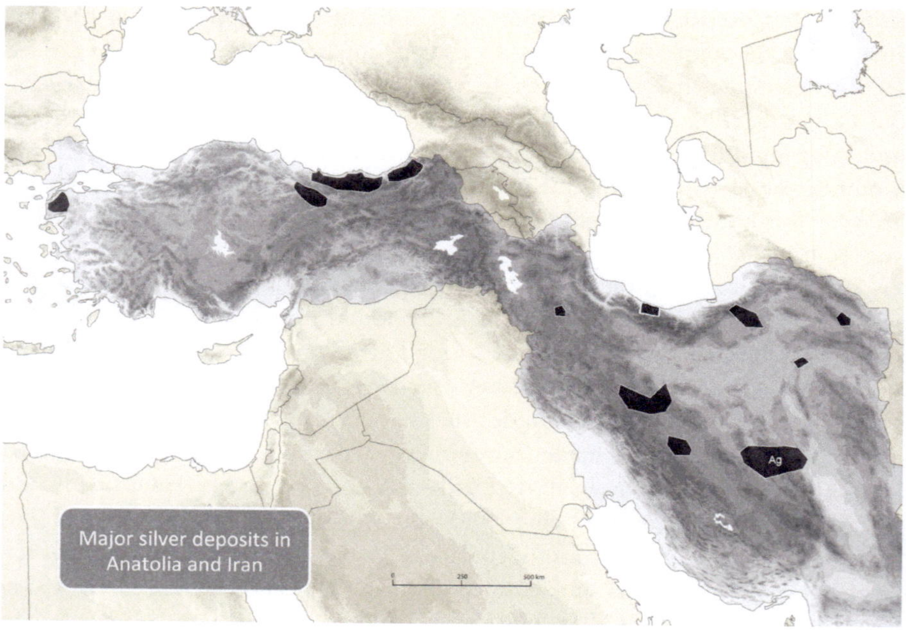

Figure 5.8. Regions with deposits of silver in Turkey and Iran. (Data based on Momenzadeh, Hajisoltan and Momenzadeh 2004; de Jesus 1978).

Analyses of iron objects have been limited because of the more extensive distribution of iron ores, and the assumption that the nearest sources would be exploited. Iron ores are more evenly spread across the earth's surface than copper (Charles 1985; Muhly *et al.* 1985; Thornton, Rehren and Pigott 2009; Waldbaum 1999), but as with copper, concentrations remain higher in the highland regions of Anatolia and Iran than in the lowlands of Mesopotamia, the Indus, or indeed, the piedmont of Turkmenistan (an incomplete distribution shown in Figure 5.9). The total number of iron objects found would in any case make it difficult to assert anything about the circulation of the metal. It is notable, however, that iron

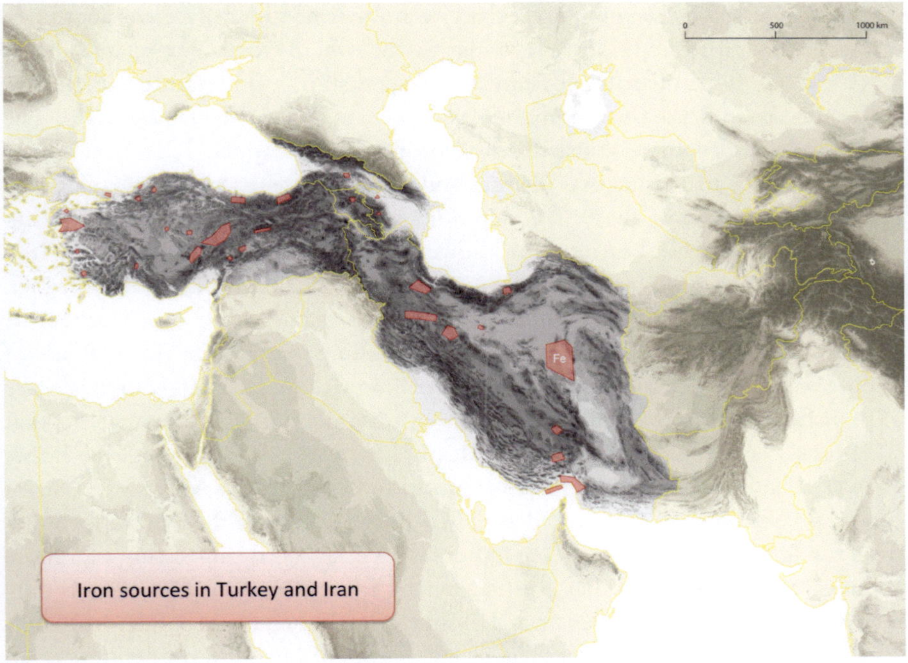

Figure 5.9. Regions with sources of iron in Turkey and Iran. (Data based on Pigott 1999a, 94).

speiss is often found at sites in Iran such as Tepe Hissar apparently in association with early silver-lead production (Thornton, Rehren and Pigott 2009).

5.2.5 Metal circulation based on geological and archaeometallurgical analyses

In 3rd and 2nd millennia BC Eurasia, only a limited range of metals seem to have been consumed widely, namely gold, silver, lead, copper and related alloys, including electrum, arsenic-bronze and tin-bronze. Iron comes into limited use in the 3rd millennium, before its more widespread adoption and application at the end of the 2nd and beginning of the 1st millennium BC (Pigott 1999a, 90-95). Of these, copper and its alloys, performed what is traditional thought of as the 'functional' aspects of metals: its relative strength and malleability facilitating the creation of tools and weaponry. But, as with gold and silver (and possibly iron), copper alloys were also used for 'high-value' items such as special drinking, eating and cooking vessels, body adornment (pins and jewellery), and a variety of other items including harness fittings, for which the 'strength' of the material was of fairly limited significance.

On the basis of the collected geological information alone, we cannot know for certain which of the mineral sources identified were known or exploited during our period of interest. Care should be taken with mapping these regions since, though some ore-areas may appear rather extensive on the map, their extent is not a direct reflection of the amount of metal which may have been extracted – indeed it is possible that small areas of intense and easy-to-access metals would be more heavily used in antiquity than large areas of sparse or difficult-to-extract ones. The direct evidence for mining activities in certain periods may confirm that a particular source was being exploited then, but it is difficult to assess the extent of the contribution of any one particular area to the overall circulation of metals. Analytical techniques, even if they cannot always identify the exact source of metals, therefore are important to identifying overall trends in circulation.

A few trends are identifiable in the archaeometallurgical literature, which have been taken to indicate the broad circulation of raw metals (see Figure 5.10). The most important of these is the distribution of arsenic-copper. There has been a strong tendency to assume that the preponderance of arsenic-bronze metal objects across western Asia in the 3rd millennium means that most copper was being sourced in the arsenic-bearing ores known from the Caucasus and western Iran (Chernykh 1992, 66; Pigott 1999a, 78-79)[61]. The question is the extent to which the arsenic content of copper-alloy objects was a result of the ores selected, or an intentional addition of arsenic. We might assume that at least the early use of arsenic was an accidental by-product of the arsenic-rich ores used (favoured because of their flexible material qualities). However, technological evidence suggests that the process of intentional alloying with pure or purified copper may be rather early (Day and Doonan 2007 passim; Thornton 2009; 2010). Analysis of Omani copper objects showed that as well as locally found copper (rich in nickel), copper from foreign sources was also being imported (Weeks 2003), suggesting a far more complex circulation than that painted by the Mesopotamian texts (which only refer to the import of copper from Dilmun – in the region of modern Oman). The origin of tin is still unclear. The recent studies of Luristani tin deposits (Deh Hosein) and nearby metal objects (Begemann *et al.* 2008; Begemann and Schmitt-Strecker 2009) remain inconclusive. A combined archaeotopogram (Figure 5.11) that highlights those areas within accessible reach of both copper and tin sources allows us to identify places in which early experimentation with tin-bronze were more likely: some of the earliest examples of tin-bronze objects from Mundigak, Afghanistan (Cleuziou and Berthoud 1982, 16) and Cilicia/Amuq (*e.g.* Seeden 1980, 7-10) may make more sense in this context.

5.3 Metal artefact types as evidence for metal flows

The distribution of metal artefact types was traditionally seen as providing the most direct form of evidence of interaction and movement of metals (*e.g.* Stronach 1957). A very brief summary of the assemblages of each region is presented here, supported by visual material, folowed by a discussion of a few artefact groups which are relevant to the study of interregional interaction.

5.3.1 Metal assemblages in Transcaucasia/eastern Anatolia, 3000-1500BC

The number of metal objects recovered from archaeological contexts of the 3rd and 2nd millennium in Transcaucasia/eastern Anatolia often seems rather small compared to surrounding regions and later periods (Yakar 1984, 78; Palmieri, Sertok and Chernykh 1993; Yakar 2002). This is despite the very early history of metallurgy in the upper Euphrates (*e.g.* early copper beads from Çayönü). Presumably this dearth is at least in part the result of limited excavations in some areas, but, equally, it also seems to be due to cultural practices that prevented the deposition of metal objects in the ground (to be discussed further below). In the

61 The levels of nickel have been taken as a marker of the 'south-east' Anatolian or Transcaucasian sources.

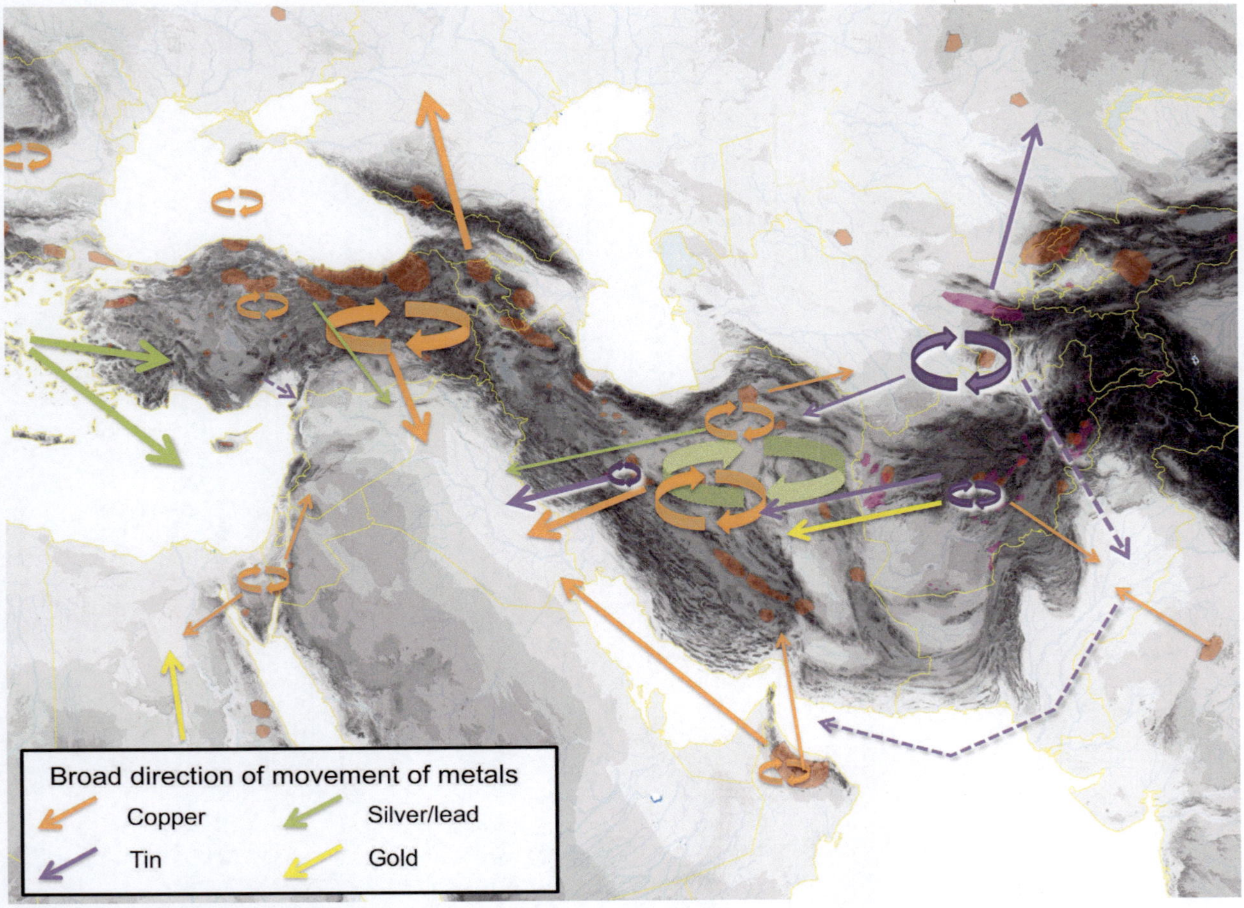

Figure 5.10. Summary of the broad direction of circulation of metals, in 3rd and 2nd millennium BC.

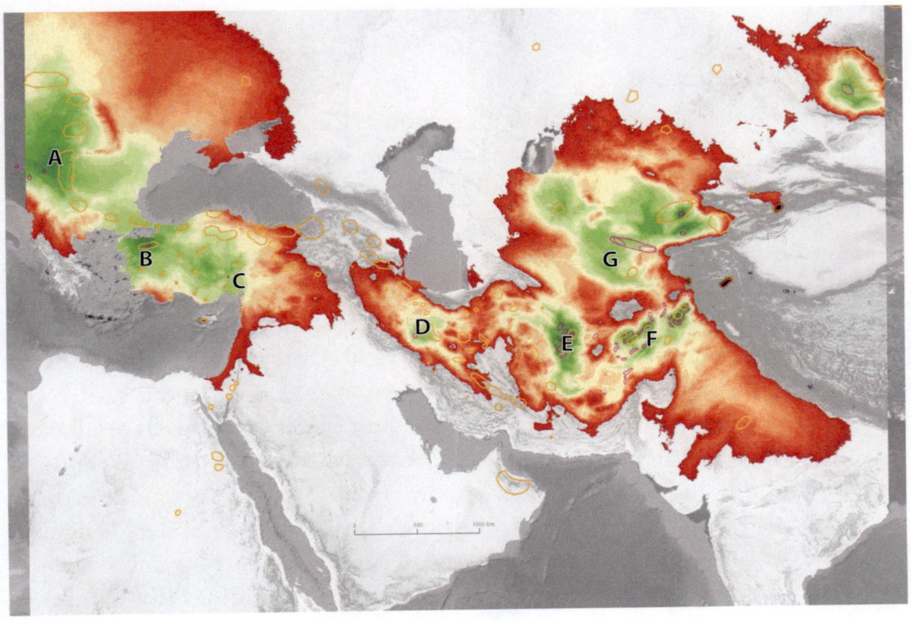

Figure 5.11. Prediction for centres of early tin-bronze experimentation – based on archaeotopogram (type B) showing sum of relative distance from copper and tin sources. In this archaeotopogram, green areas thus show regions with relatively easy access to both metals (A. Balkans; B. Marmara; C. Taurus and Cilicia; D. Luristan; E. west Afghanistan; F. east Afghanistan; G. Zerafshan and Ferghana).

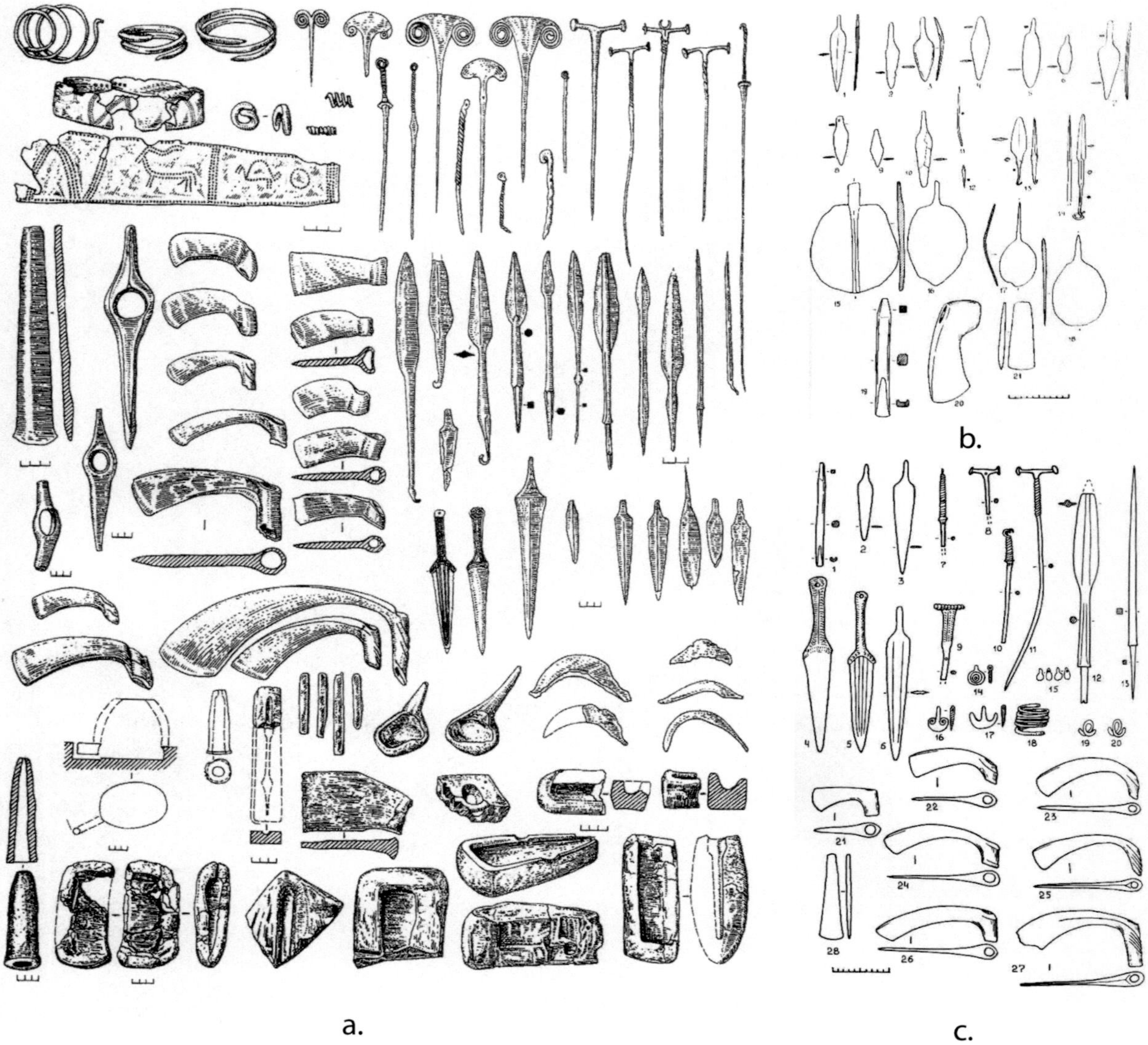

Early Bronze Age, a range of forms predominates (Figure 5.12). For example, in the EBAI/II levels at Karaz-Kahramanlar, there were "daggers, shaft-hole axes, chisels, pins, awls, some jewellery and at least one sickle" (Yakar 1984, 78; Esin 1969, 135). Some evidence for metal production and production techniques has also been uncovered at Kura-Arax/Early Transcaucasian sites: for example, crucibles at Sos Höyük (Hopkins 2003, 105). The 'royal tomb' of Arslantepe, dated to the beginning of the 3rd millennium, revealed an array of rich metal objects (Frangipane *et al.* 2001; Palmieri *et al.* 2002), including spearheads, dress pins, a diadem, earrings and bracelets, flat axes, and a dagger and a sword; and in a different part of the same site, contemporary evidence for metal production (Hauptmann and Palmieri 2000). More extensive data on early to mid-3rd millennium metallurgical traditions comes from sites to the north-east, in Transcaucasia (Kavtaradze 1999, 77; Chernykh 1992, 62). Forms include double-spiral-headed dress-pins, sickle blades, daggers, spearheads, earrings and shaft-hole axes. It is assumed that, given the similarity of ceramic traditions (the Kura-Arax

Figure 5.12. Examples of the contents of metal assemblages of eastern Anatolian sites from the 3rd millennium BC: earlier: (a) Kura-Arax (courtesy of Russian Academy of Sciences; cf. Kohl 2007, 92; Frangipane et al. 2001, 471-3); later: (b) Bedeni (kurgans) (Chernykh 1992, 105 fig. 33); (c) Sachkere (kurgan) (Chernykh 1992, 108 fig. 34). Not at uniform scale.

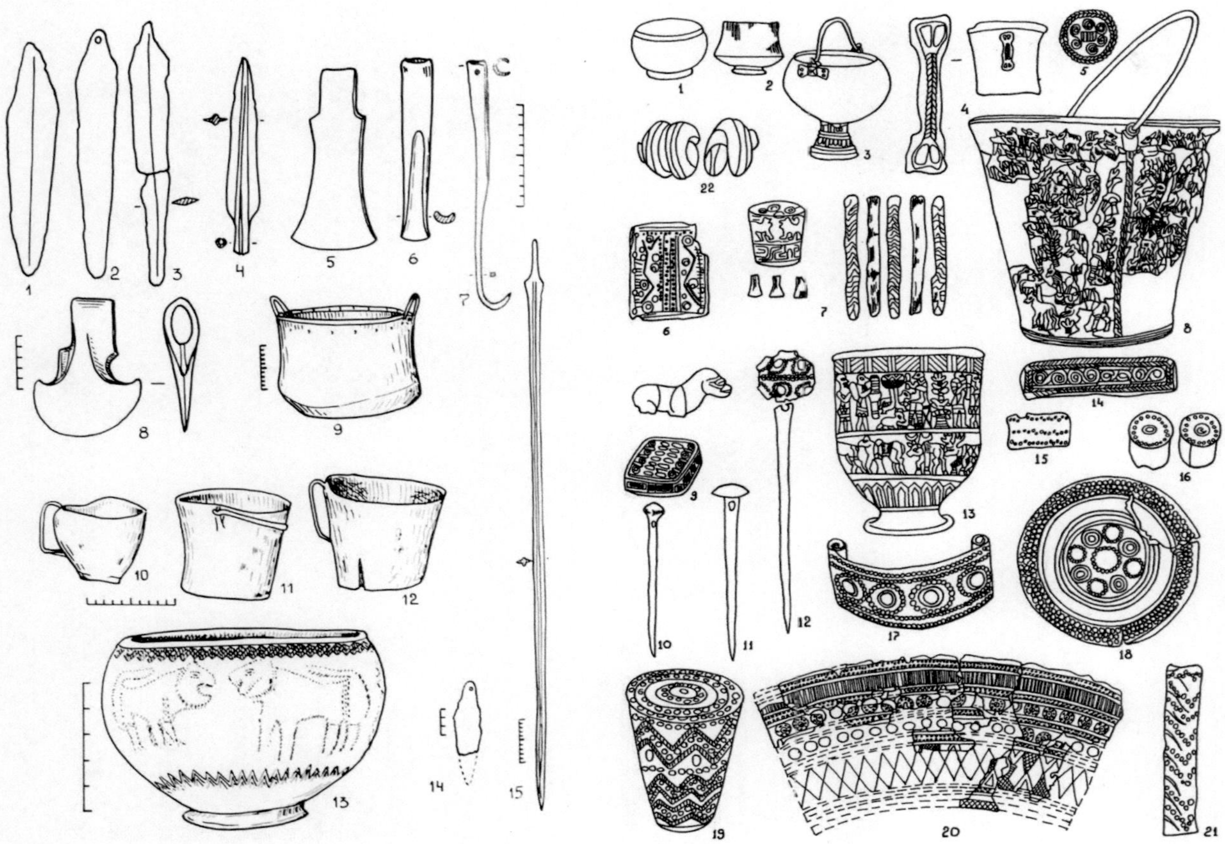

a. b.

Figure 5.13. Examples of the contents of metal assemblages of eastern Anatolian sites of the very late 3rd or early 2nd millennium BC: (a) and (b) Trialeti (kurgans) (Chernykh 1992, 112 fig. 36; Puturidze 2003, 119). Not at uniform scale.

pottery), the metalwork found at these Transcaucasian sites was also typical for much of eastern Anatolia. A similar argument is made for the later 3rd and early 2nd millennia, when substantial numbers of metal objects come from the Bedeni, Martkopi, Sachkere and Trialeti groups of burial kurgans (Chernykh 1992, 105, 108, 112; Kavtaradze 1999, 85) – although pottery traditions diverge somewhat during this period. In the Late Bronze Age (around 1600-1500BC to 1000BC) of the northern Caucasus, a vibrant metalworking tradition is recorded (Chernykh 1992, 275-295), including many finely crafted axes, swords and daggers (Figure 5.13), but similar assemblages have not been uncovered to the south.

5.3.2 Metal assemblages in western Central Asia, 3000-1500BC

In Turkmenia and western Central Asia, the distribution of metal artefacts is also somewhat uneven both spatially and temporally. The majority of the objects come from grave contexts. For example, dating to the first half of the 3rd millennium, a sizeable number of metal objects was uncovered in cemeteries of the Sumbar valley (Khlopin 1983; Chlopin 1986; Khlopin 2002). Forms include dress-pins, two-sided knives, and maceheads (Figure 5.14). For the later 3rd and early 2nd millennia, larger metal assemblages include those of northern Bactria, such as Sapalli Depe and Dzarkutan (Kaniuth 2006, 2007), the often-looted cemeteries of southern Bactria (Ligabue and Salvatori 1990), the graves of Kopet piedmont sites such as Altyn Depe (Masson 1988), and also sites in Margiana (Chernykh 1992,

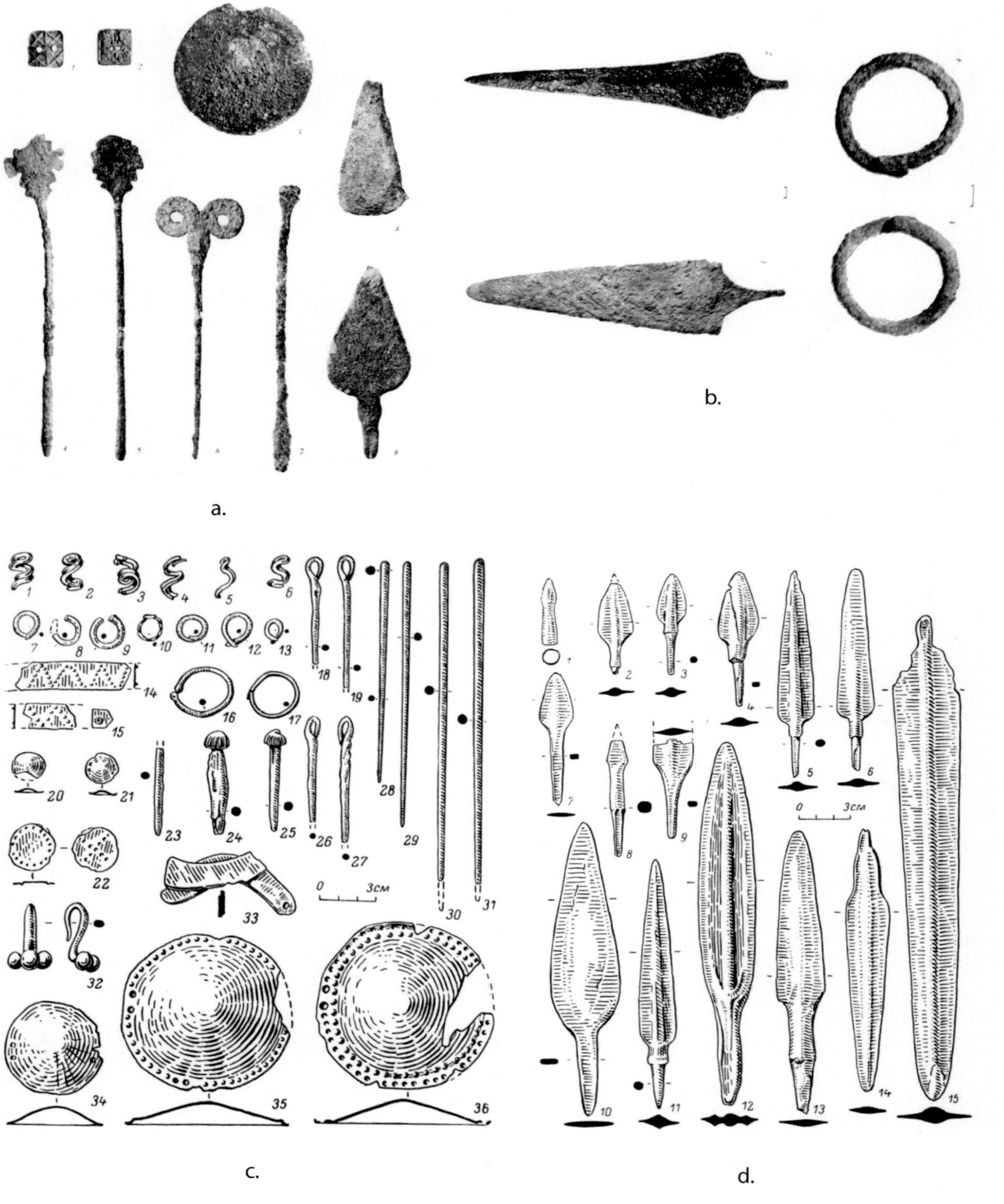

Figure 5.14. Examples of the contents of metal assemblages of western Central Asian sites of the early to mid 3rd millennium BC: (a) and (b) from Altyn Depe (Masson 1988, pls. xxxviii, xiv); (c) and (d) metal objects from the Sumbar cemeteries (Khlopin 1983, 223, 220). Not at uniform scale. (Images: courtesy of the Russian Academy of Sciences).

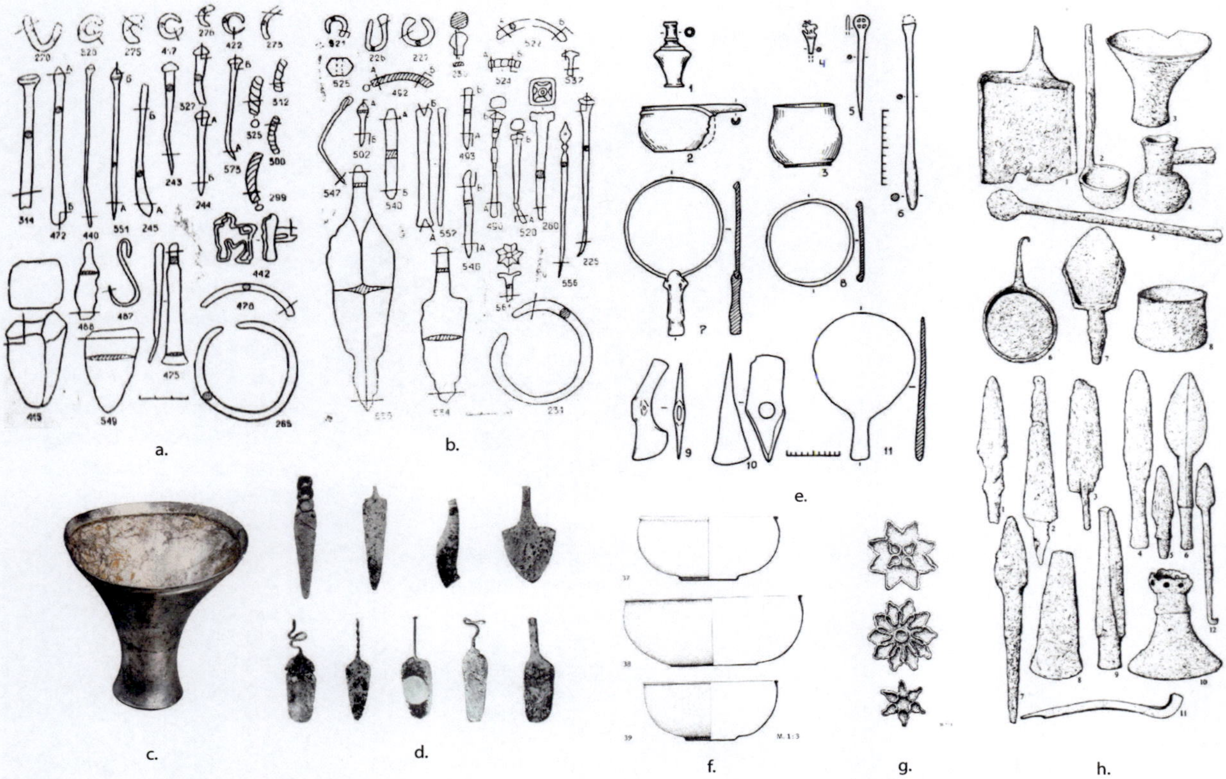

Figure 5.15. Examples of the contents of metal assemblages of western Central Asian sites of the late 3rd and early 2nd millennium BC: (a) and (b) sites in Margiana (Sarianidi 1990b, 306-307); (c) silver vessel (Sarianidi 2002b) and (d) metal blades from Gonur Depe (Sarianidi 2001, 155 pl. 20); (e) Sapalli/Bactrian assemblages (Chernykh 1992, 175 fig. 64); (f) Sapalli vessels and (g) seals (Kaniuth 2006, 81, 77); (h) various metal items, including vessels and tools, looted from graves in southern Bactria (Sarianidi 1990a, 115). Not at uniform scale.

175, 178-181), for example the necropolis at Gonur Depe (Sarianidi 2001). This later corpus includes handled mirrors, handle-less mirrors, seal-amulets, various vessel types, bottles, maceheads, shafted axes (highly elaborated), arrowheads, two-bladed knives, single-bladed knives, hooks, needles, clothes pins (with various heads), cosmetic sticks, bracelets, earrings, diadems and miniatures of various sorts (Figure 5.15). Also relevant are metal objects and metal production evidence from nearby sites such as Tepe Hissar (Pigott 1989; Thornton 2009), or to the south, as at Shahr-i Sokhta (Hauptmann, Rehren and Schmitt-Strecker 2003; *cf.* Thornton 2010), which in early periods showed strong ceramic connections with the Turkmenian sites, but during the late 3rd and 2nd millennia appear to have followed different cultural trajectories.

5.3.3 Patterns in metal assemblages

Notable in both regions, although perhaps more strongly pronounced in Central Asia, is the overall dearth of everyday metal tools in this corpus: even those weapons and tools which are recorded may often be seen as ceremonial or symbolic in some sense. Archaeologists have tended to favour the collection and publication of 'interesting' objects over the mundane, but everyday tools would be more likely to be recycled, given their relatively low value, and for similar reasons, would be less likely to have been intentionally deposited in graves. If grave goods are intended as a statement of the occupant's (or the occupant's family's) social status, such tools would make a poor statement. It is worth taking the preponderance of 'ornamental' items (personal jewellery, cosmetic items and apparently symbolic

Figure 5.16. An example of a 'Bactrian axe': intricated cast with three animal shapes (boar, goat and tiger) including silver inlay decoration for tiger's stripes; British Museum, ME 123268 (photo: © Trustees of the British Museum, #AN30790001).

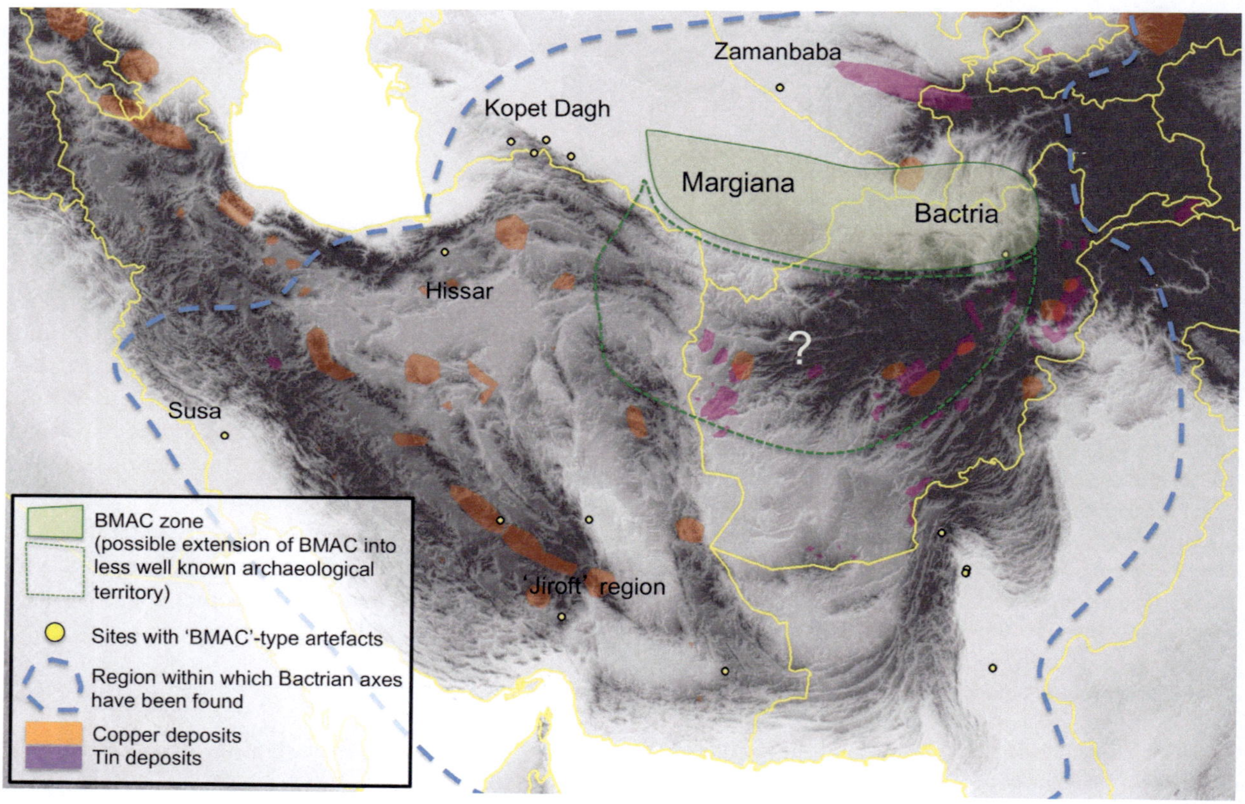

Figure 5.17. Distribution of 'BMAC' objects outside Bactria and Margiana heartland, including 'Bactrian axes' (following Hiebert 1998, 154).

weaponry and tools) seriously, therefore, and considering whether the tool/ weapon-based role of metals has sometimes been overplayed in the past: a major function of metal objects was still as a medium for social display.

5.3.4 Metal artefact types

i. Bactrian axes

Many 'Bactrian' axe-heads (an example shown in Figure 5.16) have come to light in the art-market since the late 1970s, and this lack of provenance has made it rather difficult to create a comprehensive distribution of their find spots. Large numbers have come from southern Bactria, in modern day Afghanistan, and it seems reasonably clear that they are mainly derived from the zone of the 'Bactria-Margiana Archaeological Complex' of southern Turkmenia. Additionally, however, examples of such axes have been uncovered as far afield as Susa in western Iran, and at sites in southern Iran (Figure 5.17). For this reason, these axes have been taken as evidence for very long distance interaction, and compared with other items which seem to have circulated within an 'inter-Iranian' interaction sphere during the very late 3rd and early 2nd millennia BC, first explored in detail by Pierre Amiet (Amiet 1986). Other associated items include the 'intercultural style' steatite or chlorite bowls (Kohl 1978) whose production now seems to be centred around south-eastern Iran (Perrot and Madjidzadeh 2006) (as discussed in Section 4.4.1). Hiebert, noting the 'intrusive' nature of certain burials with such Bactrian axes and associated BMAC ceramic material, which are sporadically found in southern and especially south-eastern Iran, has suggested that they may represent individual proselytizing migrants – *i.e.* itinerant preachers – perhaps associated with early Zoroastrianism (Hiebert 1998). Whilst this theory is difficult to confirm or disprove, the fact that such axes may be associated with individuals (as part of personal burial assemblages) suggests particular ways in which such material may have travelled. Rather than an overall cultural *koine* across eastern Iran, one might be tempted to see this as a very specific sub-group or diaspora, with connections to a homeland in Central Asia.

ii. Decorative metal vessels

Also associated with this 'inter-Iranian' sphere of interaction, is a collection of eclectic decorated metal vessels (Dupree, Gouin and Omer 1971; Tosi and Wardak 1972; Potts 2008a; *cf.* Aruz and Wallenfels 2003, 360-366). The iconographic scheme represented by these varied sets of vessels seems to be broadly shared across Iran (see Figure 6.27a), though some designs and types have more restricted distribution. Whilst their distribution certainly suggests very long distance circulation of vessels, it is difficult to comment on the directionality of such a circulation: it is not clear where they were being produced (whether exported from one region, or made locally in each region), and it is not clear how they were distributed. Historical texts referring to Sumerian and Elamite military campaigns into highland Iran hint at possible mechanisms of circulation, in which extensive booty was extracted or tributes paid, presumably including metals (Potts 1999, 150ff), and perhaps with other types of transfers (*cf.* Section 6.5.3).

In this context it is interesting to note a parallel situation in Caucasia – metal vessels with detailed iconographic schemes can be found in kurgans of the late 3rd or early 2nd millenium BC, though using different schemes and shapes. Metal vessels with highly elaborated decoration have been found amongst the assemblages at Karashamb, at Trialeti kurgan XVII from the Tsalka plateau (Kohl 2007, 115-116), dating to the early 2nd (or late 3rd) millennium and, from a later context,

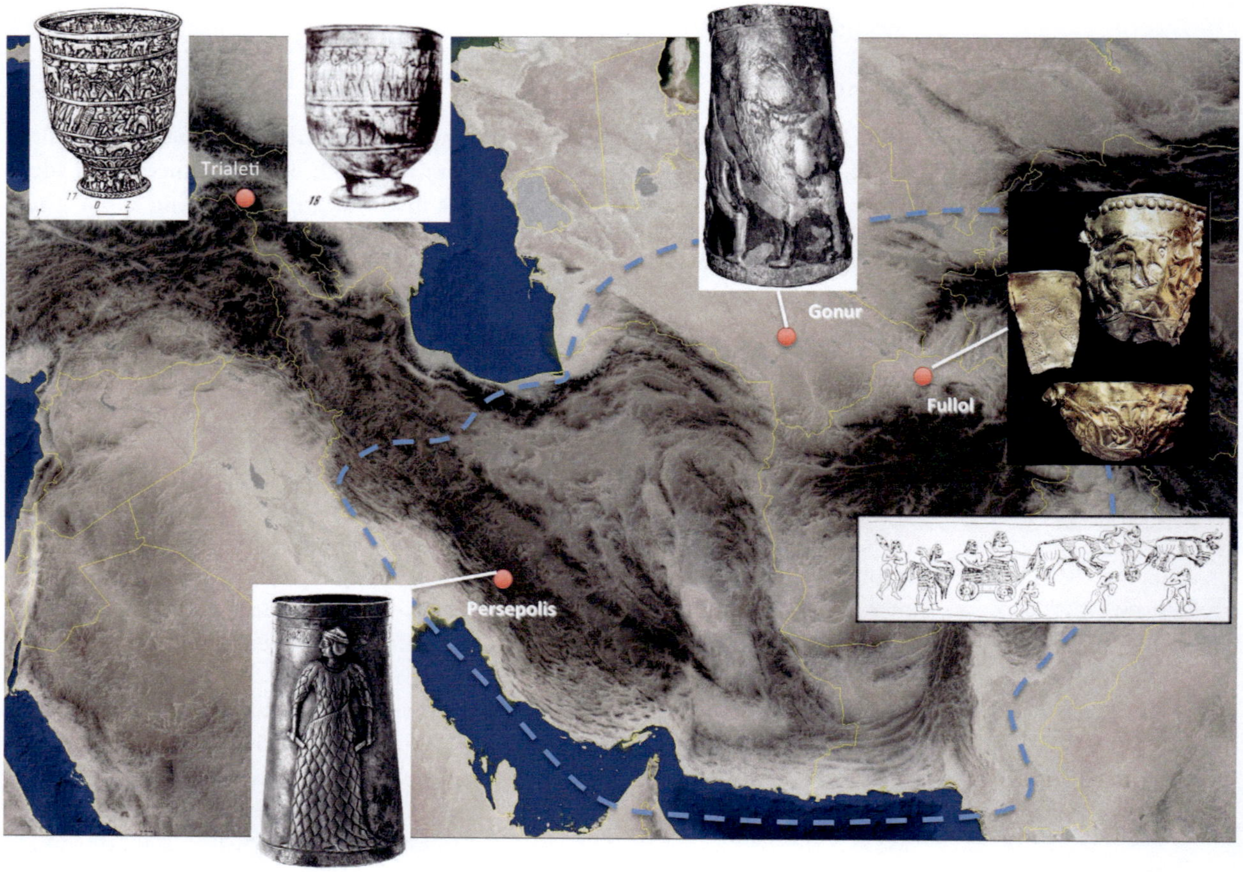

Figure 5.18. Examples and distribution of figure-decorated metal vessels, c. 2250-1750BC. Vessel photographs from: Trialeti (Kushnareva and Markovin 1994, 87[18]; image: courtesy of the Russian Academy of Sciences). Persepolis and Gonur Depe (Potts 2008a); Fullol (Ligabue and Salvatori 1990).

at Hasanlu (Rubinson 2003). Again their iconographic links to Mesopotamia and Anatolia have been highlighted, and they contrast with earlier metal vessels from the 4th millennium Maikop assemblages, but again it is difficult to assert the directionality of exchange. Were these imported vessels from the south, locally manufactured imitations of southern vessels, or local products that were mimicked in the south? Perhaps it would be more useful to view both Caucasian and Central Asian vessels as fossils of a lively circulation in images, whose axis ran through Mesopotamia (Figure 5.18), but whose inspiration was drawn from different regions (Bellelli 1989). Such images could also have travelled in other media that have not survived in the archaeological record, and their subject matter in stories, poems and songs that could have travelled even further than material objects. The nature of these depictions also remind us of their role in or reflection of the transfer of dress styles (both textiles and accessories), something which is discussed further in the next chapter (see Section 6.7).

5.4 Regional metallurgical constellations

Individual artefact type distributions show only a small part of the story, however. Perhaps more significant are the regional metallurgical constellations which emerge once large databases of metal objects, artefact types and manufacturing techniques are analysed together. E. N. Chernykh's (1992) seminal work has already been cited above on the location of raw material sources and for general perspectives on the history of metallurgy in Eurasia. One of his most significant contributions has been a synthesis of the systematic comparison and categorization of a huge number of metal objects from the former USSR and its satellite states (*i.e.* those items to which he had ease of access at the time) during what he describes as the 'Early Metal Age' (fifth to late second millennia BC). The following section relies heavily on his work to reflect on the existence of metallurgical constellations, *i.e.* geographically expansive systems of metal production and consumption.

5.4.1 Chernykh's metallurgical 'foci' and 'provinces'

Chernykh identified common patterns in the forms and metallic constituents of metal assemblages which appear to document close exchanges of ideas about metalworking, both technical and cultural, over both small and large scales. Such clusters of similar assemblages he termed 'foci' of two different types: *metallurgical* (by which he means mining, smelting and generally preparing metals) or *metalworking* (by which he means the manufacture of tools from processed metals). These 'foci' are to be defined as the metal equivalent of regional pottery production traditions – reflecting similar manufacturing techniques, similar metal forms and similar metal composition and implying interaction between sites in the region. He also noted that over an even wider area there were similarities between certain aspects of nearby 'foci' such that they can be grouped into 'metallurgical *provinces*'[62] – also implying interaction but with less intensity (in part because of the distances involved).

Partly because of the geo-political situation under which the original work for the project was done during the 1970s and 1980s, and thus the limited availability of literature and access to materials for his research group, the important regions of Mesopotamia, much of Turkey, Iran, the Indian subcontinent, Syria and the Levant, Arabia, China and many others, were originally left out of the analysis. The boundaries of several of the 'metallurgical provinces' were thus obscured or partially reconstructed. Both eastern Anatolia/Transcaucasia and western Central Asia lay on this fault-line in the analysis. Fortunately, the project continued after the collapse of the Soviet Union, extending the synthesis southward and eastward into Anatolia, Mesopotamia and the Levant (Chernykh, Avilova and Orlovskaya 2002; Munchaev 2005; Avilova 2008, 2009; Chernykh 2009).

Whilst the overall coverage of the database has been enlarged, and now claims to include at least 120,000 metal objects (Avilova 2009), the fundamental analysis of 'provinces' and 'foci' has remained broadly the same. The narrative which Chernykh and his team put forward, based on this comparison of metal objects, is broadly as follows: after some patchy experimentation with metals during the Neolithic (primarily in Anatolia and Iran), metallurgy and metal artefacts (mostly in the form of unalloyed copper objects) appeared regularly in two regions around the mid-5th millennium[63]. In Eastern Europe a clear regional

62 A pottery equivalent of which does not exist, though there are definite 'wider' patterns of pottery production styles which are rarely commented on in the literature, for example the simultaneous appearance or disappearance of painted pottery over wide areas.

63 Chernykh's Chalcolithic: partly due to the size of their field of analysis, Chernykh and his team use a very generalized version of the Chalcolithic, EBA, MBA and LBA scheme which does not always match the local sequences. For the purposes of this book, I have preferred to use absolute date equivalents as follows: Chernykh's Chalcolithic c. 5500-3500BC; Early Bronze Age c. 3500-2500BC; Middle Bronze Age c. 2500-1750BC; Late Bronze Age c. 1750-1200BC.

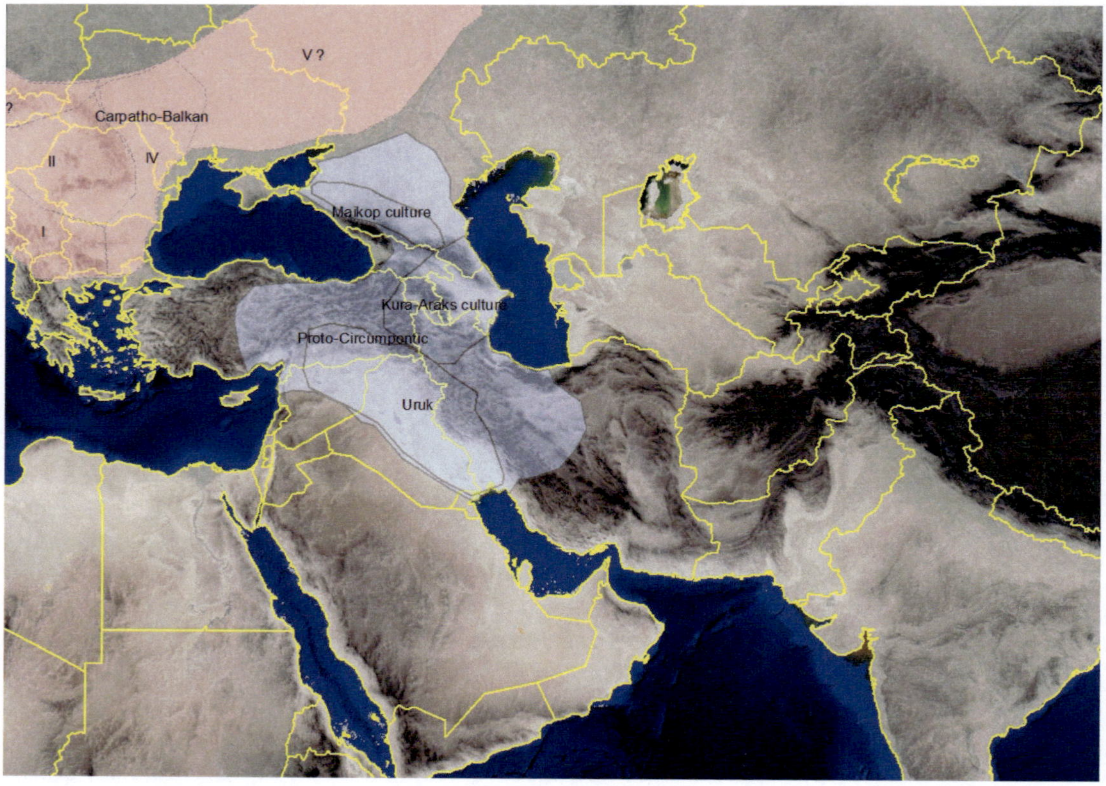

Figure 5.19. Chernykh's Metallurgical Provinces, 'Chalcolithic period', c. 5500-3500BC.

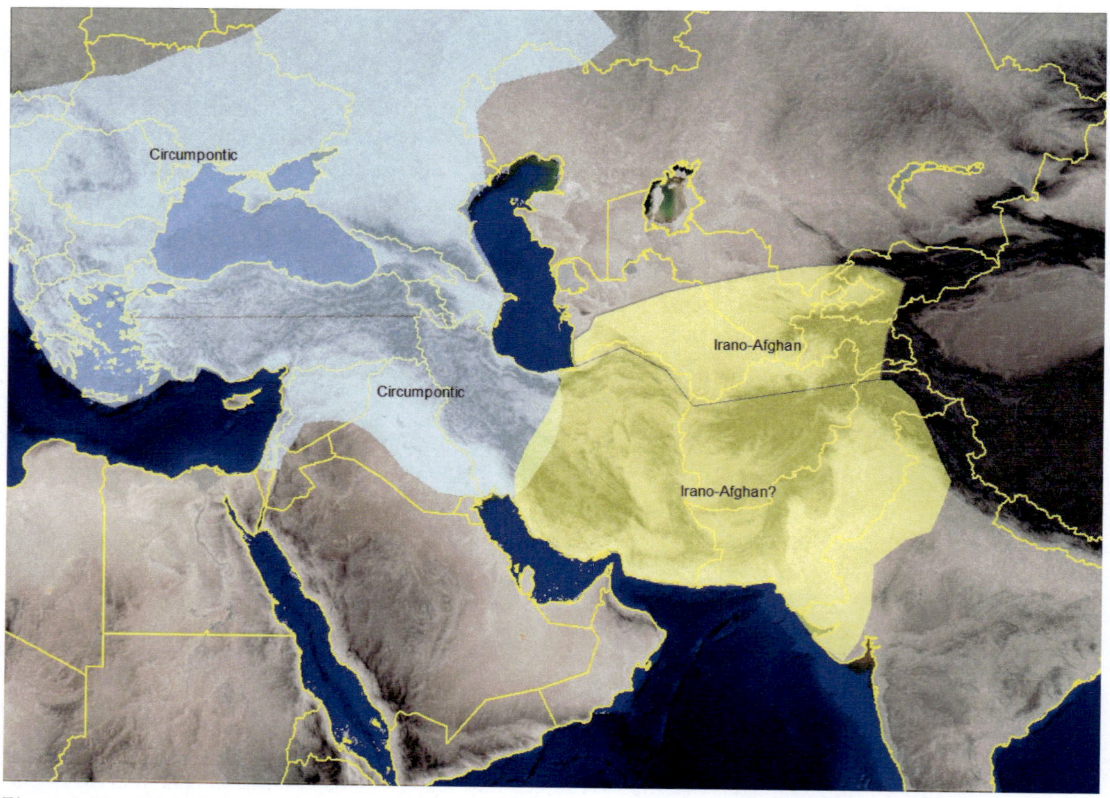

Figure 5.20. Chernykh's Metallurgical Provinces, 'Early Bronze period', c. 3500-2500BC.

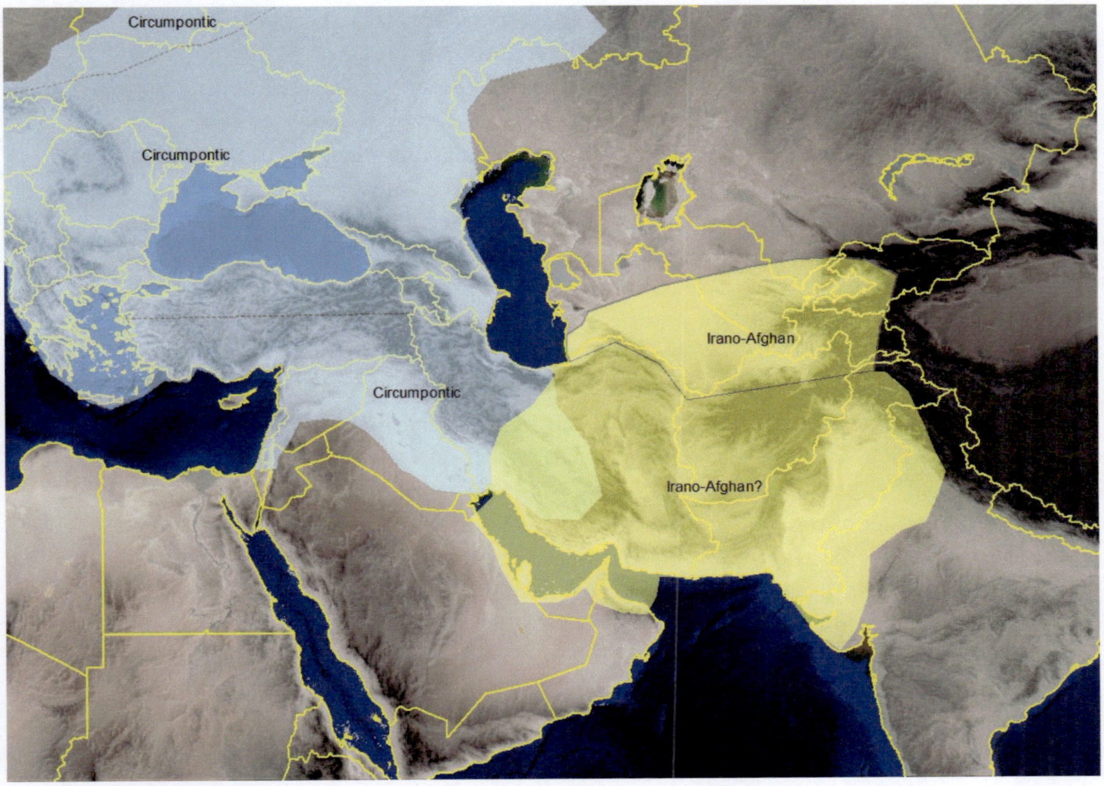

Figure 5.21. Chernykh's Metallurgical Provinces, 'Middle Bronze period', c. 2500-1750BC.

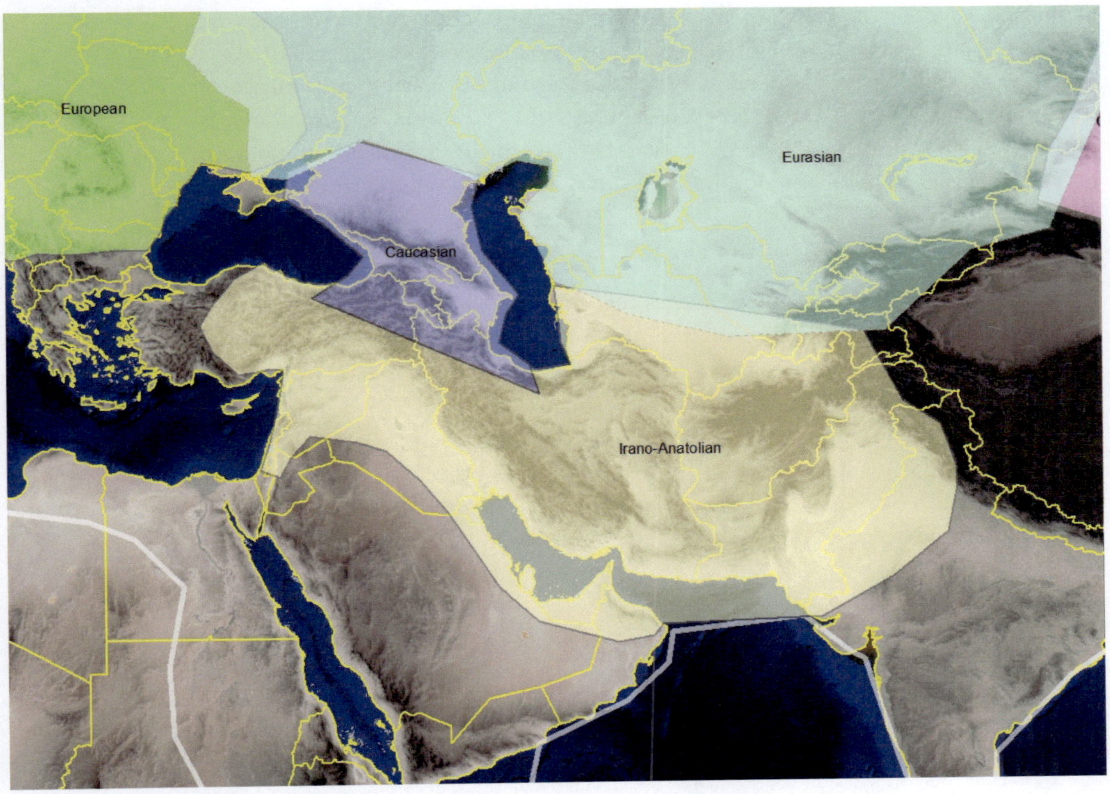

Figure 5.22. Chernykh's Metallurgical Provinces, 'Late Bronze period', c. 1750-1250BC.

tradition, defined by particular preferences for certain forms and materials, which he describes as the 'Carpatho-Balkan Metallurgical Province' (CBMP/*BKMP*),[64] forms apparently around the copper-rich Carpathians. Meanwhile in the region between Mesopotamia and the Caucasus, metallurgy develops along a different track within the early Uruk, Proto-Kura-Arax and early Maikop cultures (Figure 5.19). Around the mid-4th millennium[65], these traditions seem to have broken down or transformed, to be replaced by a new set of metal traditions, now mostly involving copper-arsenic alloys and extending over a much larger area, namely the 'Circumpontic Metallurgical Province' (CMP) (Figure 5.20). The outlines of this cultural entity or network remained relatively stable until the early 2nd millennium[66], though there were changes within the entity (Figure 5.21). Meanwhile, in other parts of the Old World, metal usage either remained very limited or else formed part of an alternative metallurgical tradition, such as the 'Irano-Afghan Metallurgical Province' (IrAfMP), whose precise outlines have not been fully investigated by Chernykh. At some point around the early to mid-second millennium BC, the system which maintained the 'CMP' must have broken down, to be replaced by a series of different entities: in Caucasia by the 'Caucasian Metallurgical Province' (CauMP/*KaMP*), across Europe by a new 'European Metallurgical Province' (EuMP/*EvMP*), from eastern Europe across the steppes and extending far to the east to the Chinese frontier by the vast 'Eurasian' (EaMP/*EAMP*) and related 'East Asian Metallurgical Provinces' (EAsMP/*DKMP*), and from Anatolia across the Near East to Iran, the Indus and southern Turkmenia by an 'Irano-Anatolian Metallurgical Province' (IrAnMP), apparently taking the place of the older IrAfMP[67], though not necessarily a descendent in terms of content. All of these provinces are marked by an increased application of tin-copper alloys (*i.e.* true-bronze) and a greater range of forms and techniques. As the relative similarities between the contents of the focuses wax and wane through time, the result is a narrative of suddenly coalescing and dissipating 'constellations' of metal cultures which sometimes cover huge areas of Eurasia, a story which is both informative and intriguing in its evocation of shifting strands of interaction between peoples living many hundreds or thousands of kilometres apart.

5.4.2 Internal mechanisms of metallurgical provinces?

Chernykh's categorization of assemblages into 'foci' and of groups of focuses into 'provinces' is of course a matter of interpretation, so the exact outlines of what constitutes a focus or a province are necessarily open to some debate or revision (see, for example, some local-level objections in Brovender 2009). Various regions remain relatively unrepresented (Chernykh 2009, 245) including, importantly, Iran (Pigott 1999b; Thornton 2009). The most serious problem with Chernykh's work is the difficulty in examining and testing the full database (and clarifying issues of dating) because 'raw' data has not been made publicly accessible – the

64 These abbreviations are sometimes used in Chernykh *et al.*'s publications – the first is the English abbreviation and the second the transliterated Russian equivalent. They will be used here only where the context is obvious.

65 The start of Chernykh's Early Bronze Age.

66 Through Chernykh's Early and Middle Bronze Ages.

67 Though, on the basis of Chernykh's data, its full extent and therefore longevity in Iran in particular are unclear. The dating that he appears to have relied on is the older Soviet chronology, which placed the Namazga Late Bronze Age around 500 years later (1600-1200) than is now generally accepted.

web could offer the perfect format for such a database. In the meanwhile, the definitions of these foci and provinces allow us to start to frame interesting questions about the exchange of both metals themselves and socio-technological ideas about them on many different levels. On a geographical level, for example, they simultaneously reveal, on the one hand, stable networks of interaction and, on the other hand, key social boundaries: how did these cultural constellations come into existence, how were they maintained and how did they disappear? To put it in more concrete terms, why is it that, for example, smiths from Eastern Europe wanted to produce metal objects with strong affinities to those made by metalworkers in Caucasia (or vice versa) during the late 4th and early 3rd millennia, whilst in other regions – for example southern Turkmenistan, which is a similar or even shorter distance from Caucasia – the assemblages are somewhat different in both form and metal composition? Chernykh perhaps prudently shies away from too much speculation on these points, though throughout the text he reminds us of the mutual importance of 'ideological' and 'technical' factors in the creation of foci and provinces. It is worth us considering some possible scenarios or mechanisms by which such provinces, or rather the relationships that they purport to represent, could have come about.

Whichever way we approach the problem, we need to evoke a dynamic model in which there must have been substantial movement between a province's constituent regions, whether by this we mean movement of people, movement of objects and materials, or, less tangibly, movement of ideas. The migrant people we would need to envisage should not be the monolithic and unidirectional hordes of traditional culture-history[68], nor a version of modern day nomadic pastoralists[69], but groups or individual crafts people moving in both directions with particular interests or motivations in maintaining cultural links for a wide variety of reasons. These groups of individuals might include permanently itinerant metalworkers or smiths, parallels for which have been recorded in recent African ethnography (*e.g.* Jeffreys 1951). Admired craftsmen, or craftsmen of admired cultures might be kidnapped or invited, or might emigrate of their own accord to neighbouring or distant locations, as is known from the movement of sculptors and artists in the classical world. Even if an individual moves only a few kilometres to the next village, if that individual's apprentices also then migrate a few kilometres, over only a few generations the knowledge of particular techniques and shapes can be transmitted over large distances without necessarily requiring the bulk of population to move in the same direction. Marriage and similar social alliance patterns can played a role in this kind of mobility and transmission of techniques[70]. Consumption may have played a central motivational role: the demand for particular objects is most likely to be the driving force behind any movement of producers (*i.e.* smiths), promoting local metalworkers to mimic shapes and materials of imported items. Local 'consumers' themselves may also have been moving as individuals or groups (though not necessarily all in the same way or at the same time), and their travels and thus relations with 'outside' groups and their objects (whether in the form of

68 Which are still sometimes called upon in Russian texts by way of explanation, or as straw-men of post-New Archaeology English texts.

69 Who increasingly seem to be wheeled in to characterize any archaeological phenomenon that cannot be 'explained' in any other way.

70 Though metalworking is often seen as male work, and relocation on marriage often associated with women, but both are simply culturally-specific assumptions, especially the second, patrilocality.

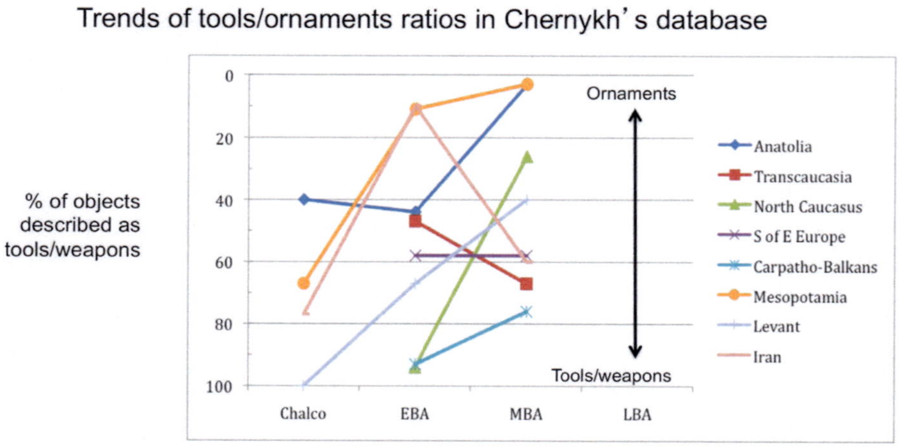

Figure 5.23. Chart showing relative proportion of items classified 'tools' vs. 'ornaments' in Chernykh's database.

gifts, trade, booty or mere distant admiration) offered as models for imitation in other regions.

It is worth reiterating Chernykh's suggested delineation between *metallurgy* and *metalworking* – the first referring to the processing of ores to create workable metals, and the second referring to the reforming of such workable metals into finished objects of various types (Chernykh 1992, 8). There are different sets of skills involved, and given the unequal distribution of mineral ores across the earth's surface, and the high cost of transporting unprocessed ore, it is likely that 'metallurgical' knowledge and activities were much more restricted than 'metalworking' ones. This has consequences for both the types of people we might envisage moving and also the materials and objects. If Chernykh's analyses, reconstruction and interpretation of the 'Circumpontic Metallurgical Province' are correct, for example, we must envisage considerable quantities of copper metal (but presumably not ore), either in finished object or ingot form, being transported from Caucasia and Anatolia over vast distances to the north and north-west around the Black Sea, presumably to be then worked to suit local needs (Chernykh 1992, 156-157). If this trade involved finished objects or moulds, it is easier for us to imagine how similar objects could be produced over such wide areas, by metalworkers copying and adapting objects and designs. The existence of standardized ingots is well established from texts and archaeological data of the eastern Mediterranean and Mesopotamia from the second millennium (Gale and Stos-Gale 1999), but though there are some hints that ingots were also circulating the 3rd millennium (as mentioned above in Section 5.2.5; Doonan, Day and Dimopoulou-Rethemiotaki 2007, 106), the evidence is somewhat less clear. Under what conditions might metal scrap or ingot-like objects actually travel is difficult to say: was it as part of long-distance 'caravans' or shipments; in a series of trade-based exchanges (as objects are sold from one village to the next); in the form of gifts or tribute as part of complex and little understood obligation and exchange networks; or indeed a mixture of all three (Renfrew 1975)? Gift exchange, even on a very small scale, must have been a very important part of the creation and maintenance of long-distance connections for both elite and ordinary people. Similarly, multi-directional cultural emulation (both 'horizontally' from elite to elite, or 'vertically' from elite to the wider population or, less often, vice versa), in which individuals or groups make use of the objects of others (by copying, buying or stealing them) to make social statements or performances about their identity, must also have had a role in the transmission of metal types and materials. It is via these sorts of mechanisms (and also through apparently unrelated and sometimes invisible media) that ideas about the 'correct' forms, materials and their 'correct' usage must have travelled.

5.4.3 Identifying trends in consumption: Chernykh's statistics on forms and materials

Though the raw data on metal finds as used by Chernykh and his colleagues have not been published, their summary publications do furnish us with a few statistics[71] that may furnish us with some insights into broad trends in metal consumption across Eurasia.

i. Functional categories: tools/weapons and ornaments

Chernykh's published statistics (summarized in Figure 5.23) divide objects in the database between 'tools/weapons' and 'ornaments' (Chernykh 1992, 150; Avilova 2008, 77). However this categorization is very problematic. Many objects could fall into either category: the weaponry of the Alacahöyük Royal Tombs might have had the form of weapons, but clearly served 'ornamental' functions in the act of their deposition if not before. The word ornament is rather passive, suggesting a lack of function, when of course such ornamental items have central social functions: signalling their consumers' and viewers' roles and status.

ii. Constituent metals

Turning to the materials from which objects in the database were made (Figures 5.24-5.26), Chernykh's figures highlight the following notable features or trends:

A. *Copper and copper-alloy objects dominate the recorded objects with some exceptions.* Assumed to be the least-valued and most 'functionally-orientated' materials, this dominance presumably does reflect a relatively high consumption of copper and copper-alloys as compared to the 'precious metals'.

B. *Gold forms near to, or more than half of, the objects for Mesopotamia, Iran and the N. Caucasus c. 3500-2500BC (the 'EBA') and Anatolia, the Aegean and S. Caucasus, Mesopotamia, c. 2500-1750BC (the 'MBA'), in the case of Anatolia apparently including more than 90% of the objects.* This is what has been described as the 'gold problem' – the sudden explosion in the use of gold in burial assemblages in many regions, which has proven difficult to explain satisfactorily (Kohl 2006, 19-20).

C. *Gold forms an important percentage of the N. Caucasus material c. 3500-2500 (the 'EBA'), but disappears c. 2500-1750BC (the 'MBA').* The high early concentration must relate to the distorting effect of the Maikop material, but the complete contrast in the following period suggests a more profound cultural shift, since the contexts of deposition for the region remain burial-related.

D. *Silver nearly always represents a very low percentage of overall metal objects, except in the Levant and Aegean, and to a lesser extent the northern side of the Black Sea, c. 3500-2500BC (the 'EBA').* Its rarity and subsequent elevation to the

71 See Appendix E for tables showing the statistics and how they were assembled.

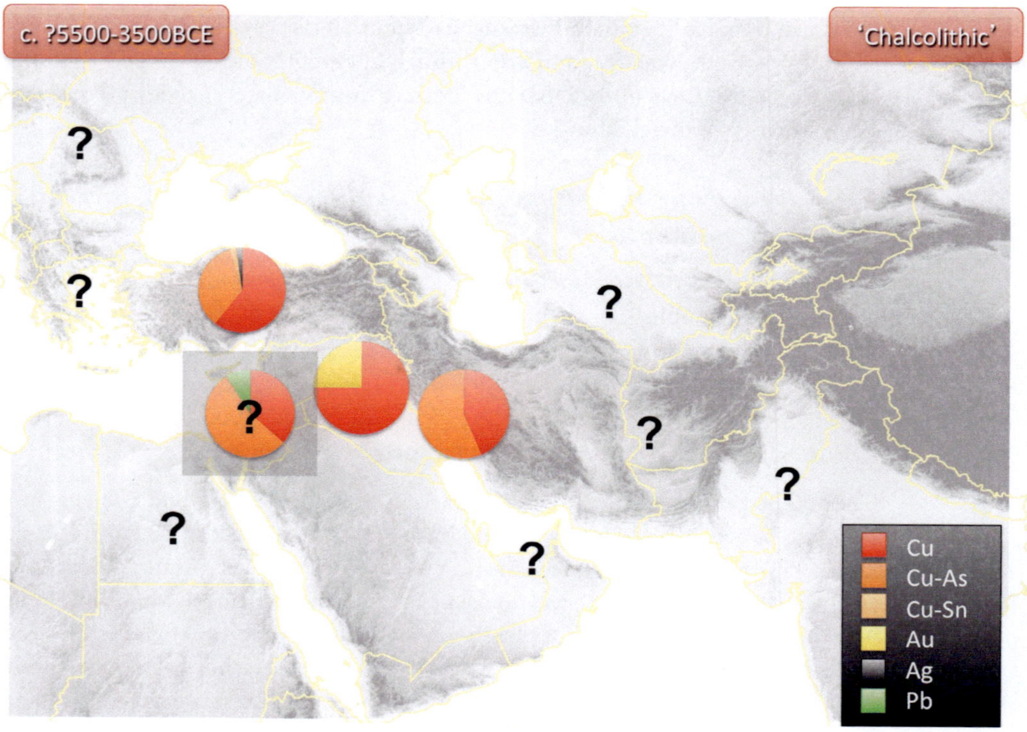

Figure 5.24. Relative proportion of copper, copper-alloys, gold, silver and lead objects during 'Chalcolithic period' from Chernykh's database. Question mark indicates insufficient data.

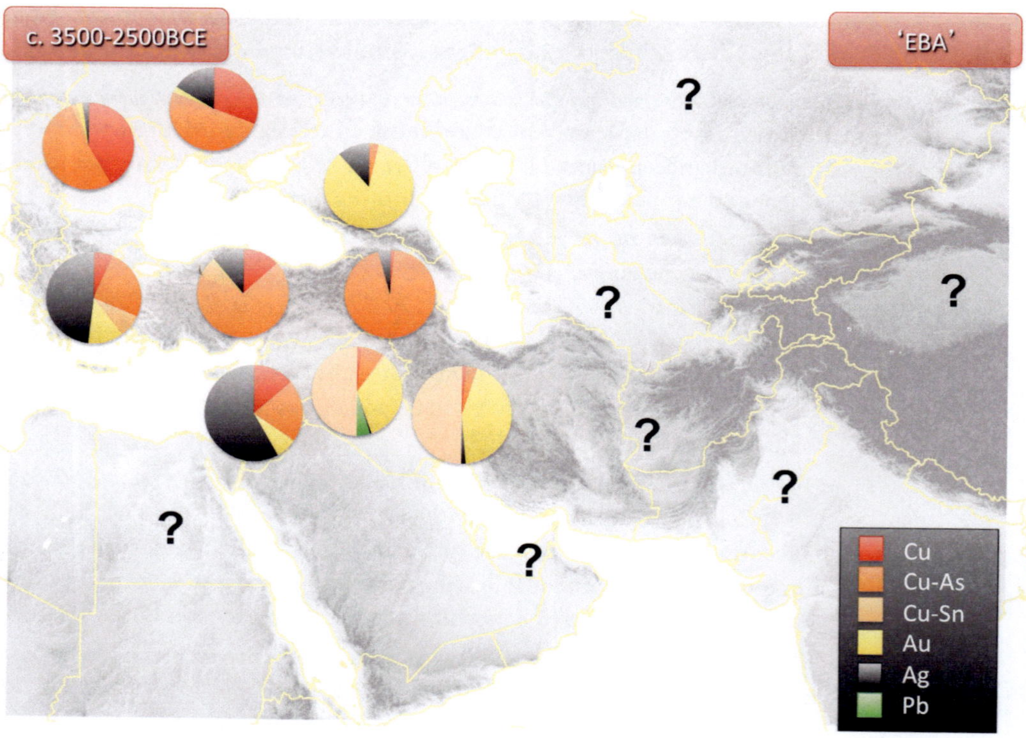

Figure 5.25. Relative proportion of copper, copper-alloys, gold, silver and lead objects during 'Early Bronze period' from Chernykh's database.

role of abstract medium of exchange in Mesopotamia may mean that silver was either less often used in actual objects (for elite items only), and/or that because it had become a kind of currency it was too valuable to put into the ground, except by groups who did not yet 'buy-into' this economic system.

E. *Lead forms a minor portion of the assemblages described here. Iron does not appear in these data, but is known from at least the 2nd millennium BC.* Both lead and iron generally form a minor footnote in discussions of metallurgy during the Bronze Age. But it seems highly likely that various groups across western Asia were experimenting with ferrous metallurgy on a limited scale from at least the second millennium, and lead metallurgy (probably mostly in association with silver) more systematically from even earlier (Hauptmann and Palmieri 2000; Hauptmann, Rehren and Schmitt-Strecker 2003; Thornton, Rehren and Pigott 2009).

Much of this data was based on visual identification of objects. It should be noted, therefore, that many museum or field descriptions of metal objects turn out to be inaccurate or over-simplified once the object has undergone material analysis (for example, objects described as gold or silver often turn out to be electrum), but such analysis has been performed on only a fraction of all known metal objects. Whilst there was clearly a great deal of fluidity between which metal could be used for particular shapes and tools, the efficacy of each to produce the desired result was clearly not equal. Gold swords may not have been very useful in battle, but gold's difficulty to source provided a social index of the power or influence of its owner, which at times surpassed that of the functionality of the sword. Likewise bronze, silver or bone pins might have been equally efficient in holding cloth together in a dress, but the ability to procure or be given silver functioned in a non-discursive manner, to place individual wearers within a social hierarchy. Thus it is obviously important to locate the ancient values of metal not in simple terms of technical function or economic wealth but also in wider cultural negotiations and performances of relative identities (power, gender, class, ritual-religious groups, ethnicity *etc.*).

iii. Alloys of copper

Making use of the object numbers provided by Chernykh (Figures 5.27-5.30), we can also perceive the following trends within copper and copper-alloy consumption (or at least its deposition):

A. *There is a broad trend from dominance of pure copper to the addition of arsenic-copper and then to tin-copper when viewed as a whole.* However, the details show a rather more complex set of alloying trajectories that do not follow the same pattern.

B. *Arsenic bronze constitutes some portion of the objects from Anatolia and Iran (Susiana) early, namely c. 4500-3500BC (the 'Chalcolithic'), with Iran showing a relatively high percentage of arsenic-copper to native copper. Despite the zero counts, Mesopotamia may also have had arsenic-bronze, but the sample is too small to rely on.* The significant use of arsenic-bronze in Iran may relate to the early exploitation of copper sources bearing ores heavy with arsenic (Pigott

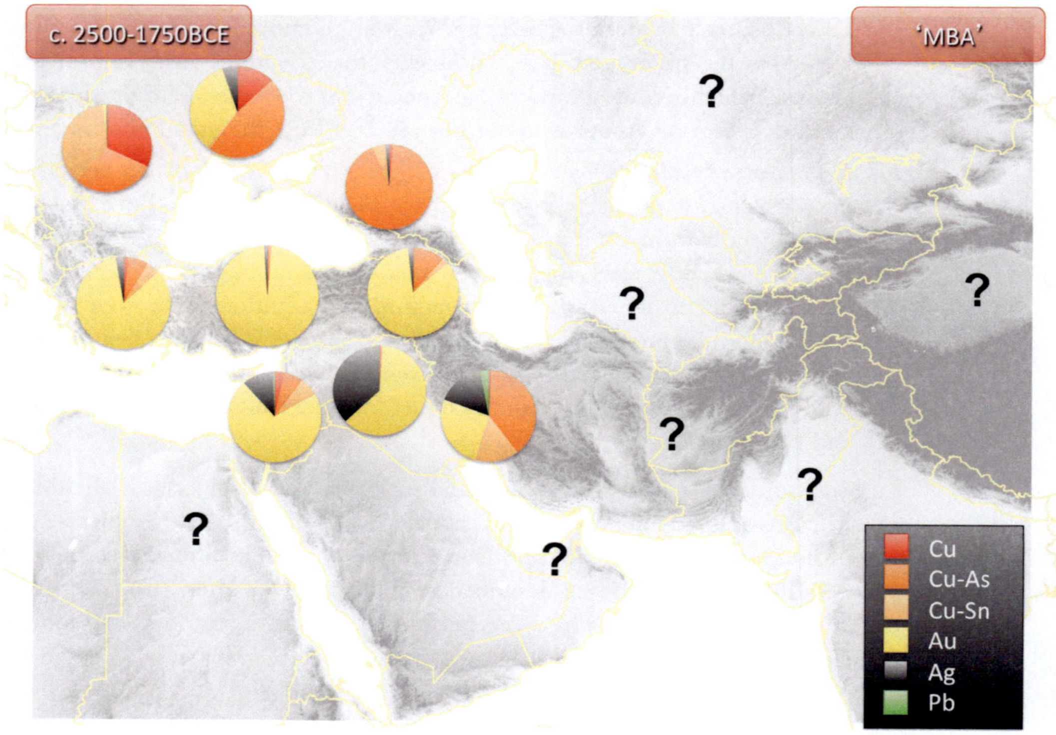

Figure 5.26. Relative proportion of copper, copper-alloys, gold, silver and lead objects during 'Middle Bronze period' from Chernykh's database.

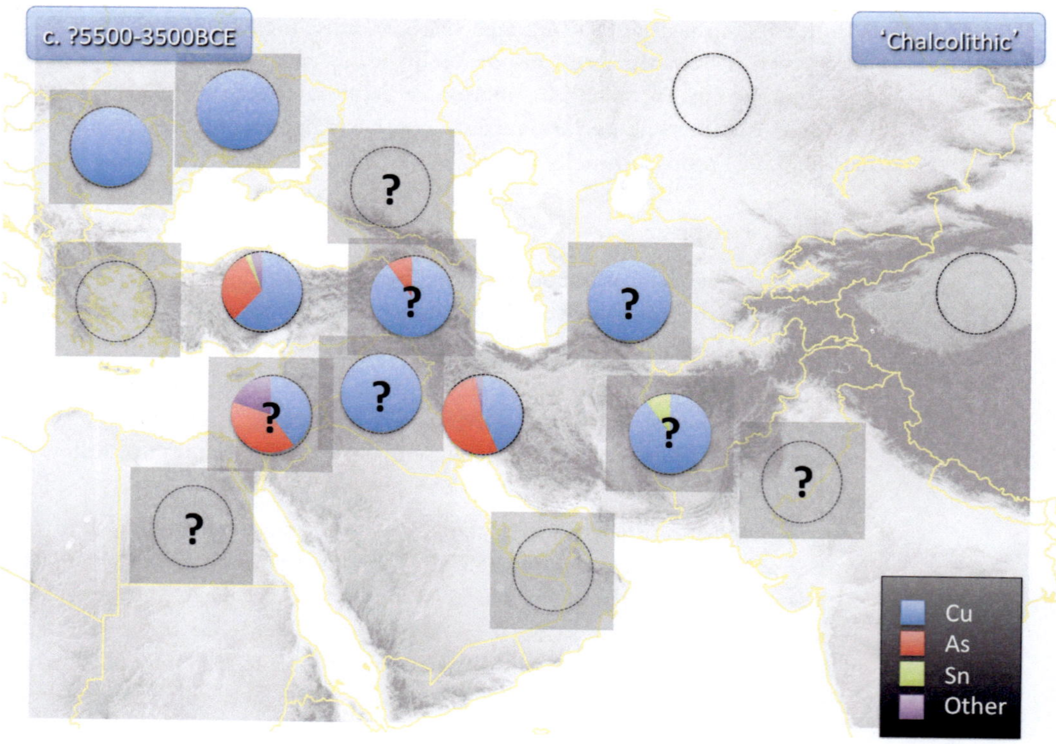

Figure 5.27. Relative proportion of pure-, arsenic- or tin- alloys used in copper objects during 'Chalcolithic period' from Chernykh's database.

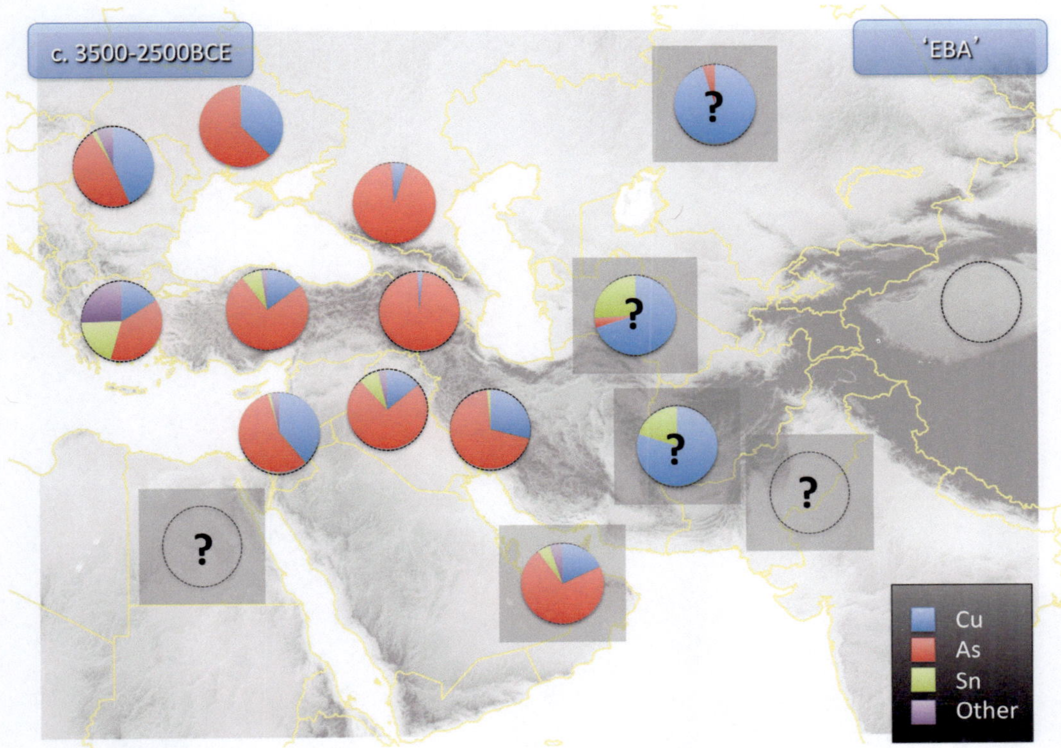

Figure 5.28. Relative proportion of pure-, arsenic- or tin- alloys used in copper objects during 'Early Bronze period' from Chernykh's database.

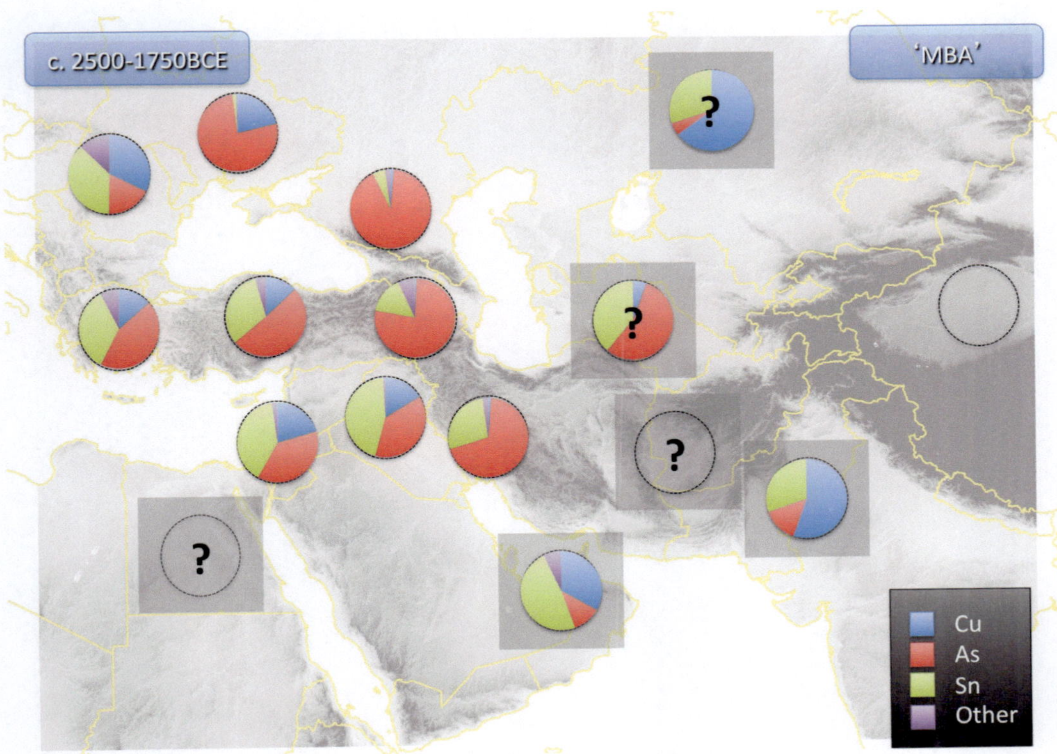

Figure 5.29. Relative proportion of pure-, arsenic- or tin- alloys used in copper objects during 'Middle Bronze period' from Chernykh's database.

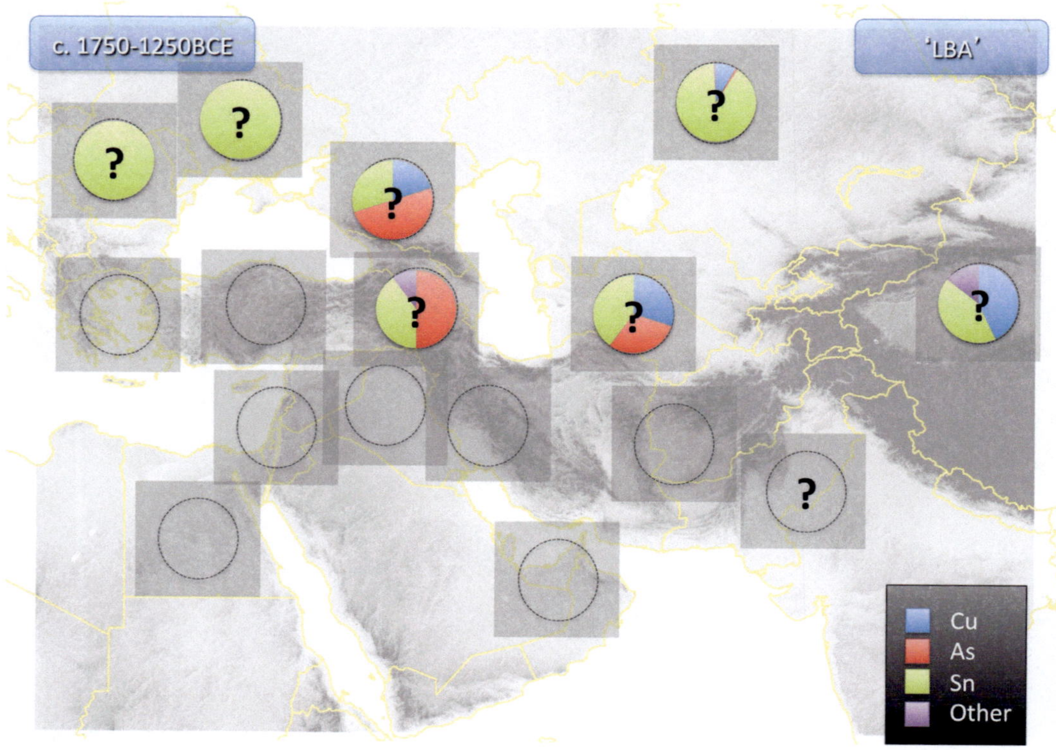

Figure 5.30. Relative proportion of pure-, arsenic- or tin- alloys used in copper objects during 'Late Bronze period' from Chernykh's database.

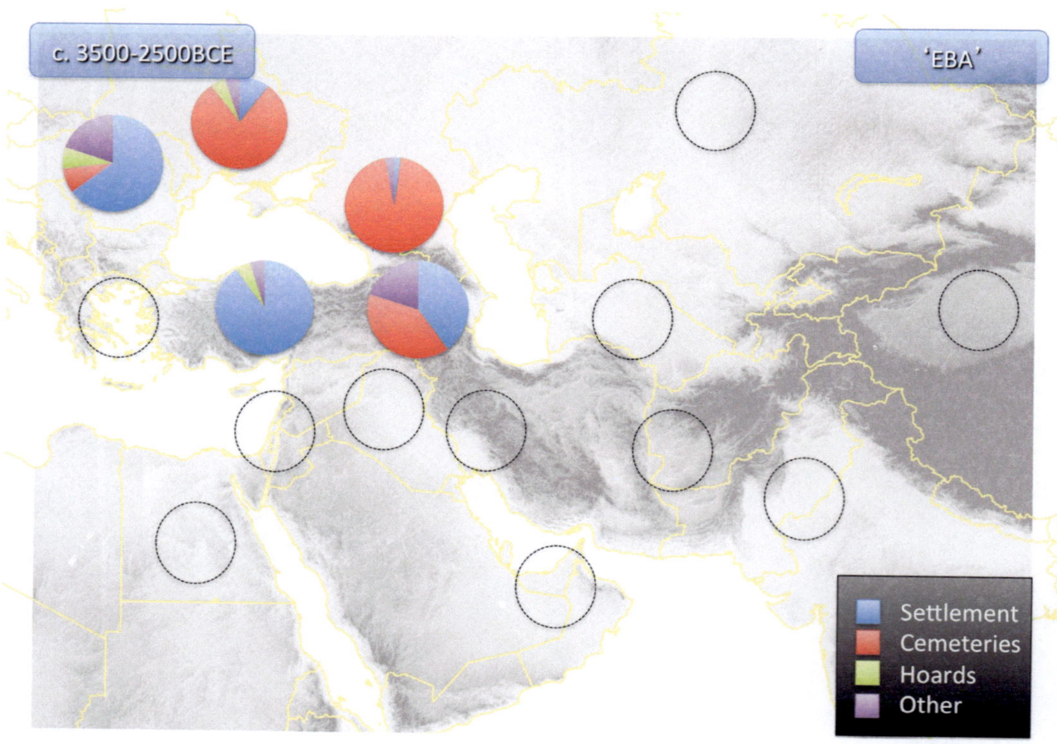

Figure 5.31. Metal objects classified by context of deposition, 'Early Bronze period', according to Chernykh's database.

1999b, 112), though recently it has been argued that the addition of arsenic to copper was very much an intentional procedure during the third millennium (Thornton 2009, 318).

C. *The limitations of these particular data actually hide the true complexity of copper-alloy technology*. Other probably intentional alloys (including those with lead and antimony, and more importantly complex combinations of tin, arsenic, antimony and lead in various amounts) form a significant part of the corpus when one looks at the detailed listings of metallurgical analyses. However, these are hidden here under simplified categories of 'tin-bronze', 'arsenic-bronze' and 'other'.

The simple view of a unilineal technological evolution in which the discovery of the functional advantages of arsenic-bronze, then tin-bronze, facilitate the replacement of one material by another, does not account for the complex patterns of metal consumption depicted in the graphs. According to these numbers, native, pure or purified copper remains a significant part of most assemblages at least to c. 2500-1750BC (Chernykh's MBA), and tin-bronze, upon which so much psychological weight is placed by archaeologists as defining a 'Bronze Age', forms less than half of the corpus in most regions at this time. The evidence for the consumption of tin in the 3rd millennium remains complex (see, for example Helwing 2009). But it seems that only in the late second millennium or early first millennium BC, did tin-bronze become a dominant copper alloy, especially on the Eurasian steppes (Chernykh 1992, 191). Rather than seeing the process as an evolution from one material to another, it may be better to think of a shift (or series of oscillations) from relatively uniform technologies to emerging diversity at different times and places: for example, from the second half of the third millennium, there was an increasing range and variation of alloy 'recipes', not shown in these simplified data, one of which was tin-bronze (Chernykh 1992, 106, 113). The range and variation of alloy 'recipes', and different metals overall (but particularly between arsenic-bronze and tin-bronze whose technical properties are very similar), may have more to do with consumer preferences for more or less subtle colour effects or exotic sourcing than technological 'advantages' (Stech and Pigott 1986, 48). These consumer preferences could then stem from a desire to differentiate taste as part of the third millennium emergence of systems of social hierarchy and, in the context of literacy and long-distance trade, an increased international awareness and desire to create ethnic and cultural differences on a large scale (both perhaps also witnessed in the expansion of textile technology across western Asia and the eastern Mediterranean – see Sections 6.4).

These patterns of diverse materials may lend support to Yener's (Yener 2000; *cf.* Thornton 2009) suggestion of the 'balkanized' nature of early metallurgy from the 5th to early 2nd millennia, by which nearby groups were using quite different metallurgical technologies, with localized cultural imperatives driving the consumption of materials, despite apparent similarities in fundamental technologies or types. On the face of it, this idea appears to run counter to Chernykh's sustained demonstration of large-scale convergence or similarities within his metallurgical provinces, but the two perspectives do not have to mutually exclude one another, since Chernykh acknowledges the variations of preferred forms and materials within provinces (as 'foci'). Combined, these approaches facilitate a shift away from a narrative of uniform technological trajectory in metal forms and metal materials towards a more complex picture in which there are periods of divergence, with 'balkanized' metal cultures, and other periods of uniformity across large-areas.

5.3.4 Contexts of deposition

The size of Chernykh's database is unparalleled, and the sample size makes it potentially very useful. The nature of archaeological survival and discovery makes statistical statements of any kind potentially unrepresentative or open to revision, however. These statistics should be treated with some caution, since they are not easily verifiable nor 'statistically significant' for many regions, and because they are incomplete (critically, even with the recent updated bulletins from Avilova 2008 and 2009 even general statistics for Central Asia, the Indus, Egypt have not been made available, and no statistics have been made available for the Late Bronze Age material, reducing the comparative power of the results for our purposes). In all the statistics cited here, we should of course also question to what extent these patterns are a result of differential deposition, and avoid the temptation to assume that they provide a representative sample of actual everyday consumption. For example, metals placed in graves or hoards may have special characteristics not representative of the field of circulating objects of any particular period. Chernykh (1992) provides some general statistics on the contexts of deposition for his database (see Figures 5.31 and 5.32), but unfortunately they were not updated in later publications so are sadly less complete than the other data. The amount of metal deposited at any one time, deliberately or otherwise, may vary according to cultural imperatives (such as the imperative to recycle or sacrifice wealth) rather than simple availability (see below for more discussion on hoarding practices), and so absolute numbers do not reflect such availability.

Ultimately, in the absence of the raw data, the 'flattened' statistics explored here show only a small slice of information, and potentially hide relationships between context, form and material that could reveal much more about the actual flows of these metal materials and types. These limitations of the current data show the urgency of creating a unified and standardized inter-regional database for more precise investigations into metal consumption.

5.5 Contexts of metal consumption

Whilst the range of metal forms and materials consumed by Bronze Age societies is important for understanding consumption and flow, it is also important to consider the social contexts of consumption, or at least as much as can be reconstructed from the textual and archaeological records. More detailed contextual comparison on social practices involving superficially similar objects allows us to see if *ideas* about the use of metal objects were also transferred over long-distances, or just the forms and materials. This clearly has important implications for the nature of interconnections and level of integration and flows.

5.5.1 Metal flows and metal values in the ancient texts

Some textual evidence for metals has already been mentioned above with regards to the sourcing of raw minerals for processing. There are a few direct textual sources regarding metals from Anatolia, primarily from the Cappadocian Old

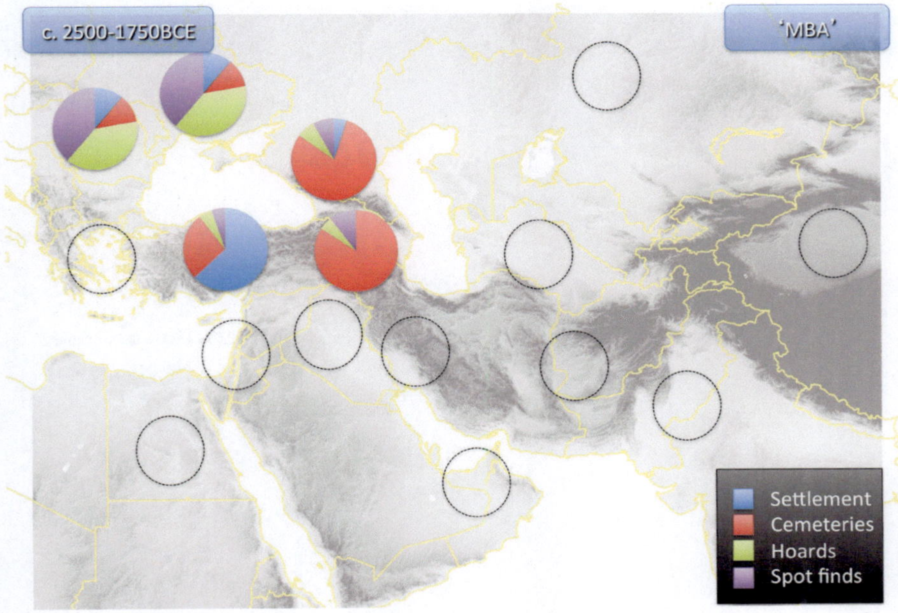

Figure 5.32. Metal objects classified by context of deposition, 'Middle Bronze period', according to Chernykh's database.

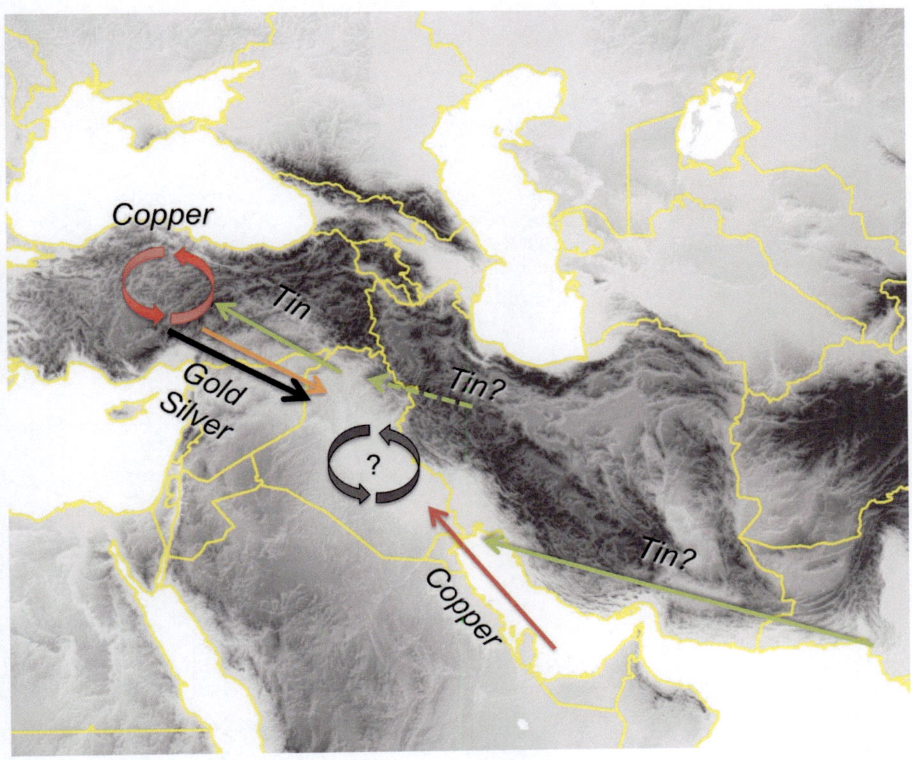

Figure 5.33. The directionality of metal exchange according to Syro-Mesopotamian texts.

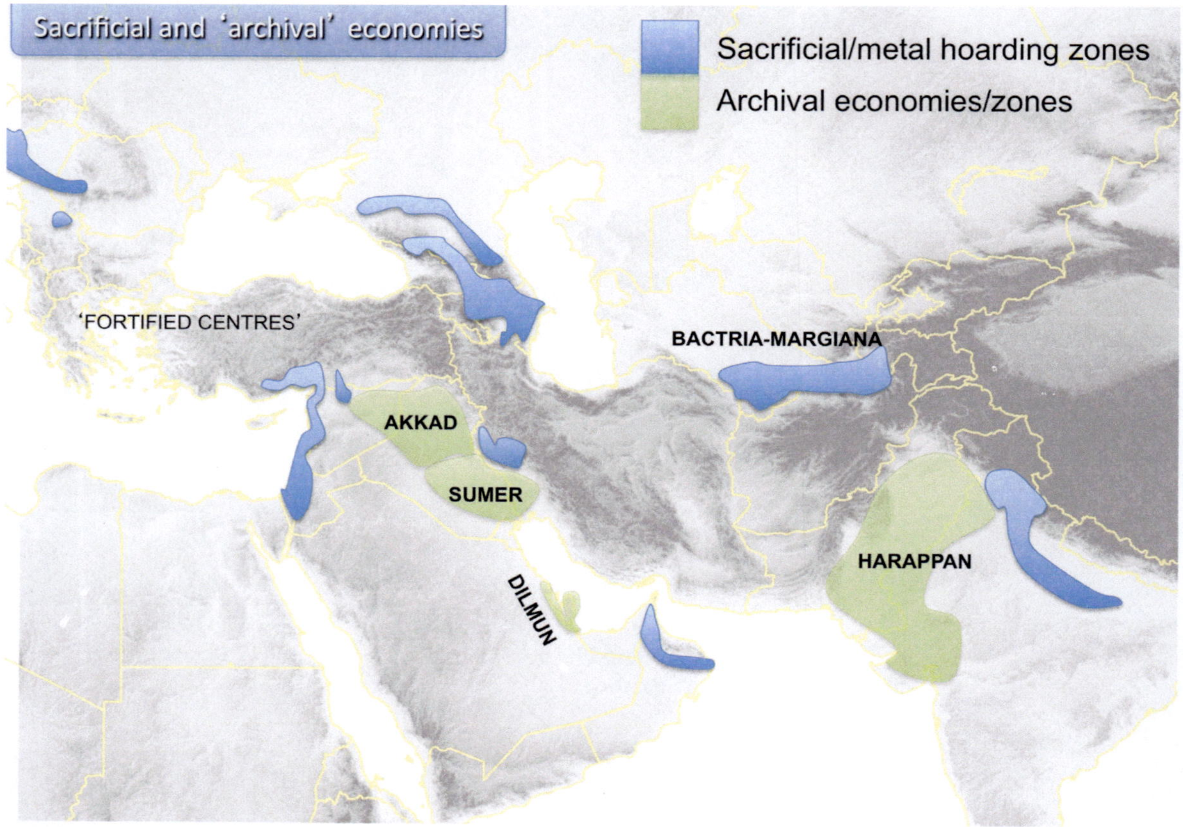
Sacrificial and 'archival' economies
Sacrificial/metal hoarding zones
Archival economies/zones
'FORTIFIED CENTRES'
BACTRIA-MARGIANA
AKKAD
SUMER
DILMUN
HARAPPAN

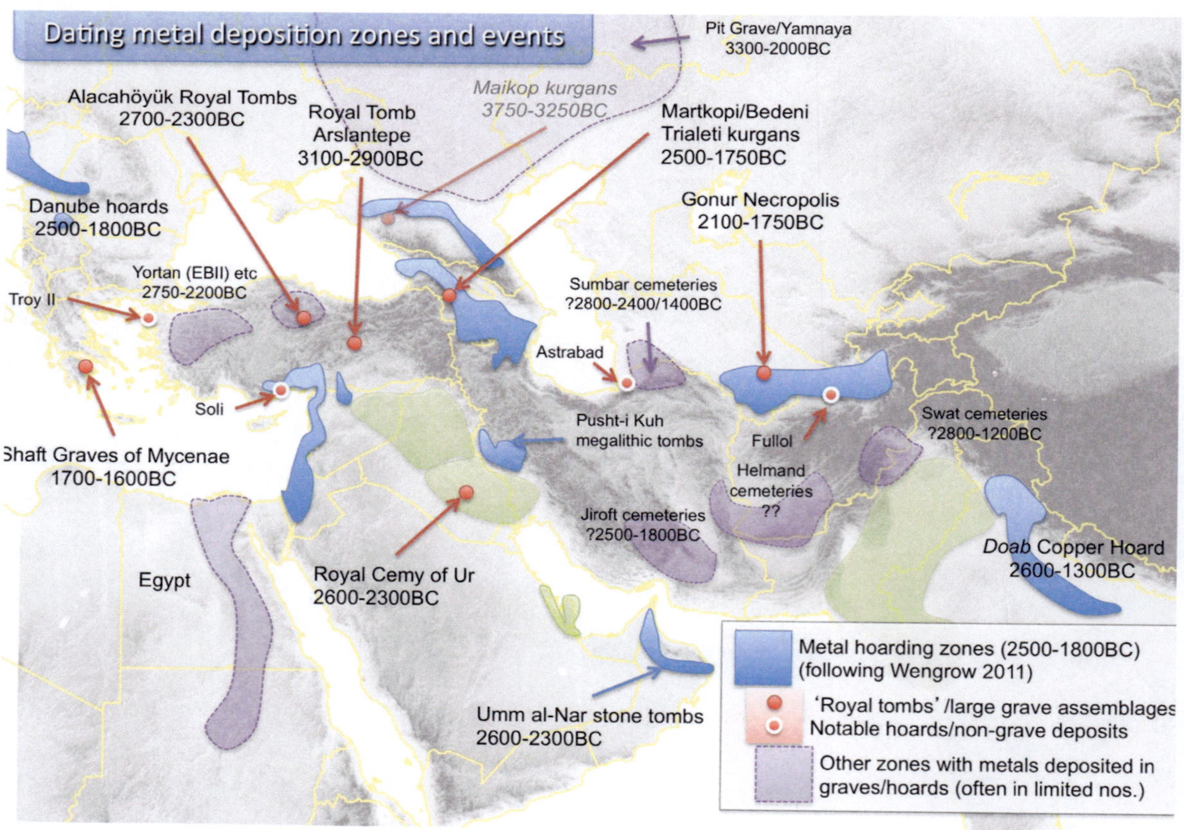
Dating metal deposition zones and events
Pit Grave/Yamnaya
3300-2000BC
Maikop kurgans
3750-3250BC
Alacahöyük Royal Tombs
2700-2300BC
Royal Tomb
Arslantepe
3100-2900BC
Martkopi/Bedeni
Trialeti kurgans
2500-1750BC
Danube hoards
2500-1800BC
Gonur Necropolis
2100-1750BC
Yortan (EBII) etc
2750-2200BC
Troy II
Sumbar cemeteries
?2800-2400/1400BC
Astrabad
Soli
Pushti-i Kuh
megalithic tombs
Fullol
Swat cemeteries
?2800-1200BC
Shaft Graves of Mycenae
1700-1600BC
Helmand
cemeteries
??
Jiroft cemeteries
?2500-1800BC
Doab Copper Hoard
2600-1300BC
Egypt
Royal Cemy of Ur
2600-2300BC
Metal hoarding zones (2500-1800BC)
(following Wengrow 2011)
'Royal tombs' /large grave assemblages
Notable hoards/non-grave deposits
Other zones with metals deposited in
graves/hoards (often in limited nos.)
Umm al-Nar stone tombs
2600-2300BC

Assyrian tablets, but none from highland eastern Anatolia or Transcaucasia, and there are also no known texts found in Central Asia, so much of the information about metals in textual sources is skewed towards the Syro-Mesopotamian world.

Various textual sources hint at the directionality of the flow of raw metals to Mesopotamia (Figure 5.33). For example, texts suggest that copper was imported from *Dilmun* (probably Bahrain), tin from *Meluḫḫa* (possibly the Indus), silver from the east (Potts 1994, 148, 156, 162) – or in the early 2nd millennium, the north-west (Dercksen 2005) – though there is debate about the exact locations referred to by these geographical terms. But there is much less direct textual evidence of metal circulation involving eastern Anatolia or Central Asia. Additionally, whilst attention has often been focussed on raw material sources, much less attention has been devoted to attempting to make sense of the patterns of distribution of metals and metal objects once they had left the source region and entered the economy. The closest textual evidence to eastern Anatolia is that of the Old Assyrian caravan trade between Upper Mesopotamia and Central Anatolia. This has been relatively well studied through the large number of texts from Kültepe (Dercksen 1996, 1999, 2005; Veenhof 1972). These texts describe an exchange system that included the bulk transport of tin from Aššur (on the Tigris) as one of the commodities exchanged for silver in various towns of central Anatolia. The tin was apparently bought in Aššur, having come originally from somewhere 'in the east'. The documents also mention payments and shipments of copper within Anatolia, apparently from the mountains toward the Black Sea. Besides the archaeological evidence for metal ingots, this is one of the clearest illustrations of an organized trade in bulk metals, exchanged by their commoditized (or 'liquid') value rather than in socially-embedded forms or exchange systems – *i.e.* in market-based exchange of materials and not in the form of finished objects, or as part of a gift exchange system. It is likely of course, that 'raw materials' were exchanged in similar ways for centuries before and after. However, it is rather difficult to say whether the picture provided by the Kültepe texts is typical of the wider 3rd and 2nd millennia in central Anatolia, or a unique formulation, since the documents appear to span a period of, at maximum, 150-200 years (c. 1950-1750BC). Other mechanisms (gift exchange, tribute, redistribution) could well have had a significant effect on the movement of metals and metal objects, as indicated by the large quantities of copper presented as 'gifts' (but in the expectation of other gifts in return[72]) between rulers described in the 'Amarna letters' (for example, the messages between the Egyptian and Alašiyan kings, see Moran 1992, 104-105, letters EA33, EA34).

Figure 5.34. Conspicuous sacrifice: metal hoarding zones (redrawn from original, Wengrow 2011).

Another theme is the relative value of different metals to each other and to other materials. The textual evidence from Syro-Mesopotamia makes it quite clear that different metals had very different values within the same socio-economic system, in texts where prices are recorded comparing particular metals and other materials. Iron, when it started being used in limited amounts, may well have been more valuable than bronze – c. 30-40 times that of silver (Dercksen 2005; S. Sherratt 1994b). It has been argued on the basis of utilitarian iron tools listed in Hittite texts that iron was not a high value material in central Anatolia during the later 2nd millennium – unlike other parts of the eastern Mediterranean (Košak 1985; Muhly *et al.* 1985, 73). But the context of references to iron in Hittite texts

Figure 5.35. Hoarding, metal burial, 'Royal' cemeteries or graves in 3rd and 2nd millennia BC. Approximate dates based on: Maikop (Kohl 2009a; Lyonnet 2000), Arslantepe (Palumbi 2008b), Mycenae (Dickinson 2002), Gonur (Hiebert 1994a), Danube (A. Sherratt 1993).

72 The effective line between commercial exchange and gift exchange is thus rather narrow.

places it alongside other valuable materials used for ritual objects: "silver, gold, lapis lazuli, rock crystal, iron, copper, bronze, alabaster and basalt". This list resembles a hierarchy in itself, and one in which lapis lazuli is therefore higher than bronze, by some way. This reminds us that the overarching category of 'metals' is a modern construction, and, though the pyro-technical or material features associated with metals might have also led ancient peoples to group them together, it is equally possible that other attributes were more relevant. Thus in the 3rd millennium the association between gold and lapis may have been more important than between gold and iron, for example (see Sections 4.3.1 and 5.2.2)[73].

Mesopotamian texts indicate that silver had a widespread role as an almost universal mediator of value by which any saleable goods could be exchanged across Syro-Mesopotamia, sometime during the 3rd millennium BC (Moorey 1999, 237). This 'monetization' appears to be associated with a package of other cultural traits: sealing of commodities, weighing systems for standardization, and recording/notation systems, *i.e.* writing. Given the price differentials between silver and tin, which drove the Old Assyrian Trade system, it is clear also that values were not constant in different regions. No doubt these differential values were based partly on the effect of scarcity, but perhaps also on differing socio-economic systems of valuation and hierarchy which may have varied widely. It is these alternative systems of valuation that may account for the very different patterns of deposition of metals in the archaeological record.

5.5.2 Metal deposition: conspicuous sacrifice, identity and systems of valuation

Excluding the impressive remains of the Royal Cemetery at Ur, it was fairly uncommon for residents of 3rd and 2nd millennium Mesopotamia to deposit large amounts of metal as grave goods or as hoards. By contrast, a large number of metal objects for certain other regions come from grave or hoard contexts. Chernykh (1992), notes that, in total, hoards and cemeteries comprise around 55% of the objects in his database compared to around 30% from settlements. In cases where spectacular numbers, or quality, of finds are deposited intentionally by ancient consumers in graves or hoards, the ritual or social significance behind such deposits must have been considerably heightened. Rich or 'royal' burials have for some time been viewed not simply as 'reflections' of the contemporary wealth (or poverty) of elites, but rather the material remains of dramatic performances of wealth (or, more properly, wealth sacrifice) which constitute a strategy used by emerging elites to establish new hierarchies. Similarly, the potential function of large metal hoards as a social strategy akin to 'potlatch', designed to manage the flow of metal wealth, was discussed by Andrew Sherratt in his study of the economic developments and interregional interactions within eastern Europe during the Bronze Age (A. Sherratt 1993). Both can be seen as examples of conspicuous sacrifice, or what Appadurai has called 'tournaments of value' – socially sanctioned or abrogated strategies designed to facilitate the competitive negotiation of status and hierarchy through the deployment of valued materials, ultimately designed to create "new paths of commodity flow" (Appadurai 1986, 56).

73 This may have changed in the 2nd millennium, when iron and gold are more often found together – for example the iron dagger with gold hilt from Tutankhamun's tomb (see *e.g.* Maxwell-Hyslop 1974, 144), or the iron battle axe with gold mount from Ugarit (see *e.g.* Waldbaum 1978, 17).

David Wengrow has recently revisited Sherratt's approach to conspicuous sacrifice on a wider scale with regards to both hoarding and spectacular burial practices across Eurasia between 2500 and 1800BC (Wengrow 2011). He notes that many of the hoarding deposits and rich burial assemblages of this time (along the Danube, the Ganges, the Levantine coast, in Transcaucasia, and in Luristan, see Figure 5.34) are located along major axes of exchange on the peripheries of 'monetized', urbanized centres. He argues that these deposits reveal a form of 'sacrificial' economy, in which social, ritual and moral values must focus on the sacrifice of wealth (such as depositing metals in the ground) to preserve social reproduction; this is in opposition to what he calls an 'archival' economy, in which values focus instead on the constant recirculation of wealth (such as recycling of metals). Wengrow focuses only on the period 2500-1800BC and does not explore the changes to the overall system; rather, perhaps for the sake of clarity, he presents a snapshot of the location of these two opposing 'ritual economies' during this particular period. His ideas could be taken further, if one looks diachronically and in more detail at the data and attempts to explain some of the complexities revealed when metal deposits are viewed through this 'ritual economic' lens (Figure 5.35).

To begin with, it is worth noting the similar but much earlier metal deposits of the Maikop kurgans (A. Sherratt 1997; Lyonnet 2007; Kohl 2009a) north of the Caucasus, dating probably to around the mid-4th millennium, which prefigure the more southerly Martkopi/Bedeni/Sachkere and Trialeti kurgans mentioned by Wengrow as examples of 'sacrificial economies' by at least a millennium. The concentration of copper ore sources in the eastern Anatolian-Caucasian region within which both sets of burials were made has provided an obvious candidate for the source of this metallic wealth: in both periods it has likewise been presumed that the builders of these impressive tombs were important agents in the distribution of metals to the north or south. Can we suggest then that the Maikop folk were also therefore acting within a similar 'sacrificial economy' – where items which have been kept in circulation in the south have greater local socio-political efficacy through funerary sacrifice?

To complicate the picture further: the Maikop kurgans are separated both temporally and spatially by groups who deposited very few metal objects in either graves or hoards. From the Kura-Arax 'culture', relatively few cemeteries have been identified, and those that have been contain relatively few metal artefacts. This of course does not mean that the Kura-Arax folk were not producing and consuming metals. Indeed it has been suggested that they were producing metal objects on a similar scale to the Maikop groups (Chernykh 1992, 73). Rather they simply did not deposit them in the ground. Following Wengrow's model, this suggests that the Kura-Arax groups did not subscribe to a 'sacrificial economy' (or at least, not one that we can recognize) – despite being located in apparently a very similar geo-economic location to both the partially contemporary Maikop and later Trialeti. But does this also mean that these groups had something like an 'archival' economy, or do we need to imagine some other form of 'ritual economy'? In the absence of the emerging urbanism that characterizes Late Uruk and Early Dynastic Mesopotamia (which are contemporary to the early part of the Kura-Arax phenomenon), it would be difficult to claim that Kura-Arax communities had any kind of 'archival' economy. It may, however, at least suggest that the Kura-Arax and Syro-Mesopotamian zones were far more integrated into an overall economic system than is often assumed.

The far-flung regions of Turkmenia also offer a more complex situation than the general model suggests. Intriguingly, supposedly opposed 'archival' and 'sacrificial' economies seem to be very closely located here. In particular, at Gonur Depe (Rossi Osmida 2002a; esp. 2002b; Sarianidi 2002a, 2006), a site of the 'BMAC' (Bactria-Margiana Archaeological Complex) dated to around 2100-1750BC, there are features of 'archival' urbanism (with monumental architecture perhaps for redistributive systems, and stamp seals) alongside metal-rich tomb assemblages, more akin to 'sacrificial' economies. The plundered graves of Bactria also show large amounts of metals being deposited in the ground (Sarianidi 1990a). There are various chronological and interpretative questions to be explored further here of course, not least how we should compare various BMAC-related metal assemblages with possibly contemporary grave assemblages to the north-east (such as at Sapalli), to the earlier sites along the Kopet Dag of the Namazga IV/V (3000-2000BC) period, such as Altyn Depe (Masson 1968, 1988; Sherratt 1997; Kircho, Korobkova and Masson 2008) – which have relatively modest grave assemblages – and to grave assemblages in the Sumbar valley (Khlopin 1983; Chlopin 1986; Khlopin 2002). We should also be reminded of the massive sacrifice of wealth represented by the Royal Cemetery at Ur and the pyramids of Egypt – both constructed within a dominantly 'archival' economy. The co-location of different types of 'ritual economies' makes the point that such systems may not be mutually exclusive either geographically or culturally. Indeed, there may have been many alternative overlapping (and potentially contradictory) value systems at work within the same regions that relate to shifting imperatives of individual agents or groups.

It is perhaps worth noting the apparently short-lived nature of 'sacrificial' phenomena relative to 'archival' practices. Perhaps we might suggest that burying large amounts of 'wealth' (in the form of metals or other precious goods) is inherently a short-term strategy since, once sacrificed, the items cannot be re-sacrificed – essentially resources are exhausted by excessive consumption. Conversely Mesopotamian communities' adherence to the 'archival economy' may simply have represented a cultural necessity in a land with no metal ore resources of its own. An alternative view comes precisely from another supposedly interstitial zone. In a complementary essay to Wengrow's, Bachhuber attempts to explain the deposition of metal objects in the Alacahöyük Royal Tombs as a measure not of the 'wealth' of the areas at that particular moment, but rather as the consequence of the "coexistence of two divergent valuations of metal in north-central Anatolia during the period of the royal tombs at Alacahöyük (Figure 5.36) – namely an indigenously constructed sacrificial value, and a liquid value ultimately derived from foreign commercial enterprise" (Bachhuber 2011, 170) – systems analogous to Wengrow's 'sacrificial' and 'archival'. He argues, on the basis of ethnographic parallels of contact between monetary and non-monetary systems of value cited by Bloch and Parry (1989), that there may have been a moral imperative behind the sacrificial deposition of metal, designed to deal with the introduction of new ideas about the value of a material which was already a culturally significant material, in order to "uphold its [local] ideological value". If correct, Bachhuber's ideas might be taken to suggest that other short periods of rich deposition of metals (or other materials) might be taken to precisely indicate zones of contact between regions with differing value systems, or at least regions in which there is struggle between competing value systems – and might also explain the apparent co-location of

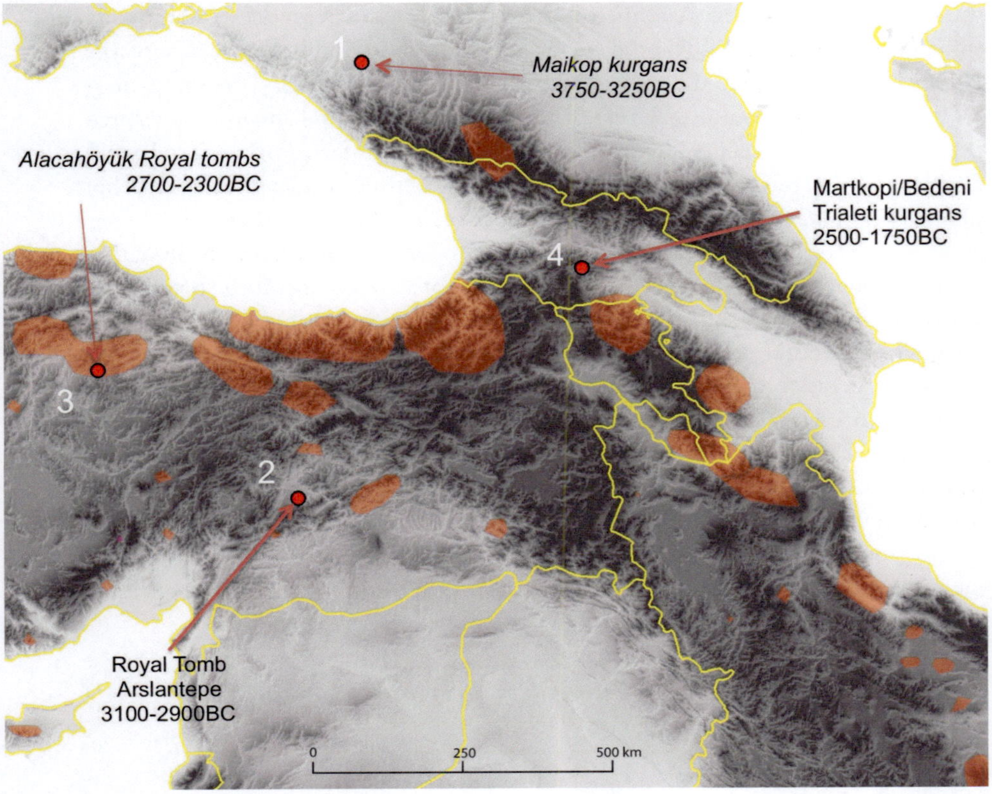

Figure 5.36. Major conspicuous deployment of metals in funerary contexts in Caucasia and Anatolia and location of major copper sources.

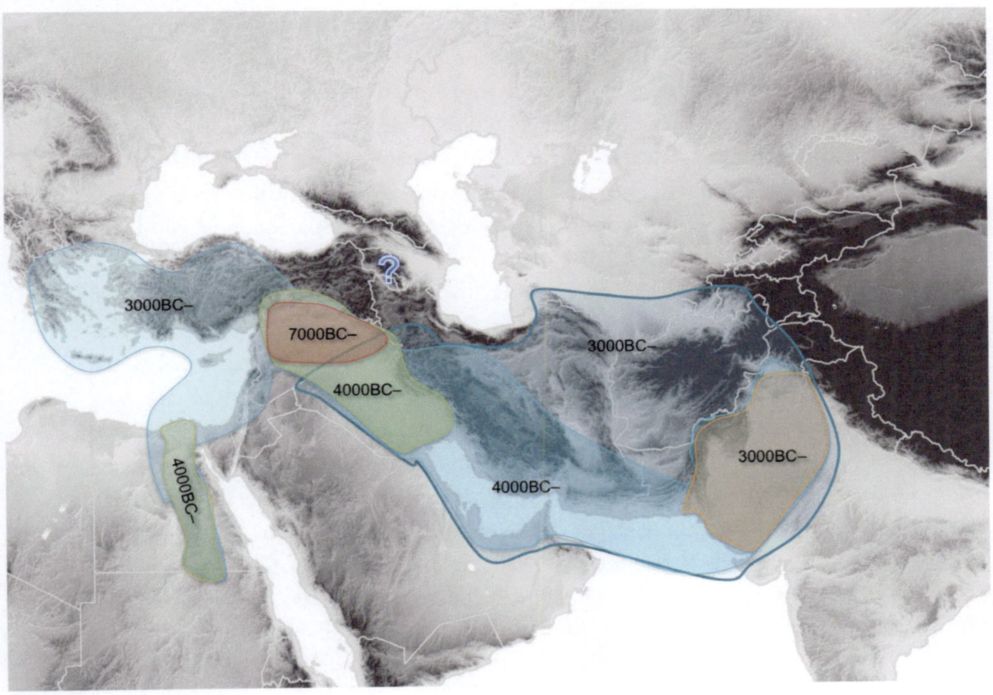

Figure 5.37. Regional adoption of sealing practices (cylinder/stamp) by date (adapted from Rahmstorf 2011).

value systems in Central Asia mentioned above. A similar model of competing value systems might be advanced regarding the location and significance of the Arslantepe 'royal tomb', dating a little earlier to c. 3300-2900BC (Frangipane *et al.* 2001), here between post-Urukian and Kura-Arax cultural spheres – although the exact nature of each valuation system is unclear. In terms of the extent and direction of flows of metals, such deposits would indicate at least a certain level of intensity of material and information flow between the regions in question.

If these interpretations of the data are acceptable, it is interesting to note that the movement of materials and metal forms did not seem to stop at the boundaries between value zones, and that exchanges were still undertaken. This means that Chernykh's metallurgical provinces or foci do not necessarily coincide with systems of values. There might be possible exceptions to this flexibility, however, in the distribution of certain forms integral to one of the value systems, for example standardized ingots, could be argued to be associated with Wengrow's 'archival' or Bachhuber's 'liquid' economy. The trouble is that ingots which made their way outside of the 'archival' system are the ones most likely to be deliberately deposited in the ground. Perhaps more productively, it may be possible to trace alternative networks and zones, such as those involving such value systems, through non-metal materials, including weights, or seals (see below, Section 5.6).

5.5.3 Materiality of metals: liquidity, partibility and origins of exchange value

The physical properties of metal were probably of vital importance to their elevation to the status of near-universal mediators of value from the 4th millennium BC onwards. Metal objects can be melted down and 'recycled' into different forms with quite different purposes of consumption, an infinite divisibility that can facilitate universal measurement (even if those measures are in flux). Contemporary with the emergence of metal as value mediator across parts of western Asia and the eastern Mediterranean in the 4th and more particularly 3rd millennia, are the creation of means of 'measurement' (such as weighing systems) and various attempts to 'standardize' material culture. For example, one of the features of wheel-made pottery, gradually ascendant in the 3rd and 2nd millennia from the Aegean to Central Asia, is its connection with mass-production and tendency towards standardization[74].

Such processes of abstraction and standardization changed the way in which economic transactions could be undertaken and hint at very intense and far-reaching interactions across a large part of the Old World. The metallic economy, weighing systems and standardized production are directly related to the way in which, as part of the process of 'urbanization', face-to-face knowledge of and obligations between individual members of small-scale societies are replaced by basic abstract yet widely-shared codes by which individuals may interact with relative strangers within much larger social configurations. In an urban setting, networks of obligation and exchange of labour and materials may be bypassed through abstract value (*e.g.* through intermediate materials such as metals). Simultaneously, the ability to redeploy this socio-material capital over longer

74 One can also compare one interpretation of the hand-made bevel-rimmed bowl characteristic of Late Uruk Sumer and northern Syria as an early attempt to standardize volume for payments in grain or bread (Goulder 2010).

distances or via indirect relationships, rather than relying on social capital derived from direct social relationships (since metal is more partible and 'mobile' than human individuals), also subsequently allows labour forces, armies and elites to become more mobile. Political entities can grow because of the communicative potential of mobile capital, people and symbols. Hence the material basis of value is intricately built into the trajectories of urban societies.

Whilst the liquidity and divisibility of metal facilitates this expansion of 'abstract value', it does not necessarily follow that metals always functioned in the same manner under different social conditions. Indeed the analysis of different deposition practices relating to metals has indicated the existence of alternative, contrasting and perhaps competing ways in which metals were integrated into social relationships between 3000 and 1500BC, sometimes within the same communities and geographical spaces. The reason for metal's role as mediator in exchange relationships should not be confused with the reasons for its own 'high value' and desirability. The ability of metal, particularly copper alloys, to make highly attractive and functional[75] objects such as weaponry (axes, daggers, swords and maceheads), should not be forgotten. Other materials (*e.g.* precious stones or textiles) may of course be 'more costly' than metal per weight, but the recyclability of metal objects can heighten their social role in the circulation of 'value' in a disproportionate way. But like other commodities, the uneven and often distant distribution of metal sources (part of their 'resistance to being owned'), at least played some part in the ability of certain individuals to monopolize or control their movement; as did the material's magical properties. Ideas of magic related to metallurgy may have been espoused by both urban and non-urban communities alike: it has long been noted from folklore studies and ethnographic accounts that smiths are frequently believed to have magical powers (Budd and Taylor 1995). These powers presumably come from their ability to transform inert rocks into metal objects, with 'technical virtuosity' (Gell 1992). Metalsmiths themselves perform complex 'magical' rituals to ensure the success of their work (see *e.g.* Rowlands and Warnier 1993). These magical properties and associations of metallurgy return the material's significance to a kind of 'transcendental' value, beyond the bounds of everyday social or economic relationships.

5.6 Metal flows traced through other materials

Actual metal objects are poorly represented in the archaeological record and, as has been discussed extensively already, their distribution in the archaeological record is not a reliable direct indicator of ancient metal distribution. Attempting to study the flow of metals from metal objects alone will thus always remain an incomplete exercise, and it is necessary to look for alternative proxy sources of information, including materials that relate to metals indirectly.

75 The 'functional' strength of copper objects versus their stone equivalents is not always clear however: "Copper as a material for tools has no great advantage over stone except that it can produce a variety of shapes and sizes of tools, for example, particularly thin and therefore sharp blades" (Ratnagar 2001, 363). Tin alloys are substantially better than either natural copper, or the predominantly low-arsenic bronze which was in use for most of the 3rd millennium for example (Northover 1989, 113-114; *cf.* Chernykh 1992, 3). Metals can, of course, be recycled into very differently shaped tools according to current needs – whereas stone tools can only be 'retouched' to form smaller tools.

5.6.1 Weighing systems and metals?

The emergence of a number of weighing systems in different parts of the Near East (Figures 4.10 and 4.11) has already been discussed in the chapter on stones (see Section 4.6). It is relevant again here because one major catalyst for the emergence of weighing systems was the facilitation of the exchange of standardized quantities of metal. As such, the full distribution of weighing systems could be taken as proxy evidence for the distribution of metals or, perhaps more specifically, for the distribution of socio-economic systems in which metals have taken on the role of liquid value mediator. The lack of indisputable weighing systems in eastern Anatolia and western Central Asia thus may be significant not only for networks of interaction (especially for metals themselves) but also for the social systems in place in both regions. It might suggest, for example, that the 'liquid' valuation of metals was relatively unimportant in the highlands of Transcaucasia/eastern Anatolia after 2500BC, in contrast to cultures to the west in central Anatolia, despite the geographical proximity to urbanizing Syrian settlements. This may also suggest that Syro-Mesopotamian metal was being sourced from more distant western and central Anatolia (where weighing systems were coming into use), rather than from Transcaucasia. However it is difficult to be confident, unless one assumes that the location of the centre of Old Assyrian trade in central Anatolia is a result of the earlier direction of interaction.

5.6.2 Seals and sealing practices?

Seals can be seen from a certain perspective as a technology designed to ensure 'quality control' and 'authentication' which also relates to standardization and impersonal exchange. The distribution of seals (Figure 5.37) is related to, but perhaps more extensive than, that of weighing systems, especially if the stamp 'seals' of the BMAC can be argued to serve a similar function to the cylinder seals of Mesopotamia (though there may be reasons to treat them differently – see Section 6.6.3). A small number of cylinder seals have been found in western Central Asia, starting from early examples at Sarazm (Lyonnet 2000), down to both seals (and some actual sealings) at Togolok-21 and Gonur Depe (Sarianidi 2002b, 276-278; Kohl 2007, 201). Whilst a few odd seals have been found, the distribution of actual sealings does not seem to extend into highland Transcaucasia or eastern Anatolia. Sporadic finds of seals in both regions[76] probably indicate their recirculation as items of exotic interest rather than their use in regular practice of sealing commodities. In a general sense, seal distribution might provide a very rough proxy for networks in which metals were circulating since they could be argued to form part of the same cultural package of standardization. Indeed, metals may well have been one of the materials which were sealed perhaps as sealed bags of metal ingots. However, the direct relationship to metals as opposed to any other commodity (some of which might be more likely to be 'sealed') seems somewhat tenuous, and it may be better to treat them as indicators of general networks of exchange that are not specific to metals.

---

76 For example, in the Caucasus: the cylinder seal from the Maikop kurgan apparently worn as a bead (Anthony 2007, 284-285); and the Mitanni seal (dated to the 15-14th centuries BC) at Gegharot (Bobokhyan *et al.* 2008).

Figure 5.38. Examples of distinctive/defining forms of the 'Anatolian Trade Network'; (a) depas amphikypellon (Şahoğlu 2005, 348 fig. 4); (b) tankards (ibid. 349 fig. 6); (d) pyxis (ibid. 351 fig. 9) (e) spouted jar (ibid. 350 fig. 8); (c) Syrian bottle (Zimmermann 2005).

5.6.3 *Metallschock*: metal skeuomorphism

One possible indication of the power of metal consumption is in the evidence for skeuomorphic features in other materials, which recall the physical properties of metals or metallic techniques. In particular pottery forms frequently emulate the shapes, decoration, colours and lustre of metal vessels (*cf.* Vickers 1989). This is not simply a lack of imagination amongst potters, but a conscious effort to draw on the aura of metal via a kind of sympathetic magic. This concept of *Metallschock* (Schachermeyr 1955), as the way in which pottery types of the Early Bronze Age seem to emulate metal aesthetics or actual metal objects, has a long pedigree and was widely recognized by early archaeologists (Forsdyke 1914; Childe 1915). Despite its long relegation to the margins, the idea is now receiving wider attention and reapplication in archaeological explanation. A finely crafted example is to be found in Nakou (2007), in which fabrics and forms of EBA pottery in the Aegean are shown to have skeuomorphic features of metal types. Here Nakou demonstrates how the identification of skeuomorphism can enlighten both the mechanism of ceramic change (as 'cheap' copies of contemporary metal equivalents), and the development of metal types and metal-working techniques for which no examples remain (as the pottery forms and colours also stand as proxy copies and indications of trends in metal designs). The question as to whether any particular ceramic ware can be said to be skeuomorphic of metal, or showing features of *Metallschock*, is always a matter of interpretation and differing extent. In some cases it may be possible to find clear metal prototypes for ceramic shapes: hence the argument for skeuomorphism is stronger (whilst we assume that the metal versions came first, we should remember that cross-craft processes are interactive and it is also possible that ceramic styles occasionally fed into metal styles). But where such direct parallels are missing, metallic features (fine walls and lips, metallic colouration and surfaces, incised, mock-'soldered' or similar decoration, 'tinny' sound when struck) may still recall prototypes that have not survived, or less tangible systems of aesthetics that recall metal as a material.

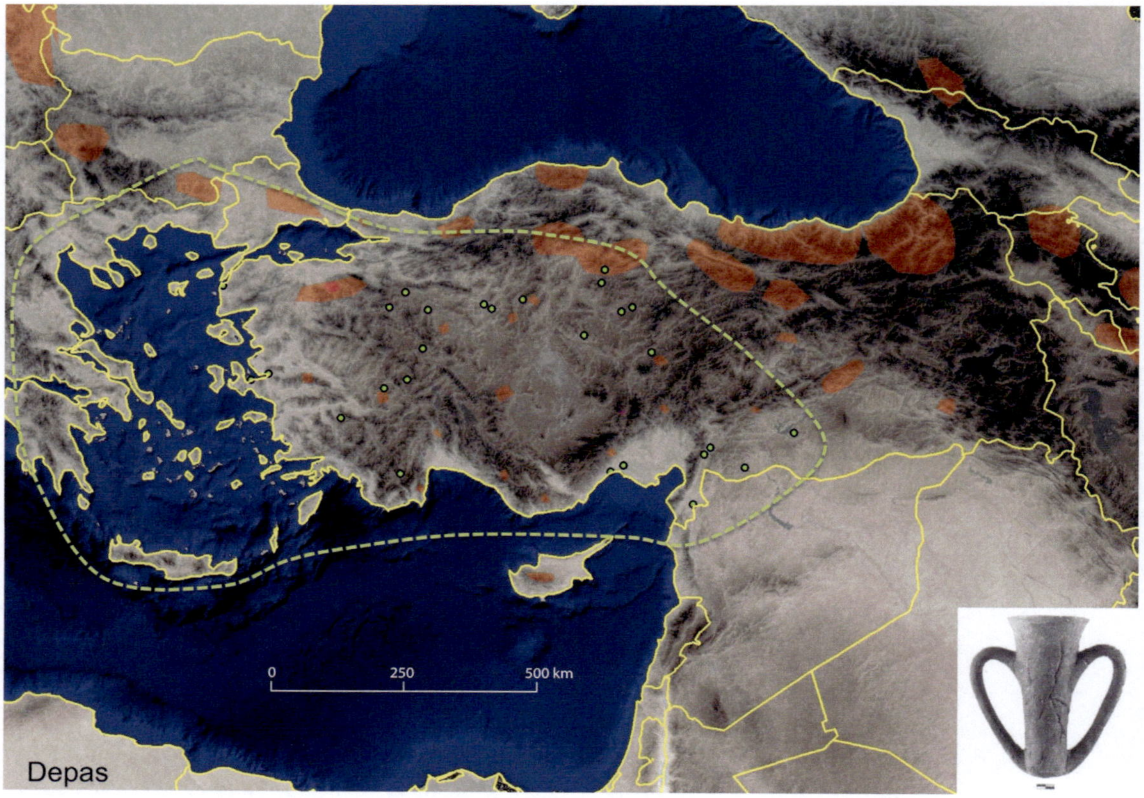
0
250
500 km
Depas

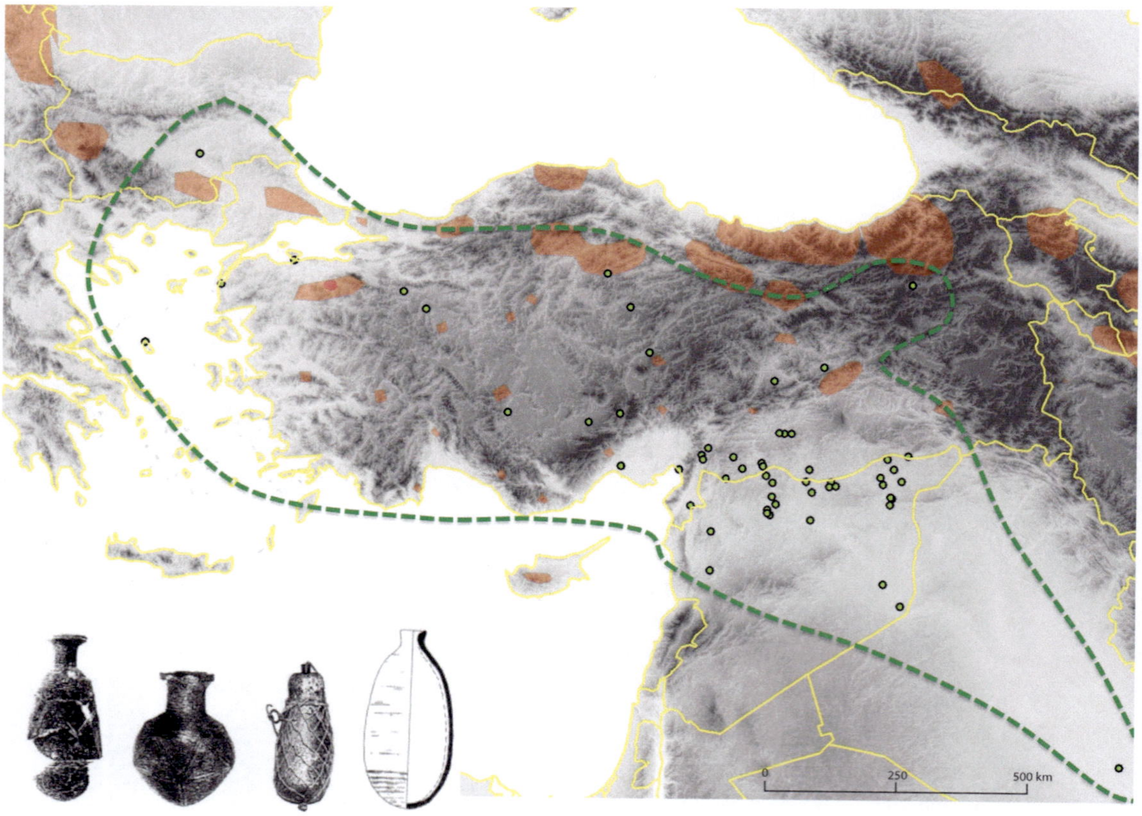
0
250
500 km

i. [Aegeo-]'Anatolian Trade Network': western and central Anatolia (2400-1800BC)

Though mostly outside our main area of interest, certainly pottery assemblages of the Early Bronze Age Aegean and western Anatolia represent a classic example by which other suggestions for skeuomorphism are measured (Figure 5.38), since acutal metallic prototypes have also found within the archaeological record. The most famous is the so-called '*depas amphikypellon*' cup[77] – a distinctively shaped vessel likely to have been used for a special form of social drinking (since the form prevents it from standing by itself – and hinting that it was passed from drinker to drinker until empty), the contents presumably alcoholic or narcotic (Figure 5.39). This is only one of a complex of chronologically related pottery shapes however, including the "tankard, two-handled cup, wheel-made plate, incised pyxis, cutaway-spouted jug and 'Syrian bottles'" which spread alongside (or soon after) the introduction of wheel-made production techniques (Şahoğlu 2005, 339), which for convenience we might label 'Anatolian Trade Network' wares as first coined by Vasıf Şahoğlu. This complex of assemblages dates from around the end of the Anatolian EBII period (c. 2400-2200BC), with a distribution that stretches from Syria to the Aegean, along what Turan Efe has termed 'the Great Caravan Route' (Efe 2002, 2007). The existence of metal prototypes for the vessels has long been recognized (Mellink 1998; Rahmstorf 2006). As well as the shapes, the clay fabrics used for these vessels tend to be 'metallic' (without painted patterns, relatively plain with slipped or burnished surfaces, with colours ranging from blacks to greys, and some reds – possibly mimicking both tarnished and untarnished silver, lead, gold and copper), and the associated introduction of the wheel to pottery production (a technology which facilitates mass-production of vessels, and greater adherence to standardized forms, based on clear prototypes), also points towards the heightened importance of metal prototypes for ceramic production. These related types are not the only ones in the Early Bronze Age Aegean to recall metalwork, and Georgina Nakou (Nakou 2007) has argued that changes in ceramic vessel forms from the Aegean and western Anatolia of this period may well directly reflect changes in contemporary metal vessel forms which have not survived in the archaeological record. Less clearly, earlier ceramic forms and fabrics in this region may also recall metals or metal prototypes: both Late Chalcolithic and Early Bronze Age ceramic assemblages from Anatolia and the Balkans (such as the Baden ware) tend to be plain, polished or burnished, in contrast to earlier pattern-painted wares, and the colours favoured are uniform blacks, greys and reds, with an increasing tendency towards thinner, lighter, walls and rims. In the absence of clear metal parallels for vessels, it is difficult to label Late Chalcolithic wares as metal skeuomorphs, but the aesthetic relationship between lustrous pottery, and the emerging importance of lustrous metals needs more analysis.

The period of 'Anatolian Trade Network' wares, however, is one in which long-distance contact between the Aegean and Syro-Mesopotamia via Cilicia and the Levant seems to have intensified, along a possible interaction route that Turan Efe

Figure 5.39. Distribution of 'depas amphikypellon' drinking vessels and apparent limits of circulation – not including the many examples from the Greek mainland (data based on Rahmstorf 2006).

Figure 5.40. Distribution of 'Syrian flask' bottles and apparent limits of circulation (data based on Rahmstorf 2006 with additions).

77 Although, as has often been noted, this term derived by Heinrich Schliemann from Homeric epic is unlikely on etymological grounds to have originally referred to the shape with which it is now associated in the archaeological literature, since it is more likely to have meant something along the lines of 'vessel-with-cups-on-both-sides' or 'vessel-with-cups-all-around' rather than 'vessel-with-two-handles'.

has called 'The Great Caravan Route' (Efe 2007). Whether or not an inland 'caravan route' or sea-based interaction (*cf.* Mellink 1998) is more important, there are clear east-west connections highlighted also in the adoption of weight-systems and sealing practices at a similar time (see above Sections 5.6.1 and 5.6.2, *cf.* Rahmstorf 2006). The combined effect of an apparent interest in standardization (weighing and sealing) alongside ceramics that mimic metalwork, seems to emphasize the importance of metals to Aegeo-Anatolians at this time – especially with regards to their relationships with peoples to the east from where some of the 'prototypes' may have arrived. It is very tempting to see this as the imprint of an explosion in the intensity of exchange of metals within and beyond this region. Tracing the details and explanations behind any such economic acceleration remains difficult. Perhaps it marks the moment when Aegean or western and central Anatolian metal sources become integrated into the consumption demands of Syro-Mesopotamia. Cyprus is also integrated into the same metalworking sphere (documented by the Philia phenomenon) from a similar period (Webb *et al.* 2006).

The particular role and distribution of individual forms thus need careful analysis: and the 'metallic' nature of the form in question is only one aspect of its importance to the flow of commodities. The depas cup, for example, though clearly with metal versions/prototypes, may also be useful as a indicator of the spread of particular drinking customs, than of the flow of metals. 'Syrian bottles' (Figure 5.40) provide another example of a complex phenomenon. The distribution of these vessels reaches from northern Syria, across Anatolia, to the Aegean (Zimmermann 2005, 2006), dating mainly to Anatolian EBIII contexts but with examples into the MBA: "The production of 'Syrian' type bottles is continued in the early and advanced 2nd millennium BC as bottles with narrow necks and protruding feet from Middle Bronze Age II contexts in Kültepe/Karum-Kaneš clearly indicate" (Zimmermann 2005, 161). They have been generally connected with Syria/northern Mesopotamia, where examples were made from a pottery fabric known as 'Syrian metallic/stone ware' (see below; Rova 1991). A number of metal versions have been uncovered in Anatolia: for example, a silver 'version' found at Eskiyapar and lead 'copies' or, rather, prototypes at Demircihöyük-Sarıket. An example was also found in eastern Anatolia at Sos Höyük (Sagona *et al.* 1996, 37, 44 fig. 10.9). The small bottles (which actually are represented by a variety of shapes) were perhaps used to transport or market some specialized unguent, perfume or oil. The rounded bases of these bottles and a couple of examples with net-pattern reliefs – suggesting the shapes were normally carried in netting (Zimmermann 2006) – certainly hint towards their role in transportation or at least an association with something that is transported from afar. If it is true that the ceramic bottles were being taken from northern Mesopotamia all the way to central Anatolia (as has been suggested, but which to my knowledge remains as yet unconfirmed petrographically or chemically), then they represent a very interesting phenomenon. The ceramic forms could be local or low-cost copies of the 'real' metal versions that were circulating within a network revealed by the distribution of the ceramic forms for a non-elite 'market'. Another possibility is that the ceramic versions of this form were merely used to transport liquids which were then transferred to the metal variants for personal use: the ceramic forms would therefore have been mere skeuomorphic marketing tools, 'fake' packaging designed to show the purchaser the content by reminding them of their 'real' bottle at home. In both senses the ceramic forms might be taken to reveal the

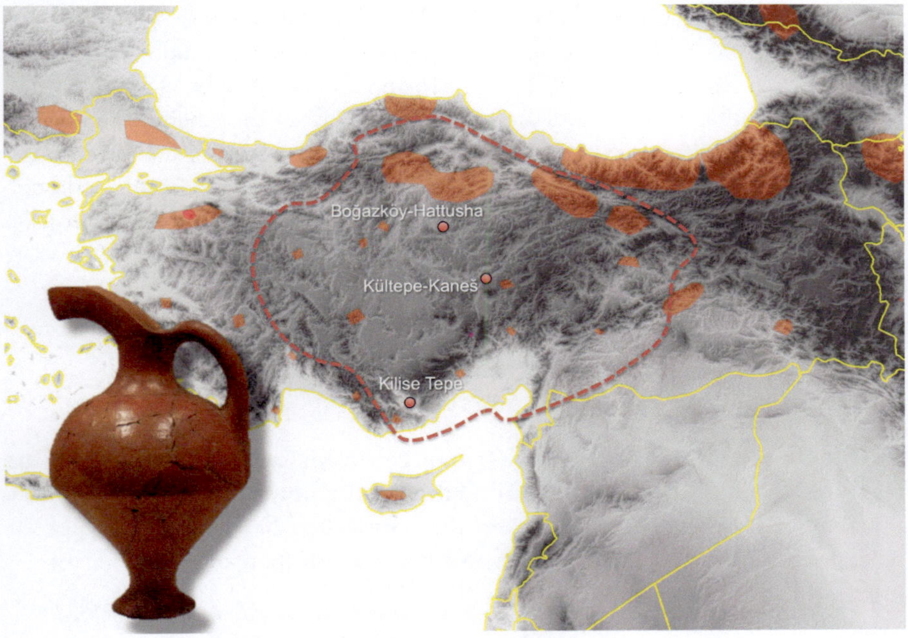

Figure 5.41. Distribution of ´Metallische Ware' (according to Kühne 1976; Kühne and Schneider 1988). Copper bearing zones are shown in orange.

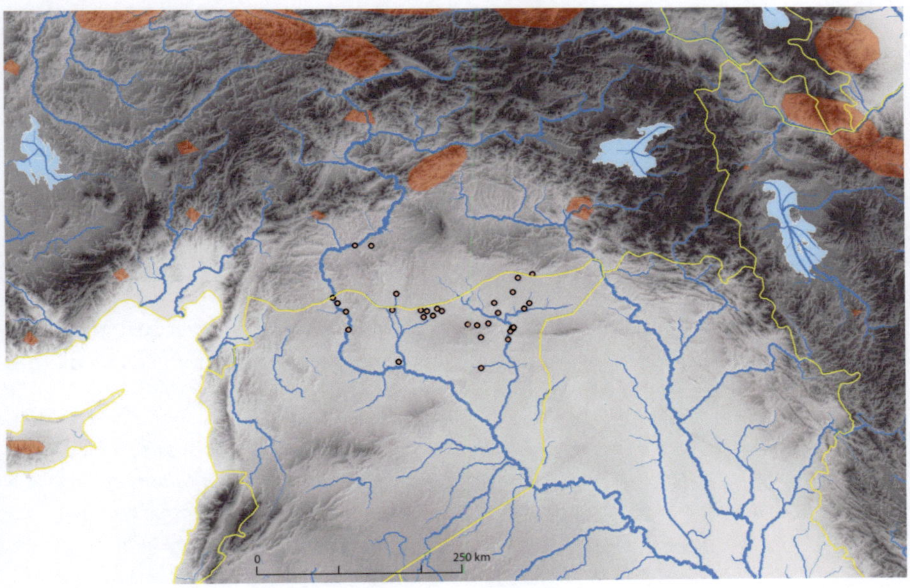

Figure 5.42. Distribution of 'Metallic ware' (according to Pruß 2000). Note the more restrictive definition of the term adopted by Pruß results in a smaller footprint. Copper bearing zones are shown in orange.

boundaries of a particular metal flow, though the intensity of that flow might be different in each case.

ii. 'Metallic Wares' in Syria and Cilicia (2700-2000BC)

In northern Syria, the initial use of the term metallic ware (or rather *Metallische Ware*) was by Kühne (Kühne 1976, 33-72) to define a set of fine wares with a copper-like appearance, identified at Tell Chuera and recognized from sites across the region[78] (Figures 5.41 and 5.42). What is known as 'Metallic ware' in northern Syria and south-eastern Turkey is also sometimes known as 'Stone ware' (*cf.* Schneider 1989), or 'North Mesopotamian Metallic Ware'/NMMW (Broekmans 2006). These wares are known mostly from the Khabur, Balikh and mid-to-upper Euphrates valleys across the north Syrian plain from around 2700-2200/2000BC. The association with a metal aesthetic is widely accepted, though the reasons for this connection are less often considered in detail. In a publication of the material from Tell es-Sweyhat, the physical characteristics of Metallic Ware are described thus:

> *[A] highly fired ware with a fine, hard fabric and dense paste, usually gray, reddish or orange in color. It has a metallic clinky ring, hence its name.* (Wilkinson *et al.* 2004, 89, fn. 58).

The distribution of the various types of North Mesopotamian Metallic Ware are relatively constrained, and its cultural longevity only around 300-400 years. Like Kura-Arax (see below), however, Kelly-Buccellati has linked the types she recognizes as ('Syrian') Metallic Ware to a north-south trade of actual metals:

> *Metallic ware originated in the vast plains of northern Syria and was transported southward at least as far as Terqa and Mari along the Khabur / Euphrates route. To the north we find it in the Keban sites, as well as in the Malatya region and now more of it is published from Kültepe. All of these northern sites are part of a trading network serving the metals trade, but which must also have transported other items, as it did later in the Old Assyrian period.* (Kelly-Buccellati 1990, 124).

In fact, for assemblages from Cilicia and central Anatolia, the term 'metallic ware' has been used to designate an apparently different set of pottery assemblages (*e.g.* Mellaart 1968). Yener and Vandiver (1993) in their publication of the metal-processing settlement of Göltepe in the Taurus mountains mention the presence of a 'clinky ware', painted or unpainted, and relate it to pottery described according to them as 'metallic ware' by Mellaart. This ware, strictly-speaking termed "thin painted metallic ware" by Mellaart (1954) is described by him as[79]:

> *[H]and made, buff or red, gritty, hard fired, often overfired and then with a blue-grey-purple colour. The surface most often has a buff or light brown colour, apricot or brown or darker purplish-black. The inside is often scored or pockmarked. The walls of vessels are often very thin, the breaks are sharp and give a metallic ring. Sometimes the ware is coarse of a brick red gritty variety, only wet-smoothed and with incised handles. This coarse ware seems much more common in Cilicia, so that the excavators at Tarsus called the ware 'red gritty ware', a term not fit to describe the finer specimens from the Konya plain. The paint is nearly always*

78 In a reanalysis of the material, Pruß (2000) claims Kühne's lack of access to the Syrian material at the time of his research meant that he grouped a series of similar but actually distinct wares (namely 'Metallic Ware', 'Stone Ware (Brak type)', 'Red Euphrates Ware', 'Black Euphrates Ware', 'Jezirah Grey Ware', 'Common EBA Wares'). The consequence is a rather unhelpful terminological confusion, and it may often be unclear what any one particular author means by the term 'Metallic Ware'. Pruß suggests abandoning the term but does not offer any useful alternative designation for the ware group he attempts to more precisely define. Perhaps it would be helpful to create a new term for Pruß's materials and preserve the general term 'metallic ware' (without initial capitals, or keep the German *Metallische Ware*) to subsume the wider group of fabric and form types.

79 Mellink (1993) suggested that Mellaart had conflated two similar (albeit related) 'wares' – referred to elsewhere as 'Konya metallic wares' and 'Cilician red gritty ware'.

matt, usually thick and splashy, designs are clumsy. The colour varies: black-purple-brown, dark red and white. Red and white occur together as at Tarsus. (Mellaart 1954, 192-193).

The influence of metal prototypes is suggested by the sharp angles (especially between neck and shoulder) the shape of some rims, the painted or plastic rivets, the thinness of the ware and the pushed up bases. (Mellaart 1954, 193).

By way of comparison, Friedman describes the ware of Göltepe and nearby as:

[H]andmade, tempered with fine quartz sand, and fired at a very high temperature. It is hard, thin, breaks along a straight fracture and makes a high pitched clinky sound when struck. Its color varies from buff to orange to brown. The vessels are usually slipped in a shade very near the color of the clay itself before firing or washed after firing. Some examples are slightly burnished." … "Anatolian metallic ware has been found at Göltepe, Tarsus, Mersin, Acemhöyük and across the Konya plain". (Friedman 1998).

iii. 'Metallic Wares' in the Levant, Palestine and Egypt (3100-2700BC)

A range of generally much earlier Levantine, Palestinian or Egyptian 'metallic wares' adds further confusion to the mix. According to Greenberg and Porat (1996), *Canaanite* 'Metallic Ware' (also sometimes termed Abydos Ware or Combed Ware) is considered highly characteristic of the EBII and perhaps EBIII of northern Canaan (equivalent to around 3100-2700BC), just before, or perhaps overlapping with, the appearance of early Khirbet Kerak wares which are related to Kura-Arax pottery (Greenberg and Porat 1996, 12). Their analysis suggests that this ware may have been made in a fairly restricted area, but then distributed in larger quantities over fairly long distances (perhaps as much as 100km).

Metallic Ware sherds and vessels appear in varying shades of red, with thin vessels verging toward buff or gray. Occasionally, thicker vessels appear in gray or brown, indicating firing in reducing conditions. Metallic Ware is brittle to the touch, sounding a distinctive ring when struck. In section, it is generally uniform in appearance, rather coarse grained and evenly fired, with gray cores appearing mainly in large vessels. […] The range of vessels produced in Metallic Ware answered all household and small-industrial needs, with the notable exception of open-flame cooking, for which Metallic Ware is ill-suited. (Greenberg and Porat 1996, 6).

Potter's marks are common, particularly on the bases of platters and jugs, and on the shoulders of jars. […] Metallic Ware duplicates, almost in its entirety, the Early Bronze repertoire of non-Metallic ceramic assemblages (Greenberg and Porat 1996, 10).

Greenberg and Porat note the typological and petrographic similarities of their material to 'Brittle Orange Ware' from the Amuq G-J (Greenberg and Porat 1996, 17), but emphasize that the fabric is from a different source. In their analysis, they draw a distinction between their studied material and the "metallic vessels appearing from Sidon north, as far as Ras-Shamra and Sukas" made in a different fabric, "though of similar quality" (Greenberg and Porat 1996, 11). Similarly

Miroschedji draws a distinction between their material and another fine EBII ware, with a "chocolate brown" colouring known from across the southern Levant (de Miroschedji 2000a, 321). Meanwhile, Greenberg and Porat distance the Canaanite pottery from the northern Syrian Metallic Ware (as discussed above), though without specifying why, though the dating of each is somewhat different.

> *The roughly contemporaneous highly fired Stone Ware of northern Syria and Mesopotamia (sometimes termed 'metallic ware'; see, e.g., Schneider 1989) is completely unrelated to [south Levantive] Metallic Ware.* (Greenberg and Porat 1996, 18).

iv. Terminologies and 'metallic wares' as supra-cultural phenomena

The attempts to fix the definitions of these various pottery wares are an understandable reflex of pottery analysis which attempts to divide and delineate particularly for creating chronological sequences, in order to date sites, and also for localized provenance-based questions (*e.g.* whether particular wares are imported to the site or locally produced). However, especially with the limited availability of good colour photographs, diagrams and descriptions, it is rather difficult for anyone interested in these different wares to make any sense of the relationships between them. Clearly a systematic comparison between the regional pottery variants needs to be undertaken to assess the true relationship (if any) between these different assemblages[80]. Though one might argue that all this is simply a matter of terminology and clarity of inter-communication amongst pottery specialists, there are larger historical issues at stake.

Firstly the pottery terminology (and its confusion) seeps into the interpretation of sites and cultural changes: that some types are labelled 'true' metallic ware and others excluded has consequences for the types of analysis which can be done on a regional basis. The parcelling up into discrete named 'Wares' means that 'Brittle Orange Ware' seems very distant from 'Metallic Ware' or 'Stone Ware' even if there were complex and significant interrelations between these imposed categories. Certain ceramic assemblages which may have metallic relations are thereby less obviously associated with inter-craft transmissions because of an accident of terminology[81].

Secondly an important point about wider trends in pottery traditions is thereby missed as a result of the chronological drive of pottery analysis: even if the term may often be misleading, the fact is that during the 3rd millennium, over a wide area – albeit at slightly different times – many ancient peoples were interested in consuming pottery vessels with strong metallic associations. Whether these associations were simply as cheaper/'bourgeois' alternatives to difficult-to-secure metallic prototypes, or merely attempts to draw on the aura of the metallic aesthetic, cannot be said with any certainty. It is somewhat tempting to infer that the range of colours and differential treatment of the various wares[82] reflect

80 This is of course a huge task which clearly cannot form part of the current study, though something like the ARCANE chronological/material database project may help to facilitate this in the near future – http://www.arcane.uni-tuebingen.de/.

81 And the contrary may also be true: some 'metallic wares' might be completely unrelated to metals.

82 Compare, in particular, the terminological confusion within Syria Metallic wares, as described by Pruß (2000): perhaps the different colours of the pottery are a result of emulating alternative patinas of metallic originals.

a similar range among, or various local or temporal trends in, copper, bronze or other metal vessels (*cf.* Nakou 2007, 235-236).

This is not to say that different regions did not have very different priorities in terms of the shapes made, which reflect regional cultural aims. Equally, local traditions and availability of clays probably dictate the exact fabrics used to create a metallic effect.

> *"In Anatolia, the most common form of metallic ware is the jug. In northern Syria jars and bowls are more frequent, in the Amuq the bottle dominates, in northern Palestine bowls and platters are more familiar, and in southern Palestine and Egypt the most ubiquitous form is the storage jar. The surface treatment also differs from region to region. In Anatolia the jugs are usually painted whereas the bottles from the Amuq are often slipped and burnished. In northern Syria there is no evidence of either paint or an applied slip and in Palestine and Egypt the storage jars are combed in the leather-hard stage prior to firing and usually treated with a thin coat of lime plaster or wash after firing"* (Friedman 1998).

However what we may be missing by focussing on the microscopic aspects for the purposes of chronology and provenance is the trace of an aesthetic which was being transmitted and maintained *supra-culturally*, rather as culinary traditions may be shared and transformed over an invisible macro-scale (Fuller and Rowlands 2011)[83]. There must have been a clear imperative in certain social contexts to use such metallic inspiration, even if these contexts were not the same in different regions. It is also worth pointing out that all these "metallic wares" existed contemporaneously alongside pottery traditions which do not necessarily recall a metallic aesthetic.

In fact these examples appear to fit into a much larger and longer-lived ceramic *koine*, which starts in the Late Chalcolithic (c. 4000-3500BC onwards). Many pottery fabrics and decorations of the 3rd millennium could be compared to a metal aesthetic, especially when compared to the painted pottery of earlier periods (such as those of Halaf, Hacılar, or Sialk), and the more colourful or ornately decorated painted pottery characteristic of, say, the 2nd millennium Aegean (Minoan and Mycenaean painted pottery). The lack of archaeological discourse on these long-term ceramic trends in recent literature may be explained by either the trend towards specialization or because such features seem too obvious. This 'over'-conspicuity (like the wearing of a high-visibility jacket) tends toward an effective invisibility: the explanation and mechanism behind such widespread similarities in cultural practices thus remains unexplored.

5.6.4 Metallschock in western Central Asia

i. The transition to Namazga V/VI (2500-1500BC)

The emergence of a mid-to-late second millennium 'Anatolian Trade Network', with the inspiration behind ceramic forms being metallic, is curiously paralleled at a similar time in distant Turkmenia. Before around 2700-2500BC, the local

83 Of course, the comparison with culinary traditions may be apt when one considers that the majority of pottery production over prehistory has been as tools within human food production and consumption systems. This is not to say of course that particular pottery styles always equate to particular fashions in food, but a relationship must have existed.

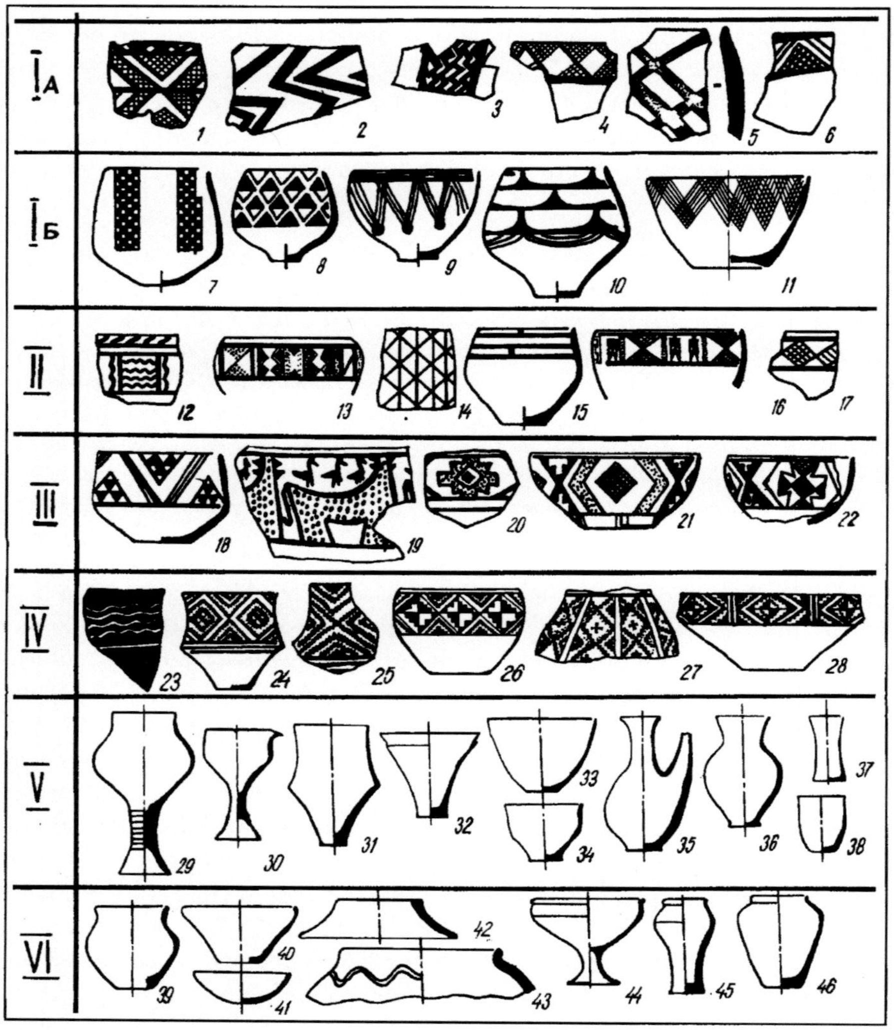

Figure 5.43. Namazga pottery sequence schematic (after P'yankova 1994).

pottery was dominated by painted wares, normally defined within the Namazga I-III sequence. However, from the Namazga IV period (referred to sometimes as the beginning of the Central Asian Early Bronze Age, and currently dated to around 2700-2500BC), the ceramic assemblages of the Kopet Dag and surrounding regions change substantially (Figure 5.43):

> *Pottery, which during the Eneolithic period was manufactured exclusively by hand, gradually begins to be made on a potter's wheel. At Khapuz-depe, the lids of certain pots were wheel-turned as early as the lowest Namazga IV layers, and later this technique becomes more and more dominant. At Namazga-depe itself, wheel-made vessels were common in the late Namazga IV layers. With the introduction of this technique, the forms of the pots become more sophisticated and ambitious compared with those of the Eneolithic culture. Painted decoration gradually dies out, becoming careless and inchoate in the late Namazga IV period, and disappearing completely in Namazga V. Two-tiered potters' kilns, unknown in the Eneolithic period, also appear.* (Masson 1968, 180).

Figure 5.44. Examples of Namazga V ceramics from Altyn Depe (Masson 1988, pl. VI; image: courtesy Russian Academy of Sciences).

The disappearance of vibrant painted decoration by the beginning of the Namazga V period (c. 2500BC), and the introduction of the potter's wheel, both point toward another case of '*Metallschock*'. As with the Aegeo-Anatolian case, the changing pottery evidence is accompanied by other features sometimes claimed as 'Mesopotamianizing':- metal stamp 'seals' appear in the Late Namazga IV period (c. 2700-2500BC), though whether their function was the same as seals in Mesopotamia is still an open question; and the monumental architecture of Namazga V Altyn Depe has been compared, perhaps optimistically, to Mesopotamian ziggurat-temple architecture.

The dramatic disappearance of painted pottery traditions has been considered merely a collateral casualty of mass-production, standardization (in Masson's view, a continuation of the unification of pottery styles of late Namazga IV) and urbanization – its relationship to a metal aesthetic rarely acknowledged (Figure 5.44). The dating of the beginning of this Namazga *Metallschock* is significant, of course. In particular the period of very late Namazga IV/Namazga V has been associated with the intensification of exchange between Mesopotamia and the east: the circulation of 'intercultural style' vessels from Iran (Jiroft) to Mesopotamia, the textual and archaeological suggestions of links between the Indus civilization

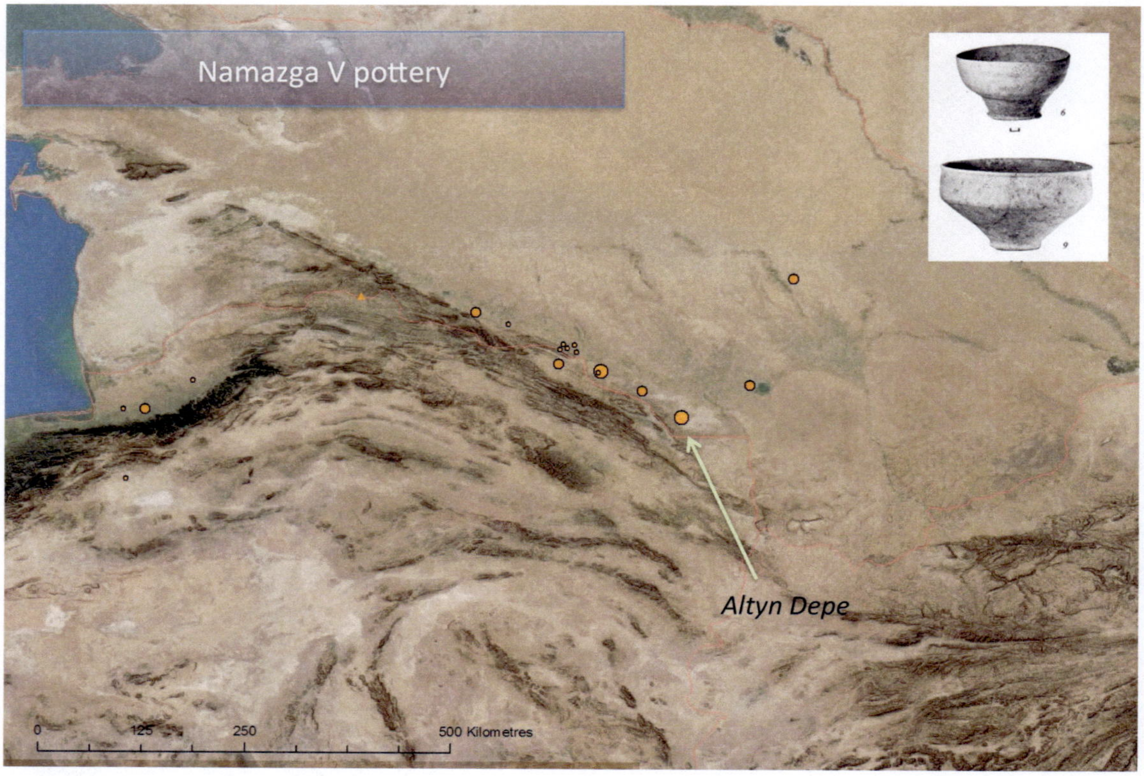
Namazga V pottery
Altyn Depe
0
125
250
500 Kilometres
a

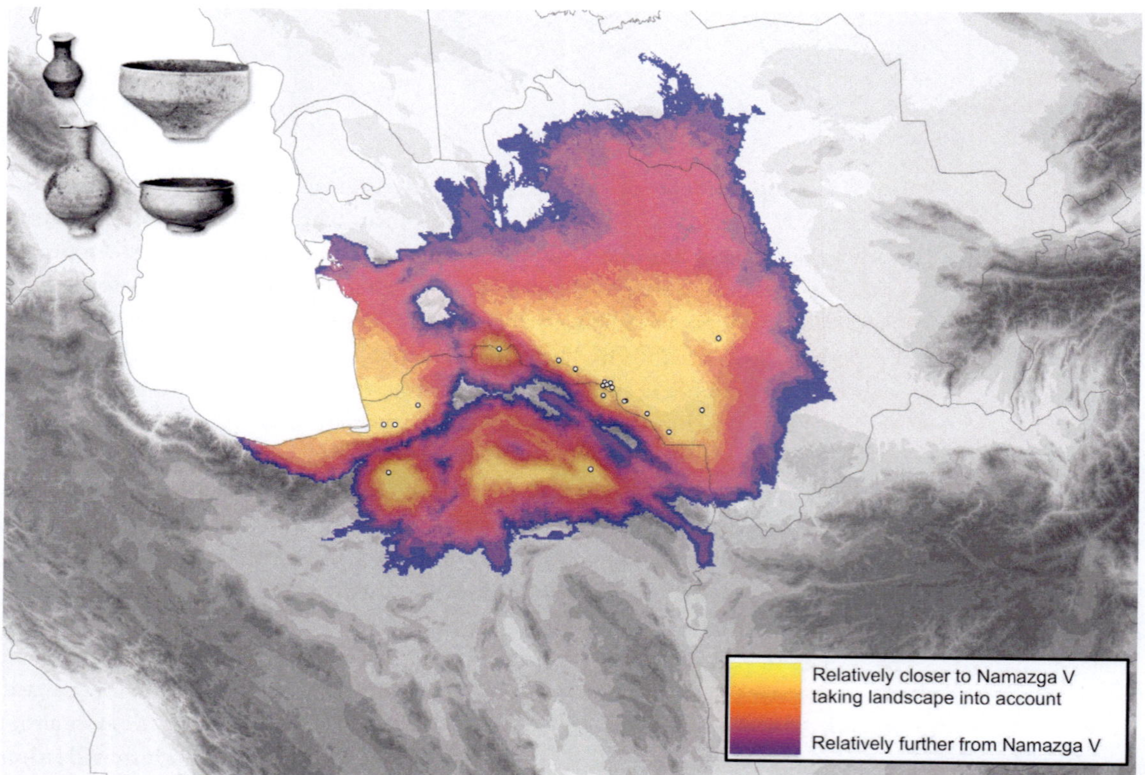
Relatively closer to Namazga V taking landscape into account
Relatively further from Namazga V
b

Figure 5.46. Metallic vessels from the 'Royal Necropolis' of Gonur Depe, Turkmenistan next to a ceramic vessel from the same site (after Sarianidi 2006).

Figure 5.45. Distribution of 'Namazga V' style pottery in southern Turkmenia and north east Iran (data based on Kohl 1984 with additions), and (b) Namazga V cultural-geographical zone (archaeotopogram 'type A2' using distribution of Namazga V pottery finds).

and Mesopotamia. When also paralleled with the expansion to the west, identified by the 'Anatolian Trade Network' and 'Eastern Mediterranean Interaction Sphere', it becomes clear that the second half of the 3rd millennium BC was a period of dramatic extension and intensification of interaction across a very wide area.

The broad decorative scheme of pottery assemblages in Turkmenia (Figure 5.45) continues into the Namazga later V/VI and BMAC periods. Examples of silver and gold metal vessels, found in graves of the later BMAC period (2100-1800BC, equivalent to late Namazga V or VI) at the Gonur Depe necropolis, show the exact same forms as contemporary pottery (Figure 5.46). The thin walls and rims and the very 'clinky' sound of the fabric of pottery vessels are in any case very metallic, and the light sandy to white colours similar perhaps to polished silver or gold – in contrast to the darker colours of Anatolia. Note, however, that the Namazga V/VI pottery is less lustrous than the Anatolian wares. As in Anatolia, these metal vessels have often been seen as high-class or ritualized 'copies' of ceramic types. The opposite possibility – that the ceramic-versions in graves are ritualized 'stand-ins' for the preferred metal versions – is not normally considered, perhaps because archaeologists found the ceramic versions first.

Again it is tempting to see the metallism in the pottery assemblages as a reflection not only of a centrality of metals in Namazga V and later BMAC culture, but as an index of metal circulation, or rather its intensification, since we know metals were already in use during the Namazga IV period. Bertille Lyonnet recently proposed that the emergence of the Bactria-Margiana Archaeological Complex might actually be explained by a relationship with the circulation of tin

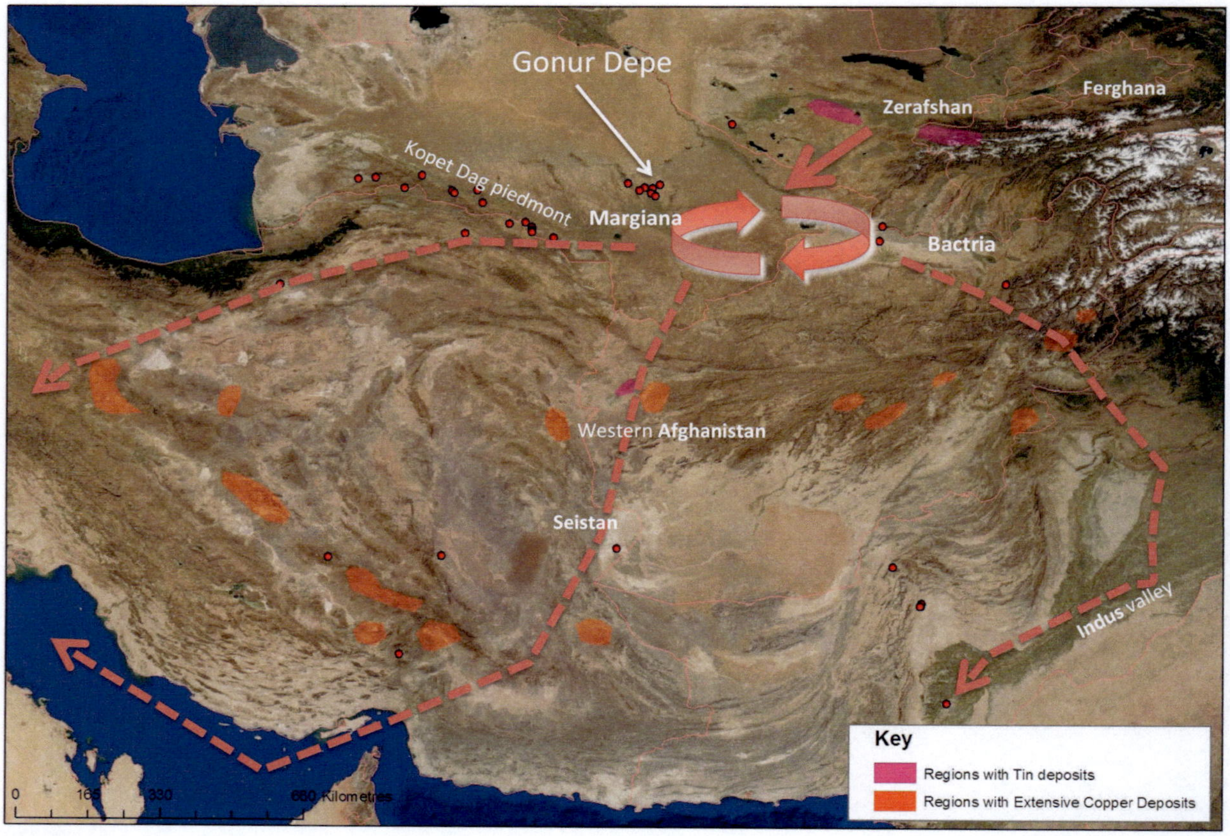
Gonur Depe
Zerafshan
Ferghana
Kopet Dag piedmont
Margiana
Bactria
Western Afghanistan
Seistan
Indus valley
Key
Regions with Tin deposits
Regions with Extensive Copper Deposits
0 165 330 660 Kilometres

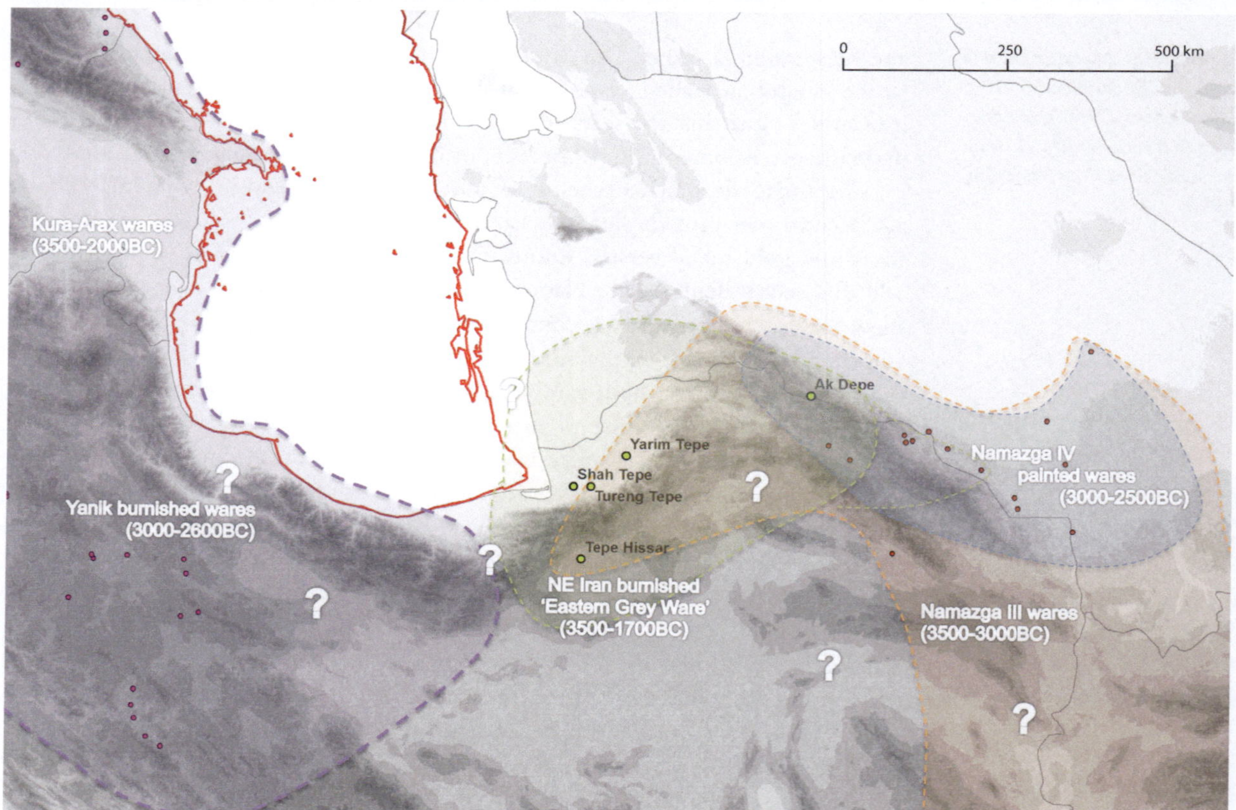
0 250 500 km
Kura-Arax wares
(3500-2000BC)
Ak Depe
Yarim Tepe
Shah Tepe
Tureng Tepe
Tepe Hissar
Namazga IV
painted wares
(3000-2500BC)
Yanik burnished wares
(3000-2600BC)
NE Iran burnished
'Eastern Grey Ware'
(3500-1700BC)
Namazga III wares
(3500-3000BC)

(Lyonnet 2005)[84]. Noting the range of evidence, and intensity of interconnections between this region and the west between 2100 and 1750BC (a similar period to that in which the Old Assyrian merchants were exchanging silver for tin and textiles at Kültepe), she suggests that the rich sources of tin in nearby Zerafshan valley might have been traded through sites like Gonur Depe, in a similar way to the Old Assyrian Trade at Kültepe-Kaneš (Figure 5.47). The relatively low use of tin in uncovered BMAC copper objects (*cf.* Terekhova 1990) does not necessarily contradict this idea, if tin were a plentiful resource, more valuable in exchange than in local consumption (this compares favourably with the situation in the Aegean – which as a major source of silver has nonetheless yielded few silver objects):

> *The location of BMAC, right in the nest of the tin resources of Central Asia, on territories that had never been settled before, its formation and rise exactly at the time when tin becomes extremely prized in the West combined with the evidence of exchanges with these regions, all this leads to propose that this tin-trade can explain most of the 'original' features of the BMAC, even if there are no proofs of tin on the sites.* (Lyonnet 2005, 197).

The Late Bronze Age of Central Asia (Namazga VI / Takhirbai, c. 1750-1500BC) is also characterized by emerging links between the north-east and south-west of the Karakum, illustrated for example by fragments of Andronovo pottery found in the last phases of sites in southern Turkmenistan (*e.g.* Sarianidi 1990b, 259; Hiebert 2002a, 244), and the emergence of the 'Zamanbaba' culture (in an oasis close to the Zerafshan valley), which potentially might be linked to an intensification of tin exchange in this region.

Figure 5.47. Hypothetical geographical role of BMAC communities in the circulation of tin in late 3rd and early 2nd millennia BC.

If, as Lyonnet argues, goods were being sealed in BMAC centres for export, and those goods included tin (or lapis – as she also suggests), we may well wonder about what goods were returning in the opposite direction. Using Kültepe-Kaneš as a model, it would be easy to speculate on a number of possibilities, including arsenical-copper (from which BMAC copper objects are made), textiles and silver. There may be a clue in the colour oppositions in the ceramic assemblages of Anatolia and Central Asia (dark blacks/greys/reds : light creams/white/pinks): Anatolians wanted darker vessels, used and tarnished through circulation, Central Asians wanted light vessels, kept clean and shiny. This all depends on whether the skeuomorphism we see is designed to mimic rare or exotic items, or reflect the common or everyday. In periods of intense international exchange, exoticism seems to be a more likely motivation (the exchanges themselves a part of this preference), and so ceramic skeuomorphism may have been designed precisely to mimic materials which were exotic and difficult to obtain: tin-bronze in Anatolia, or untarnished silver in Central Asia.

Kaniuth (2007) has pointed out that, while Margianan and southern Bactrian material includes little tin-bronze usage, the Sapalli culture in northern Bactria (southern Uzbekistan) shows relatively high percentages of tin-bronze usage: around 25% of copper objects being tin-bronze on current analysis. Rather than seeing this high tin-bronze as simply a consequence of the proximity to the Zerafshan tin sources (cited above) and to tin-bronze using Andronovo groups

Figure 5.48. Distribution of Hissar IIB-type burnished grey wares ('Eastern Grey Wares') and relationship with Namazga III and IV, and Kura-Arax pottery assemblages. Red line shows probable contemporary Caspian coastline.

84 The idea that the Zerafshan/Ferghana tin deposits may have provided a "strong incentive for the merchants from southern Turkmenia" was noted at least as early as 1972 (Masson and Sarianidi 1972, 128).

to the north (although both are relevant), he believes that it reflects a poverty in other (more valuable) materials:

> *The types manufactured from gold in these regions compare well with the range of tin-bronze objects from Northern Bactria, being mostly vessels and personal ornaments. Seen from this perspective, the higher frequency of tin bronzes within the metal inventory of the LB I stage of the Sapalli Culture need not be explained by privileged access to a highly desirable raw material. Instead, tin bronze appears as a substitute for the true luxury metals on the market, silver and – most importantly – gold.* (Kaniuth 2007, 33).

The evidence is clearly still equivocal: following Kaniuth's arguments, one could suggest that BMAC communities were sufficiently wealthy in gold and silver to make tin replacements unnecessary, but that they may still have been involved in the tin trade to the west. In either case, the connection of the BMAC to tin does not, on its own, explain the emergence of the metal-inspired pottery of mid 3rd millennium Central Asia.

ii. Iranian Grey wares (3300-2500BC)

One possible forerunner or model for Namazga V 'metallism' in the ceramic repertoire may come from fairly nearby, in the form of North-East Iranian (or Hissar II) grey wares. The term 'grey ware' is even more over-used than 'metallic ware' and there are, in fact, a number of different types of grey-wares in the archaeological record of Iran (which may or may not be 'genetically' interrelated, see Piller 2003). The 'grey wares' in question here (Figure 5.48) are those typical of north-east Iran (at sites like Tepe Hissar, Shah Tepe and Tureng Tepe) from around 3300BC to at least 2500BC and perhaps as late as 1700BC (Deshayes 1972)[85]. It is tempting, given the region's association with innovative metallurgy including the production of silver, lead and even iron-speiss (Thornton, Rehren and Pigott 2009), to compare the aesthetic qualities of these grey wares to metals (silver and/or lead might have provided the colour models). Similar grey wares turn up ('intrusively' in the context of the Namazga sequence though it is not clear whether they were actually transported or manufactured locally) in Namazga IV contexts at Ak Depe and Kara Depe on the Kopet Dag piedmont (Kohl 1984, pl. 7.a, b).

The distribution of these wares may also show us the boundary between two 'metallurgical provinces'. During the 3rd and 2nd millennia, eastern Anatolia and western Central Asia had very different metallurgical traditions and systems: eastern Anatolia appears to be a part of the 'Circumpontic Metallurgical Province' whilst, in Turkmenia or western Central Asia (or at least the southern region represented by the Namazga and related cultures), the metallurgical tradition appears to adhere to the 'Irano-Afghan Metallurgical Province' whose south-western boundaries lie somewhere in Iran. At present our general knowledge of sites in northern-central Iran is rather poor. Any border or transitional area between the two areas has been made obscure by the same mid-twentieth-century AD political map that prevented Chernykh from including much material from south of the USSR. However, the distribution of ceramic assemblages from each region, particularly those of a skeuomorphic nature, might well point to the location of

85 And which replaced the earlier painted repertoire of Hissar I (Schmidt 1937).

Figure 5.49. Examples of Kura-Arax/Early Transcaucasian Culture (ETC)/Khirbet-Kerak ceramics: (a) Shengavit, Armenia (Burney and Lang 2001 [1971], pl. 35); (b) Sos Höyük, Turkey (courtesy A. Sagona); (c) Karnut-Shirak, Armenia (Kushnareva and Markovin 1994, pl. I[2]; image: courtesy of Russian Academy of Sciences); (d) Tell Tayinat, Amuq, Turkey (courtesy of the Oriental Institute of the University of Chicago). Colours are not calibrated.

such a boundary: starting on the boundary between the dark Kura-Arax pottery to the west, and the light Grey Wares and Namazga V plain wares to the east – the two pottery constellations perhaps mimicking two very different metallurgical aesthetics, and, we may speculate, marking the borders between 'Circumpontic' and 'Irano-Afghan' metallurgical provinces.

5.6.5 Metallschock in Eastern Anatolia?

i. Kura-Arax ware and its relations to metallurgy (3400-2000BC)

The Kura-Arax phenomenon or Early Transcaucasian Culture is a distinctive assemblage assigned to the Late Chalcolithic and Early Bronze Age (Bedianashvili 2008; Burney 1989; Burney and Lang 2001 [1971]; de Miroschedji 2000b; Edens 1995; Frangipane and Palumbi 2007; Greenberg and Goren 2009; Heinsch and Vandiver 2006; Iserlis 2009; Kelly-Buccellati 1990; 2004; Kiguradze and Sagona 2003; Kohl 2001; 2006; 2009b; Kushnareva and Chubinishvili 1970; Magomedov 2006; Mohammadifar, Motarjem and Khorasani 2009; Motzenbäcker 1996; Palumbi 2003; 2008b; 2008a; Paz 2009; Philip 1999, 2000; Rothman 2003a; 2003b; 2006; Sagona 1984; 1993; 2000; 2004; 2006; Shimelmitz 2003; Summers 2004; Thornton 2009; Trifonov 1994). Kura-Arax pottery is of particular significance because of its relative unity and extensive distribution, and because it forms such a central part of the record of east Anatolian and Caucasian evidence from the mid 4th millennium, into the mid 3rd millennium possibly continuing on into the beginnings of the 2nd. The phenomenon is primarily defined by its distinctive black or combined red-black burnished pottery assemblage (Figure 5.49). This characteristic pottery was apparently made locally and not traded, but the same or similar manufacturing techniques are adopted over a very wide area and maintained over a considerable period of time to produce the distinctive burnished vessels (*e.g.*, see Iserlis *et al.* 2010).

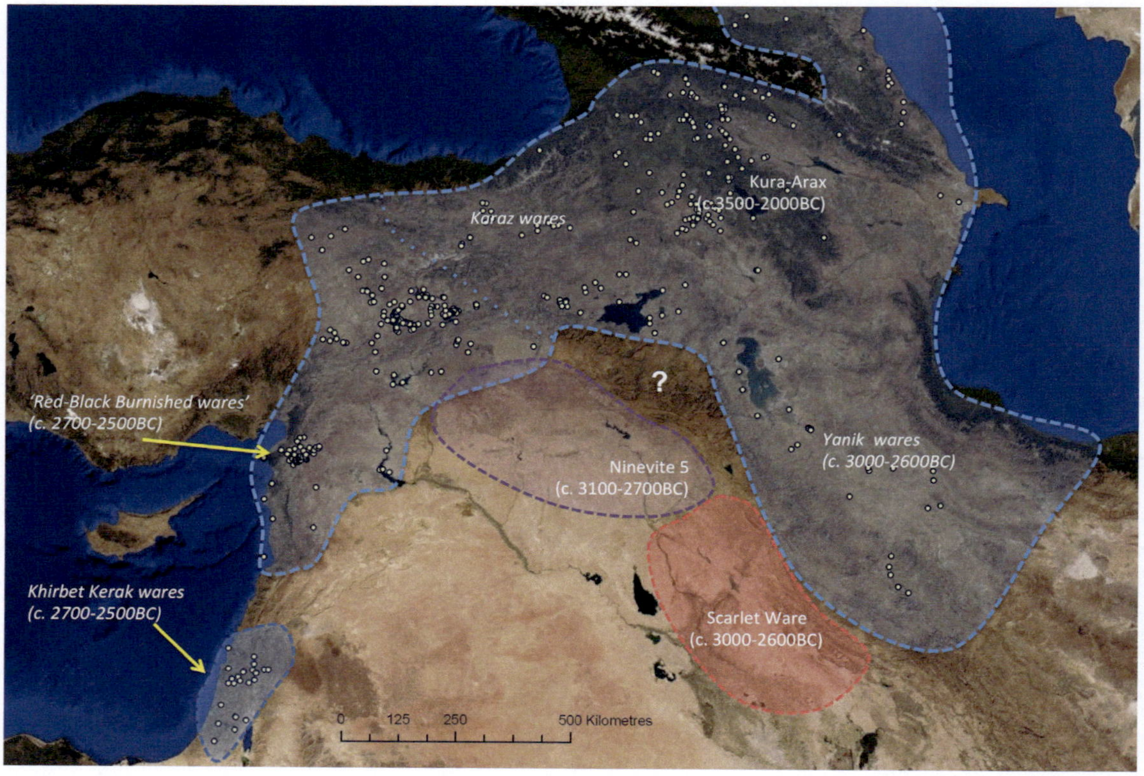
Kura-Arax
(c.3500-2000BC)
Karaz wares
?
'Red-Black Burnished wares'
(c. 2700-2500BC)
Ninevite 5
(c. 3100-2700BC)
Yanik wares
(c. 3000-2600BC)
Khirbet Kerak wares
(c. 2700-2500BC)
Scarlet Ware
(c. 3000-2600BC)
0 125 250 500 Kilometres

a

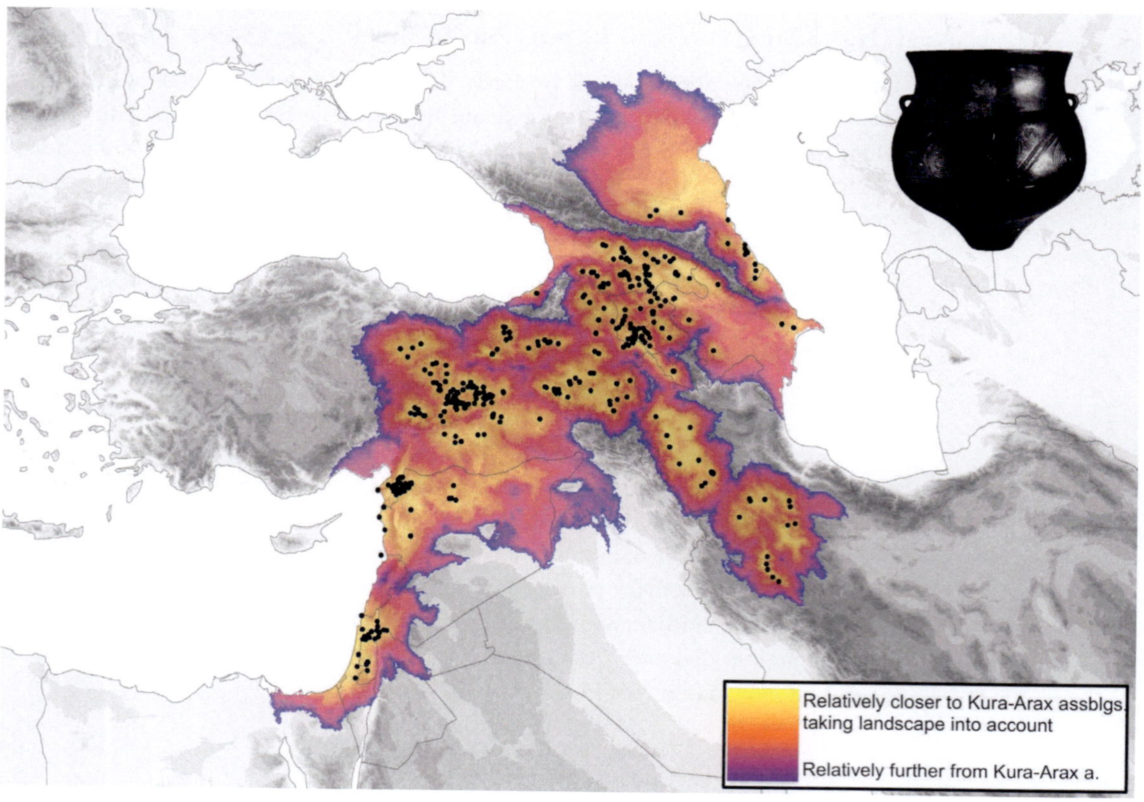
Relatively closer to Kura-Arax assblgs. taking landscape into account
Relatively further from Kura-Arax a.

b

Several terms and variant spellings are used almost interchangeably to label this pottery and the phenomena it is taken to represent: Kura-Arax, Kuro-/Kura-Araxes, Early Transcaucasian Culture/ETC, Karaz, Red-Black Burnished Ware/RBBW and Khirbet-Kerak. In fact there are several sub-regional variants of the pottery, with the style developing through time (Palumbi 2008b, 205-207, 311), some variations involving minor local differences in decorative elements such as handles (*e.g.* Summers 2004), some the configuration of the red/black colour scheme on vessels (Palumbi 2008b, 301), whilst others involve the overall proportion of form and vessel types (*e.g.* Greenberg 2007). The phenomenon is normally divided into three stages following Sagona (1984), labelled KAI, KAII and KAIII periods, with gradual variations in forms and amount/type of decoration through time. On current evidence, the earliest 'Proto-Kura-Arax' pottery forms appear to have developed in central and southern Transcaucasia (sometime in the 4th millennium), but their distribution extended from the late 4th and early 3rd millennium to wider Transcaucasia, eastern Anatolia, western Iran, and parts of the upper Euphrates[86] (beginning of KAII). In the first or second quarter of the 3rd millennium (around 2800BC), a further dramatic extension of similar wares is seen in certain areas of the Levant including the Amuq and Palestine, known there as Red-Black Burnished Ware (RBBW), or Khirbet-Kerak ware respectively (Philip 1999; de Miroschedji 2000b; Philip 2000; Greenberg and Goren 2009). After around 2600BC (KAIII), the distribution of these wares contracts, disappearing from the Levant, the Upper Euphrates and north-west Iran. A simple point distribution of findspots of Kura-Arax related pottery (Figure 5.50) can thus give a misleading picture of the unity of the phenomenon and the social networks which were involved in its creation, though more detailed dating data could improve this picture substantially. In Eastern Anatolia and Transcaucasia, at the sites of Sos, Pulur-Ömertepe, Karaz-Kahramanlar near Erzurum, or Shengavit in Armenia, this pottery makes up the dominant ceramic evidence for late 4th to 3rd millennia and perhaps later. Meanwhile in other areas, for example at Arslantepe near Malatya, Tell Mozan/Urkesh in northern Syria (Kelly-Buccellati 2004) or Tell Bet-Yerah/Khirbet-Kerak in Palestine (Greenberg *et al.* 2006; Greenberg 2007), the equivalent assemblage represents only one type amongst a range of others. The sheer longevity of Kura-Arax-like pottery, and our limited ability to date sites and material from the mostly survey-based evidence, also mean care should be taken in over-generalizing the contemporaneity (or otherwise) of all sites within the phenomenon. Despite this, there remains considerable value in studying these regional variants together. Their similarities hint at very complex long-distance connections, and a variety of explanations for the emergence and duration of this pottery repertoire (and its assumed cultural or ethnic carriers) have been put forward, with little real consensus and few persuasive arguments.

Figure 5.50. Distribution of Kura-Arax related wares. (a) Regional sub-types (Khirbet-Kerak, Yanik and RBBW) wares marked by blue zones (data based on Sagona 1984 with additions); alongside contemporary Ninevite 5 and Scarlet wares (based on Roaf 2000). (b) Kura-Arax cultural-geographic zone: archaeotopogram (type 'A2') based on the distribution of all Kura-Arax related wares. See Appendix C.1.2 for full database. Note: density or proportion of finds are not taken into account.

Though it is rarely mentioned in the literature, many of these unpainted and highly lustrous vessels (particularly those from the later periods) could be argued to recall metalwork, or at least a metallic aesthetic, suggesting the emulation of tarnished silver (hence the black lustre) or polished copper (hence the bright red) vessels. The patterning on the vessels certainly would have been possible to achieve

86 Like most large-scale cultural phenomena it would be a mistake to assign a single pure or original 'centre'. Giulio Palumbi has recently argued for the hybrid nature of the classic red/black Kura-Arax pottery as an amalgam of Caucasian pottery shapes and east Anatolian red/black colour schemes (Palumbi 2008b, 205, 311).

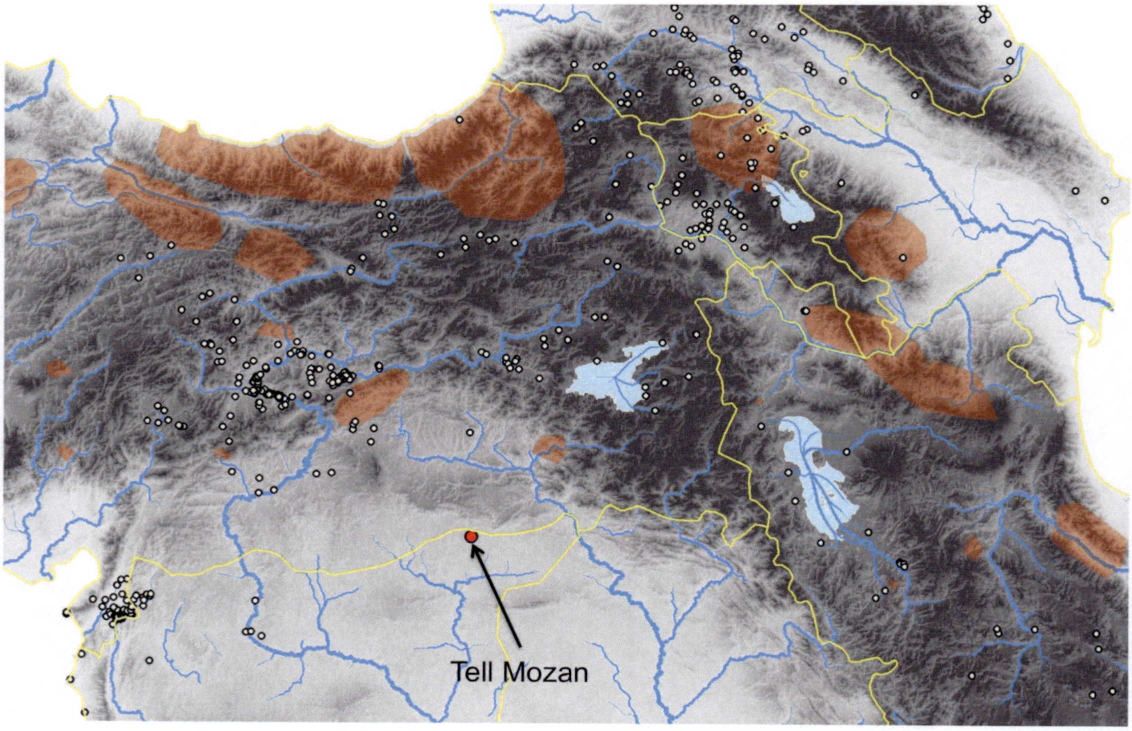

Figure 5.51. The relationship between Kura-Arax communities and the metals trade between Anatolia, the Caucasus and north Syria, including Tell Mozan. The location of sites like Tell Mozan, on the shear-lines between the northern Syrian plain and the foothills of eastern Anatolia may have placed it on a frontier between the urbanized south where metal was in great demand and a zone to the north with smaller-scale communities who had direct access to metal ores.

in metal form. Similarly, the shift from an exclusively black or dark brown/grey colour scheme to the red/black colour scheme of later assemblages from Caucasia to the Upper Euphrates might be taken to reflect a shift in the preferred colours of metal vessels that have not survived, and the same might be proposed for the gradual modifications in decorative features of the vessels.

However, the total corpus of Kura-Arax metal objects is very low despite the assumed level of metallurgical production for this region during this period (for reasons outlined by Chernykh and cited above). Suitable metal prototypes have not been identified, though metal vessels of the nearby Maikop culture at least prove that such vessels were not unknown in the region (see Kohl 2007, 83, fig. 3.13). There is certainly localized evidence for metalworking at certain Kura-Arax sites (*e.g.* the crucible at Sos Höyük) but it is missing from Kura-Arax-like assemblages in north-west Iran such as Yanik Tepe (Summers, pers. comm.). The support for this idea is therefore circumstantial: the extent of copper, silver and other metal sources across the Caucasus and eastern Anatolia has already been mentioned, and the distribution of early or Proto-Kura-Arax sites hugs the Georgian-Armenian highlands around which significant arsenic-copper sources are to be found (see *e.g.* Schachner 2002). The metallurgical connections between the slightly earlier, northerly Maikop culture and southerly Uruk assemblages, support the idea that metals (both forms and materials) were circulating through this eastern Anatolian/Transcaucasian region, from at least the early 4th millennium, even if they have left no direct traces. Later Martkopi/Bedeni and Trialeti tomb assemblages (3600/3300BC onwards) also support the idea of an advanced metalworking industry, including examples of metal vessels (Figures 5.13, 5.18) – though these do not look like contemporary pottery, perhaps because they show rather unique narrative decorative schemes.

There are alternative possible interpretations of the Kura-Arax pottery aesthetic; not least is the idea already mentioned (see Section 4.5.1) that the aesthetics of obsidian, as a valued lustrous and black material might have played an equally important role. Processed leather might provide another model upon which the Kura-Arax aesthetic could be based (though some other material would be needed to provide the framework for vessel forms), the intensive working of leather needed to provide a smooth black lustrous surface perhaps mimicked in the extreme level of burnishing and surface treatment for Kura-Arax wares[87]. The likely role of Kura-Arax communities in metal procurement and metal exchange still seems to make the idea of a *Metallschock* in the ceramic assemblages the most plausible, however.

Whether they were only producing enough for internal social relationships, or whether such metals made their way into 'international' exchange networks, is difficult to prove. The distribution of Kura-Arax-types hugs the highland fringes north of the Mesopotamian region, described as marking the 'Outer Fertile Crescent' (Kelly-Buccellati 1990, 120). This distribution does suggest that there was a relationship of interaction between the urbanization north Mesopotamian plain, and the highlands of Eastern Anatolia.

At Tell Mozan in northern Syria, Kelly-Buccellati (1990) has taken the evidence for extensive metalworking at the ancient settlement (thought by the excavators to be the historical town of Urkesh) and the co-occurrence of red-black burnished wares of 'Early Transcaucasian' inspiration as evidence for a northern trade of metals from the Ergani copper sources (Kelly-Buccellati 1990, 123). Thus the location of Tell Mozan on the "economic shear lines" between north-south and east-west routes was part of its urban success (Figure 5.51). Indeed in a more recent article, she also attempted to link the origins of Late Bronze Age 'Hurrian' identity and culture, for which Urkesh is said to be the capital, with the same Early Transcaucasian Culture, in this case via the similarities in designs of hearths or rather andirons, suggesting a cultural continuation of import rites or rituals of (perhaps familial) commensality (Kelly-Buccellati 2004). Whilst the connections between Kura-Arax/ETC culture and 'Hurrian' ethnogenesis remains a more equivocal question, Tell Mozan's suggested role as metallurgical powerhouse drawing raw materials from the less urbanized 'periphery' (Eastern Anatolia) and acting as gateway market to the local 'centre' (in this case the cities of northern Jazira) has potentially wider-ranging consequences. This picture follows a broadly 'world-systems' approach (*cf.* Algaze 1989) to inter-regional relationships, and paints the bearers of Kura-Arax/ETC culture in the north as passive participants in the metal exchange or simply as labourers seeking metal ores rather than active consumers or flamboyant artisans in their own right. However, it again raises the role of north-south relationships based on metal exchange, and though there are

87 The economy of Kura-Arax communities has often been linked to (nomadic) pastoralism, with cattle and sheep forming part of the faunal evidence, so leather might have been a prized or socially significant material suitable for imitation in ceramic. Interestingly, Kura-Arax pottery and other objects have been found in association with ancient salt works at Duzdaği in the Nakhchevan region (Marro, Bakhshaliyev and Sanz 2010) – salt being a chemical resource which might have cuisine-related, metal-industrial or indeed leather-working functions. Perhaps the cultural interaction between groups in Transcaucasia who used leather or obsidian inspired pottery and those groups to the south-west with red/black pottery and more interested in metallurgy because of the demand from Mesopotamian consumption resulted in a fusion of aesthetic sources – the relationship to any of the original inspiring materials perhaps lost within the long history of the pottery usage.

clear and relatively impermeable cultural boundaries in place to prevent certain kinds of exchange and preserving the cultural integrity of each zone (Kelly-Buccellati 1990, 123), the potential for metal to cross this threshold seems very high.

5.7 Summary: metal flows and cultural-economic trajectories

This chapter has examined various different strands of evidence for the flow of metals during the 3rd and 2nd millennia BC. The uneven distribution of metal ores has been shown to have been a major incentive for inter-regional exchange: metals, particularly tin and gold, appear to have been circulating over extraordinarily large distances (over networks that, in some instances, echo that of precious stones like lapis lazuli). Chernykh's macro-synthetic work has also highlighted the scale of technical connections: both the content of assemblages (types) and fabrication techniques seem to have been shared over wide regions, apparently grouped into technological systems which Chernykh called 'metallurgical provinces'. The reasons for the desirability of metal have been discussed (its dual liquidity, strength and associated aesthetics) but whilst metals and templates were circulating very widely, the evidence also shows that the manner of the consumption of metals differed on a local and regional level. In certain regions, this evidence (or indeed negative evidence) appears to suggest the increasing 'metallification' of certain economies: where metal becomes central to social and economic systems of value. These patterns provide indicators for – as with stone – variant value systems in which 'ritual-economic' or 'moral' rules promote either the recirculation (Wengrow's 'archival' economy) or destruction (Wengrow's 'sacrificial') of metal 'wealth'. That these variant systems do not appear to correlate neatly with the 'metallurgical provinces' suggests an extremely complex and dynamic set of interlocking or overlapping economic systems across the Near East during the 3rd and 2nd millennium BC providing support, perhaps, for the 'balkanized' model of metal consumption as well as production.

These variations in the deposition of metals, dictated by cultural practice, and the general lack of visibility of metal in the archaeological record (due to recycling and corrosion) still means, however, that it is difficult to reconstruct clear maps of the changing routes of the flows of metals. One way we might overcome this invisibility of metal flows is through engaging proxy evidence of metal movement: here, a sample of pottery assemblages (both shapes and fabrics) which appear to document cross-craft influence from metals have been analysed to create a more detailed footprint of the likely networks by which metals must have circulated. In our casy-study regions, Namazga V/VI and Kura-Arax wares represent examples of such 'metallic wares' – apparently documenting an increasing impact of metals – and in both instances the probable increasing global flow of certain metals through each region (tin and copper respectively) thus appears to have had strong local effects. The manufacture of 'metal wares' perhaps indexes another type of 'metallification' where value is increasingly measured with metal – albeit in regionally divergent manners. But the examination of cross-craft proxy evidence has other consequences. Because pottery assemblages are so densely distributed, they may offer a much more sensitive indicator of the connections between places and communities. All pottery assemblages have the potential to map social networks through which other materials, including metals, travelled. However,

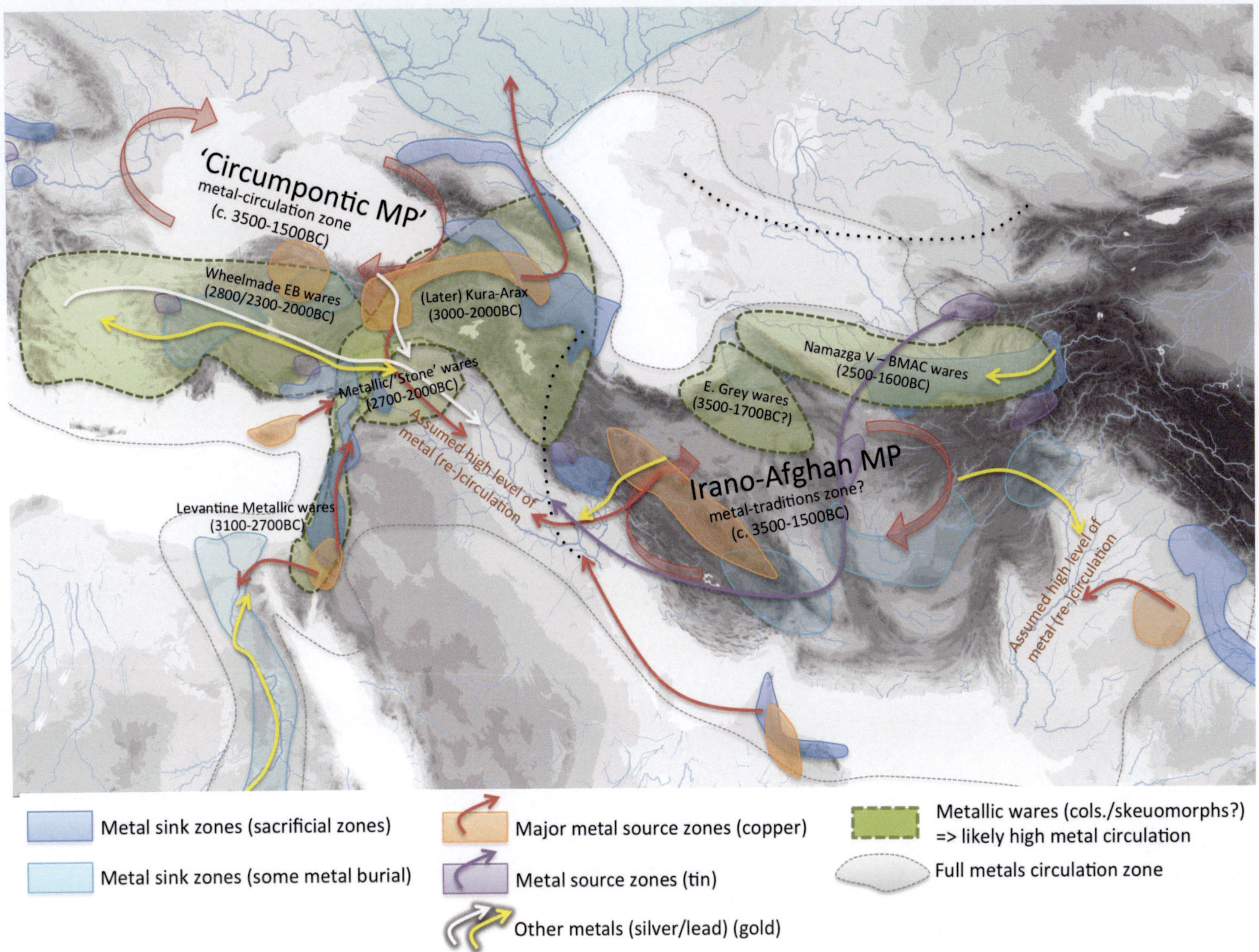

Figure 5.52. Summary of distribution data on metals (copper and tin sources zones, likely major flow directions of copper, silver and tin, metal 'sink' zones including hoarding and burial zones, metallic ware zones) over the 3000-1500BC period.

this potential is likely to be particular strong in the case of 'metallic wares' where there are overlapping technical procedures (or more general bodily praxis) between metallurgy and pottery manufacture (such as burnishing : polishing) and aesthetic consumption (such as drinking/eating in similarly shaped or decorated vessels), that connect the materials more closely in the minds of their makers and users.

Chapter 6

Mapping Material Flows: Textiles and Patterns

6.1 Introduction

The previous chapter traced the various sources of information about the flows of metals in the Near East during the 3rd and early 2nd millennium BC, both direct and indirect. Both geographical inequalities arising from geological distribution of raw sources and variations in patterns of consumption were highlighted as incentives behind the creation of metal flows. The evidence for different traditions of metal deposition or recycling and indications of cross-craft interaction in pottery were also taken as indices of varying attitudes to metals in social, moral and 'ritual-economic' terms. This chapter sets out to examine the evidence for the flow of textiles, clothing (styles) and related visual patterns in a similar way: assembling both direct and indirect evidence for the flows of textiles (including from textual records, information about production techniques, depictions or hints of clothing, dress accessories and apparently textile-inspired patterns on other media). Given that textiles are highly perishable, very little direct evidence of their forms (and thus of their movement) has been preserved. This means we are forced to rely on proxy and indirect evidence to an even greater extent than for metals, which also requires us to search for clues or potential models far outside of our case-study areas. The reader should not therefore be surprised to find jewellery and beads discussed here, since though the materials from which they were made are mostly metal or stone, they are also vital as indices of common clothing style and, therefore, of the likelihood of moving textiles.

Much of the literature devoted to textiles has been devoted to tracing the evolution of textile and weaving techniques rather than the flow, movement or exchange of cloth or styles within certain periods. The main reason for this, is the extremely limited corpus of textile fragments or impressions in other materials distributed across western Asia, alongside the better recovered but sometimes less thoroughly investigated objects of textile technology: particularly spindle whorls, loom weights and needles/awls. Since it is the demand for desirable materials that drove ancient economies, however, contemporary exchange must form an essential part of the 'evolution' or history of textiles and dress.

6.1.1 A definition of textiles and its relationship to similar materials

It is worth establishing what is meant here by the term 'textiles'. In most studies the term is used to denote *woven* items, constructed from vegetable or animal fibres which are inter-woven together to form flexible fabrics which may be used to drape over (or be fitted to) a variety of objects including, most obviously, human bodies (predominantly in the form of clothing or bedding), but also animals, vessels, floors, walls, furniture and so on. In the ancient world, yarns were made from fibres of plants (such as hemp, flax/linen, cotton), animals (especially wool, most often from sheep, or mohair from goats) or even insects or molluscs (silk, byssos, respectively). Many such fibres can be dyed or painted with a range of different colours, a fact which allows an endless variety of patterned fabrics to be made (by weaving different coloured threads, painting on plain fabrics, embroidering coloured threads, or combining multiple coloured fabrics together). Textiles may be used for clothing, wall-hangings, cushion covers, beddings, floor-coverings/ carpets, ceilings, tent-coverings, product wrapping, flags/banners, protective-cases, sails, *etc.* Baskets can be seen as a kind of sub-category of textiles in which the woven 'yarns' are normally larger plant fibres or wood.

It should be remembered however that woven textiles are related functionally and/or technologically to a whole range of materials including non-woven 'fabrics', including furs, felt or feathers, and other malleable 'sheet-like' or fibrous materials such as leather (from animal skins), ropes/strings (threads made from animal hair, vegetable fibres) – all of which may sometimes be used interchangeably (or even combined) with woven textiles in certain contexts, particularly clothing. Though there is no clear English term to subsume all these materials into a single category (as noted by Harris 2008), it is clear that they are closely inter-related, posing similar visibility problems to the archaeologist, as well as problems of potential context of use.

In this chapter the focus will remain, however, on woven textiles and/or those fabrics which show distinctive patterns, motifs or aesthetics. The basis for this is partly pragmatic (archaeologists have been more interested in woven textiles and hence there is more that can be said about them), but it also based on the idea that woven and decorated textiles frequently formed the sort of high value commodity likely to be transported long distances, and represented a vital medium for 'symbolic' negotiation of social identities, particularly through human clothing and the display and emulation of desirable colours, motifs and materials, but also in other contexts (wrapping of goods, decoration of architectural spaces and dressing of animals). Whilst there are obvious functional aspects to the use of textiles as clothing, modern anthropological accounts of textiles show that clothing is a central location for identity politics: particularly of gender, social status, morality, modernity and ethnicities (*e.g.* Weiner and Schneider 1991). Both the flexibility of textile shapes and patterns and their location on the bodies of human actors allow them serve both practical and semiotic purposes.

Textiles, particularly of the woven or painted kinds, are likely to have provided an extremely important medium for the circulation of patterns and motifs, especially over long distances. The prime motivation behind the creation of high value textiles is the aesthetic (visual and tangible) effect upon the perceiver. Patterns, motifs and images, whether woven, embroidered or painted, are often used to enhance the visual effect of the cloth itself and increase value, as did the visual or other qualities of the fibre. Indeed placing cloth within a social setting, it is difficult to untangle cloth from these aesthetic qualities. Whilst woven textiles allow an almost infinite range of designs, certain weaves promote certain geometric patterns over others, such that it is sometimes possible to suggest patterns in other media as having come from textiles (either cloth or basketry). Whilst patterns and motifs may jump from medium to medium, textiles remain important (and a likely source for many patterns), because their portability allows the patterns to travel much further than via other objects.

6.2 Textual evidence for textiles and dress

Ancient texts have traditionally provided one of our most direct pictures of the overall structure and organization of Bronze Age textile production, trade and movement. Various textual records give information about the production, buying, shipping and selling of textiles, though the information is restricted to urban centres such as Ebla, Mari, Ur (Ur III period), Kültepe-Kaneš and Acemhöyük, and in Egypt, at Ugarit, in central Anatolia (Hittite texts), and in the Aegean (the Linear A and Linear B records). The potential of these sources is considerable for textiles in general, and the corpus so large that it is impossible to deal with everything, but, as we will see, there are limitations as to the kinds of information which can be extracted and this frames the understanding of the data. Here then I will simply try to summarize the data that relate to the movement of textiles/ textile styles, their exchange and any indications of patterns and colours that will help in our reconstructions of these interactions. The following summary is aided considerably by a recent publication of a collection of studies on textile terminologies in the Near East (Michel and Nosch 2010) that focus on ancient texts referring to textiles.

6.2.1 The Ebla archive (late 3rd millennium BC)

The Ebla royal archives consist of a large number of clay tablets, dating to the middle of the 3rd millennium, which consist of different types of predominantly economic texts, including monthly accounts of textile deliveries, and the destination of gifts from the king or palace bureaucracy either to foreign allies or to workers of various statuses, apparently as payment for services rendered. The archivists kept various records of the assignment of goods and resources to different people to facilitate the production of textiles (for example of water for dyeing). The texts mention both male and female weavers[88], producers of felts and textile dyers (Biga 2010).

Despite the considerable detail on terms and production, building an overall picture of interregional movement and exchange of textiles from these texts is not straightforward. There is little detail in the directionality of exchanges in the Ebla texts. Instead the texts are mostly concerned with the distribution of textiles *within* Ebla, or at least the Eblaite state. There are a few geographic clues: some textiles are said to have come from the cities of Mari and Armi (or were perhaps made in the style of these cities, see discussion below). Accounts of textiles sent to nobles or functionaries of other cities are recorded, for example, to representatives or kings of places such as *Gú-da-da-núm, I-bu-íb, Bur-ma-an,* and *Ur-su-am* (Biga 2010, 161-163). The exact location of many of cities mentioned is unknown or uncertain however. Some textiles appear to have been sent by intermediary messengers, via merchants or given on the occasion of visiting delegations (as 'diplomatic gifts'). The texts show that the state or palace of Ebla owned or had access to large numbers of sheep for wool, but also imported wool from elsewhere

88 This fact might be used to counter the automatic assumption that it is always women who were responsible for textile production, particularly spinning and weaving. Fulling and dyeing are also recorded as being carried out by men in the Linear B tablets of the late 2nd millennium BC. However in a large percentage of ancient texts of the 3rd and 2nd millennium, women are noted as the ones responsible for textile production. Barber (1994) argues that this relates to the compatibility between textile work and childcare.

(Biga 2010, 151-152), including Mari, Armi and Kish. Wool as a raw material also circulated within Ebla itself:

> *The wool was given to women, including the queen and the queen mother to make inter alia fabrics, blankets, pillows, cushions, carpets, or ropes; to workers to make fabrics; it was used to purchase goods at fairs [...]; it was given as payment to several categories of works; and it was given to functionaries responsible for the equipment of donkeys and equids and chariots of the members of the royal family and the vizier.* (Biga 2010, 151).

Locating some of these cities (Kish and Mari, for example) is reasonably straightforward (Figure 6.1), but historical geography can be a very imprecise science, and many place names remain un-locatable. There is little mention of exchange outside of urbanized contexts, but this perhaps not surprising since Ebla was geographically and culturally at the centre of an urbanized heartland. While quantification is difficult, the impression is one of considerable movement of textiles, both raw materials and finished textiles, between countryside and urban centres, and between neighbouring towns and cities.

6.2.2 Neo-Sumerian/Ur III/Third Dynasty of Ur texts (c. 2100-2000BC)

Figure 6.1. Some identifiable places named in the Ebla archives that are associated with the distribution of different styles of textiles (Biga 2010).

The Ur III period is characterized by the domination of the state of Ur over much of the southern Mesopotamian region, Elam, and parts of northern Mesopotamia and Syria. The number of tablets that has been uncovered from this period is staggering: Pomponio (2010) modestly estimates a total of at least 75,000 from the sites of Ur, Girsu, Umma, Nippur, Puzris-Dagan and Irisagrig, for example (see map Figure 6.2). Many of these texts remain unpublished, untranslated or unanalysed, however, so the scope for future research remains very large. Unfortunately the fragmented location of the tablets[89] and the recovery of many finds from uncontrolled excavations mean that it is difficult or often impossible to identify textual 'archives' which would otherwise aid comparative diachronic and synchronic research (Steinkeller 1982). These factors, and the very size of the potential corpus, have had a certain paralysing effect on the ability of researchers to make synthetic accounts. Some general comments on the nature of textile production can be made, based on research so far undertaken.

The picture from the Ur III texts is one of massive centralized industry involving many people in the production and export of textiles:

> *The textile industry employed many thousands of female workers ... belonging to the lower stratum of the Neo-Sumerian society and in addition to them female prisoners, provided by the military campaigns of the Neo-Sumerian armies, young castrates, and children were employed in the 'houses of weaving' (e-us-bar).* (Pomponio 2010, 187).

Figure 6.2. Maximum extent of Ur III political control, with sites that have yielded substantial Ur III textual archives.

89 Fragments and whole tablets are to be found in hundreds of private collections and smaller museums as well as the major national museums. This problem may slowly be overcome by modern database projects like the CDLI (http://cdli.ucla.edu/), which attempt to bring together texts from as many collections as possible.

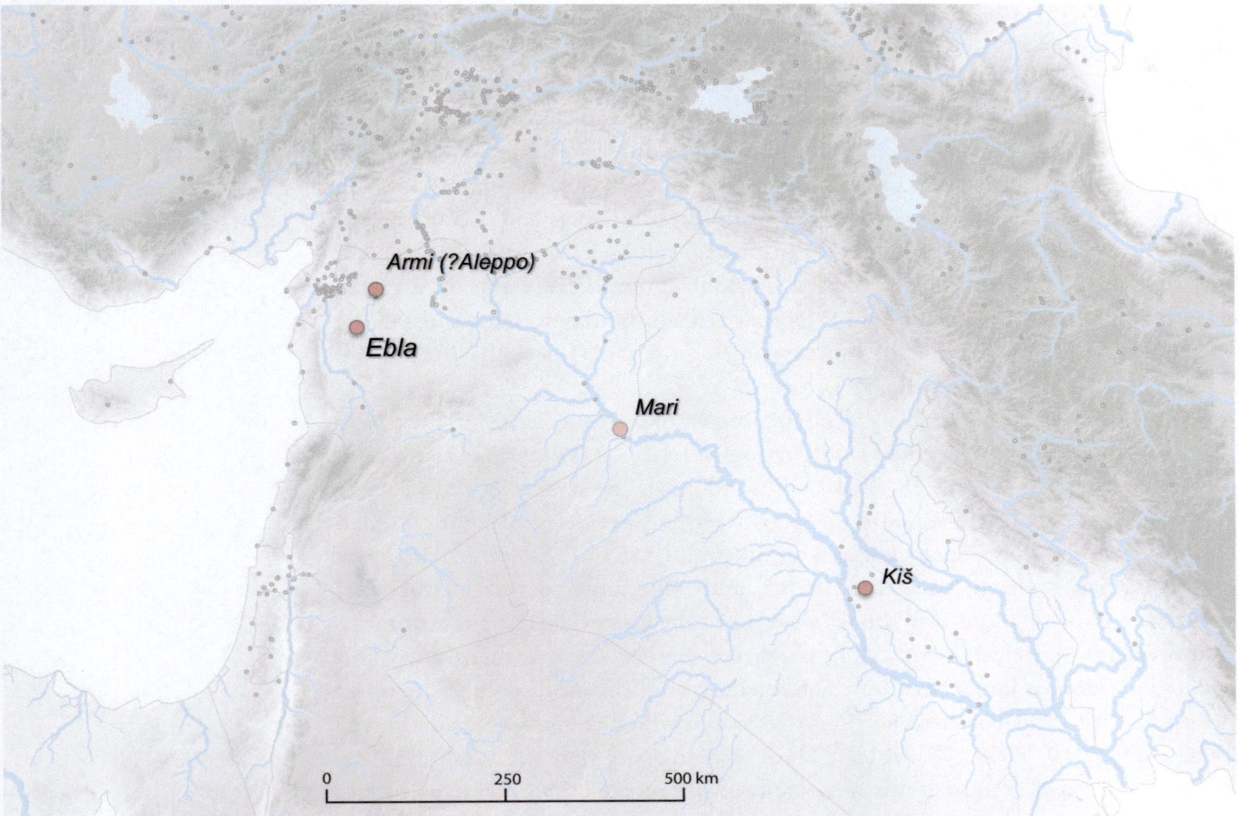
Armi (?Aleppo)
Ebla
Mari
Kiš
0
250
500 km

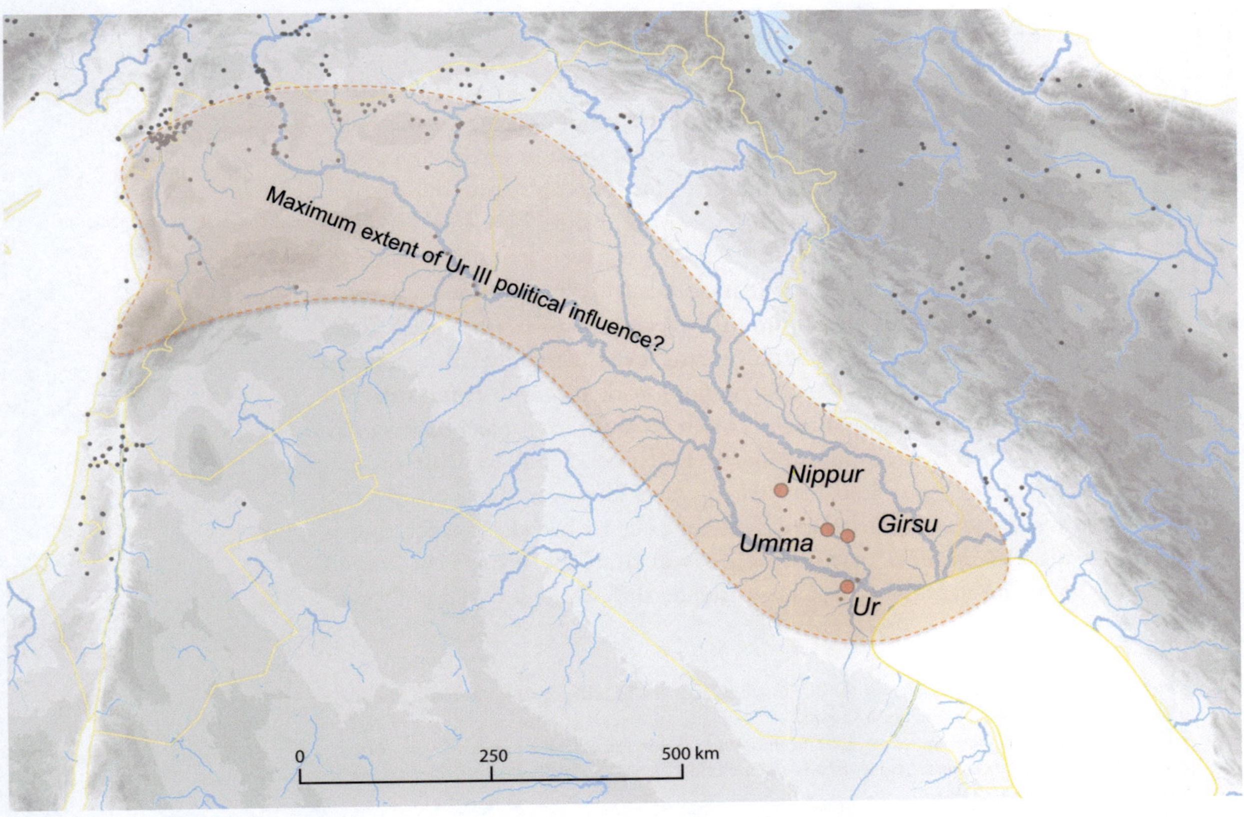
Maximum extent of Ur III political influence?
Nippur
Girsu
Umma
Ur
0
250
500 km

Many Ur III texts were, in effect, the account books of industrial work groups: the inputs/outputs (materials, textiles, wages, prices) and remainders/debts of a set period. Prices of textiles are apparently mentioned rarely (Pomponio 2010, 193), but various qualities and attributes are described, and clothing and cloth are differentiated. Again wool appears as the most commonly listed fabric, and weights of wool are one of the types of commodities listed in the accounts, sometimes with prices. "Linens, in contrast, were luxuries fit for dressing kings and divine images" (McCorriston *et al.* 1997): "flax cultivation was a minor enterprise in ancient Mesopotamia, and probably accounted for well under 10% of the total textile production achieved there (though even that is a substantial amount considering the enormous scale of the textile industry during the Ur III period)" (Reade and Potts 1993, 104; *cf.* Waetzoldt 1983).

The 'wool of mountain sheep' is apparently of highest value: 9 shekels as opposed to 4-6.5 for unspecified wool, or goat wool 1.5-2 (Pomponio 2010, 194-195). According to Waetzoldt (2010), "the colour of the textiles is almost never mentioned. We therefore suppose that most textiles were made of light-coloured wool". There are occasional mentions of apparently 'white', 'black', 'red/brown' and 'yellowish' – often for specific items, and with some sort of hierarchy of value, though which hues and shades were meant by these translated colour terms cannot easily be determined. 'Multi-coloured' textiles are also mentioned, but again "[w]e do not know whether the Sumerian word for multi-coloured (*gun-a*) meant two- or three-coloured fabrics or fabrics in which colours were combined" (Waetzoldt 2010, 202). Waetzoldt (2010) also suggests that the texts could be taken to suggest that most wools were naturally pigmented, and that dyed fabrics were extremely rare.

6.2.3 Old Assyrian texts: Cappadocian/Kültepe tablets (c. 1950-1750BC)

Tablets uncovered at the site of Kültepe-Kaneš in Anatolia[90], dating to the Old Assyrian period (c. 1950-1750BC), refer to a large-scale trade in textiles between Aššur and central Anatolia. A detailed linguistic investigation into the textile terms used in the Assyrian texts has recently been published by Cecile Michel and Klaas Veenhof (Michel and Veenhof 2010), who have attempted to draw out as much information as possible about materials, textile qualities and, very usefully, geographical terms. It is clear from this research that there is a great deal of uncertainty about the meaning of certain words, even if their relationship to textiles is clear from the context. The reconstruction of meaning from ancient vocabularies is fraught with danger, since it is often necessary to project backwards from periods in which the meanings are better understood, but by which time considerable linguistic change could have taken place. However, more positively, assuming the translations that Michel and Veenhof offer are broadly correct, there are certain trends and patterns that can be gleaned from the texts. First, wool again appears to be the material traded most frequently[91], though it is not always

90 *Kaneš* being the Old Assyrian name for the city in which a *karum* (trading centre) was based, the site known today as *Kültepe*, near modern Kayseri.

91 A large proportion of wool seems to have been sourced from *Purušhaddum*, recently argued to be located in the Konya-Isparta region (see Barjamovic 2011), and subsequently circulated across Anatolia (Wisti Lassen 2010b).

explicitly mentioned. According to Michel and Veenhof, the term '*kutânum*', the textile product that is recorded most frequently in the texts, appears to refer generally to a type of woollen cloth (and not to linen as previously proposed etymologically, though *cf.* Wisti Lassen 2010a, 275-276). Various different qualities of woollen material and textiles in general are recorded, as indicated, for example, by special requests by traders or comments about the lack of a certain quality of wool. Both Anatolian and Assyrian wool is mentioned, both taxable by the local Anatolian authorities, and exchanged for silver, with Anatolian textiles apparently always cheaper than Assyrian imports (Michel and Veenhof 2010, 226). Linen cloth is also mentioned, but apparently more sparingly.

The geographical aspects of the texts are particularly relevant to our investigation because they provide pointers to explicit movement between places. There are a large number of textiles with associated place attributions, including textile names qualified with a so-called 'nisbe' suffix (*e.g. Abarnīum* meaning 'from the [unidentified] town of Abarn', *a-li-ú-tum* possibly meaning 'from/made in/in the fashion of the city/Aššur', *Alkuaīum, Gasurīum* from Gasur=Nuzi=Yorghan Tepe?, *Hahhīum, Malku(w)aīum, Suseīum, Sarzuaium, Šilipka'um, Šurbuīum, Takkušta'um, Talhatīum*), all of which appear to have be exported from Mesopotamia to Anatolia; or names designated with a '*ša*' prefix (*e.g. ša Akkadīē* meaning 'of the Akkadians', *ša Šubirim* 'of Šubarum/Šubartum' a 'Hurrian'-speaking area to the north of Aššur, *ša Ālum* 'of the city' *i.e.* of Aššur, *ša Apum* 'of Apum/Tell Leilan', *ša Hahhum, ša Haqqa, ša Nihriya, ša Qaṭṭara, ša Tuttul, ša Zalpa etc.*); and in some letters, explicit sources for textile cargoes. The recording, in certain texts, of caravan routes taken by merchants and the amount of time required to travel between locations facilitates a rough reconstruction of the historical geography of south-central Anatolia. Thus some of these named places may be assignable to particular regions or known ancient sites, but many others can only be very vaguely located. A summary based on Michel and Veenhof's analysis is shown in Figure 6.3 and Figure 6.4. It should be borne in mind, as these authors note, that a geographically tagged textile does not necessarily mean that the material or item was actually still made there. These place-name affixes do not appear to differentiate the exact relationship of the textile to its eponymous town – they could have been bought there, made there, simply be 'in the style of' the named place or else call artificially upon the aura of famous places as a form of ancient marketing. The classic examples of modern 'denim' (coming from the French *de Nîmes*) or 'Oxford shirt' are illustrative: whilst the type of material may originally have been made famous by the production (or wearing) in a particular locale or region, the geographic name can stick long after production (and fashion) has moved elsewhere. Equally, whilst certain textiles are mentioned as coming from certain towns, it is not always possible to say whether they were actually manufactured there, or whether they were simply traded through the location mentioned.

There are few mentions of possible colours in the texts: yellow, white, red, black, 'dyed', 'dirty', 'pure/clean'. Michel and Veenhof note that colours are never specified in the caravan reports. This gives the impression that colours were of limited importance to the Assyrian traders, which contrasts with a more extensive range of 'quality' qualifiers ('of royal class', 'of very fine quality', 'of fine quality', 'of inferior quality').

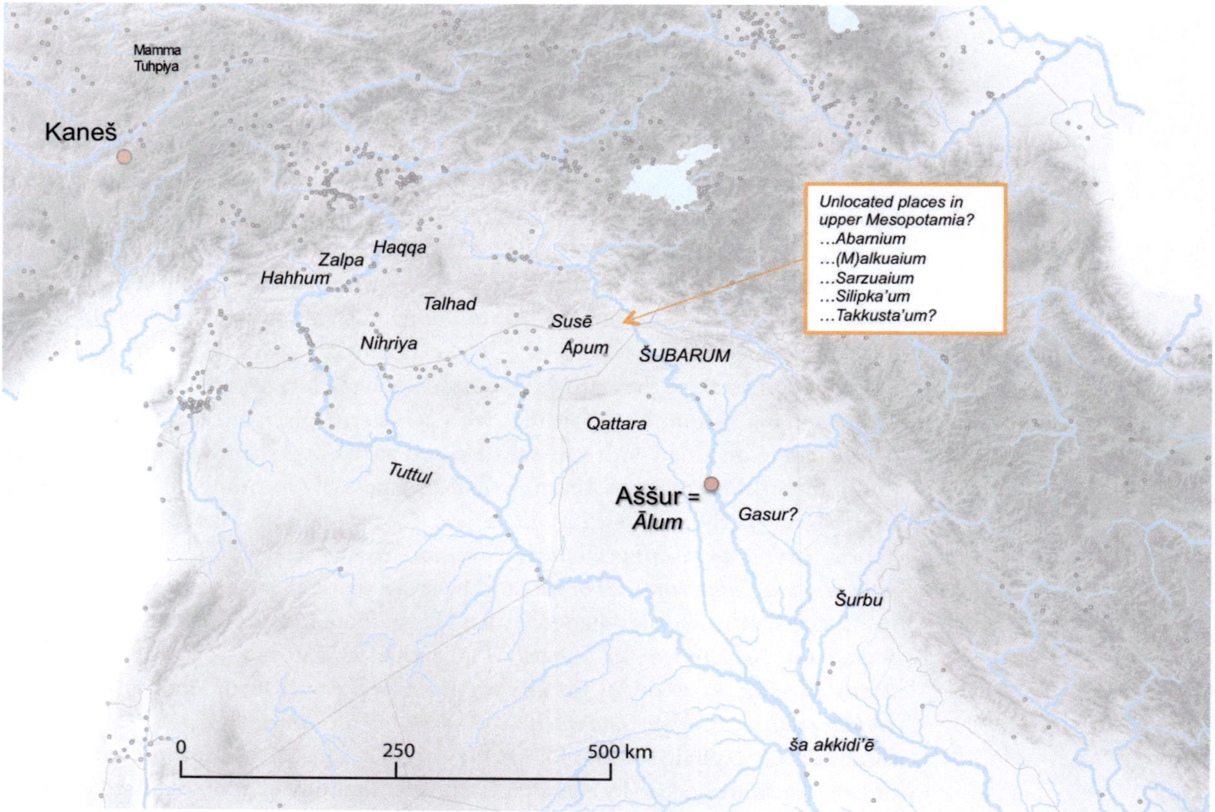

Figure 6.3.

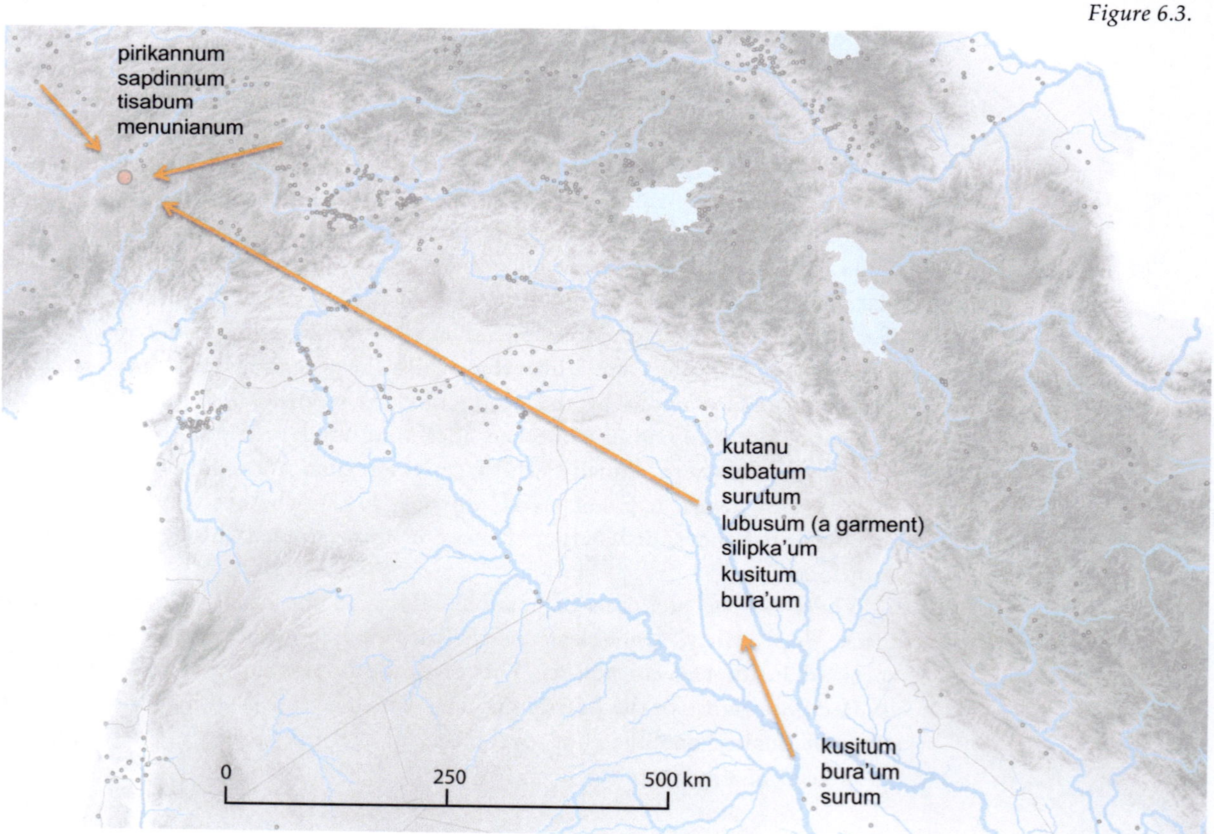

Figure 6.4.

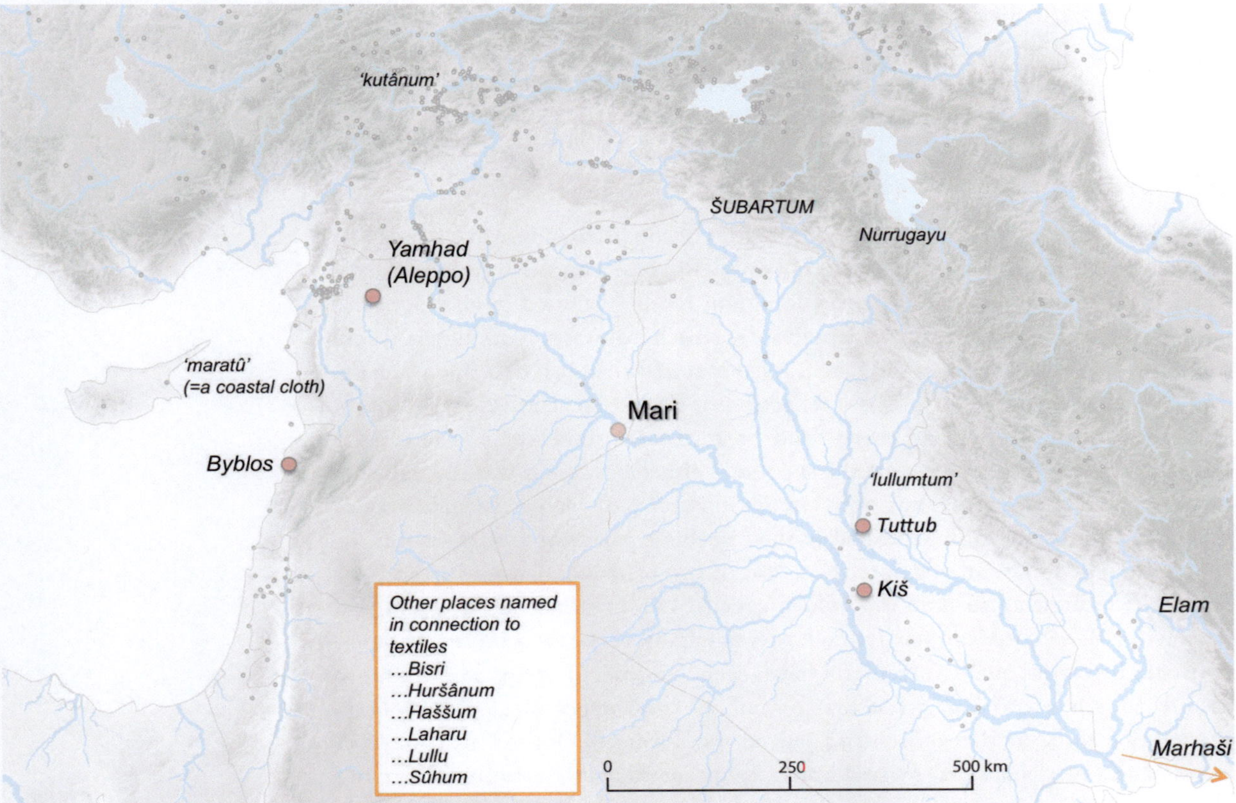

Figure 6.5. Places named in the Mari archives associated with the distribution of or with different styles of textiles (following Beaugeard 2010).

Figure 6.3. Places or place names mentioned in the Kültepe tablets associated with origins or styles of textiles (following Michel and Veenhof 2010, 224).

Figure 6.4. Kültepe textile types and their geographic origins or associations (following the analysis of Michel and Veenhof 2010, 218-226).

6.2.4 The Mari archives (c. 1800-1750BC)

The archives of Mari consist of between 12 and 15 thousand texts, many of which still remain unpublished and unanalysed. The publication of the archives so far in the series *Archives royales de Mari*, includes a catalogue of textile and clothing terms compiled by J.-M. Durand (2009). Anne-Claude Beaugeard (2010) notes that the archives include few or no terms referring to the production of textiles, but rather mostly reflect the accounts of textile repositories and the allocation of different fabrics and clothes from the palace to various recipients. She also notes that there appears to be a strong association between certain textiles (and thus presumably of styles or traditions of textile manufacture) and geographic or 'ethnic' identity (see map Figure 6.5). Comparable to the Assyrian terms found at Kültepe, many geographic/ethnic labels can be identified: *e.g. ša bišri* 'in the style of Bishri', *elamûm* 'Elamite style', *gublâyum* 'of Byblos', *huršânû* 'of Huršânum', *huššîtum* 'of Haššum', *iamhadu* 'of Yamhad', *kiššihhu* 'of Kish', *laharû* 'of Lahurû', *lullûm* 'of Lullû', *maratu* 'of the Levantine coast', *nurrugayu* 'of Nurrugayi in north east Mesopotamia', *paramšitum* 'of Marhaši in Iran?', *sûhum, šubarûm/ šubarîtum* 'of Šubartum', *tuttubum* 'of Tuttub'. It is interesting to note the wider geographical spread of the places mentioned than is the case with the Old Assyrian texts, indicating, perhaps, the extent of regular circulation of textiles and trading contacts in each region. Again, it is not always possible to determine whether the particular cloth listed was actually sourced from the place in question, or whether it was made elsewhere (perhaps even locally) but in the style of the place referred to (see the discussion of denim above). '*Kutânum*' textiles, the most commonly listed item in the Cappadocian texts (see above), are also mentioned – though whether there is any particular geographic implication, or whether it simply refers to a type of cloth which happens to be more popular in the north than in the middle Euphrates, remains moot. Most textiles again appear to be woollen, though some, perhaps small, linen items are mentioned (Beaugeard 2010, 289).

6.2.5 Textual evidence and its limitations

Some general points can be made about textiles in the urban contexts of the Near East of the late third and early second millennia BC based on these various textual sources. First, wool appears to be the dominant fabric discussed in all places and periods for which these texts are applicable, which contrasts with Egypt where we know, from both texts and well-preserved material remains, that linen was the dominant textile. Whether this is a reflection of the actual relative consumption of different materials during this period or a selective tendency of the texts is difficult to say. Linen was recorded less often – either because it was too valuable or not valuable enough – thus only circulating above or below the networks of exchange and distribution described in the respective palace or *karum* records.

If, though, this is an accurate reflection of the primacy of wool in the Near East, it is interesting to note that whilst specific colour terms can be identified in most of these corpora their frequency is generally very low. Other attributes, particularly cloth 'quality' seem to have been considered more important to record. This is surprising, given that one of the advantages of wool over linen is supposed to be its dye-ability, and indeed the dyeing of cloth is mentioned in some of the texts (at Ebla, for example). Egyptian characterizations of 'Syrian'/ oriental dress (for example, in the 19th century wall paintings of a tomb at Beni Hassan, Goedicke 1984) also suggest the prevalence of multi-coloured textiles. There are various possible explanations for this apparent paradox. Dyed fabrics may have remained extremely expensive items, especially compared, for example, to naturally coloured wools, and hence restricted in number and distribution. Early wool may have been less easily dyed than modern wools. Another possible explanation is that colours (or indeed patterns) were implicit in other attributes: for example that textiles were coloured or patterned according to traditionally understood place names. The Old Assyrian tablets from Cappadocia offer the clearest example of an organized trade in textiles, but the occurrence of textiles with geographic or ethnic associations also strongly supports the idea that textiles were circulating widely (in order for such associations to become recognized). The intensity of this interaction is difficult to assess since, as we have already noted, it does not necessarily follow that a *particular* geographically associated item of textile was made there. Presumably, however, most such styles did indeed originate in the place by which they are named and were made popular by the movement of people wearing such clothes, or selling them abroad.

To finish this review on a cautious note: whilst the texts may provide considerable information, it is difficult to judge how representative the data are. The danger with these texts is, of course, the potential to be blinded by detail. Each of the above textual archives covers a relatively short, non-overlapping, period of time. Each has its own focus: for example, the Old Assyrian records from Kültepe provide much detail on internationally traded textiles (but less on their form or usage), whilst the Ebla royal archives provide more detail on the conditions of production and the circulation of wool and textiles within the state itself (for example which cloth types are given to which types of individuals). Given that all the texts come from specialized contexts (in the hands of literate scribes, in the service of the palaces or traders), we may even question how representative they are of local textile circulation. Are there other flows of textiles that would not be recorded in the texts (household production, gift giving), which were as important as or more so than the economies actually described? Certainly the details of the organization of textiles and their exchange are unlikely to be representative of the situation in all parts of the Near East. North-east of Central Anatolia, no surviving texts of this period have been uncovered – detectable literacy appears only to have arrived in eastern Anatolia and Transcaucasia with the Urartu in the Iron Age. The cultures of Central Asia in which we are interested are very far away from any historical sources referring to textiles. Thus, while the information from ancient texts provides important insights into regional developments in textiles, for these non-literate zones we will ultimately need other types of evidence to reconstruct even the most basic of information about textile materials and movements.

6.3 Direct textile evidence: fragments and impressions

Direct evidence of ancient textiles and yarns comes in 5 forms, listed here in order of decreasing preservation:

1. Very well preserved items, full garments of items of cloth, preserved in extremely dry or wet anaerobic conditions which prevent the decomposition of the organic textile fibres (*e.g.* in Egypt, the Tarim basin or in waterlogged bog-sites of Europe).
2. Small fragments of threads, normally preserved due to the toxicity (to the organisms which normally decompose organics) of those materials in close proximity to the original textile (*e.g.* metal), preventing decomposition.
3. Mineralized fragments of cloth or thread whose organic components have been transformed by materials in close proximity to the textile (*e.g.* metal), or chemicals in the ground. These are often called 'pseudomorphs'.
4. Impressions of cloth, the shape of the fibres visible in other materials, such as clay, earth, ceramic or plaster, through close proximity in manufacture or in deposition, but the original organic textile fibres having completely decomposed.
5. Chemical residues of textiles, particularly dyes.

6.3.1 Preservation, contexts and materials

The distribution map and table (Figure 6.6) presented here shows as many published textile pieces, fragments and impressions that appear to be datable from the 4th millennium to the beginning of the 1st millennium BC within our broad region of interest as could be assembled for this work. This list is still not comprehensive, however, since textile fragments and impressions, despite their exceptionality, remain poorly published. A particular problem is the issue of textile preservation: in older excavations, conservation science was at a relatively early stage, and much important textile evidence was rapidly destroyed immediately after its discovery

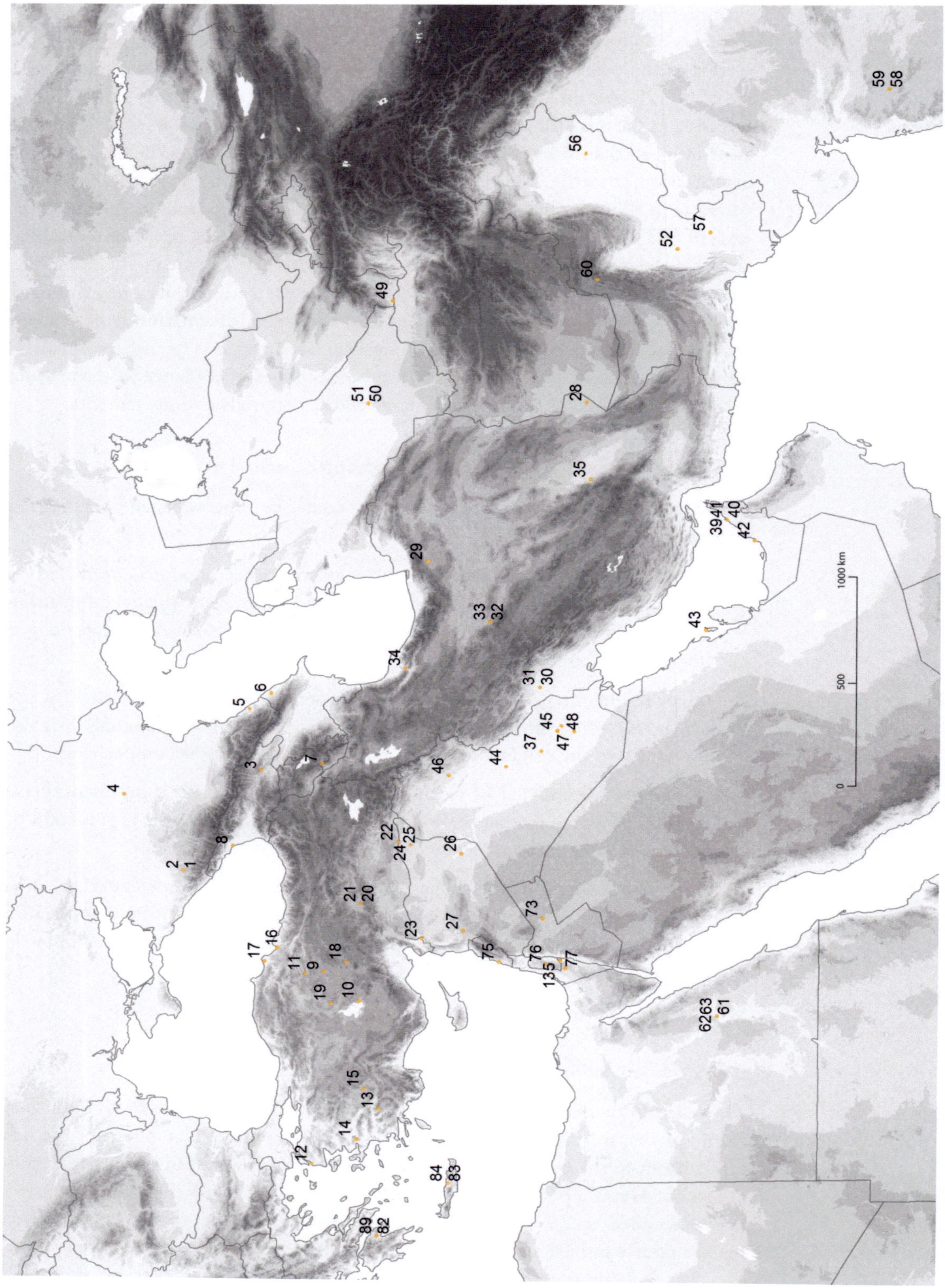

Figure 6.6a. Distribution of published textile fragments, impressions and threads dated from the 4th to the 2nd millennium BC. See Appendix C.1.3 for key to numbers.

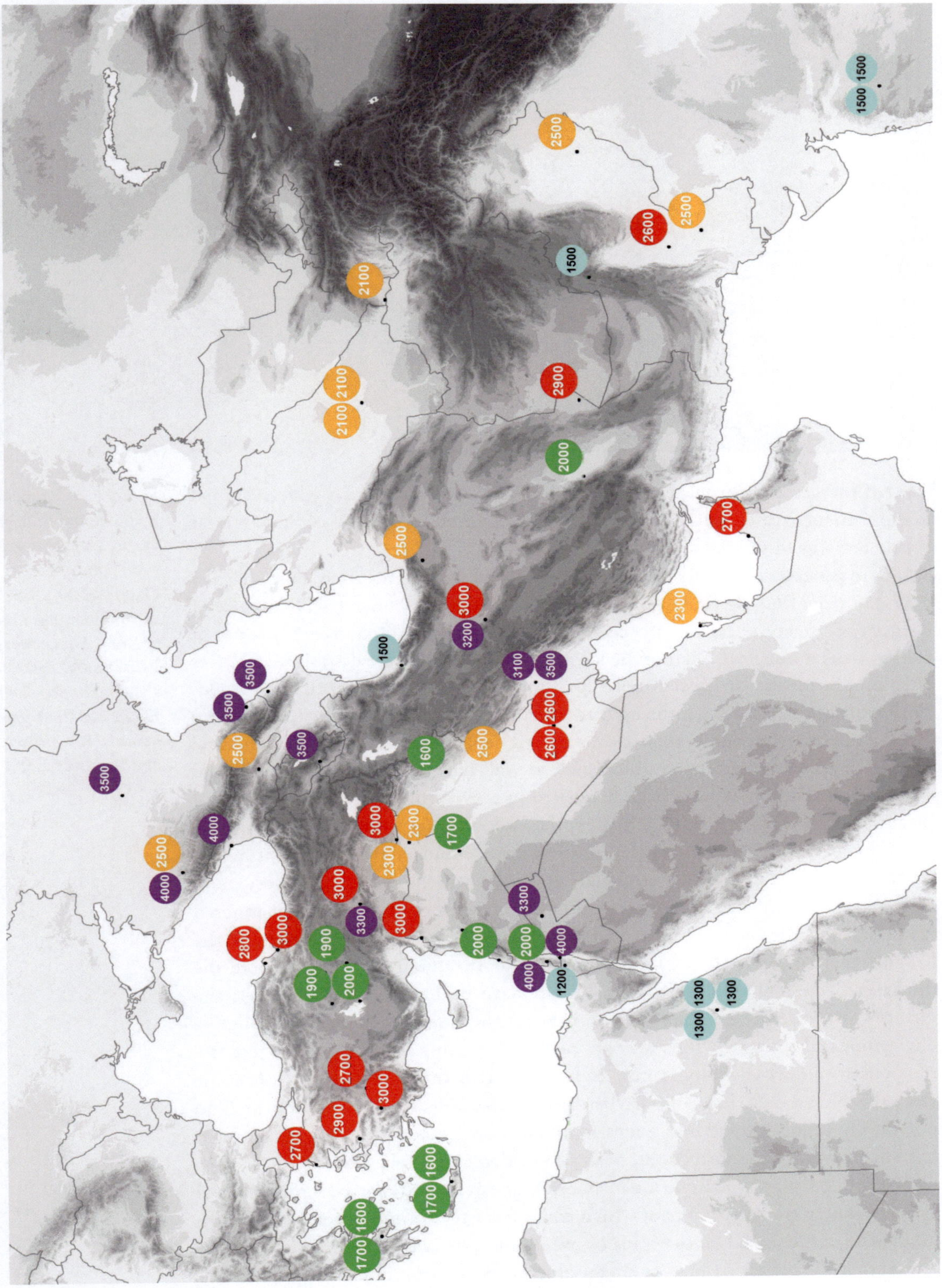

Figure 6.6b. Broad earliest dates for published textile fragments, impressions and threads (cf. Figure 6.6a). Concentrations or regional differences of date presumably reflect the level of archaeological investigation or preservation rather than ancient patterns of consumption.

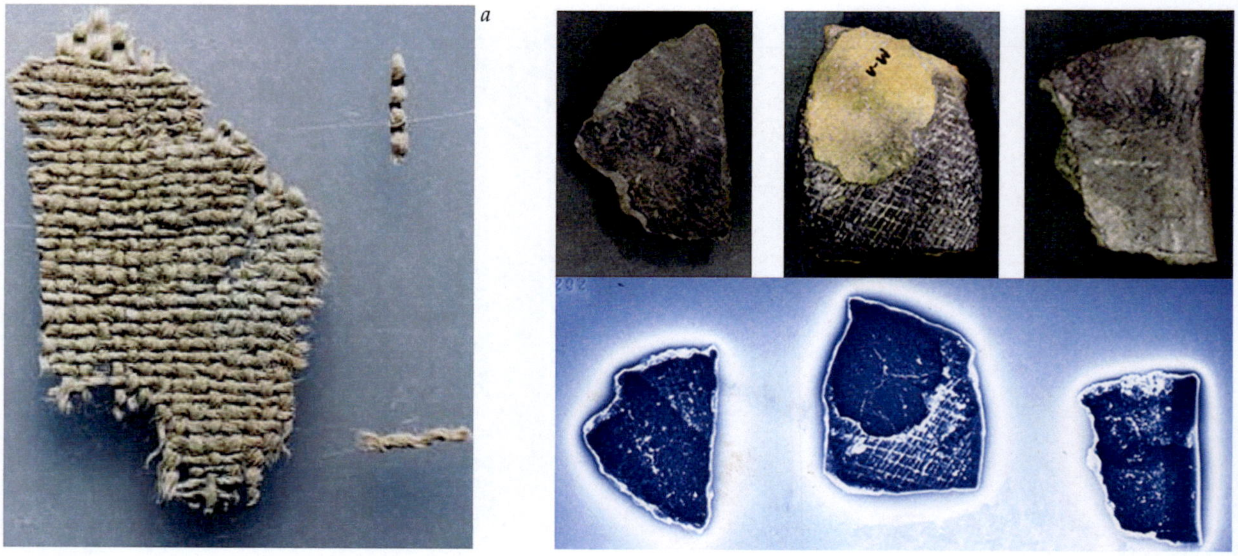

Figure 6.7. Examples of textile remains: (a) photograph of tiny fragment of linen from necropolis of Gonur Depe (Tsareva 2007, 330-331); (b) xeroradiographs of fragments of Kura-Arax pottery from Velikent and Serkertepe showing impressions of coarse textile apparently used in manufacture process (Heinsch and Vandiver 2006).

(see *e.g.* the descriptions of textiles from the Royal Cemetery at Ur which "turned to dust" upon excavation). Even today, there is limited knowledge of how to treat textile remains in the field.

Some general comments on this corpus can be made. First, it is clear that the vast majority of textile finds are from burial or mortuary contexts. The nature of mortuary contexts means that the corpus we have may be limited to certain kinds of textiles – depending on whether the kinds of goods which were deposited in graves were in general use or special items for funerary deposit, *e.g.* specialist 'shroud' cloth or death clothing. Cloths used to wrap or protect other goods (particularly metals) also survive better than cloths placed separately. 'Household', domestic or even high-value 'public' textiles (such as wall-hangings, carpets, and normal everyday clothing) may have been deposited in graves only occasionally as part of displays of wealth sacrifice (*e.g.* in Egyptian tombs or in the Royal Cemetery at Ur). The geographical distribution of direct textile finds appears to be primarily related to climatic and local contextual conditions which impact on preservation: thus Egyptian tombs are famously well endowed with cloth, whilst well-watered Anatolia has produced very few and very fragmentary samples. Curious is the low number of reported textile finds from the Turkmenistan region:- here the desert environment of the Karakum might have been expected to increase the chances of textile preservation. That this is not the case may be a consequence of environmental changes, which may have led to increasing aridization only recently (Cremaschi 1998). Equally it may indicate that in this much flatter landscape than the Nile valley, the contrast between wetter/irrigated lands and arid (and potentially textile-preserving) zones is much more gradual. Burials may simply have been sited in less arid zones, nearer to settlements.

The bulk of all textile finds can be put into one of two categories: impressions of cloth found on pottery (many of which come from the Caucasus and its northern hinterland), or original cloth scraps or pseudomorphs (where the textile has been remineralized) preserved next to metal objects (or occasionally bone). Our case-study regions yield a small range of examples. A tiny fragment of cloth from the necropolis at Gonur Depe (Figure 6.7a), in plain weave and apparently made from linen, was found by the leg bone of a body from burial 2380: "the cloth was folded together several times and then wrapped round a bronze 'ladder'" (Tsareva 2007, 330-331). The piece was also a green colour, although this is likely to have been a post-depositional process of leaching from the copper. Impressions of textiles are occasionally pretty much the only evidence we have. In Transcaucasia and eastern Anatolia so far very few surviving textile fragments from the 3rd and 2nd millennia have been identified. The use of textile to actually produce Kura-Arax pottery, as documented by a xeroradiographic analysis of some pottery fragments from Velikent or Serkertepe (Figure 6.7b), is therefore quite significant – although what exactly 'textile-formed' meant in practice remains a matter for debate (Heinsch and Vandiver 2006).

The kinds of information that can be extracted from such tiny scraps or impressions and 'pseudomorphs' in other materials is normally quite limited: it may be possible to identify the weave (plain, basket, twill *etc.*), and perhaps the fibre type (wool, linen) and direction the fibres were spun (S or Z) – with the naked eye or through the microscope. Actual preserved fibres can be tested through microscopic inspection or chemical analysis to increase the reliability of the fibre identification[92]. From the current corpus of data, we might be tempted to assume, then, that linen was the most popular or common material in use during the 3rd and 2nd millennia BC), with an occasional but increasing use of wool which had started in the 4th or 3rd millennium, but this seems likely to be merely a bias in preservation. Experiments suggest that flax fibres are more likely to be preserved next to metals (through mineralization) than woollen ones, so the weighting of this corpus is unlikely to be very representative (Janaway 1989). Cotton also makes an occasional appearance – in the Indus, where it is thought to have been first systematically cultivated, in the desert of eastern Jordan (Betts *et al.* 1994) and, surprisingly, north of the Caucasus (Shishlina, Orfinskaya and Golikov 2003).

6.3.2 Textile patterns and colour

Patterned fabrics, or rather fragments or impressions that show any kind of evidence of patterning (particularly in colour), are the exception in this corpus (their distribution shown in Figure 6.9). Given the primary importance of patterns to the desirability of textiles (and their ability to convey semiotic messages), this considerably hampers our ability to put even this small corpus into social context, or trace the geographical or social meanings of such patterns. This paucity is partly because the majority of fragments are too small to show any patterns, but also because the techniques of dyeing or decoration used are unlikely to have left any evidence (especially in impressions, unless the weave itself showed some three-dimensional pattern, with or without associated colour). The fact that the most commonly preserved fragments in the archaeological record are linen cloths used to wrap metal objects may also play a factor. Since linen is more difficult to dye than wool, if decorated it would often be painted – a decoration even less likely to be preserved than coloured dyes. We also should keep in mind that the extent to which patterns were used in textiles may have varied considerably through time, for both technological (availability/knowledge of dyes, dye-ability of the fabric) and socio-economic reasons (cost, sumptuary, moral or 'aesthetic' restrictions). Egyptian clothing (of which we have a relatively good corpus of large

92 This is a recent development in research, however, and relatively rarely used.

fragments or full garments) seems to have been relatively free of patterns for most of the 3rd and 2nd millennia. Whether this is simply because linen, as a favoured fabric, was difficult to dye, or, perhaps a more compelling idea, that conventional Egyptian aesthetic ideas demanded the limited use of patterns on clothing, is hard to prove.

Some textile patterns are also formed by the nature of the material and weave used. Twill is a good example of such weave-based patterning, because the nature of twill weave provides such a distinctive diagonal patterning to the cloth it produces. The distribution of twill seems to have been limited to certain areas (see inset of Figure 6.8). The earliest direct remains of twill weave are known from Alişar (some fragments of linen attached to bone, in a burial apparently dated to the 4th millennium[93]), from Martkopi in Georgia (particularly a woollen fragment from a kurgan perhaps dated to the late 3rd or early 2nd millennium BC, see Karaulašvili 1979, pl. III.1-2; Barber 1991, 168), and much later from the Tarim basin (including various items dated probably after 1200BC, see Good 1998). Better known examples are of course known from central Europe during the Hallstatt period, dating to c. 1200-800BC (Barber 1991, 186-195). A pseudomorph left on a copper axe-head from the 4th millennium levels at Susa may represent an alternative (non-twill) example of weave-based patterning if Barber's (1991) interpretation is correct – *i.e.* as a weave in which a fine warp is interwoven with weft fibres of varying coarseness. This would give the effect in the finished cloth of a striped pattern. The descriptions of the textiles uncovered from the Royal Cemetery of Ur (c. 2500-2400BC), suggest other forms of weave-based patterning: a "finely woven cloth with a diagonal rib" (Woolley 1934, 238) – representing either a twill weave or an appearance of diagonal ribs achieved using simple plain weaves (Barber 1991, 164).

Figure 6.8. Distribution of plain and twill weaves 4th-2nd millennium BC; (inset) visual difference between plain and twill weaves

Based on the relevant direct evidence, a range of dyeing colours was known during the 3rd and 2nd millennia BC, including shades of red (including brown, orange, purple), blue, yellow and green (Barber 1991, 223-225). One example of a coloured textile was found in a kurgan (burial mound) at the village of Novosvobodnaya (previously known as Tsarskaya), which included "two garments: an undergarment of linen-like fabric 'brightly decorated with purple color and covered with red threads in the likeness of tassels' ... and an overgarment of fluffy (or downy ...) yellow cloth ornamented with narrow black stripes forming a close-set and regular plaid" (Barber 1991, 169). The kurgan is dated to the mid- or late 3rd millennium by Barber, though may be earlier. Here colours appear to be used in relatively large blocks, rather than as detailed patterns (though the original size of the garments is difficult to tell from the published fragments).

One of the few potential examples of a substantial piece of patterned textile from outside Egypt also comes from a burial mound north of the Caucasus, this time Kurgan '9/7' of Tri Brata, near Elista in southern Russia. In a report of the kurgan's excavation by Soviet archaeologists, Sinitsyn (1948) claims that one corner of a patterned rug was preserved as an impression in clay at the edge of (and

93 The item is quoted as coming from a 'Chalcolithic' burial. Note however that the Alişar stratigraphy and dating scheme is rather poorly understood and care should be taken with dating information from the site.

Figure 6.9. Location of sites yielding textile fragments with identifiable distinctive patterning (red points).

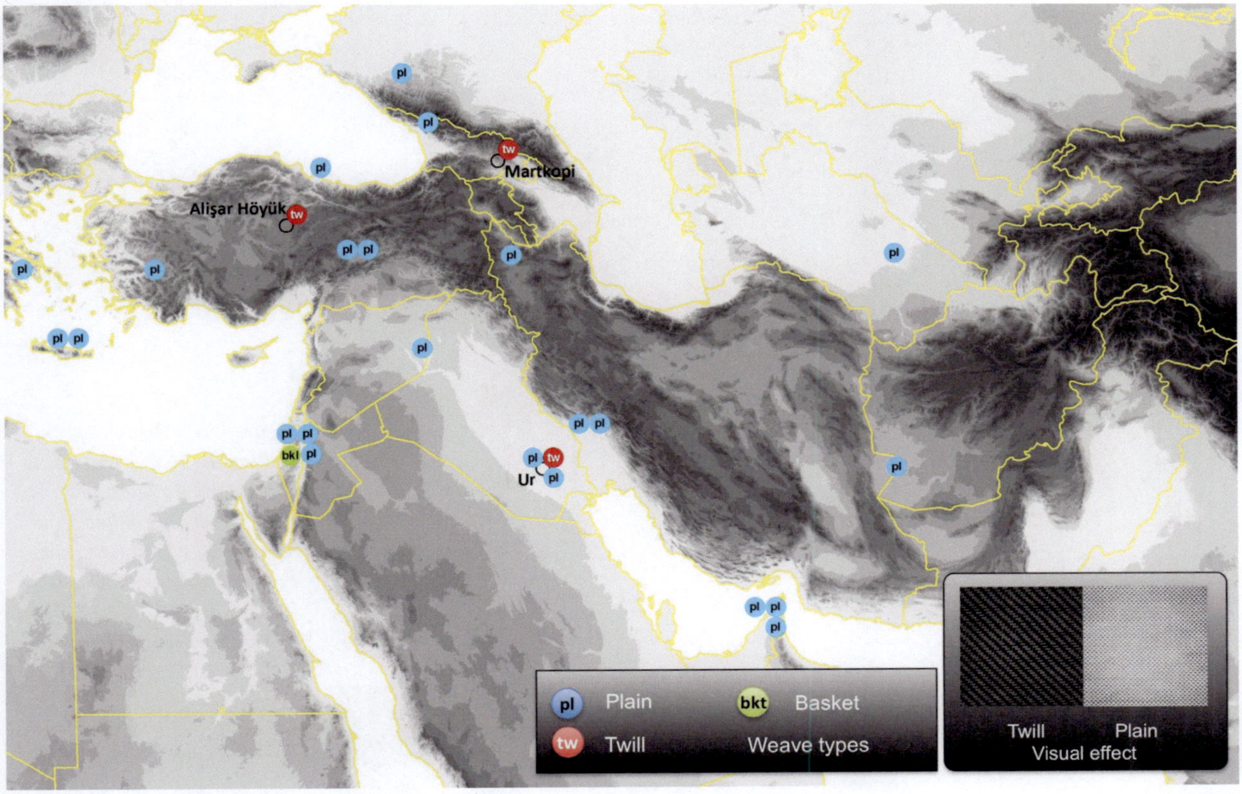
Martkopi
Alişar Höyük
Ur
Plain
Basket
Twill
Weave types
Twill
Plain
Visual effect

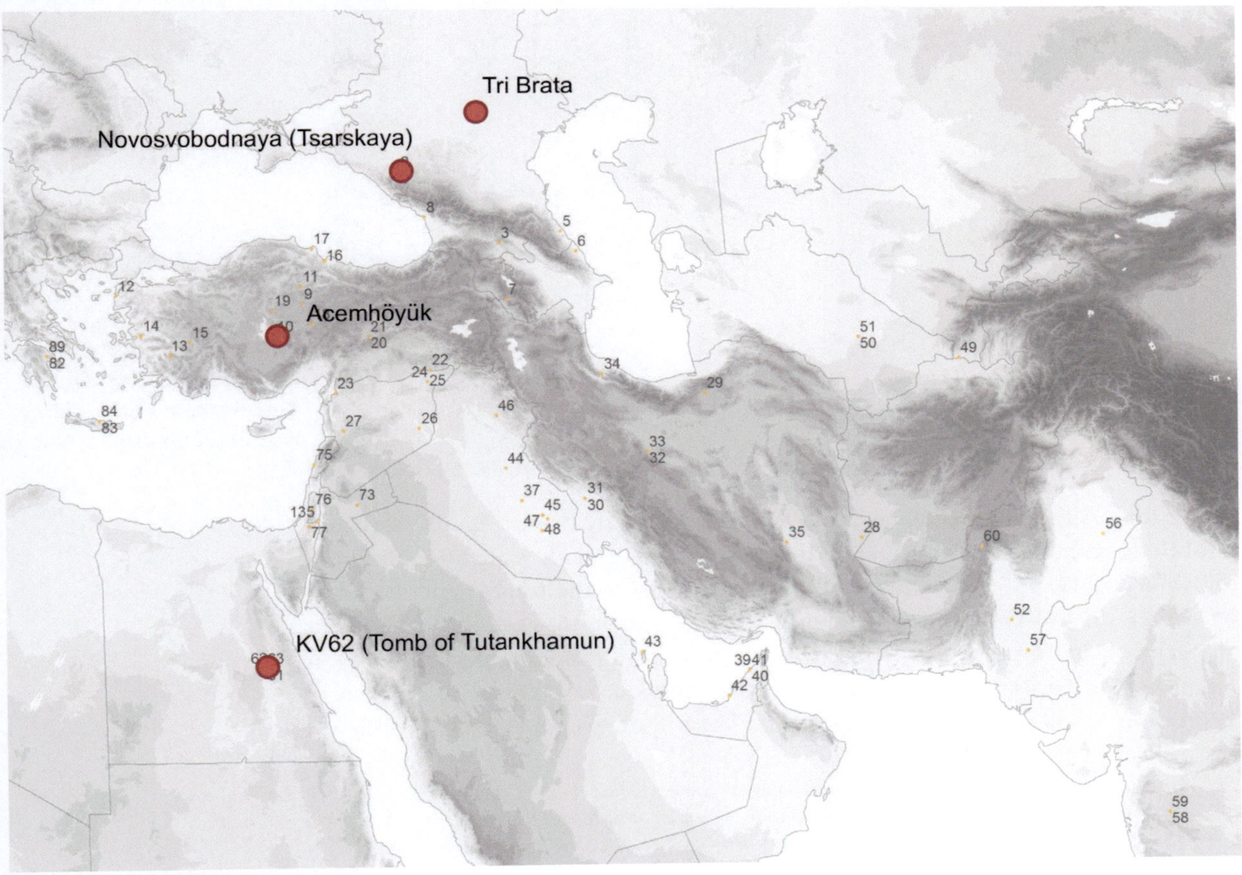
Tri Brata
Novosvobodnaya (Tsarskaya)
Acemhöyük
KV62 (Tomb of Tutankhamun)

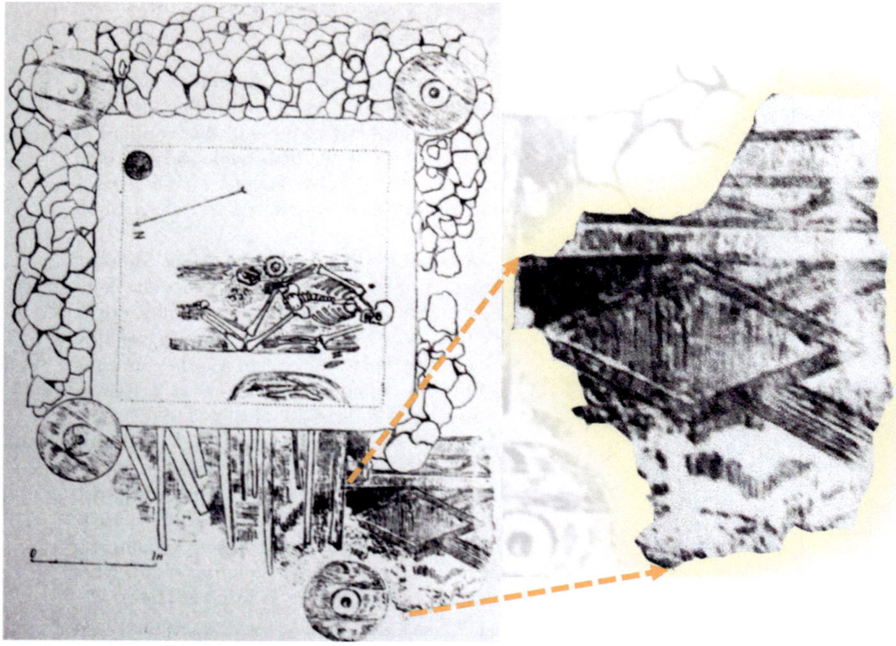

Figure 6.10. Reproduction and highlighted enlargement of 'carpet'-like impression found in the burial pit of Kurgan 9 at Tri Brata near Elista, dating c. 2600-2200BC (Sinitsyn 1948, 152; cf. Barber 1991, 169-170).

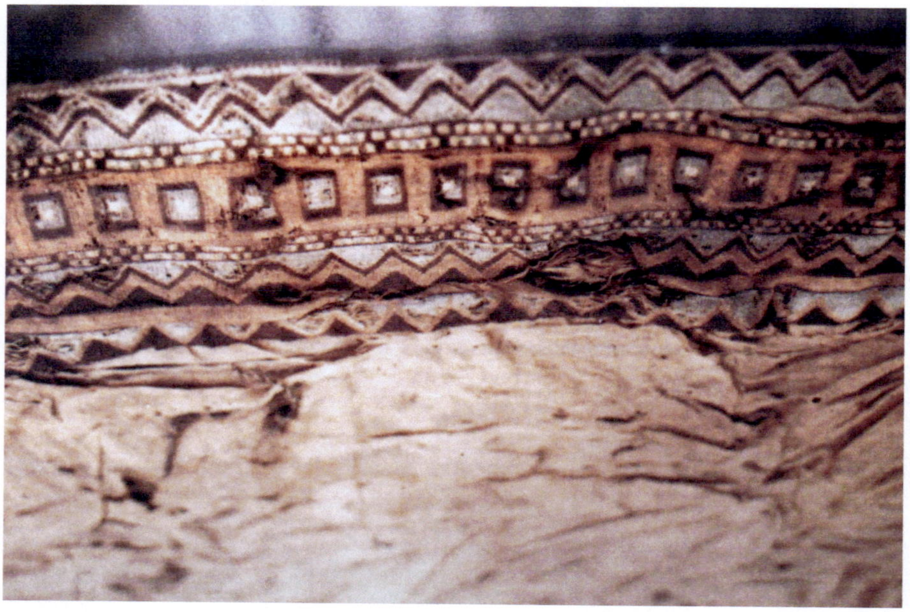

Figure 6.11. Detail of the patterned edge of the so-called 'Syrian-tunic' found among the grave offerings of Tutankhamun's tomb, dating c. 1323-5 BC (after Vogelsang-Eastwood 1999, 82).

on the top of) the burial chamber[94] and dated to "the Early Bronze Age". Natasha Shishlina's research dates this tomb to the Yamnaya period, c. 2600-2200BC (Shishlina 1999, 133; Shishlina 2008). The drawing of the tomb (republished by Barber 1991, 169-170; and Gimbutas 1956), if shown to scale, illustrates an apparently large-form pattern including lozenges, triangles and framing bands (see Figure 6.10). Barber suggested that the textile technique most likely to leave an impression would be tapestry, since it would leave "quite palpable discontinuities

94 The technique of placing textile 'roofing' over a burial chamber before closing (which is what might have happened here) is known from a much later period in north-east Arabia (Haerinck 2002) although there is presumably no direct connection between these.

between the color zones of the design", in contrast to other possible techniques such as pile knotting, in which each "knot tends to be of the same size regardless of color" (Barber 1991, 170) and hence would not produce an impression[95]. Sadly, there is no photograph showing the actual weave, knots or thread of this supposed rug, and no further data about the impression, so that it is easy to question the authenticity of the report. Given that most textile patterns are not detectable in impression form, could this imprint instead be more simply explained as that from some other material (*e.g.* wood or basketry) more likely to leave such an impression?

Slightly later than our main period of interest, but nonetheless providing a vivid illustration of the potential for textile hybridity is a tunic from the tomb of Tutankhamun ('KV62' in the Valley of the Kings). This garment (from the 18th dynasty, dated to around 1323-5BC), included finely embroidered hem-panels (see Figure 6.11), "edged like the rest of the tunic in a lavish array of patterned, warp-faced bands" (Barber 1991, 161; *cf.* Vogelsang-Eastwood 1999, 82). Given the general history of Egyptian textiles as understood from the larger corpus (with its relatively minimal patterns and very limited use of embroidery as a technique), the tunic is already unusual. However, the style and content of the iconographic motifs depicted in these hem-panels (made as separate panels sewn onto the tunic), is arguably closest to 'Syrian' (or Syro-Mesopotamian) models. All of this emphasizes the panel's 'foreign' character, whether or not it was the cloth or the cloth-makers that were imported from the north-east. However, the tunic is also embroidered with Tutankhamun's name and the ankh symbol, showing that the garment and decoration are designed to act within an Egyptian cultural *milieu* for one individual. The 18th dynasty (and particularly the Amarna period) appears to be a unique era of considerable change within Egypt, presumably driven through an intense cultural, political and economic interaction with the Eastern Mediterranean and Syro-Mesopotamian world at this time. It is perhaps significant that it is also in the 18th dynasty that an exceptionally large number of pottery vessels are painted, compared to the very plain way in which pottery is treated throughout the majority of the 3rd and 2nd millennia BC in Egypt. This tunic, and other 'foreign'-style pieces in Tutankhamun's wardrobe may represent an unusually eclectic range by Egyptian standards. However, it reminds us of the potential for 'mixing and matching', and the adaptability of imported textile for local needs (in opposition to the rigidity of 'ethnic' clothing, for example).

6.3.3 Comments: textile remains and textile routes

Any attempt to reconstruct the historical flow of textiles needs to explore the corpus of direct evidence, both fragments and impressions in other materials. Unfortunately, whilst an essential piece in the jigsaw, the nature of the corpus – extremely fragmented (both geographically and temporally), biased (where linen is over-represented by processes of decomposition) and fragmentary (in terms of preservation) – makes the data unsuitable for the reconstruction of routes or contemporary flows of materials on their own. It is thus necessary to place these direct fragments and impressions into the context of material sources and indirect evidence.

95 Dyes might have been responsible for the patterned remains, as Barber recently pointed out to me in a personal communication.

6.4 Production of textiles: evidence for materials and technologies

To understand the circulation of textiles it is of course very useful to look at the sources of the raw materials, particularly those plants and animals that have traditionally been the source of textile fibres and coloured dyes until the invention of synthetic materials during the 20th century AD. Additionally, some of the technologies of fibre/dye processing and textile production do leave traces in the material record that provide additional information about traditions of manufacture and thus the movement of technologies.

6.4.1 Distribution of raw materials: fibres

The reconstructed 'natural range' of plants like flax and cotton at first exploitation or domestication and their subsequent migrations through time can help to localize the economic sources of materials: this is something which would help macro-economic reconstructions, such as world-systems based theories, that are based on unequal distribution of resources and the dislocation of wealth production. Archaeological identifiers of such 'sources' are extremely difficult to find, however, for two main reasons. First, the distribution of many relevant plants and animals can and has changed substantially during the history of human exploitation and the available data is very patchy. Second, the species themselves have changed physically and genetically through time, such that it is not always possible to compare like with like. All reconstructions are therefore extremely partial, and open to revision. Various types of fibres can be used to make textiles and each varies in specific sources and physical characteristics[96]:

Figure 6.12. The origins of flax and the extent of its cultivation by the 3rd millennium BC, (following Zohary and Hopf 1988, 122; Barber 1991, 250, top).

Flax, hemp, jute and *nettle* are the most commonly cited plants which yield 'bast fibres' used to make textiles, the most significant being *linen* from flax. Evidence from fibre impressions from central Europe suggests that twined bast fibres were known and perhaps widely used from very early on, at least as far back as the Upper Palaeolithic (Adovasio, Soffer and Klima 1996). The 'natural' distributions of the plants in question are geographically very wide, and the history of their earliest exploitation remains difficult to access. Local climatic and soil conditions affect the growth of each individually, something that may have affected ancient populations and their respective abilities to produce sufficient fibre. Flax is considered one of the earliest cultivated plants (Zohary and Hopf 1988, 122). The growth of flax for fibres appears to have been easier in relatively temperate climates, but irrigation techniques may have allowed cultivation of domesticates in hotter climes (Reade and Potts 1993, 103-104), such that by the 3rd millennium BC it was under widespread cultivation (see Figure 6.12). Whether the ability to grow and harvest particular plants in a region necessarily implies their exploitation there, is, of course, impossible to assert definitively without botanical and textile evidence. Suitable palaeoenvironmental studies are

Figure 6.13. Summary showing the likely geographical variations in the use of different textile fibres during the 3rd millennium BC.

96 See also Barber (1991) for a general introduction to the domestication of plants and animals associated with textile fibres.

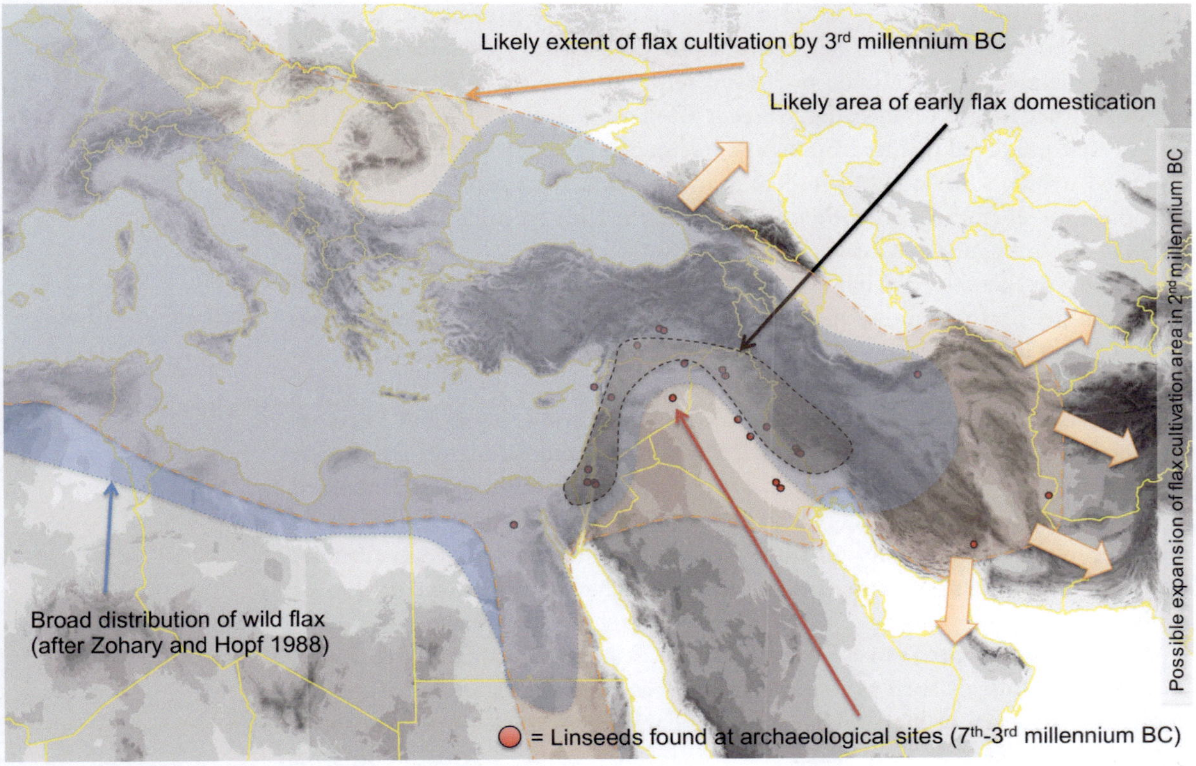

Likely extent of flax cultivation by 3rd millennium BC
Likely area of early flax domestication
Possible expansion of flax cultivation area in 2nd millennium BC
Broad distribution of wild flax
(after Zohary and Hopf 1988)
= Linseeds found at archaeological sites (7th-3rd millennium BC)

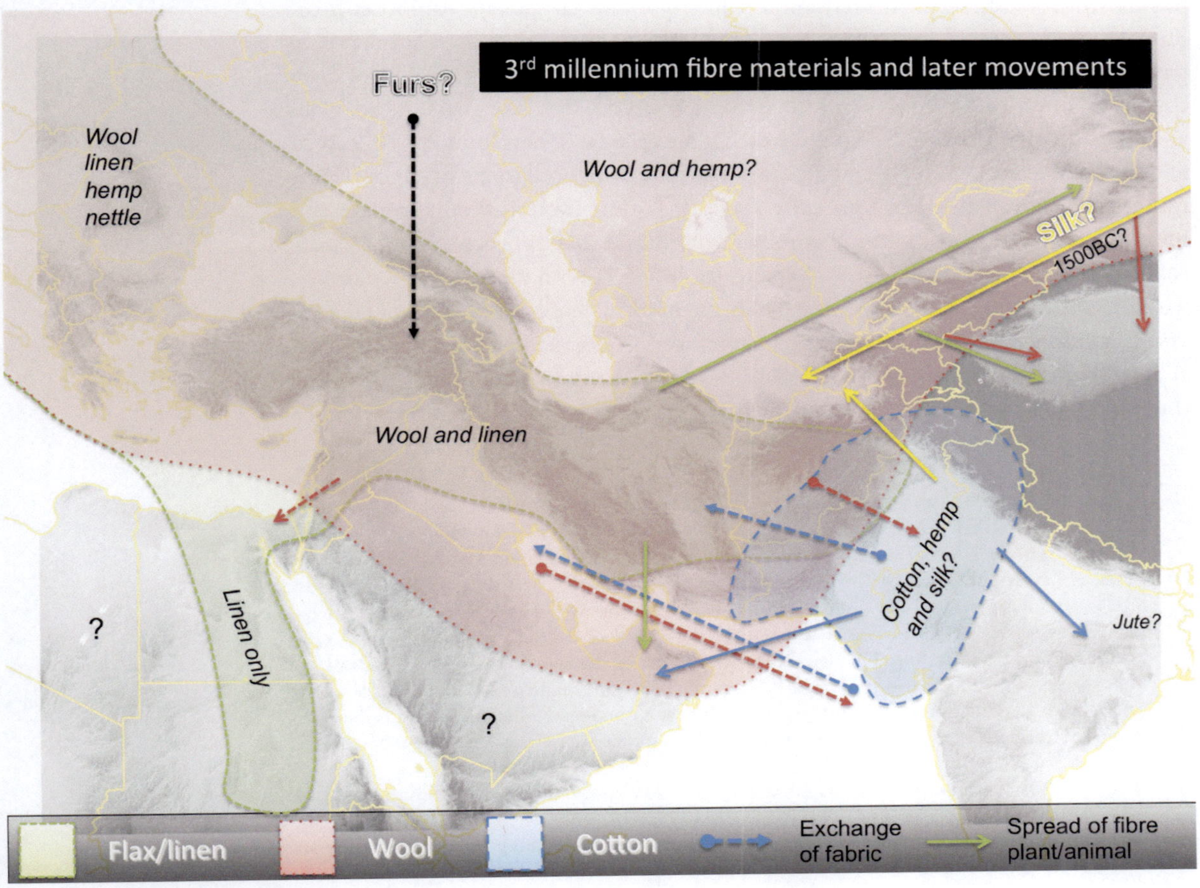

3rd millennium fibre materials and later movements
Furs?
Wool
linen
hemp
nettle
Wool and hemp?
Silk?
1500BC?
Wool and linen
Cotton, hemp and silk?
Jute?
Linen only
?
?
Flax/linen
Wool
Cotton
Exchange of fabric
Spread of fibre plant/animal

geographically uneven, but even where flax seeds have been found[97], it is not always certain that the plant was being exploited for fibres over its oil[98]. Flax may be processed to make linen or linseed oil, though today different cultivars produce either longer fibres or larger seeds (Zohary and Hopf 1988, 122). The earliest well-preserved textiles from Egypt are linen, and, except for a few, perhaps imported items in the late 2nd millennium, the plant fibre seems to have been used there for all clothing and other textiles until the later 1st millennium BC (see *e.g.* Hall 1986; Vogelsang-Eastwood 1999).

Cotton is a vegetable fibre but not 'bast' because its fibres come from the seeds of the cotton tree or shrub, rather than the stalk of the plant. There are various species of cotton, but it has been argued that the earliest cultivation or domestication was undertaken in the Indian subcontinent (Gulati and Turner 1929; Fuller 2006; 2008), though the picture remains unclear (*cf.* Hutchinson, Silow and Stephens 1954). Cotton plants require very particular climatic and geological conditions and it seems that the distribution of cotton cultivation was very limited until at least the 1st millennium BC (Alvarez-Mon 2005, 4-5), perhaps exclusively to the Indus region. Finds of cotton fibres (such as at Novosvobodnaya, see below) have thus been taken as strong evidence of its exchange rather than spread of cultivation.

Sheep and goats appear to have provided the majority of animal-based fibres, though other animals such as *camel* may also have provided alternative sources locally or for special or high-value exchange. The earliest domestication of sheep and goats is believed to have taken place somewhere in the Zagros mountains, but as Andrew Sherratt stressed some time ago, the large-scale exploitation of domesticated animals' 'secondary products' (such as wool or milk) appear not to have been part of the initial domestication process (A. Sherratt 1981). Fibres from the first domesticated sheep seem likely to have been hairy, like goats, rather than woolly, with the expansion of wool-bearing sheep a later development (Halstead and Isaakidou 2011, 67). Sheep hair may only have been used to make felt or fleeces rather than woven textiles at an early stage. It is thus more difficult to date and identify the regions from which woolly sheep originated and expanded. Despite some potential problems[99], the characterization of population kill-off patterns based on the data of sheep bones has been used at various sites (see Halstead and Isaakidou 2011, 68) to attempt to track the emergence of wool exploitation archaeologically. In theory, a large number of old sheep in a population suggests they were being kept for their wool (or milk) rather than for meat. This evidence

97 Indeed, McCorriston (1997) argues that it is more likely that flax seeds are preserved in processing the seeds, rather than when extracting the fibre. Flax seeds have been found in archaeological contexts from the 7th millennium onwards, see Figure 6.12 (Zohary and Hopf 1988, 124-125; Reade and Potts 1993, 102; McCorriston *et al.* 1997, 519). In later ('post-domestication') periods, little attention has been paid to the distribution of such evidence so our knowledge is rather patchy. Exceptionally, such evidence is used to highlight the arrival of flax in a region: for example, at Loebanr 3 in the Swat valley of modern north-west Pakistan, the earliest identified linseeds in the Indian subcontinent are dated to 'Period IV', around 18th-15th centuries BC according to Stacul (1987), though perhaps earlier given the general shift in chronologies. The temptation is therefore to suggest the cultivation of flax for linen was part of the western imports of this 'migration period', to replace cotton, but the aim of cultivation could have easily been for linseed oil.

98 "Berger (1969) stresses that the geographical range of flax cultivated for the manufacture of oil [...] is greater than so-called 'fibre flax'" (Reade and Potts 1993, 102-103).

99 Including the difficulty involved in differentiating faunal remains of sheep vs. goats and hairy vs. woolly sheep (see Marom and Bar-Oz 2009).

suggests that widespread exploitation of wool started at latest during the 4th millennium BC. During the late 3rd and 2nd millennia BC, if the contemporary textual evidence is to be believed, cities had access to many hundreds or thousands of sheep in the plains of Syro-Mesopotamia. Sheep and goats are also well adapted to mountainous environments, and we might expect the highlands of eastern Anatolia and western Central Asia to be well provided: the faunal evidence from Early Transcaucasian/Kura-Arax period Sos Höyük (Piro 2009) indeed shows an emphasis on sheep and goats, though it is unclear whether wool was the aim of herding practices. Large-scale shepherding practices often involve considerable transhumance, *i.e.* movement of animals over long distances seasonally[100]. This 'source mobility' makes reconstructing flows of wool particularly complex.

The same 'source mobility' is of course true of animal skin based products, such as *furs, leather and fleece*, from *cattle, sheep, horse, pig* and a whole range of wild animals. Given the range of possible animals from which leather can be obtained, it is difficult to apply any kind of geographic boundaries to potential sources, even if such products were not consumed in equal proportions by different peoples.

Silk (from the cocoons of certain species of moth, see Good 1995), originally thought to be exploited exclusively in China until the 1st millennium AD, now appears to have been known in both China and India by the 3rd millennium BC, though the direct evidence of production or use is very sparse. Some very early remains of silk fibres have been found at Sapalli Tepe, in Central Asia dating to the local 'Middle Bronze Age' (Good 1995, 964), c. 2500-2000BC, and more recently at Chanhu-Daro on the Indus, dating to around 2450-2000BC (Good, Kenoyer and Meadow 2009). This last example overlaps in date with the earliest direct evidence for silk in China. As Irene Good has pointed out, however, there are a number of species of silk moth that could have been exploited at different times, each with a very different 'natural' range (Good 2011, 220). The presence of silk in the Indus during the 2nd millennium BC thus does not necessarily imply that the silk came from China. We may speculate, though, whether the knowledge of silk extraction could have been exchanged between these regions, given the near contemporaneous evidence in both regions. Silk was produced in Central Asia in large amounts by Sogdian farmers during the early medieval period, and later in Byzantium (Barber 1991, 32).

Based on current archaeological and textual evidence, other fibres like *byssos* (from certain species of shell fish, see Dalley 1991) were either unknown or used only to a very limited (and localized) extent during the 3rd and 2nd millennia BC in the Mediterranean or western Asia.

Fibre type can provide a very crude indicator of fibre provenance (see Figure 6.13): the ecological range of wild and domesticated species of bast-fibre plants (flax, cotton-tree/bushes) offers a large but distinct area for the source of certain fabrics. This gives a broad indication of the movement of textiles in regions where the identified fibre is extremely unlikely to have been harvested locally: for example, the linens from Tell Abraq (Reade and Potts 1993) and the possible cotton-thread from a fragment apparently uncovered in a kurgan at Novosvobodnaya (Shishlina, Orfinskaya and Golikov 2003, 334) are likely to represent imports of some sort,

100 This is something that could now be studied through an analysis of strontium isotopes in tooth enamel, (for example of techniques, see Meiggs 2009).

but the exact sources of the fibres cannot be directly identified at present[101]. In the case of Tell Abraq, a highland Oman or Iranian source for the linen has been suggested (Haerinck 2002, 246-247) on the basis of proximity, but it is difficult to exclude more distant sources completely. In an analysis of 3rd millennium textiles from Shahr-i Sokhta, Irene Good has suggested that some of the net- or string-based textiles may represent the packaging from various imported products, partly on the basis of the fibres from which they are made (Good 2006, 202-203). Most of the textile evidence from the site was made from sheep's or goats' wool or hair. However, a few fragments or fibres of "sunn hemp" (*Crotalaria juncea*), which only grows in tropical regions, and of linen, which is difficult to grow in the ecological zone in which Shahr-i Sokhta is situated, may have travelled from the Indus and Mesopotamia respectively. Given the site's apparent involvement in the procurement and production of lapis lazuli, this is of course perfectly possible, with textile bags from distant locations used to transport lapis or other commodities. But it is rather difficult to be more specific: the linen fragments in particular could have been sourced from a very wide area. Theoretically some of the tools for processing plants like flax into yarn may be recognizable in the archaeological record, but, although some work has been done on objects from Egypt (where models and depictions of processing are common, as well as the tools themselves), little large-scale synthesis has been undertaken to track the extent and development of these technologies.

6.4.2 Distribution of raw materials: dyes and colours

Our current knowledge about the full range and source of colours and textile dyes during the 3rd and 2nd millennia BC remains for the most part conjectural. Relevant chemical or physical analysis has rarely been undertaken on this topic, at least partly a result of the low number of textile finds and thus few samples. This is a shame because, given the specificity of certain dyes and mordants (because of botanical/faunal or geological limits) and thus their potential geographic sources, research on this topic could have the potential to offer insights into textile interaction networks.

Textual evidence, such as that cited above, gives us a few clues about the colours used, but, given the problem with translation of these ancient words, we cannot be sure what shades these really represent, or even that the broad meaning can be relied upon (both because of linguistic change, and because of different cultural colour codes). Barber has catalogued what she judged to be the most likely range of colours used, and the order in which they become available, with some suggestions for the possible chemical (both organic or inorganic) sources from which such colours could be made, and what processes or additional substances ('mordants') would be needed to 'fix' the colours (Barber 1991, 223-246). For example, red dyes may be produced from a variety of sources: kermes[102] (from the *Kermococcus vermilio* insect), a variety of other insects, iron-bearing mud or other geological dyes, a variety of plants. One better studied example is the colour

101 Recently attempts have been made to 'provenance' textiles using strontium isotope techniques (Frei *et al.* 2009). This is a very interesting potential method, though the interpretation of individual results should be treated critically, as with other isotopic techniques of tracing movement in the archaeological record.

102 Kermes from the same root as the word 'crimson' in English and 'kırmızı' in Turkish, and *vermilio* of course yields the word 'vermilion' in modern English.

produced by extraction from the shells of certain sea snails. The rich purple or red dye, later known variously as 'Tyrian purple', 'imperial purple' *etc.*, made from the crushed shells of *Bolinus brandaris* (previously *Murex brandaris*), appears to have been used to make extremely high value cloth and clothing from at least the 2nd millennium BC (Schneider 2011)[103]. In fact, "[v]arious colours were possible based on the species utilized, amount of sunlight and air, weave and type of textile employed. The colour varied from a pale pink or rose to a dark violet or black-purple; our modern idea of 'purple' is only one of the possibilities" (Reese 1987, 203). The distribution of the marine snails from which the dye can be extracted is of course restricted to coastal locations around the Mediterranean (see Figure 6.14), most famously to the northern coast of the Levant, from where the Phoenicians successfully exported dye in the 1st millennium BC (Dalley 1991, 123-124). Murex shell middens from this region attest the size of the industry, and similar evidence of shells has suggested that murex-based dyes were being made in the Aegean from at least the mid-2nd millennium (Reese 1987). The appearance or remains of textiles dyed with this substance thus shows the movement of either the dye, dyed yarn or more likely dyed cloth. For example, in a richly furnished tomb dating before 1350BC, at the inland city of Qatna (Tell el-Mishrife), traces of a distinctive purple dye from murex shells were identified chemically in the soil, and must have come from a garment or cloth dyed with this colour – which presumably had been imported to the site in either cloth or yarn form. Subsequent microscopic examination of the soil also revealed dye-based pseudomorphs of the cloth, showing tapestry weave (James *et al.* 2009). Given that the source of this dye is definitely coastal, this represents a clear example of the movement of textiles or dyes, and hints at important exchange relationships at least at the regional level. But we cannot identify the full biography of the cloth, and it is impossible to say whether the cloth in question came from the 'nearby' coast of the Levant or further afield. In the absence of comparable examples, it is nearly impossible to reconstruct any more details about the mechanisms of movement.

6.4.3 Spinning and spindle whorls

'Spinning' is the process by which groups of raw fibres are spun tightly together to make yarn, which is longer and stronger than the original fibre. There are different methods of spinning, depending on the specific fibres, but in the ancient world the process was normally done by hand using a spindle (a sort of rod to wind the yarn onto) and whorl (a small weight for balance and spin). One variable relevant to the movement of textiles is the direction of spinning: to the left or right, or 'Z' or 'S'. Certain fibres are spun more easily in one of these directions (for example linen is spun more easily in a S-direction, hemp in a Z-direction, whilst for wool it makes no difference) and because different fibres were available in different regions, long-lived cultural traditions in spin direction appear to have been set up from early on and carried into other fibres for which the same spin direction was either unnecessary or even counter-intuitive (Tsareva 2007, 332; Barber 1991, 39-50) (see Figure 6.15). It might therefore be tempting to use this idea to trace imported yarns, but we have too little information about whether such broad generalizations are sustainable.

103 A related species, *Hexaplex trunculus* (previously *Murex trunculus*) was apparently used to make 'royal blue', and other dye-shades appear to have been collected from related species.

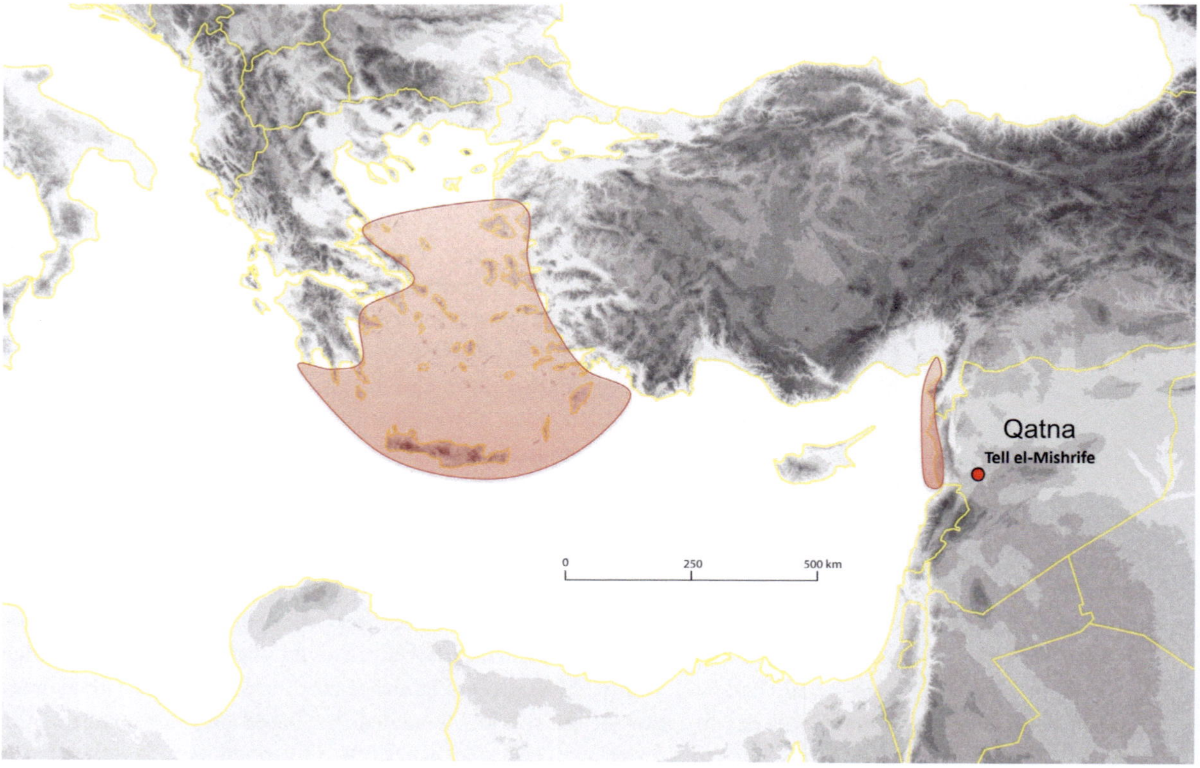
Qatna
Tell el-Mishrife
0
250
500 km

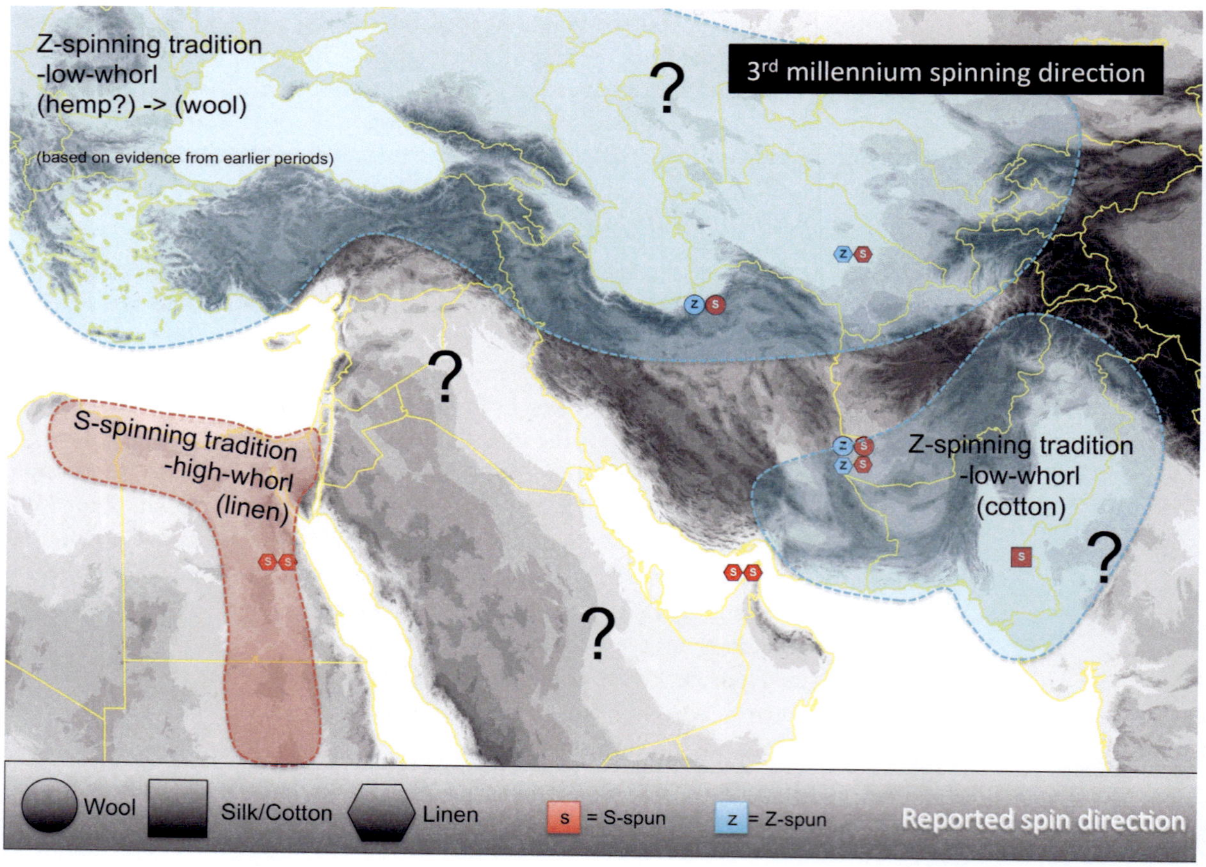
Z-spinning tradition
-low-whorl
(hemp?) -> (wool)
(based on evidence from earlier periods)
3rd millennium spinning direction
?
S-spinning tradition
-high-whorl
(linen)
Z-spinning tradition
-low-whorl
(cotton)
Wool
Silk/Cotton
Linen
S = S-spun
Z = Z-spun
Reported spin direction

Whilst spindle shafts may be made of wood or bone[104], and therefore prone to poor survival in the archaeological record[105], whorls are often made from stone or baked clay. Spindle whorls are a ubiquitous item of many assemblages from the Near East from the Chalcolithic onwards, and their numbers increase exponentially in the Early Bronze Age in many regions. Spindle whorls display clear variations in functional morphology and in decoration, some of which have technological or cultural significance, and this is also true in our regions of interest. The weight or size of the whorl appears to have a relationship to the material and thickness of the desired thread. Cotton, for example, requires particularly small whorls compared to wool or linen and these may act as indices of cotton production. However, whorls have received limited large-scale comparative investigation or synthesis[106]. More often their presence is used to indicate zones of use within a particular site: domestic vs. industrial, female vs. male. A survey of the primary literature shows that in eastern Anatolia and Transcaucasia, a variety of bone spindle whorls (and occasionally possible spindles) from Chalcolithic and Early Bronze Age contexts related to the Kura-Arax ceramic assemblages have been uncovered, whilst clay and stone whorls are well known in high concentrations from most sites of central and eastern Anatolia during the 3rd millennium[107]. At Arslantepe, a change in the weight of spindle whorls in period VIB (3000-2750BC) has been interpreted by the investigators as indicating a change in the thickness of fibres used, which may indicate a shift to wool production (Frangipane *et al.* 2009, 16-17). Whorls are also widespread in Central Asia from at least the 3rd millennium, examples appearing in the publications of most major sites[108]. Unfortunately larger catalogues and comparative typologies that might make some sense of the different morphological features remain lacking and must await future research.

Figure 6.14. Regions likely to have been producing murex-based dyes during 3rd and 2nd millennium BC.

An attempt to study spindle whorls on the macro-scale was undertaken by Barber (1991) who tried to create a spatial and temporal distribution map of a particular form she calls 'hollow whorls' (see Figure 6.16). She suggested that the concave shape would allow the spindle to be used both as spindle and as shuttle for weaving (so removing the 'rewinding' step from the process) – in a manner of weaving which did not require a vertical loom (hence the low correlation with loom weights). This is supported, she argues, by the number of such whorls found together: the weaver would have to have many spindles to make this practical. However, Jack Davis (1986, 98) has argued that this hollowing may represent a technical feature rather than a cultural one – designed to reduce the risk of whorl breakages. If we assume these temporally disparate items are indeed related, these whorls may index both a technique of spinning and weaving and, in all likelihood, a type of cloth or style whose origins may have been in Central Asia in the 5th or

Figure 6.15. Distribution of textiles whose spin-direction (Z- or S-) has been reported; overlain by coloured zones showing possible different spinning traditions (as argued by Barber 1991, 65-68, 250).

104 As seems to have been the case in the Early Bronze period (late 3rd millennium) at Ebla (Peyronel 2007).

105 Though spindles or combined spindle and whorls are occasionally preserved in high value materials, like metal, *e.g.* in tomb L8 (and possibly tomb H) of Alacahöyük (Koşay 1951, 168-9, pl. 197; *cf.* Barber 1991, 60-61), Horoztepe (Özgüç and Akok 1958, pl. VIII), Karataş (Mellink 1969, 323, pl. 74.23), Hissar IIB (Schmidt 1937, 120, pl. 29, H2171).

106 One recent and exceptional example is that of Good (2012).

107 For example, Karaz (Koşay and Turfan 1959, 411); various sites in Armenia (Badalyan and Avetisyan 2007, 31, 329); Sos Höyük (Sagona, Sagona and Özkorucuklu 1995, 213 fig. 13.6), Arslantepe (Frangipane *et al.* 2009).

108 For example, the Sumbar valley cemeteries, dating from the late 4th to the early 2nd millennium, have yielded a variety of stone spindle whorls in some of the tombs – *e.g.* from the late 4th-early 3rd millennium graves: tomb nos, 12, 13, 14, 21, 257, 262, 265 (Khlopin 2002).

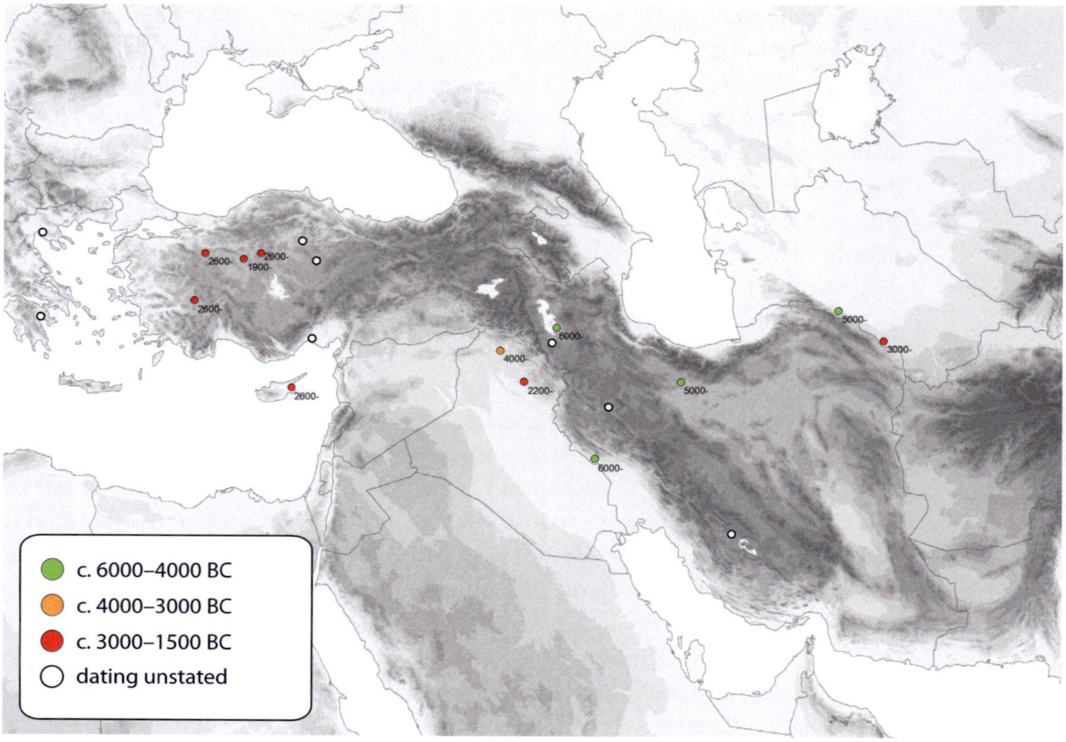

Figure 6.16. Distribution of 'hollow-shaped' spindle whorls with approximate earliest possible date (following Barber 1991, 391-392).

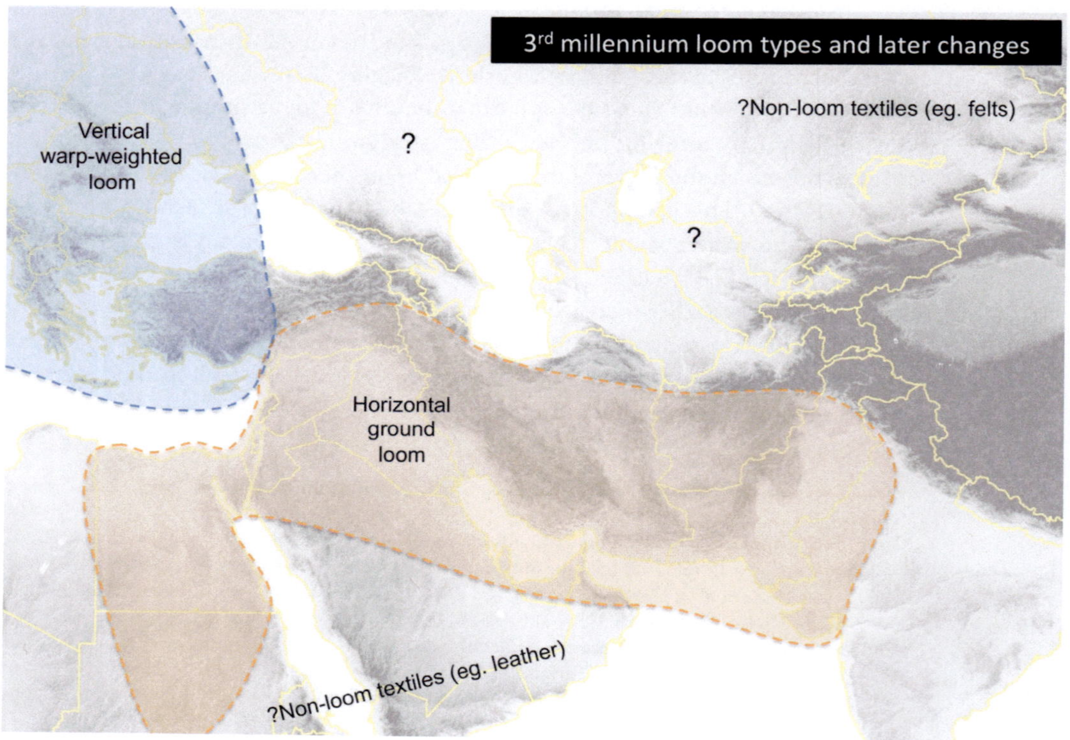

Figure 6.17. Distribution of different loom-types during 3rd millennium BC and after (following Barber 1991, 250, centre map).

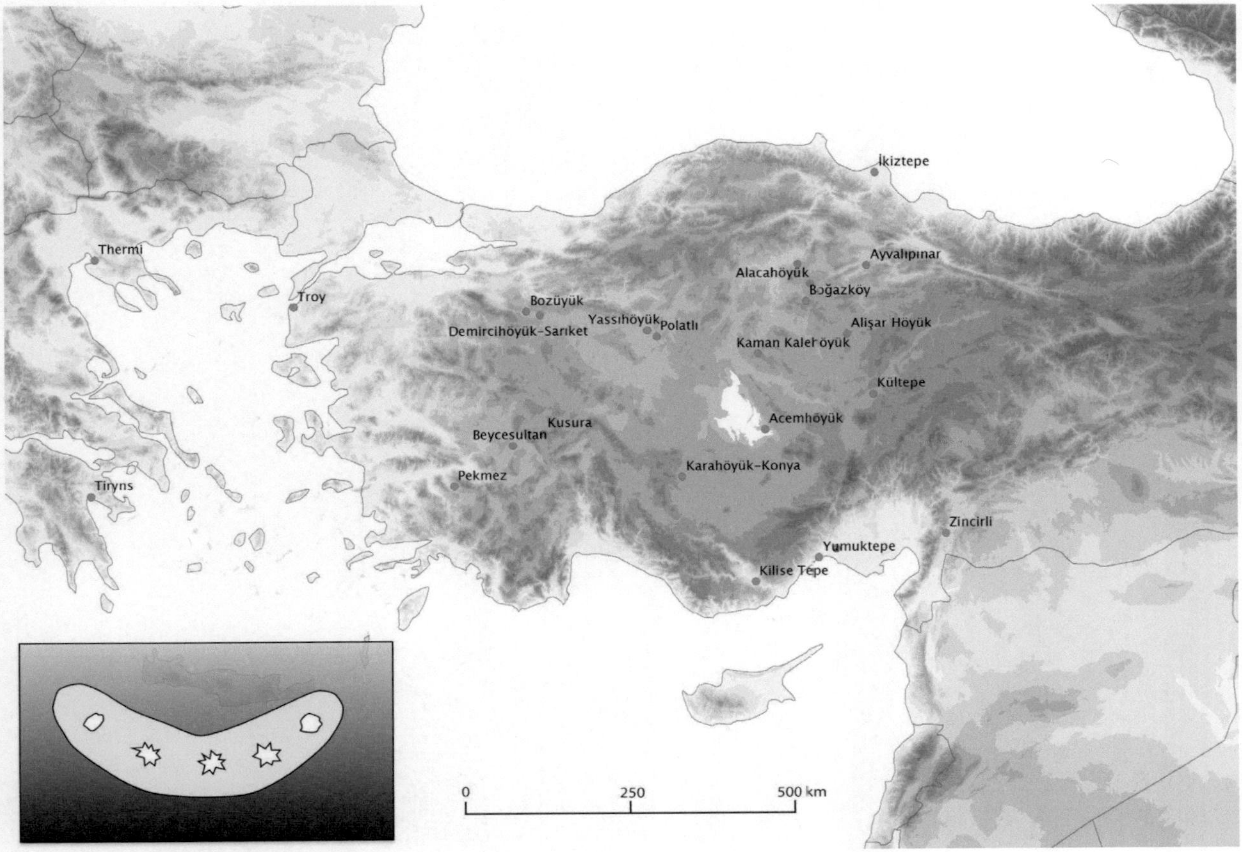

Figure 6.18. Distribution of crescent-shaped loom weights of the early 2nd millennium BC (data based on Wisti Lassen 2013a).

4th millennium, but which by the Late Bronze Age was popular as far as Central Europe.

6.4.4 Loom types and loom-weights

To produce woven textiles, some form of loom is required to hold the 'warp' yarns in place, and allow the weaver to weave the 'weft' yarns (Barber 1991, 79-125). There are two main types: horizontal, ground-based looms (which because they were made only of perishable materials tend not to leave any trace in the archaeological record but from pictorial evidence are assumed to have been dominant in Mesopotamia and Egypt) and the more complex vertical and warp-weighted looms (which do leave traces in the form of loom weights, particularly if the loom is abandoned suddenly, though it is not always possible to assert whether particular objects are indeed loom weights). The type of loom affects the type of textiles that may be produced: the vertical loom, for example, makes it easier to produce 'tapestry'-style weaves (in which 'weft' yarns are interwoven individually). Loom-weights from the 3rd millennium have been unequivocally identified at Arslantepe (Frangipane *et al.* 2009) but are harder to locate in the publications of smaller excavated sites of Transcaucasia. The situation is similar in Central Asia where there are relatively few mentions of possible loom-weights, though the 'stone pocketbooks' (whose chronological range stretches from the Namazga III to BMAC periods) could possibly have been used as loom-weights (Hiebert 1994a, 157). Unfortunately, as for spindle whorls, detailed research and synthesis on a

Figure 6.19. Depiction of twill fabrics (as 'herringbone pattern') on 'Anatolian-style' seal impressions from Old Assyrian Kültepe: (a) CS 826 (Özgüç 2006, pls. 84, 278); (b) CS832 (Özgüç 2006, pls. 85, 279).

cross-regional basis remains limited. One exception is, again, the work of Barber (1991, 299-303) in studying the expansion of the warp-weighted loom (on the basis of the distribution of loom-weights) from south-east Europe and Anatolia to the east (see Figure 6.17), and the subsequent emergence of 'doughnut' shaped weights in the late 2nd and 1st millennia BC. As with spindle whorls, she proposes that the spread of loom-weights reflects the migration of women, bringing with them their weights, or at least their knowledge of the construction of a loom (for example from Anatolia to the southern Levant in the Middle Bronze Age). Whilst this offers one scenario[109], alternative mechanisms of the spread of such cultural knowledge are equally imaginable: technology transfer always requires some kind of movement of people but the desire to produce or consume certain types of new textile may have driven adoption as much as in-migration. At the very least, such commonalities do indicate inter-cultural interaction.

The difficulty for wider synthesis is partly the ability of excavators to feel confident that particular objects are indeed *loom*-weights as opposed to non-textile related weights, since their forms do not have to be very distinct. Loom-weights, like spindle whorls, are often made of stone or baked clay, but we should also bear in mind that they may have been made in a variety of archaeological invisible forms, for example sand-filled bags, known from recent Iranian tradition

109 We can imagine women travelling for marriage purposes, for example, which might explain 'intrusive' textile technology in regions where other material culture remains 'local'.

(Hiebert 1994a, 157), or simple stones of more or less the same size and weight. Unlike spindle whorls, loom weights appear to be unlikely to have been placed in tombs as grave offerings: perhaps not surprising when one considers the full size of a loom and the likelihood that looms would be shared items (with the potential to be recycled) and much less personal to the deceased. Where sets of weights have been found grouped together in the same context and appear to have formed part of a collapsed loom, our confidence is obviously higher. For example, the investigations at Arslantepe (Frangipane *et al.* 2009 and mentioned above) have allowed a reasonably detailed diachronic reconstruction of changing loom-weight shapes from the early 4th to the early 2nd millennium BC. Such detailed technological studies are essential and, if the results of studies from different locations were synthesized, we would also be able to reconstruct contemporary technological (and therefore cultural) differences, which might give us more clues about textile interaction spheres.

Recent experimental research by Agnete Wisti Lassen into crescent-shaped objects from western Anatolia (see Figure 6.18), often assumed to be loom-weights, has shown that the mechanics of mounting a loom using these weights gives the ability to weave four-thread layers with only two rows of weights, and thus shift instantly between 2/2- and 3/1-twill, tabby (plain), panama and pattern weaves (Wisti Lassen 2013b). As already noted above, the earliest direct examples of twill fabrics are from Anatolia and the Black Sea region during the 4th millennium BC (see Section 6.3.2). These crescent-shaped loom-weights appear to be dated later, to the 2nd millennium BC (though similar crescent-shaped weights are also known from Neolithic contexts in Eastern Europe – see Barber 1991, 100) when Anatolian weavers appear to have wanted to be more flexible in the types of weaves they produced. The distribution of these loom-weights becomes more significant with regard to dress style, when seen within the context of glyptic (seal-based) art styles. A distinctive style of dress with diagonal lines is shown in early 2nd millennium 'Anatolian style' cylinder seals (Figure 6.19), which seems likely to represent twill as a distinctive, perhaps ethnic, type of clothing (Wisti Lassen 2013a), also visible on some of the lead figurines (or their moulds) from the same period (*e.g.* Emre 1971, pls. V.2, VIII.2, IX.1a, XI.1). This contrasts with other figures depicted in glyptic, which show plainer clothing, or 'flounced' garments (linked to Mesopotamian representations of clothing, see Section 6.5.1 below). Whatever commodity these images were designed to seal, the fact that a distinctive dress is shown in these images that were designed to ensure authenticity of the commodity, suggests a very self-conscious assertion or construction of dress identities.

Additionally, some scholars have suggested an association between twill and Indo-European (both the language group or its postulated users), mainly based on the similarities between Celtic plaids or tartans (*i.e.* traditional Scottish and Irish cloth), the Hallstatt textiles and the geographically very distant evidence from Tarim mummies (Barber 1999; Mallory and Mair 2000). The spatial location of the earliest twill (Anatolia and the Black Sea region), and the broad region of the contentious reconstructed 'homelands' for the language group (see Anthony 2007; Renfrew 1990; Sherratt and Sherratt 1997) happen to coincide, which, it could be argued, offer some kind of support to this view. But whether one can make a link between a weave and a language group given the huge temporal and spatial gaps between the few highly localized scraps of evidence is extremely doubtful.

Leaving aside the general question as to whether there ever was an original group of Indo-Europeans, and whether one can trace them archaeologically, it is not clear exactly what kind of mechanism could produce the strong association between a language or ethnic group (or even a whole language-ethnic continuum) and one particular weaving style. Whilst the direct association between language groups and weaving styles seems somewhat implausible, certain social networks could have provided common paths for both. Tuck (2006), for example, argues for a synergy between song and weaving as repetitive rhythmical activities which might offer a mechanism to explain how musical (and possibility linguistic) ideas and textile patterns might travel along the same routes.

6.4.5 Specialized tools and their distribution

Other specialized tools for textile production (for processing fibres, weaving, dyeing and sewing) may be even more difficult to identify, and suffer from the same low level of comparative synthesis as loom weights and spindle whorls. It is worth mentioning one example from the Central Asian corpus, namely certain blade tools from the Sumbar valley cemeteries (Khlopin 1982), whose distinctive shape matches that of the tool used by modern pile carpet makers (Figure 6.20 inset). The dating of these cemeteries is between the 4th and early 2nd millennia BC, though these blades appear in contexts apparently dating to the local 'Late Bronze Age', *i.e.* the early 2nd millennium. If this identification is correct, it suggests that textile manufacturers of this region were able to make 'pile-carpet' type fabrics (and perhaps carpets themselves) from an early period. Similar knives have also been uncovered previously at Shah Tepe (level II) and Tepe Hissar (level IIIB) – around 2300-2100BC – see map on Figure 6.20. Further investigation needs to be undertaken to see whether these are exclusive to this area alone, or whether they can be identified elsewhere. The distribution of such specialized tools may indicate differences between zones of cultural knowledge (or practices?), which in turn reflects on the potential for textile exchange[110].

6.4.6 Comments: raw materials, production tools and identities

The evidence for changing raw materials and weaving technologies appears to represent more than just a technological change. The form and range of weaving tools may index ethnic, cultural or regional identities (and their distribution in the archaeological record, a trace of the movement of the people – perhaps primarily women – who used them), and this is due both to the particular skills required to make textiles (which take time to learn and are acquired at the same time as other identities are forming), and as a consequence the particular styles of textiles which particular technologies can produce come to stand as identity markers themselves – particularly ethnic or regional ones. Meanwhile the appearance of fibrous materials far outside of their normal cultivation zone (and presumed areas of production) points towards exchange of cloth over large-distances. A few broad directions have been mentioned (*e.g.* the possibility of cotton being imported from the Indus to eastern Iran), but our knowledge is so incomplete that these can only remain tentative suggestions. Since preserved remains of textiles are so

110 Additionally, in the context of the modern value of carpets, we are prompted to wonder whether such textiles could have formed items of export to the urban societies to the west.

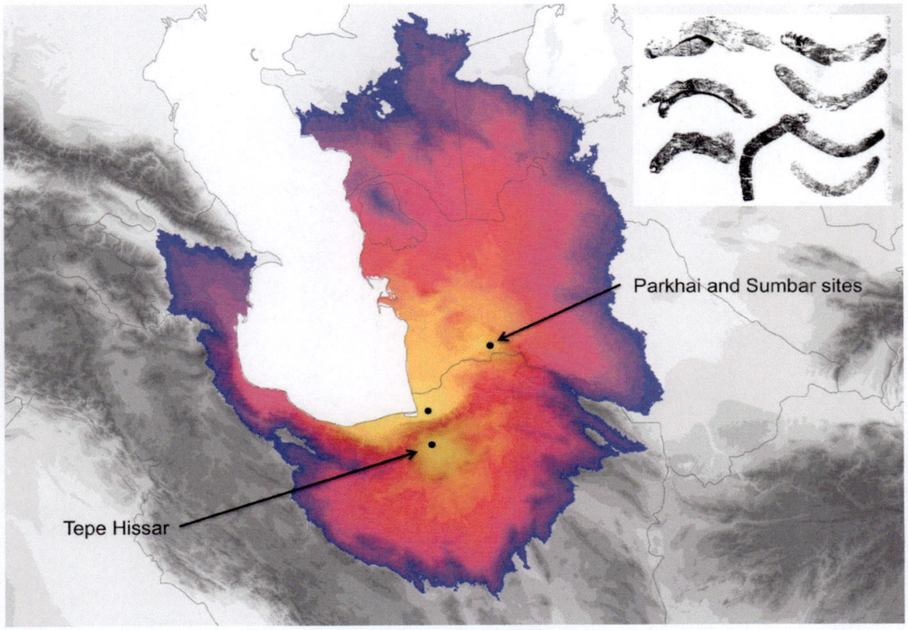

Figure 6.20. Distribution of 'pile-carpet-makers' knife (red dot), overlain onto a type 'A2' archaeotopogram showing the accessibility of these sites to surrounding regions (distribution data from Khlopin 1982; inset photo: courtesy of Russian Academy of Sciences).

rare in the archaeological record, linking them to their raw sources still remains a partial picture. We must thus turn to more indirect data: depictions of cloth and patterns in other media.

6.5 Indirect evidence: depictions of dress and clothing

The evidence from depictions of clothing on figures in wall-paintings, figurines, figures in relief (on stone or other materials) and decorated objects (like metal vessels and seals/cretulae) have long served as the primary means of reconstructing the clothing of the Near East. They also provide indications of hair-styles, jewellery and other bodily adornment.

6.5.1 Figural representations of dress from Mesopotamia, Egypt and the Aegean

In Mesopotamia, in the early part of the 3rd millennium, the so-called Early Dynastic period, most (male) statues are shown unclothed above the waist, with long skirts/kilts, often with long tassels (see Figure 6.21a). From the Akkadian period (from around 2300BC), men are shown with pleated or more flexible cloth draped over one or both shoulders (Figure 6.21b; Strommenger 1962, fig. 61b). After this time, fleecy garments are depicted, normally full-body covering, with tufts suggesting a woollen fabric of some kind. From the Old Babylonian/Old Assyrian era (the beginning of the 2nd millennium), more complex hems and edges are depicted (Figure 6.21c and d; Margueron 2004, 441-442), apparently culminating in the elaborate dress shown in Neo-Assyrian depictions of the 1st millennium. Though patterns are often depicted (see for example, the 'dotted' dress of some Elamite statuary, perhaps indicating sewn-on jewellery, Pittman 2003), there is fairly little evidence for colours because of the low number of surviving figural representations in wall paintings from this region. Colours in surviving wall-paintings were in any case subject to various processes of decomposition

Figure 6.21. Clothing styles in Mesopotamia: (a) Early Dynastic period unclothed above waist, skirt, tassels; (b) mid-late 3rd millennium pleated/flexible cloth over shoulders, Khafejah (image: the Oriental Institute of the University of Chicago); (c) and (d) Old Babylonian/Assyrian – complex hems and edges (Margueron 2004, 441, 442).

Figure 6.22. 'Foreigners' in Egyptian depictions: individuals wearing patterned dress in wall-painting from a tomb at Beni Hassan, Egypt, 12th dynasty, c. 1990-1780BC (photo: the Metropolitan Museum of Art, Rogers Fund, 1933, 33.8.17).

which means they may not represent the 'original' colour. Glyptic art also shows features of dress, shown schematically, and mostly in keeping with the evidence from statues. Nakedness, or lack of dress, appears to have negative connotations of shame or defeat, given the contexts of its depiction, such as in the Royal Standard of Ur (Asher-Greve and Sweeney 2006).

In conformity with the vast majority of textile finds from Egypt, wall paintings and statues mostly depict Egyptians in simple plain white clothing, presumably linen (see, for example, plates in Shaw 2000, esp. opp. 129). Various types of garments are depicted (Vogelsang-Eastwood 1993), including some patterned

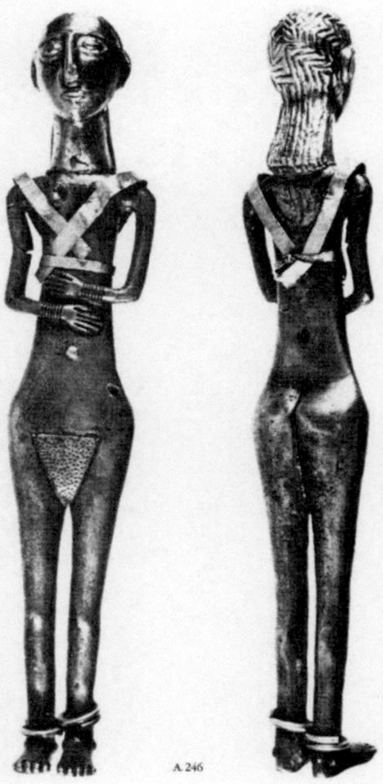

Figure 6.23. Silver figurine with gold head from Hasanoğlu, in north-central Anatolia, dated to late 3rd millennium BC (Edgü 1983, 111 cat. A.246).

a

b

c

Figure 6.24. Depictions of various textile fabrics and clothing in seal impressions, from Kültepe, showing a wide variety of fabric and styles: (a) CS704 (Özgüç 2006, pls. 68, 246); (b) CS767 (Özgüç 2006, pls. 76, 263); (c) CS770 (Özgüç 2006, pls. 77, 264).

examples. For example, a net-like dress is sometimes shown (*cf.* comments in Section 6.7.2), whilst high-ranking individuals or gods may be depicted with colourful patterned garments (*e.g.* as illustrated in the 'Harris' papyrus held at the British Museum), at least during the New Kingdom, *i.e.* after 1600BC, whose patterns are very likely to derive from the arrangement of coloured beads (see Riefstahl 1944, 1, 11-12, figs. 11 and 12). Depictions of patterned weaves, like the direct example of Tutankhamun's 'Syrian' shawl already mentioned above (Figure 6.11), appear to index 'foreignness' quite clearly. The 12th dynasty/Middle Kingdom (1991-1783BC) paintings from cave-tombs at Beni Hassan (Figure 6.22), for example, show a series of figures wearing highly colourful patterned clothing, who are normally identified as 'Bedouin' coming from the Levant or Syria (*e.g.* Goedicke 1984). By way of comparison, similarly intricately decorated and multi-coloured dresses supposedly representing 'Syrians', 'Libyans' and 'Nubians' are shown on faience tiles from the temple of Rameses III at Medinet Habu, dating to c. 1180BC (for photograph, see Shaw 2000, 321). These appear to offer a striking insight into the dress colours and patterns of Syro-Mesopotamia at the beginning and end of the 2nd millennium.

The figures shown in Aegean wall-paintings, such of those at Knossos, Thera and Pylos (dating to the mid-to-late 2nd millennium), are often shown with quite complex patterned dress[111]: we may speculate that these were of tapestry weave made on a warp-weighted loom. In Crete, apparently male figures are often shown wearing a simple loincloth, while female figures are shown with open-bodiced dresses with flounced tunics. Colours shown include reds, greens, blues and yellows. Paintings which appear to be of a similar or related style to the Aegean ones have also been uncovered at Alalakh/Tell Atchana, Qatna/Tell el-Mishrife, Tel Kabri and Tell ed-Dab'a (for example, see Winter 2000; Niemeier and Niemeier 2000; Morgan 1995). This does not necessarily mean, as has been suggested, that Aegean fresco painters were travelling around the eastern Mediterranean on commission[112]. Instead it may suggest an intensive Middle to Late Bronze Age circulation of textiles, which inspired the depictions we see in mural form (S. Sherratt 1994a).

6.5.2 Figural representations of dress in Anatolia

In highland eastern Anatolia there are no significant depictions of dress (and indeed, few figural images as a whole) until the Urartu period, well into the 1st millennium BC. For clues we can thus only consider examples from nearby central Anatolia. A metal figurine from Hasanoğlu, apparently dating to the 3rd millennium, may show fabric straps (Figure 6.23), though the overall effect of the piece is one of nakedness, as is true of most figurines found during this period (see discussion below). The idea that the particular manner of depicting certain figures on so-called 'Anatolian-style' cylinder seals of the early 2nd millennium BC with herringbone-patterns as a representation of twill (see Figure 6.24), has already been mentioned (see above Section 6.4.4). In fact the extraordinary variety of clothing styles (or at least of the patterns shown) shown in such seals is quite striking and hints that there was considerable scope to play out or fix identities through clothing in the early 2nd millennium. It is difficult, however, to say what exactly would have been read into such differences: class or status, distinctive religious or craft roles or perhaps, more politically, some form of geographic, kin-based or 'ethnic' affiliation.

---

111 Note, however, that care should be taken with the colours and patterns from wall-paintings – our interpretations often depend very heavily on 'reconstructed' versions rather than on the remaining fragments (or their chemical indications of the *original* colours).

112 Even if we assume that the frescoes were considered valuable, there is no clear evidence to say that such individuals had to be from the Aegean, or were of Aegean 'ethnicity'. If the paintings are seen as simply stand-ins for wall hangings (*cf.* the Mari investiture scene, which has tassels hanging from the bottom), then it is more important to establish the system of movement of the textiles: and the style may have been a shared one within the east Mediterranean *koine*, rather than necessarily emerging from one cultural or production location (S. Sherratt 1994a).

6.5.3 Figural representations of dress in Central Asia

In western Central Asia, there are few depictions of clothed people. Most figurines are shown apparently unclothed (though see note below Section 6.5.4). One exception is shown in the form of so-called 'Bactrian composite figurines', which do seem to depict a distinctive dress style (for examples, see Ligabue and Salvatori 1990, 246, 242, 241; Amiet 1986, 330; Francfort 2003, 53). Carved from different coloured stone (green chlorite for the dress, white limestone for the head and arms), these figures are always shown seated. Precisely dating these composite figurines, and plotting their archaeological distribution, is unfortunately very difficult because most of them have entered into the public domain through the art market, decontextualized and lacking true archaeological provenance. Given the diversity of details and mixed stylistic attributes, and their high price on the art market, it may also be possible that some are forgeries. The few figurines with good provenance do seem to be related to the 'Bactria-Margiana Archaeological Complex', which dates them to the beginning of the 2nd millennium (notably later than the earliest similar depictions of dress in Mesopotamia). The clothes depicted always cover the full body, except for the head and arms, and tend to show triangular tufts or wavy segments, perhaps showing a fleece. Their social function, as with most prehistoric figurines, is difficult to identify. A very similar type of seated figurine, with the same dress, is depicted on a silver vessel with a Proto-Elamite inscription (see Figure 6.25a), recovered near Persepolis in south-west Iran though without good archaeological provenance (Potts 2008a, 165-171; see also Section 5.3.4.ii), and in the form of a silver pin from Gonur Depe, tomb 3220 (see Figure 6.25b). The appearance of necklaces, belts and bracelets may suggest that the spaces on the stone figurines may have originally included such adornments, perhaps removed as valuable metal extras.

This type of dress has come to be known as a *kaunakes*, though the label is somewhat anachronistic (Kawami 1992, 9). Fleece-based dress is of course widely known from Mesopotamia, though the type depicted in Central Asia appears to be distinctive in its 'all-body' nature. Kawami points out that these garments are normally assumed to be formed of "leather with the fleece left on and perhaps dressed or trimmed in a decorative way" – which would mean that these are not woven textiles at all – though she notes that an "artificial fleece", made by pile-technique, is also possible. Another silver vessel of the same form as the Persepolis one, also from Gonur Depe, tomb 3220 (Sarianidi 2006, 236-7; Figure 6.25d), shows two camels whose neck hair is depicted in a very detailed way, which interestingly, is executed in the same manner as the tufts of the *kaunakes*. This may indicate one of a number of things: that the *kaunakes* is indeed a 'natural' fleece, perhaps even made from camel wool[113], or, by contrast, that the camels were 'dressed' in the same way as these female figurines. Animals with similarly decorated fleeces are not unknown in Near Eastern imagery[114], and would presumably indicate some special civilized status of the depicted animal (as deities, or as sacrifices). In

113 The reported presence of camel wool amongst the textile fibres of Shahr-i Sokhta might support this idea (Tosi 1983b, 163), though camel hairs are not mentioned in a summary report on the textiles from this site (Good 2006).

114 For example, the fleece of the 'Ram Caught in a Thicket', one of two goat figures from the older Royal Cemetery of Ur, depicts the fleece in a very similar way to the way human figures are clothed (Woolley 1934, 121, 264, pl. 87).

either case, the use of animal-like fleeces hints at a symbolic relationship between possibly 'mythical' human and animal characters.

Potts has recently argued that the Persepolis silver vessel discussed here seems, in the light of the Gonur Depe vessel and the distribution of the composite Bactrian figurines, most likely to have been made in Central Asia, and this, alongside various strands of evidence, including the association of the Linear-Elamite text and the importance of Bactrian camels on the Gonur Depe vessel, makes the identification of the Bactria-Margiana archaeological complex with the textually attested region of Šimaški a strong possibility. He argues that the subsequent conquest of Susa by a group from Šimaški, and the presence of BMAC objects in discrete locations across Iran, may represent the political takeover of an "'*ethno-classe dominante*' that moved from its original homeland and established a state over another people (Susians, Elamites and Anšanites)" (Potts 2008a, 193). He further suggests that "it was this infiltration of south-western Iran by the Šimaškians which accounts for the iconographic influence of the Bactrian-Margiana seated females in *kaunakes* on 'Anšanite' style glyptic at Tal-e Malyan and Susa". Though very persuasively presented, the arguments for this integration of historically-cited places with certain aspects of the archaeological evidence remains circumstantial. Potts attempts to establish an archaeological identity for the historical Šimaškians by linking them to the BMAC material (or, vice versa, an historical identity for the archaeology). In this scheme the 'old-fashioned' *kaunakes* dress appears to be interpreted – at least by the late 3rd or early 2nd millennium – as a distinctive ethnic marker, within a field of competing contemporary regional ethnic identities. The danger of course – though Potts himself is careful not to fall into such a trap – is if this interpretation becomes circular: *i.e.* if this dress style is now assumed to be a Šimaškian/BMAC ethnic dress (albeit possibly a special ritual one), so that other finds assumed to index Šimaški presence elsewhere.

6.5.4 Flat figurines in Central Asia and Anatolia: civilization, dress and nakedness

Figurines are treated separately here because they form a distinctive category of representation in the 3rd and 2nd millennia, found in both Central Asia and Anatolia, and many regions beyond, but for the most part executed with minimal or no clothing visible. This is in contrast with the bulk of Mesopotamian 'high-art' imagery, for example. Figurines are of course known in the archaeological record from the Upper Palaeolithic onwards, and widespread during the Neolithic and Chalcolithic in our regions of interest. However, what is most interesting about figurines in the 3rd and 2nd millennia is the emergence or adoption of the 'flat figurine' across a very wide area, in contrast to the prevalence of the much more three-dimensional styles which predate them.

In southern Central Asia, flat figurines replace earlier crescent or 'boomerang'-shaped figurines, and are associated with the emergence of plain Namazga V/VI pottery and dated accordingly between the mid-3rd and early 2nd millennium BC (Rossi Osmida 2007; Masson and Sarianidi 1973). These figurines are generally made from terracotta, with great effort spent on differentiating headgear or hair-styles. Sexual features are often clearly marked, and are in some cases exaggerated. Legs are often unrepresented, at least in the case of apparently 'female' figurines. Some figurines appear to show indications of jewellery or personal adornment

Figure 6.25. Depictions of the Central Asian/Iranian kaunakes dress: (a) on a silver vessel which came to attention (though is not necessarily originally sourced from) near Persepolis (Potts 2008a, 167, 168; photo: Ebrahim Khadem Bayat/Instituto Nacional de Antropología e Historia, Mexico); related objects from the Margiana area, including (b) a pin head from the Gonur Depe necropolis and (c) a Bactrian composite figurine (Sarianidi 2002b, 231, 140); and (d) a similar metal vessel depicting camels with similar type of 'fleece', from Gonur Depe (Sarianidi 2006, 236-237).

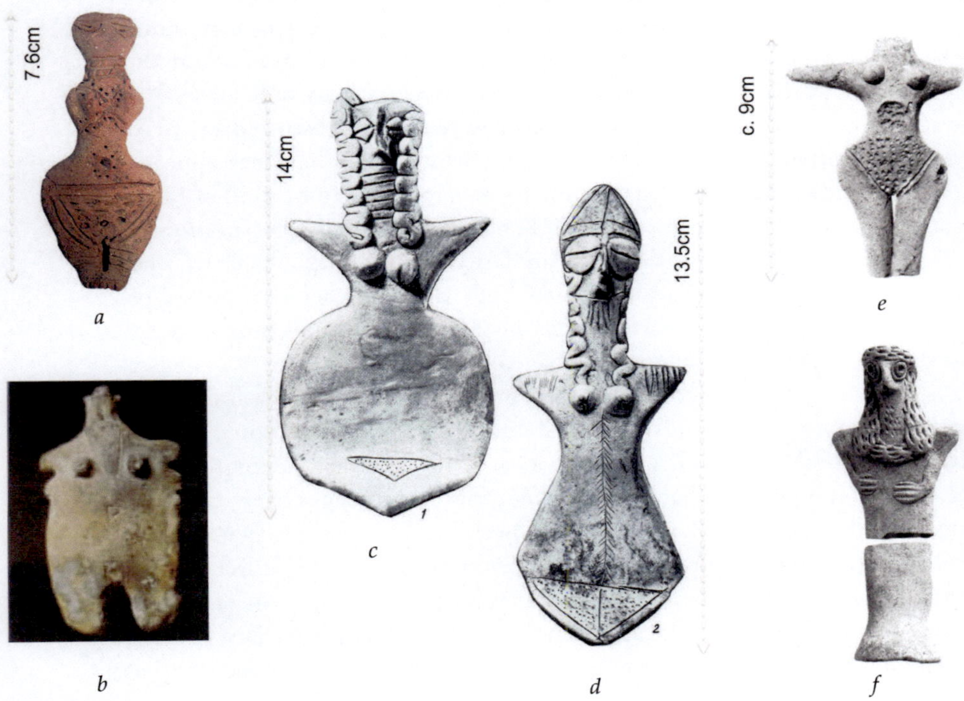

Figure 6.26. Examples of Bronze Age 'flat figurines' of variant dates: (a) Demircihöyük, western Anatolia, (oda 9, H yapı kat), EBA I, c. 2300BC (Baykal-Seeher and Obladen-Kauder 1996, 364, Taf. 118.8; photo: Aydıngün 2005, 97); (b) Shengavit, modern Armenia, Kura-Arax contexts, early 3rd millennium (Simonyan n.d.); (c) and (d) Altyn Depe, Turkmenistan, late 3rd millennium (Masson and Sarianidi 1973, pl. 1; images: courtesy Russian Academy of Sciences); (e) Tell al-Judeidah, in the Amuq, T.JD.X 2152 (Badre 1980, pl. XXIII.18); (f) Tell Selenkahiye, Syria, SLK 75-H 27a (Van Loon 2001, 6.390, pl. 6.10).

around necks, arms and head, although actual clothing is difficult to identify. Some also display marks, which have sometimes been associated with a kind of runic or proto-writing (perhaps related to Elamite or Indus), but they may also indicate scarification or tattooing of 'magical' patterns (see Figure 6.26d).

In western Anatolia, flat, abstract figurines appear a little earlier, perhaps in the first half or middle part of the 3rd millennium (Aydıngün 2005), with perhaps related kinds of figurines continuing to be made into the 2nd millennium (*e.g.* see Aruz and Wallenfels 2003, 257)[115]. There are several different variant styles (see Figure 6.26a), some far more abstract than others, co-existing with more three-dimensional figurines. Some early forms may perhaps be distantly related to Cycladic forms (Renfrew 1969). Anatolian flat figurines are made in a greater variety of material than the Central Asian examples, they are far from identical. Interestingly, however, clay-based flat figurine types have been found in a wide number of places between the Aegean and central Asia: in Cyprus (Orphanides 1983; Bolger 1996), in Syria (Moorey 2003; Badre 1980; Marchetti 2000), in Iran (Tosi 1983a; Shirazi 2007) and in the Indus (Yule 1985; Clark 2003; 2012). Each version has slightly different emphasis but ahre certain characteristics, particularly exaggerated headdress or hairstyle, scarification, jewellery or 'beaded' dress (see below Section 6.7.2) without obvious reference to 'normal' clothing. The earliest examples start to appear in the 4th millennium but the greatest concentration is in the late 3rd millennium BC. In Transcaucasia only a small corpus of human figurines or fragments have been identified (see the example from Shengavit, Figure 6.26b) despite its intermediate geographical position between Central Asia, western Anatolia and northern Syria where figurines are very common. It may be that they made use of such figurines, but they were more commonly made in materials that have not survived. Alternatively, it may indicate that the cultural interconnections of eastern Anatolia were strongly focussed on north-south routes, rather than the east-west routes which these figurines appear to trace. When put together the distribution appears to imply some kind of loosely integrated but very widespread pattern of hand-scaled figural representation (see map Figure 6.27). The overall effect of similarity is intriguing, especially where alternative modes of representing the human body, quite distinct from these types, existed concurrently. Can this pattern of more or less synchronous formal changes in figurines be more than just a remarkable co-incidence?

It is difficult to imagine exactly what mechanism could have linked such dispersed groups of communities. We should perhaps assume some form of concrete medium through which aesthetic ideas about figural representation could be transferred. It is possible that actual flat figurines were being actively exchanged (or moved with their owners) over large distances, but were made in materials that have not survived in the record (*e.g.* metal, wax or wood). Certainly different forms of metal figurines[116] and moulds[117] for figurines are known from parts of the Near East, though they are widely spaced chronologically and geographically. Concrete religio-philosophical ideas or conventions about ritual meanings (especially those

115 Also notable is that certain 3rd millennium figurines can be compared to the naked figure shown between two satyr-like mythical creatures on a seal from early 2nd millennium Kültepe, see Figure 6.24a.

116 For example at: Tell Brak (*e.g.* Aruz and Wallenfels 2003, 257, cat. 163a), Alacahöyük, Hasanoğlu (see *e.g.* Aruz and Wallenfels 2003, 256, fig. 75), Tell al-Judeidah (Braidwood and Braidwood 1960, 300, 305, pl. 60) and İkiztepe (*e.g.* Bilgi 1984; 2001; *cf.* Zimmermann 2007).

117 For example at: Sippar and Titriş Höyük (see Aruz and Wallenfels 2003, 257, cat. 163b and 163c)

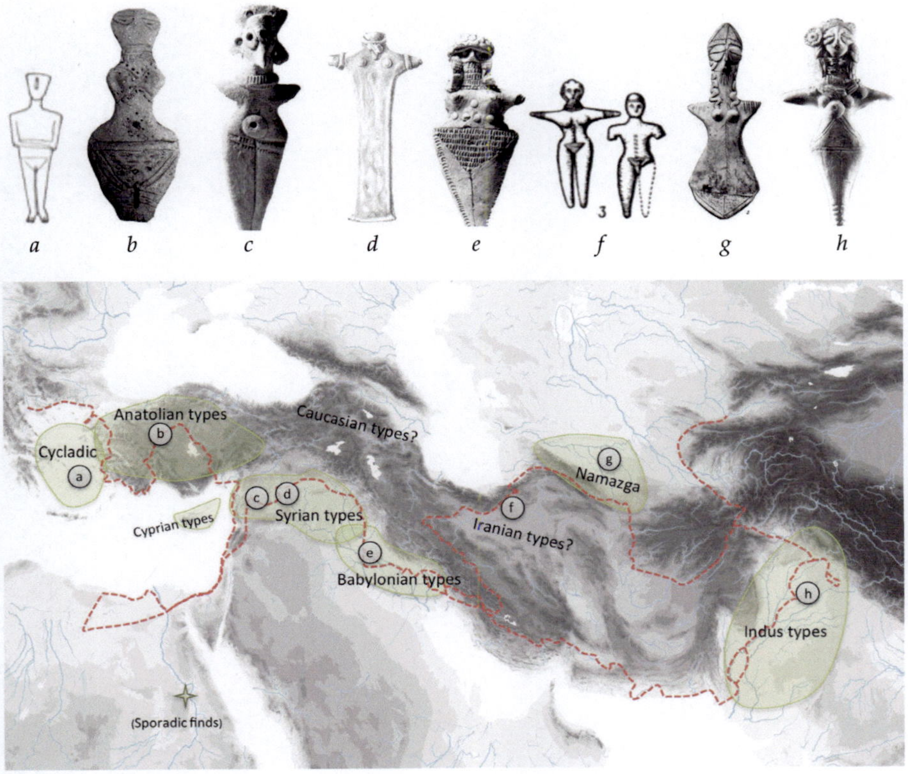

Figure 6.27. A selection of flat-figurine types from the 3rd and early 2nd millennia BC: (a) Renfrew 1969, 28; (b) Aydingün 2005, 97; (c) Feininger 1960, 114; (d) Braidwood and Braidwood 1960, 469; image courtesy the Oriental Institute of the University of Chicago; (e) Feininger 1960, 115; (f) Masson and Sarianidi 1973, 46; (g) Masson and Sarianidi 1973, pl. 1; images courtesy of Russian Academy of Sciences; (h) Feininger 1960, 117. Additionally: rough zones of stylistic variants types; overlain on supposed itinerary of Alexander the Great's campaigns in the east (cf. Chapter 2).

relating to depicting bodies) may have travelled with such figurines, though each region exaggerating features important to locals, much as later colonial religious movements borrow the cultural frameworks from the societies into which they enter.

Further, whilst most of these figurines have come down to us apparently in a state of near-nakedness, we should remember that this may be a property of archaeological preservation. Some figurines may have originally been painted to show dress, tattoos or other body decoration (*cf.* the Cycladic figures)[118]. Others may have actually been 'clothed' using textiles. Modern toy dolls invariably come with clothes[119] and idols in various ritual traditions must be clothed like their human or divine counterparts. By way of comparison, figurines and idols from recent Indian religious practice are often dressed and undressed as if animate people or gods during everyday ritual (*cf.* Waghorne 1994), and similar rituals may have been undertaken in Classical Greek/Roman temples[120]. In this sort of animistic cultural context, depicting clothing in the main medium of the figure

118 The Cypriot 'plank' figurines of late EB/MBA date (see Bolger 1996, 370 fig. 3 top) are in contrast to many other flat figurine types are painted with what look like textile patterns.

119 And 'reflecting' our modern consumerist ideals, dolls like *Barbie* often have multiple interchangeable clothes and accessories!

120 References in early Greek texts attest the treatment of 'gods' (meaning their representations as idol form) in sanctuaries as being bathed and clothed (Simon and Hirmer 1985, 283): "[t]he laughter-loving Aphrodite, went to Cyprus to Paphos, where is her demesne and fragrant altar. There the Graces bathed her and anointed her with [365] immortal oil, such as gleams upon the gods that are forever. And they clothed her in lovely raiment, a wonder to behold" (Homer, Odyssey 8.362); "Then with sacred cries they all lifted up their hands to Athene; and fair-cheeked Theano took the robe and laid it upon the knees of fair-haired Athene, and with vows made prayer to the daughter of great Zeus" (Homer, Iliad, 6.303).

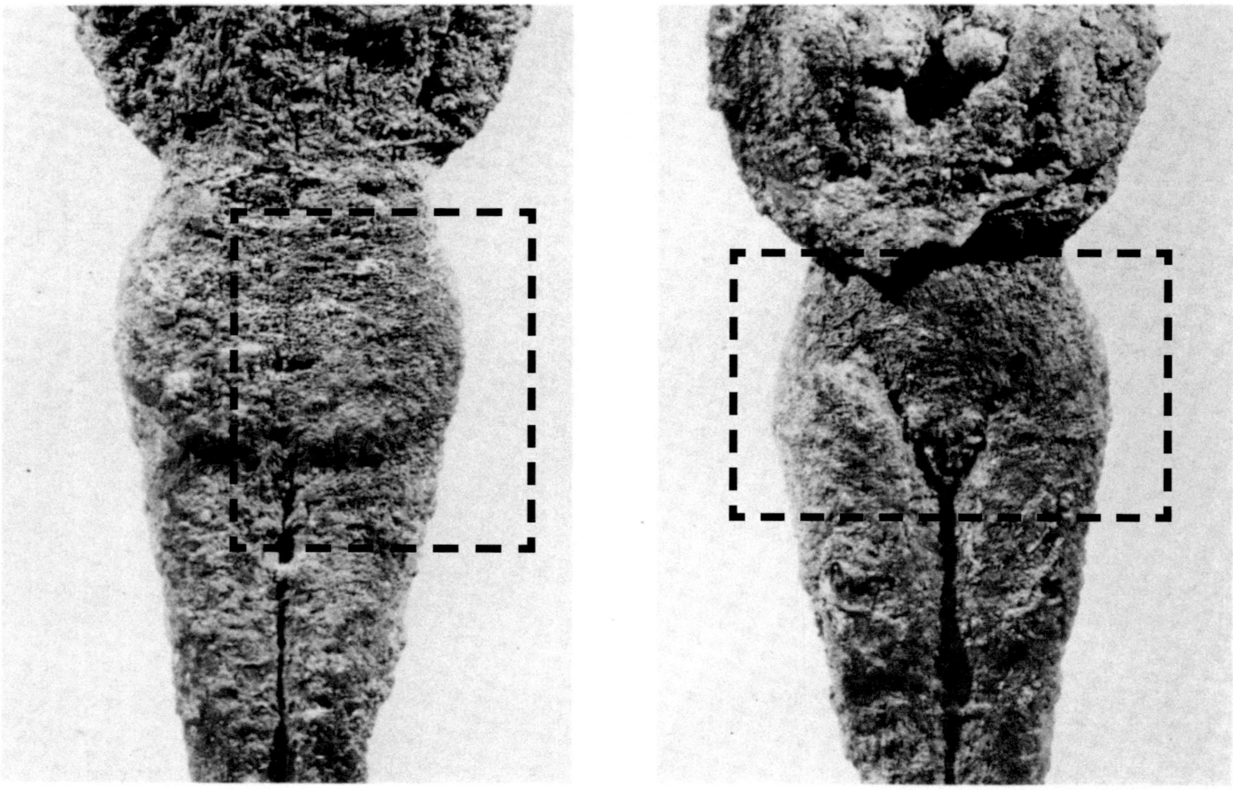

(*e.g.* sculpting clothing in clay) is actually a rather strange thing to do: it implies that the statue or figurine is considered a lifeless and inanimate symbol, or simply a *trompe l'oeil*, instead of being treated as an animate being of veneration itself. There is some evidence which may support such the idea that figurines were 'clothed' by real textiles: a couple of the copper figurines from the site of Tell al-Judeidah in the Amuq, dated to the Amuq G or H phases (probably around 2800BC), showed the impression of textiles as pseudomorphs (Braidwood and Braidwood 1960, 300, 305, pl. 60), shown in Figure 6.28. These fabrics could simply represent protective wrapping into which the figures were placed, since they appear to have been intentionally cached, but equally they could have been part of the figure's dress. It is only because these particular figures were made from metal – in fact a very early tin-bronze – that any indication of textiles survived (see Seeden 1980, 7-10, for detailed critique on dating and composition).

Figure 6.28. Detail of lead figurine, one of a cache uncovered at Tell al-Judeidah in the Amuq valley and dated to the Amuq G phase (c. 4000-3000BC), that shows the impression of a textile weave which may have been used to dress or simply wrap the figurines (Braidwood and Braidwood 1960, 300, 305, pl. 60; photos: courtesy of the Oriental Institute of the University of Chicago).

6.5.5 Comments: clothing styles and movement

The depictions described above have various limitations for the reconstruction of dress styles. First we are often unable to judge the fabrics that were actually used to manufacture the textiles shown. More seriously we have little idea how representative the images are of fashions over society as a whole. For example, it is assumed that a large proportion of Mesopotamian figures represented deities of some sort or their worshippers (as with the small statues from Babylonian temples) – though presumably such dress reflected contemporary elite clothing. With increasing frequency through time, apparently 'secular' images of powerful rulers or ancestors become more common, presumably designed as tools of propaganda to increase the status or aura of the signified individual through the association with divine images and media. Given this, it seems likely that the kinds of fashions and garments shown in other media, like finds of real clothing preserved in tombs, are likely to be rather exceptional, most often indicating the special role of the characters shown (as king, priest or deity) rather than everyday or widely-used garments. More importantly, though, can this evidence of clothing styles from depictions tell us about the movement of cloth, clothing style, or indeed raw materials and textile techniques?

On a general level, the depiction of clothing shows the geographical extent over which particular styles of clothing were known and worn, or at the very least considered suitable dress to indicate generalized characters (such as foreigners), or important figures or deities. Where the same clothing style is depicted in different media within the same region, we may assume that garments themselves were also circulating in this region – although it is very difficult to measure the intensity of this. Since textiles are often associated with ethnicity in ethnographic literature (and since the geographical affiliations of certain fabrics or garments are mentioned in the historical texts), it has also been common to attempt to provide ethnic labels for certain depictions. Whilst there were undoubtedly certain clothes that provided such 'ethnic'-like identities for their wearers, we should take great care in uncritically attributing labels picked up from ancient texts to the material culture from other regions. Likewise, the distribution of the *depiction* of clothed figures may be misleading. It may not indicate one-to-one the distribution of the particular clothing style, since some regions may have used similar clothing but simply had no tradition of depicting such figures.

Ultimately, it should be remembered that the distribution of the depictions of all the known clothed figures from the 3rd and 2nd millennia is fairly restricted both temporally and spatially (shown as blue zones in Figures 6.29 and 6.30). The lack of direct textile finds of figural depictions in 3rd millennium Anatolia and early 3rd millennium Central Asia presumably does not mean clothes were not worn, so we must turn to other types of indirect evidence.

6.6 Indirect evidence: textile-related patterns in other media

Whilst the representations of garments on human figures may provide relatively direct if partial evidence for clothing, patterns on various other media, including non-figural features of wall-paintings and painted or incised decoration on pottery, may also offer insights into textile patterns. Given the absence of actual surviving examples of patterned textile in most regions it is impossible to ever assert definitively the textile origin of such patterns. Apparent reflexes of textile aesthetics are then, like most instances of suggested skeuomorphism, open to debate. Given that textiles are an extremely portable resource, however, it is often likely that, even if the 'original' source of a particular pattern or motif is not exclusively textile (something which is any case impossible to prove), it is often the textile versions that facilitated the actual transfer of the patterns from one region to another. As such, the patterns and motifs discussed below at least hint at extensive inter-craft transfer of patterns and techniques which in themselves document the extent of certain exchange networks related to the often nebulous concept of artistic or cultural 'influence'.

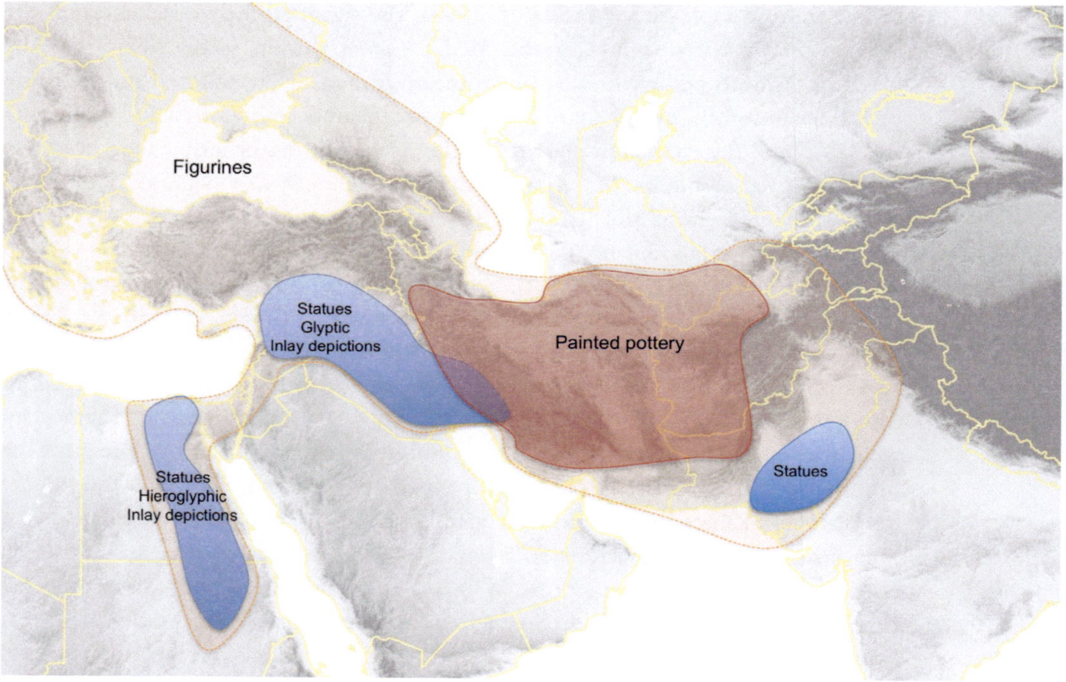

Figure 6.29. Geographical distribution of classes of evidence indicating clothing during the 3rd millennium BC.

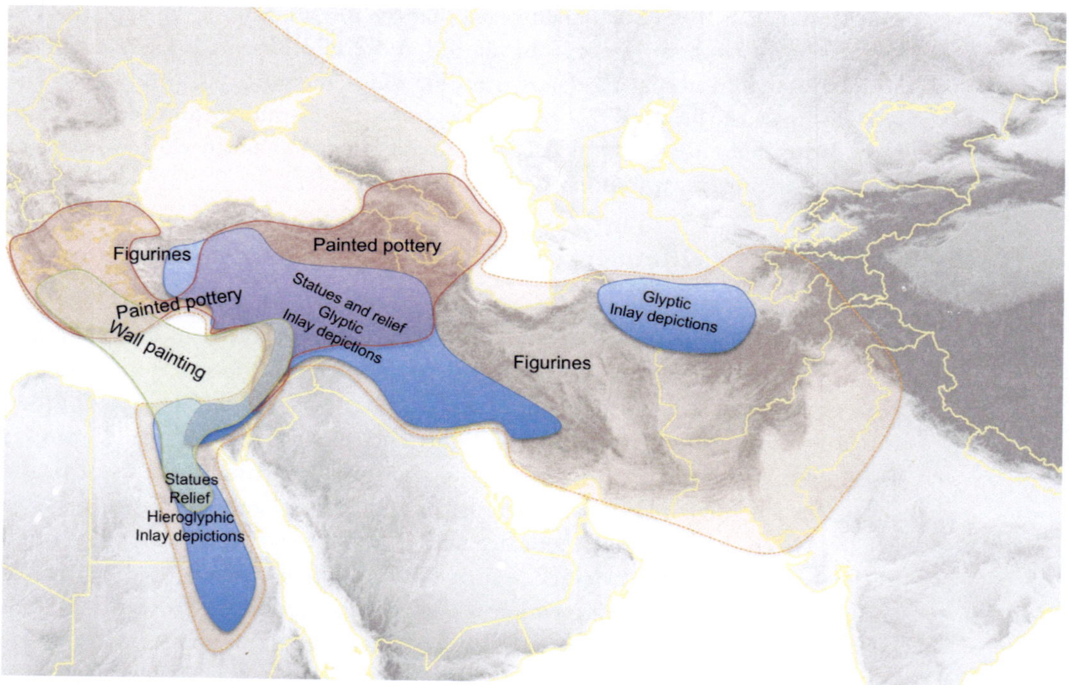

Figure 6.30. Geographical distribution of classes of evidence indicating clothing during the 2nd millennium BC.

Figure 6.31. Painted patterns from the Aegean: Late Minoan I examples on ceramics (after Lacy 1967, 99, fig. 43).

6.6.1 Textile patterns and motifs in the Aegean, Egypt and Mesopotamia

Fragments of wall-paintings from the Aegean and the Levant show many patterns which could very easily have come from, or be transferred to, textile media. The floral and geometric designs (see Figure 6.31; see also Donahue 2006, 89) found on both the pottery and walls of 'Minoan' and 'Mycenaean' palaces dating to the late 2nd millennium BC have long been compared to textile patterns, for example (Shaw and Laxton 2002). Sue Sherratt (2008a) has suggested that the wall-paintings may in fact be 'stand-ins' for real, and much more expensive, textile wall-hangings when not required (*e.g.* in hot summer months) or where it was too expensive to deploy them (*e.g.* in tombs). The same could therefore be said for wall paintings found elsewhere in the eastern Mediterranean and beyond, such as those at Mari which have hoops at the top of the designs which might indicate the manner in which the depicted textile was hung (see Figure 6.32). In Egypt, walls and ceilings of tombs are often painted with geometric designs that could easily have had textile antecedents (see Figure 6.33), at least some of them may be related to this same east Mediterranean *koine* (Shaw 1970; Riefstahl 1944, 43-44). Such patterns turn up as early as 12th dynasty contexts (*i.e.* the early 2nd millennium) and with particular intensity during the 18th dynasty. This is, of course, particularly interesting in light of the general plainness of textiles used for Egyptian clothing.

Related but somewhat different from the 'Aegean' or east Mediterranean paintings are those of Nuzi (Yorghan Tepe), dating to the mid-second millennium. These patterns (see Figure 6.34a) are believed by the excavators to represent small framing borders which appeared at the top of otherwise plainly painted walls. According to Kelly-Buccellati (1996), similarly-styled geometric motifs to those found on fragments of wall-paintings may be identifiable in sealings from Tell

Figure 6.32. Painted patterns from wall-paintings uncovered in Cour 106 of Zimri Lim's palace at Mari, and thus dated to early 18th century BC: (a) reconstructed colour image (Parrot 1958, pl. A); (b) enlarged detail of 'hooks' or 'tassels'; (c) photograph of the wall as excavated (Parrot 1958, pl. viii).

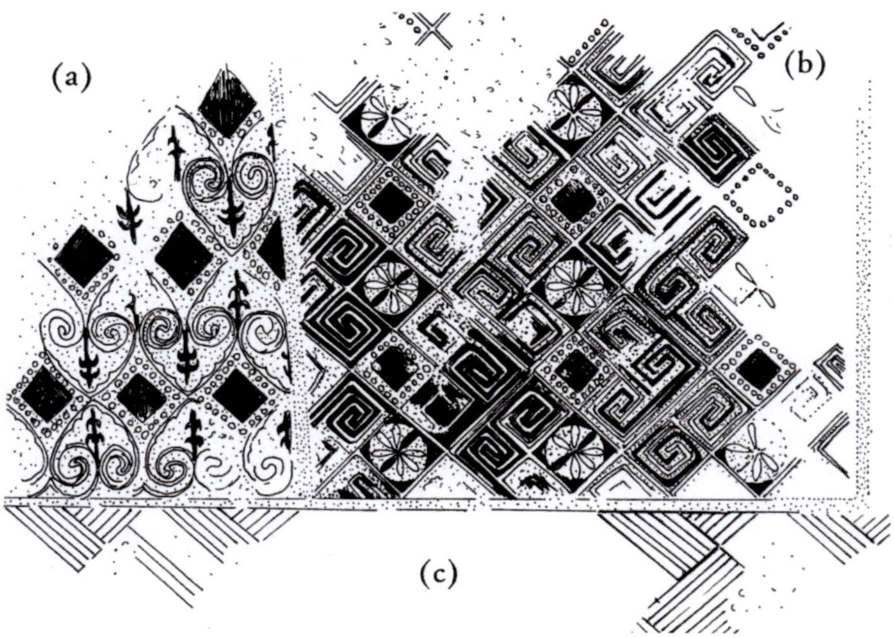

Figure 6.33. Painted ceiling of Egyptian Tomb of Hepzefa at Assiut (Shaw 1970, pl. 5, fig. 4).

Figure 6.34. *Painted patterns from Nuzi on (a) wall-paintings (Starr 1939, pl 129) and (b) ceramics (redrawn from Cecchini 1965).*

Mozan (Urkesh) and Tell Brak that date up to a millennium earlier. She thus argues for a long continuity of artistic traditions in the region of northern Mesopotamia: "what unites the northern artists over the span of about a thousand years, roughly from the prominence of Urkesh to Nuzi, and singles them out from their southern counterparts, is their innate appreciation for strong geometric designs and the possibility of using disarticulated human and animal heads as part of the overall compositions. Geometric frames and borders are basic to their sense of design…" (Kelly-Buccellati 1996, 265). Such frames, borders and strong geometric designs are typical of textile traditions, and therefore the mechanism of such a continuity (or at least 'apparent' continuity) may well be related to the continued use of typical textile motifs and styles from the 3rd to 2nd millennium in these different media. Traditions may have been created in the context of the central importance of patterned textiles to local identities in Assyria or northern Mesopotamia and their export to other regions. A patterned frit threshold (Figure 6.35) found at Mari has been claimed to represent the design of some kind of early carpet (Dalley 1991; *cf.* also much later 1st millennium thresholds Albenda 1978). The fact that later 1st millennium thresholds of metal have sometimes been found, does not

Figure 6.35. Patterned door-threshold in frit from Mari that may have been based on a rug design (after Dalley 1984, 52, fig. 24).

exclude an originally textile-based origin. Urbanized nomadic elites might, for example, 'convert' the rugs they used in tents to a different medium once settled, as flashy indices of their risen status.

The end of the 3rd millennium and much of the 2nd millennium is characterized by widespread adoption of re-assertion of painted pottery traditions with geometric or floral designs, which conceivably could have their origins in textile forms and motifs[121]. Painted Aegean pottery from the 2nd millennium is well known (see Figure 6.31), and the potential connection of its motifs to textiles has been noted before (Barber 1991, 354). Widely distributed in Syria, Cilicia, central Anatolia and the Levant (see Figure 6.36) is also a range of painted wares of varying dates, including: Upper Euphrates Painted wares, Gelinciktepe-wares (Marro 1997); Alishar III[122] and Cappadocian wares (Emre 1966), Ciradere and Intermediate Wares (Öktü 1973); Levantine painted wares, Syro-Cilician wares and Khabur wares (Bagh 2003). Though the designs are quite different from the Aegean ones, there are possible affinities between these decorations and basic textile designs: many have some kind of patterned band, often around the neck, showing triangles, zigzags, lattices or tassels, and motif-styles typical of netting, embroidery or woven patterns. Many of these wares are also bichrome or polychrome – a strong indication that the pottery may be mimicking other materials because of the difficulty in adding two or more colours to a pot.

Dating slightly later to the mid- to late 2nd millennium but sometimes described as being related to the earlier Khabur ware, 'Nuzi wares' show complex and dramatic geometric and floral patterns and bordering (see Figure 6.34b) – which are suggestively textile-like. Interestingly, there are "no two examples of Nuzi Ware with exactly the same white painted design, nor are there two sites

121 Painted pottery was of course common in earlier periods, especially during the later Neolithic: examples include Halaf, 'Ubaid, Susianan, Hacılar and Bakun wares. However, sometime during the 5th or 4th millennium, unpainted or plain wares became dominant.

122 Note that the name of the type site 'Alishar' is now normally spelt *Alişar* (following modern Turkish orthography) while the 'Alishar III' assemblages continue to be referred to using the Anglicized spelling.

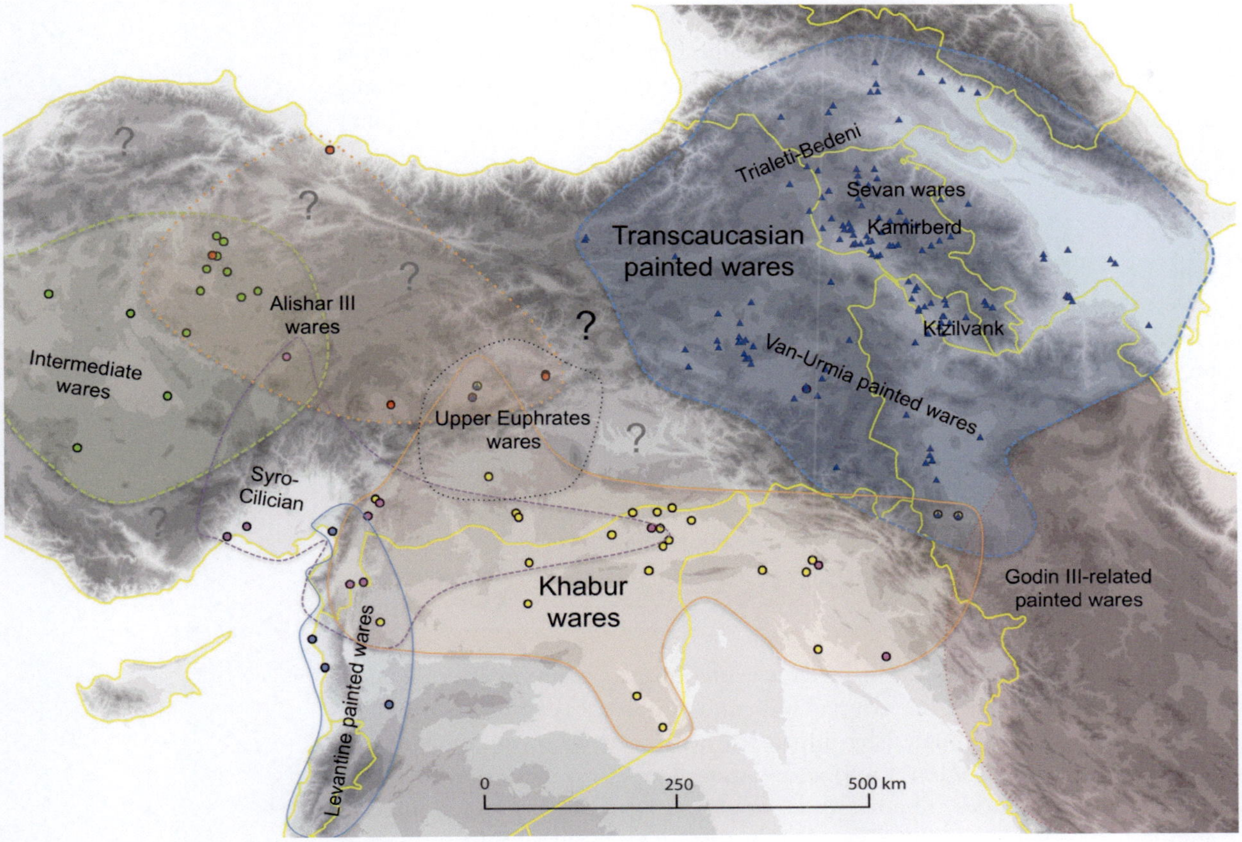

Figure 6.36. Distribution of emerging painted pottery traditions in late 3rd and early 2nd millennium BC, including 'Transcaucasian Painted Wares' (Özfırat's 'Aras' wares). See Appendix C.1.3 for database.

where the Nuzi Ware decoration is conceived in exactly the same way … [so] it is probable that not one but several diverse influences contributed directly or indirectly to the Nuzi style of decoration" (Stein 1984, 27) – something which is also typical of pre-modern textile production. We may propose that this pattern reflects an association between the pottery and *localized* textile traditions or even *particular* textile pieces, and not just generic patterns and motifs[123]. Considerable archaeological effort has been spent on attempting to understand the origins and relationships between all the different painted wares of this region, normally formulated as how one pottery type influenced the development of another despite some variations in vessel form across different regions. If the pottery is viewed differently, however, as merely the fossilized refractions of textile designs that were circulating within these networks, such fluid and indistinct interrelations become easier to understand. Whilst we should pay careful attention to the specific history of each of these assemblages, it is curious that painted pottery should have become so widespread from the end of the 3rd and more particularly in the 2nd millennium, precisely when we believe, on the basis of textual evidence, that the exchange of textiles across the Near East and eastern Mediterranean region had become a large-scale industry.

If we accept this basic premise, then the proliferating painted designs also allow us to identify both small networks of intense exchange involving textiles (*i.e.* those regions with related designs), and larger networks of interaction as sets

123 This individuality of design is also paralleled by Cretan 'Kamares ware' dating to the Middle Minoan period.

of nested communities. In this way, we may begin to construct textile provinces and foci in a similar way to Chernykh's metallurgical groupings. However, more work needs to be done to integrate these patterns with the distribution of and variation in textile technologies – which, as we have seen above, we still know very little about.

6.6.2 Patterns and motifs in Transcaucasia/eastern Anatolia

In the area of Transcaucasia and eastern Anatolia, there are very few examples of recovered wall-paintings from any period, and none dating to our main period of interest. The best known are those found inside the Uruk temple and later alongside doorways of the 'palace' at Arslantepe, dated to the late 4th millennium BC. The figures are in a style that strongly recalls embroidered red lines of thread on a white field (Figure 6.37), which may mean the paintings were temporary 'stand-ins' for valuable wall-hangings, as has been suggested for the Aegean (see above Section 6.5.1). This evidence is more significant if, as McCorriston has proposed, the Uruk 'expansion' was strongly related to what she calls the 'fibre revolution': a shift from linen to wool as the dominant textile fibre, and concomitant 'extensification' of agro-economies via shepherding (McCorriston *et al.* 1997).

For the 3rd and 2nd millennia, ceramic evidence again seems to offer the most promising source of patterns. Early 3rd millennium pottery assemblages for much of this region were dominated by unpainted burnished wares, particularly the Kura-Arax wares (see Section 5.6.5) and without any obvious relation to woven textiles. At the end of the 3rd millennium or beginning of the 2nd millennium (in common with other areas of the Near East at this time), the archaeological record reveals the introduction of a range of interrelated painted wares in Transcaucasia, including 'transitional' Bedeni and Trialeti wares, and a group of assemblages known variously as Aras Painted, Kizilvank, Van-Urmia, Sevan-Uzerlik or Yayla wares (Özfırat 2001). To refer to the whole constellation of these wares, which are distributed over a wide area (Figure 6.36), the term 'Transcaucasian Painted Wares'/TPW has been coined here. Where not the result of illicent excavations, much of this pottery seems to have been placed in mortuary contexts, and, given the relative paucity of settlement evidence, it has often been concluded that the producers of these objects were transhumant or nomadic pastoralists (Sevin 2004)[124].

It has also not gone unnoticed that these painted pots carry design features that strongly evoke textiles (Belli and Bahşaliyev 2001, 62, 73). Whilst acknowledging real regional variations (Özfırat 2001, 105, 17-26), there are some distinct similarities in the choice of design that include square 'fields' as backgrounds to the paint, geometric patterns (triangles, borders) with concentric circles, tassels, netting and occasional 'embroidery'-like figures (see Figure 6.38)[125]. If

124 There are serious problems with these labels of course, since pastoralism and nomadism are very general terms and hide a multitude of different possible economic and subsistence lifestyles, with very different consequences and level of integration with surrounding groups.

125 Similar patterns to which can also be seen on other painted pottery traditions – *e.g.* those mentioned above or the EBA painted wares of Luristan dating to the late 4th or early 3rd millennium BC (*e.g.* Haerinck 2011, 92, 97) – albeit in different configurations. These commonalities presumably indicate common medium of inspiration (*i.e.* textiles or baskets) rather than any direct cross-fertilization or genetic link between the pottery styles themselves.

Figure 6.37. The schematic figural wall-paintings from Uruk period Arslantepe (Frangipane 1997, 66 fig. 15), with strong resemblance to patterns produced by textile designs (Photo: courtesy of the archive of the Italian Expedition in Eastern Anatolia, Sapienza University of Rome).

Figure 6.38. Examples of Aras, Yayla or 'Transcaucasian Painted Wares'-style painted pottery (following Özfırat 2001).

we accept the basic idea that these designs represent textile patterns transferred to a different medium, then it seems very plausible that the distribution of these pottery patterns reflects in some sense the ancient distribution of otherwise invisible textile patterns. The fragmentation of styles (especially when compared to the strong uniformity of Kura-Arax pottery) suggests, however, that, at least for the Transcaucasian example, this is not a bulk trade dependent on economies of scale and abstract trading relationships (although the original textiles might have entered into elite luxury exchange networks) but rather the movement took place within craft networks dependent on local or regional networks of marriage, clan and kinship relations – perhaps akin to how carpets were made in Anatolia until very recently. It seems likely then that there are regional communities who were exchanging motifs and techniques within strong local networks, whilst they were also linked into wider but weaker networks of aesthetic and technological possibilities.

The more difficult question to answer is why textile patterns should be applied to pottery at all. It could be a response to the possibilities offered by new techniques of dyeing and weaving wool (*cf.* Sherratt 1983) – which facilitate new forms of asserting identity and difference through clothing in an increasingly internationalization context. The context of finds of these TPW wares may provide a clue to such a reading of their significance, and perhaps also an explanation for the wider trend towards painted pottery beyond Anatolia. Except for a few settlements (such as Kyultepe in Nakhchevan), where these painted pots are found in large numbers, the majority of the TPW wares come from tombs (and sadly, in many instances, from already looted contexts). On current evidence, each pot's decoration appears to have been unique, and it is tempting to assume that this in some sense reflects the uniqueness of the person whose tomb the pot furnishes: indeed, individual motifs and their combination may have provided direct clues to the contemporary viewers of these pots about the owner (perhaps specific kin, craft or ethnic affiliations). They may also have mimicked (or fossilized in paint) individualized cloth bands or belts worn by the owners.

In support of this idea is the use of hems or belts to sign contracts when seals were not available in northern Syria, attested most clearly in texts from the Neo-Assyrian period but probably with a much longer history (Dalley 1991, 125), which vividly demonstrates the relationship between patterned textile strips and individual identity in an economic or exchange context[126]. This is not to say that all the different painted pottery types of this period were necessarily used as markers of *individual* identity: they may also have represented 'corporate' groups (clans, tribes, ethnic groups) – although the Middle Bronze emergence of richly furnished graves such as the Trialeti or Martkopi/Bedeni graves may suggest a breakdown of corporate kin groups into class or house-based groups. However, the phenomenon does suggest that the use of textiles and textile patterns were becoming increasingly important in establishing interrelationships between individuals and groups in a much more serious manner than before.

126 Irene Good (2007) places the association of these individualized cloth with the emergence of the concept of 'the individual' to the time of Hammurabi, in the mid-2nd millennium BC, but the 'individuality' of pottery designs in late 3rd and early 2nd millennium BC assemblages which appear to reflect textile designs may be argued to place the beginnings of such a conception much further back in time.

6.6.3 Patterns and motifs in western Central Asia

As in Transcaucasia and eastern Anatolia, very few fragments of wall-paintings from Central Asian sites have been published. A couple of exceptions, from Anau and Yassi Depe (now Ulug Depe), are dated much earlier than the 3rd millennium BC. Interestingly, the one photographed example from Yassi Depe shows patterns which are very similar to the designs found on painted pottery of the contemporary Namazga I (see Figure 6.40). Painted pottery again forms the main source of potential textile motifs in western Central Asia, albeit from an earlier period. It has long been commented that the Namazga I-IV pottery (see Figure 6.41) has strong affinities to textile patterns (Pinner 1982; Masson 1992, 231), with some describing Namazga III assemblages as 'tapestry ware' (Good 2006). The distribution of these patterned wares in fact reaches out across eastern Iran and Afghanistan (see Figure 6.42). In fact, painted wares – some perhaps related to basketry – have a long tradition across wider Iran (for example the Fars, Bakun, Yahya or Sialk sequences) stretching back into the Neolithic[127].

The idea that motifs and patterns may reflect textile patterns has rarely been taken any further to consider what the motivations for such a transfer of patterns might signify. As with the Anatolian examples cited above, we may postulate that expressions of ethnic or individual 'identity' may be one of the motivations behind the transfer of motifs to pottery. However, most of the evidence for painted pottery comes from settlement areas, not tombs like the Transcaucasian Painted Wares. It is thus much more difficult to relate these early painted designs to some wider trend, such as the same notions of personhood suggested by the Babylonian texts (which are of course at least a 1000 years later).

These Central Asian painted wares disappeared rather rapidly after the mid-3rd millennium, to be replaced with plain wheel-made pottery, more likely to have metal antecedents (see Section 5.6.4). However, it has been suggested that some of the motifs, such as the crosses, appear to survive in the form of 'seals' (Figure 6.43), like those of Altyn Depe (Masson 1988, pl. XLII and XVI). Central Asian seals, like other BMAC-related objects, actually turn up sporadically over an extremely wide area (Hiebert 1998), as far as Malyan, level III (Amiet 1986, fig. 52) – where fragments of wall-painting with cross-designs have also been identified (see Figure 6.44a) – and Mari (Beyer 1989; see Figure 6.44b), dated, perhaps over precisely, to 2225BC. At Gonur Depe, these seals often appear in graves somewhere near the waist, neck or wrist of the deceased (Sarianidi 2001, 195), perhaps suggesting they were part of a belt or personal ornament. If these were indeed used as 'seals' as well as dress adornments, then it may be that their usage was not far off the use of hems in Babylonia, and the continuation from earlier motifs on pottery is thus highly suggestive.

6.6.4 Comments: patterns as stamps of personhood and individual identity

We should take care of course not to get carried away with the idea that painted pottery somehow simply equals textile designs. There are many examples of painted pottery in the Near Eastern archaeological record which are not likely to

127 This is also true outside of Central Asia during the Neolithic and Chalcolithic: *cf.* Halaf pottery in northern Mesopotamia, or Hacılar in western Anatolia.

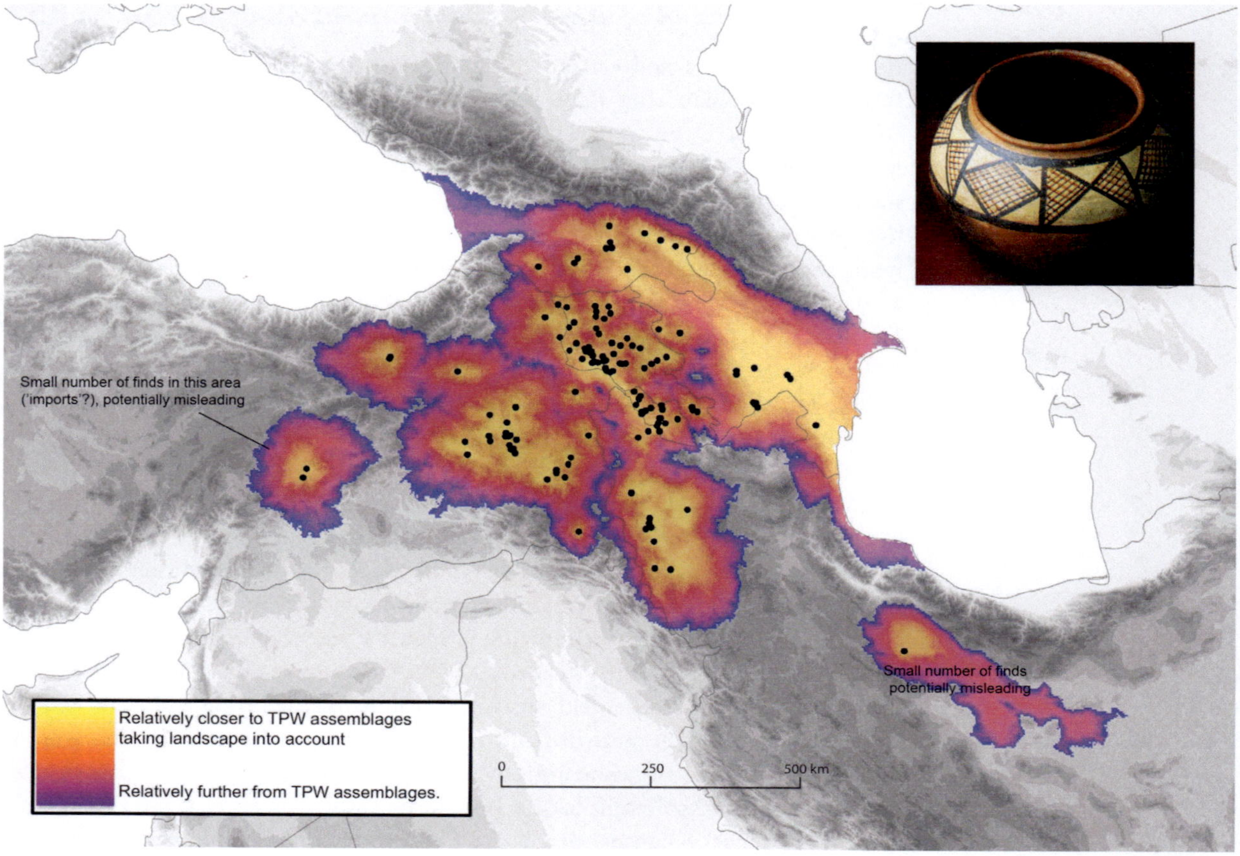

Figure 6.39. Transcaucasian Painted Wares cultural-geographic zone: archaeotopogram (type 'A2') based on the distribution of Transcaucasian Painted Wares related wares (using data from Özfırat 2001).

bear any relationship to woven textiles: the painted wares of Hacılar levels III/II for example, with their very different rounded swirling patterns might suggest a relationship to, for example, body paint rather than textiles or basketry[128]. However, the overall trend of emerging painted wares is suggestive of a new or renewed engagement with patterns in textiles as important cultural markers. Though difficult to prove, the distribution of pottery wares apparently inspired by textiles might reflect the distribution networks of certain styles of real textiles themselves, since the textiles were the more mobile medium. The data from patterns and motifs in other media have led us to the use of cloth representing not just generalized roles, but also individual identity to be used in contracts, display ownership and confirm authenticity. It is well known of course, that textiles can function to display, reflect or construct both 'ethnic' and individual identities, but the tradition of archaeologists sometimes wanting to use pottery assemblages to represent such groups begins to make more sense if such textile patterns were being transferred to pots. Perhaps, on occasions, through the use of distinctive motifs copied from identity-laden or 'ethnic' dress, pottery styles indeed did represent emergent and ever-changing 'ethnic' or other similar corporate groups.

128 Though the earlier Hacılar IV/V wares are in fact more textile-like in style – perhaps related to basketry, as proposed by the excavator (see Mellaart 1958).

Figure 6.40. Wall-decoration from Ulug Depe (formerly Yassi Depe) with patterns reminiscent of contemporary Namazga I pottery designs, and –potentially– textile patterns (after Hiebert and Kurbansakhatov 2003, 20).

Figure 6.41. Examples of Namazga IV-style painted pottery (from Masson and Merpert 1982, pl. 4b; photos: courtesy of Russian Academy of Sciences).

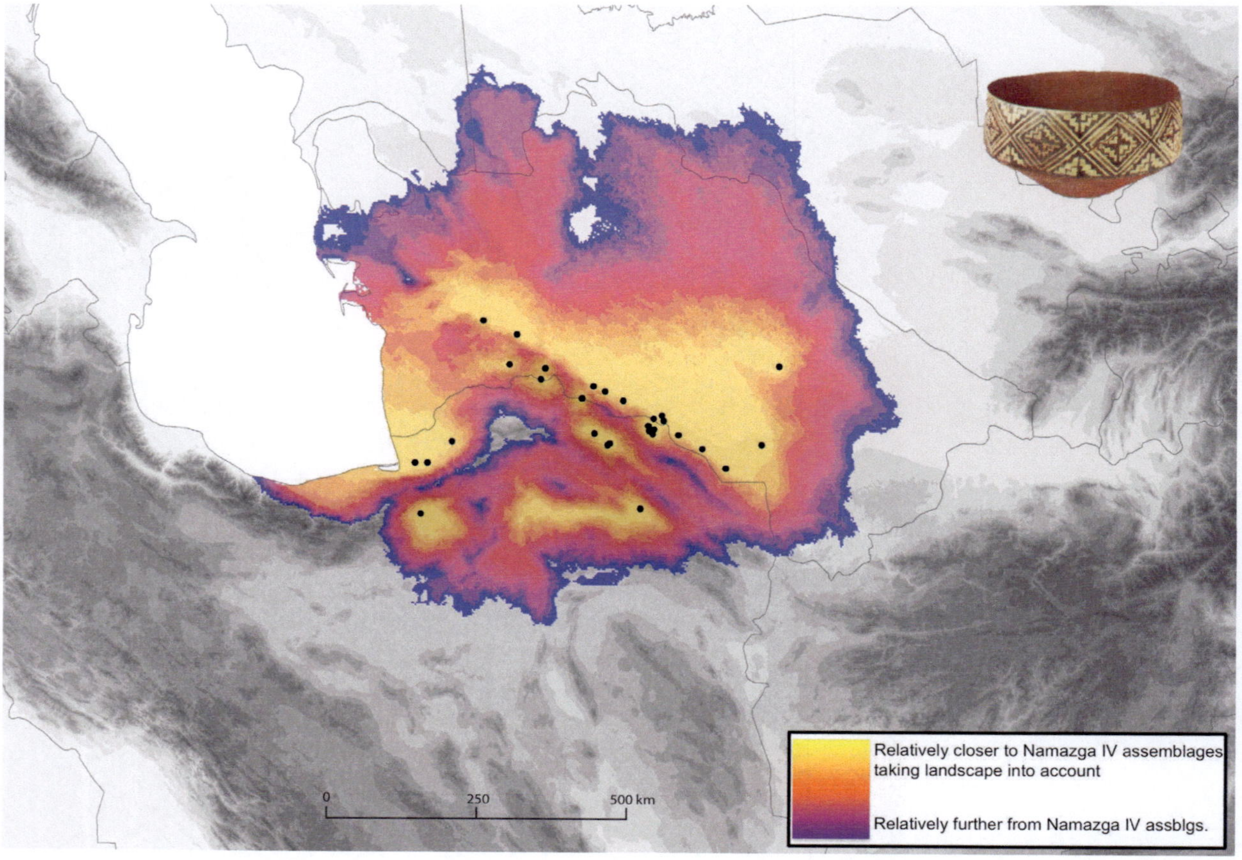

Figure 6.42. Distribution of Namazga IV-style painted ceramics in Turkmenistan and north-east Iran plus Namazga IV cultural-geographic zone (archaeotopogram type 'A2').

Figure 6.43. Central Asian seals: (a) cross-motif seals and impressions from Altyn Depe, Namazga V period (Masson 1988, pl. XVI); (b) figural/iconographic seals from the Gonur Depe necropolis, BMAC period (Sarianidi 2001, pl. 8).

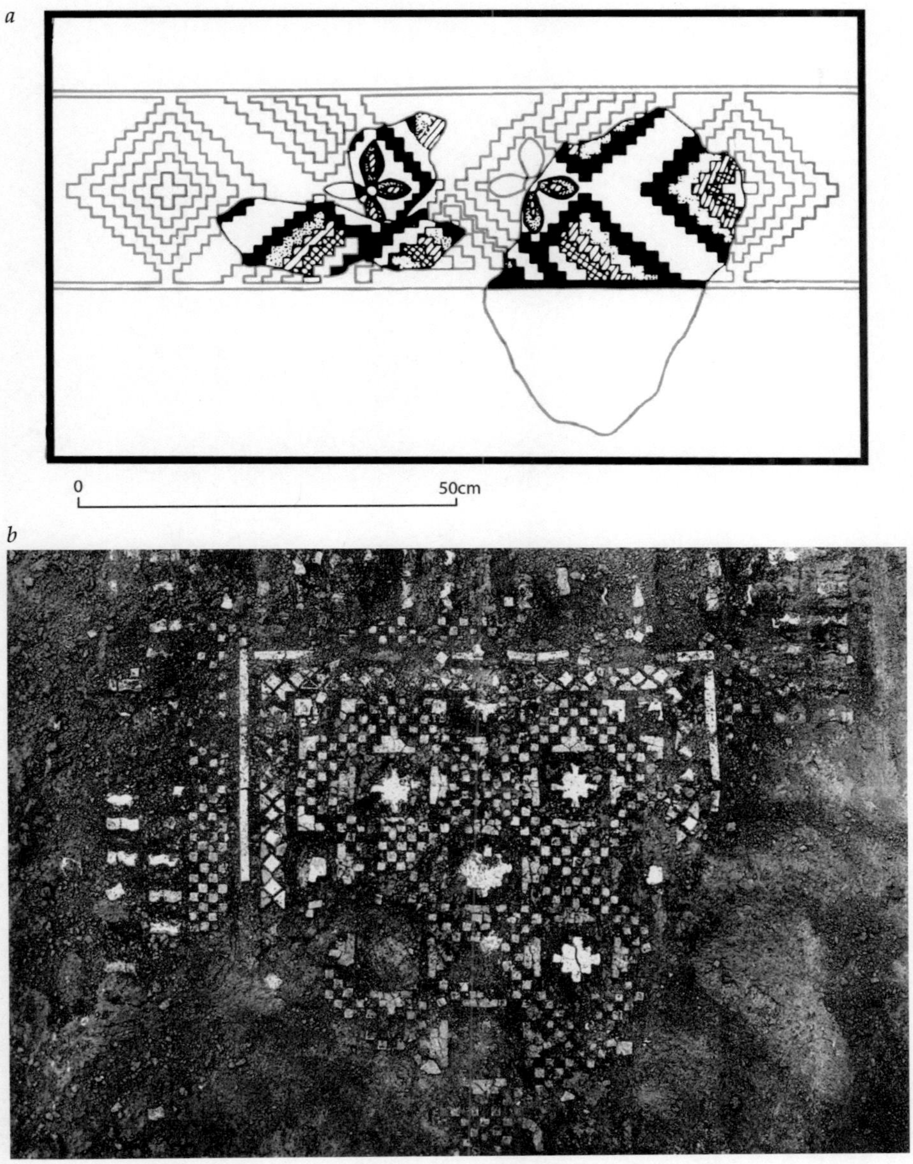

Figure 6.44. Concentric cross-pattern in various media (a) fragment of patterned wall-painting from Tell Malyan, Iran (drawing: Janet Nickerson; cf. Amiet 1986, 262, fig. 52); (b) similar pattern in a mosaic from Mari (after Parrot 1959, pl. xxxiv).

6.7 Indirect evidence: dress accessories

Another category of indirect evidence, rarely considered in the context of textiles but which may well indicate the extent and movement of dress styles, is that of dress 'accessories' such as jewellery and beads, or essential items for dressing, like dress pins. The middle of the 3rd millennium in particular seems to represent a period of elaborate and cosmopolitan accessorizing: the gold crown from Queen Puabi's tomb at the Royal Cemetery at Ur, with its leaf patterns apparently of the sissoo tree, which grows in the Indus, a good example of the international flavour of dress fashions (Tengberg, Potts and Francfort 2008).

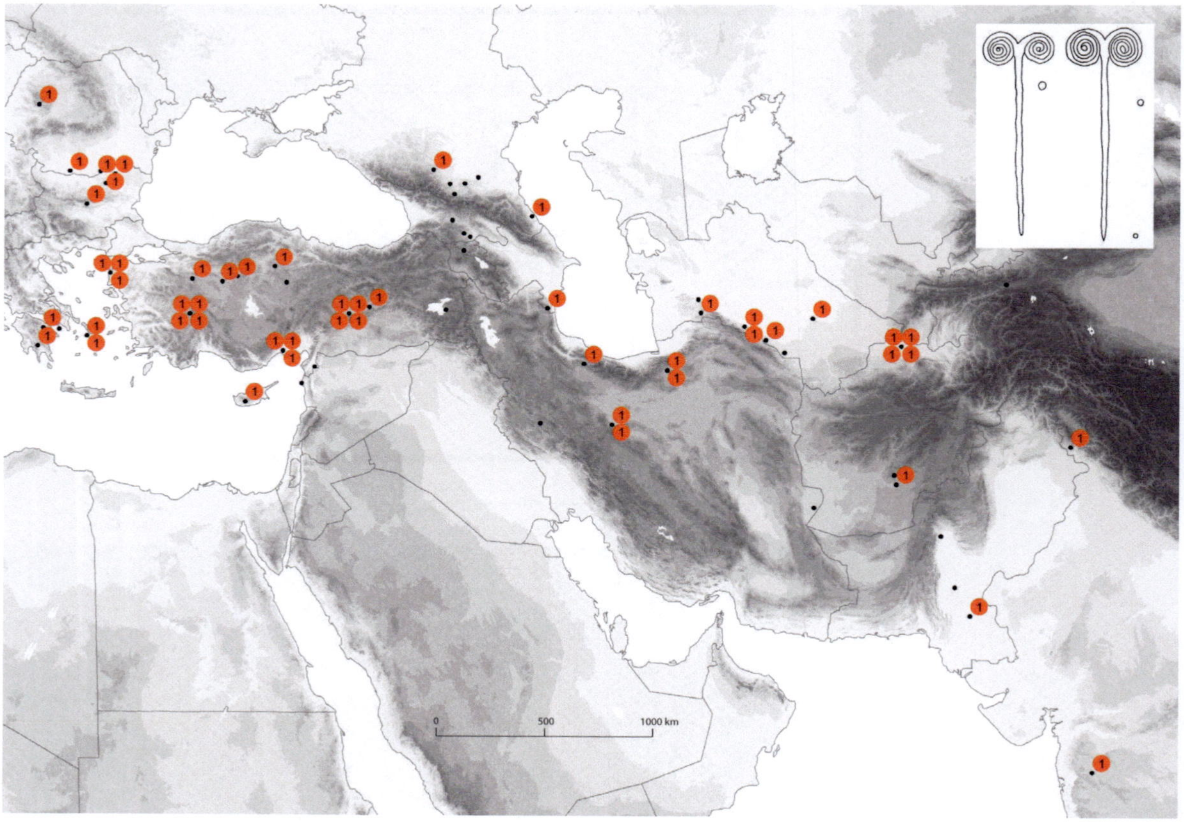
0
500
1000 km

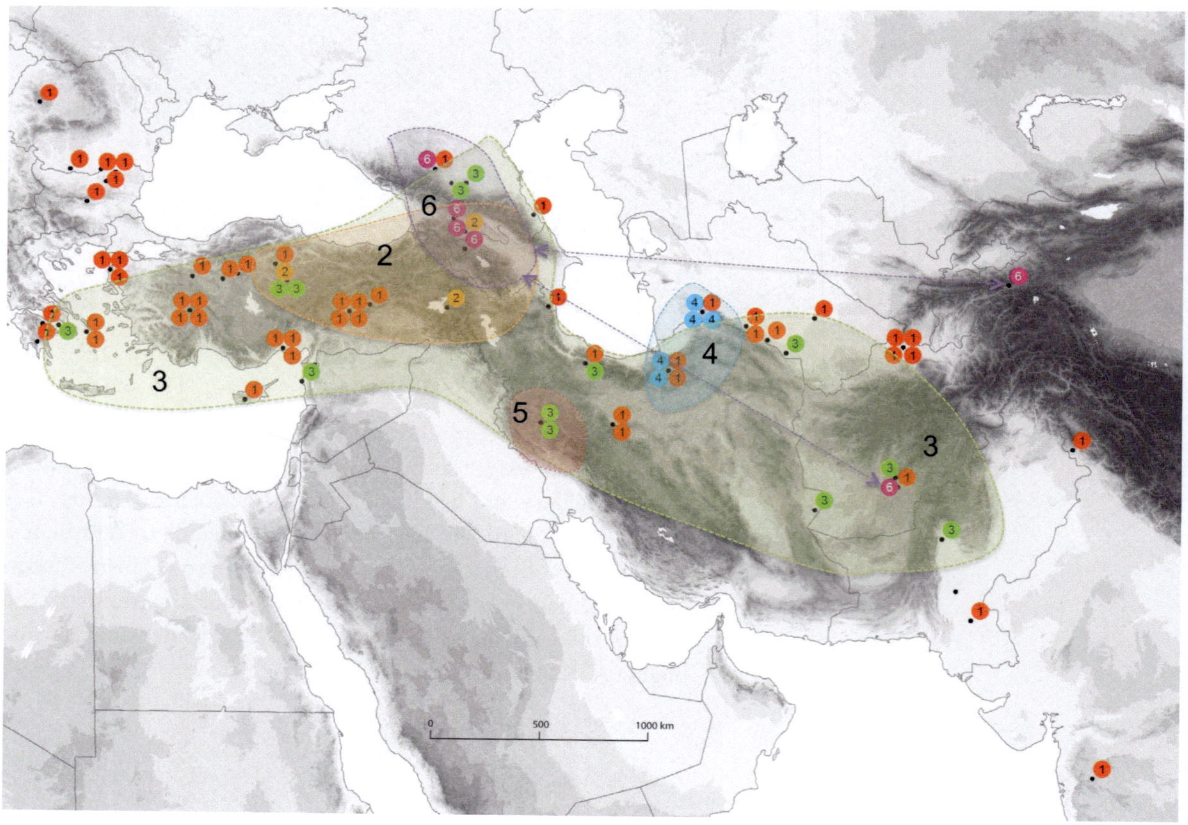
0
500
1000 km

6.7.1 Pins and clothing styles

Dress pins of various types are known from across the Near East during the 3rd and 2nd millennia. Particularly elaborate examples are known from Caucasia, especially during the 2nd millennium (see Motzenbäcker 1996). One type of pin that stands out as particularly interesting because of its very wide distribution is the double spiral-headed pin (Huot 1969; Sagona 1981; Huot 2009). A useful database and typology of these pins was established by Huot (1969) and recently updated (Huot 2009). Huot highlights the stylistic variation within this category of pin, categorizing them into a number of different 'sub-types'. This variation suggests that there may be no simple explanation of the overall distribution. Huot's (2009) 'type 1' pin, for example, has the widest geographic distribution, stretching from the Aegean to the Indus (see Figure 6.45) – a picture in miniature of the international connections of this period (compare, for example, Rahmstorf 2010b). These pins appear to reference a distinctive and long-lived formal motif, which had wide currency across the Near East – also appearing, occasionally, as decoration on Kura-Arax pottery (see *e.g.* Sagona 1984, fig. 120 cat. 191-193). It is likely that the pin itself indexes a certain style of clothing: something that needed to be pinned together, of course, but also, potentially, a particular pattern or dress style that signalled distinctive identity. Whilst one could argue for independent invention, especially given the lack of clarity in dating these objects, the overlapping distribution of the different types makes this difficult to believe.

Figure 6.45. Distribution of Huot's 'type 1' double spiral-headed pins (data from Huot 2009; inset Huot 1969 fig. 1[5, 6]).

As Huot highlights, one very curious fact is the complete lack of any double-spiral pin finds from Mesopotamia (see Figure 6.46). There is no obvious explanation for this, since, given the wide distribution, and other pins discovered in Mesopotamia, they must have been familiar. The context of most archaeologically uncovered examples suggests, however, that this may relate to burial customs – that Mesopotamians did not consider this item suitable for burial with the dead, unlike many 'peripheral' groups. This might simply be a consequence of Mesopotamian attitudes to metals (as discussed in Chapter 5), but equally it could reflect attitudes towards dress style.

6.7.2 Beads, patterns and beaded dress

A similar example to the spiral-headed pin comes in the form of the distribution of certain flat, tubular beads and quadruple-spiral beads (Rahmstorf 2010b, 280-281). As with the spiral-headed pins, the flat-tubular (see Figure 6.47) and quadruple-spiral beads (Figure 6.48) have variant shapes that could be taken to argue independent invention. However, the strong chronological parity and the context of long-distance exchange of other items (such as lapis) makes this harder to sustain. Rahmstorf considers both classes of jewellery as "by-products of the contacts of the elites in the third millennium as in the case of tin and lapis lazuli" (Rahmstorf 2010b, 281).

Figure 6.46. Distribution of all double spiral-headed pins (according to Huot 2009's typology). Some types are much more geographically constrained than others (data from Huot 2009). See Appendix C.1.3 for database.

Whilst there is often a tendency for beads to become decontextualized in archaeological publications, various pieces of evidence suggest that 'beaded' dress was an important part of the repertoire of 3rd and, particularly, 2nd millennium clothing. In Egypt, different types of beaded dress are well known from both

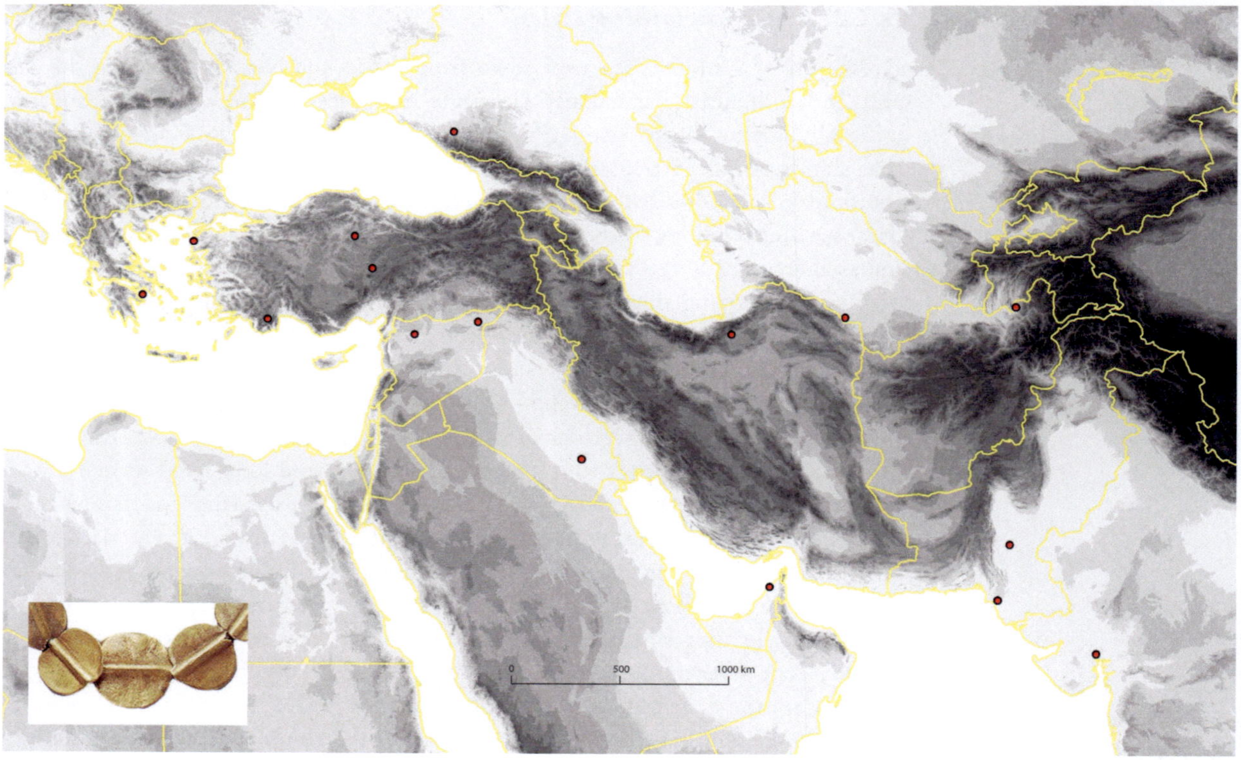

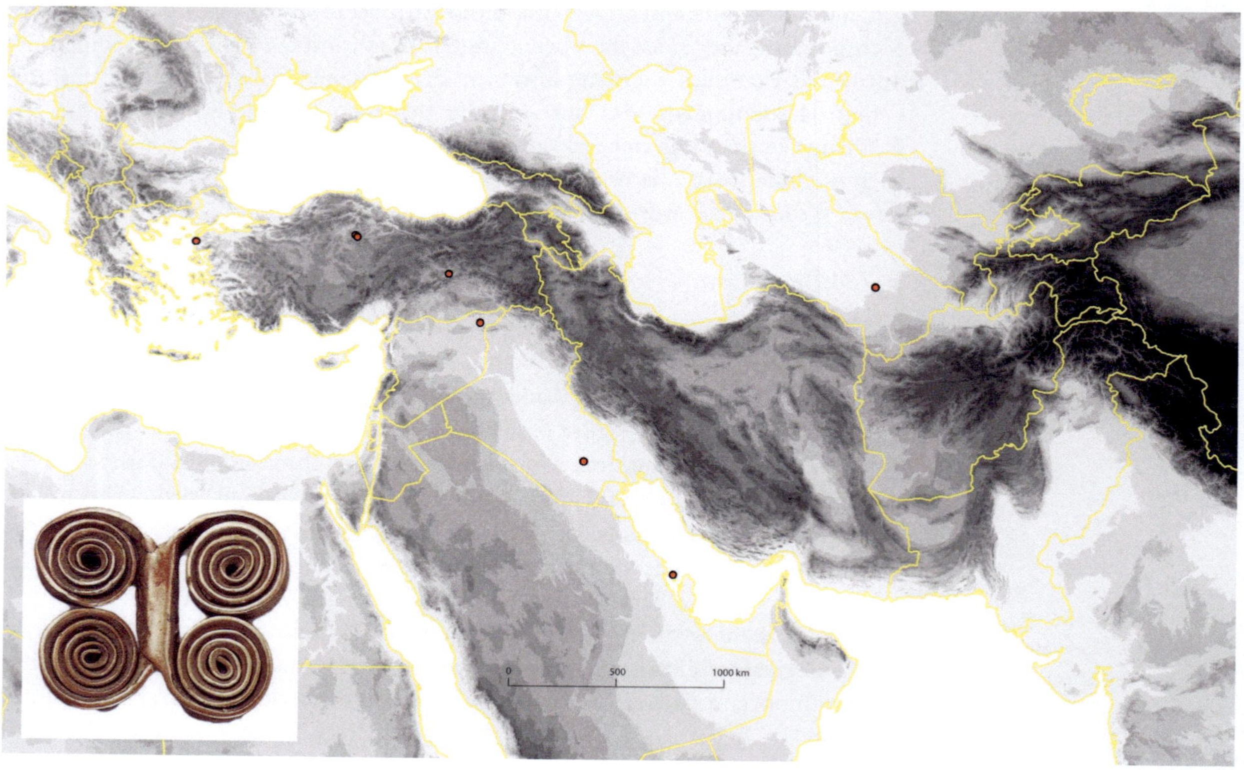

surviving pieces (an example from the Petrie Museum shown in Figure 6.49) and depictions (Riefstahl 1944, 11-13). The coloured patterns adorning many figures and some New Kingdom (mid-to-late second millennium BC) pottery types[129] (Aston 2003, 154) appear to represent complex beaded pieces made from precious stones such as lapis lazuli, carnelian and turquoise or else feathers and flowers (Figure 6.50; see Chapter 4). We see indications of similar 'beaded' dress in Anatolia. Barber has persuasively argued that a collection of gold beads uncovered at Troy with a group of loom weights (in Room 206, level IIg, dated now to the mid-third millennium BC) represent the remains of an unfinished beaded cloth, left on the loom at the moment of destruction (Barber 1991, 171-172). There is a similar example from the Middle Bronze period of Old Assyrian colonies in central Anatolia, c. 1900-1750BC, uncovered in a building at Acemhöyük (Özgüç 1966). The building, described as a palace by the excavators, was destroyed by a conflagration, which, as well as preserving some impressions of its wooden support structure in vitrified mud-brick, also preserved fragments of cloth, described as white linen with tiny beads of lighter and darker faience, sewn together with gold thread (Figure 6.51). Barber (1991, 171) argues that clear patterns formed by the beads can be perceived in the remaining fragments: "one can still make out a pattern of stacked chevrons and part of a swastika or meander". Here the use of two types of blue faience, although impressive, may have acted as a cheaper alternative to lapis lazuli and turquoise. Woolley argued that many of the beads around wrists and waists from the Royal Cemetery (dating to the mid-3rd millennium) were actually sewn into garments rather than representing separate bracelets, belts or necklaces (Woolley 1934, 239; Aruz and Wallenfels 2003, 112-113). Given the combined value of fine cloth and precious stones or metals, bejewelled items like this may of course have represented clothing of only a particular social stratum, for example courtly dress, or indeed for very particular occasions – the wearer's death being only the final dramatic example. In these instances, we do not know the origin of the cloth itself, but we do know that beads of lapis, carnelian, gold, and faience did have distant sources, and these composite pieces hints at social values which emphasize the role of exotic materials in establishing and signalling social roles, at least in death. However, clear evidence for these kinds of beaded dress have not been uncovered in our case-study areas of Transcaucasia/eastern Anatolia or western Central Asia.

Figure 6.47. Distribution of flat circular beads with tubes. Inset: detail from string of gold flat circular beads with bisecting tube from Troy IIg, room 253, c. 2500-2300BC (Aruz and Wallenfels 2003, 263, cat. 167). See Appendix C.1.3 for database.

6.8 Summary: motivations and means of textile-based interaction

This chapter has catalogued and analysed evidence about textiles and their circulation from texts, fragmentary remains of textiles themselves, archaeological remains of textile production (including the raw fibres, spindle whorls and loom weights), pictorial depictions of clothing, apparently textile-inspired patterns in other media and finally dress accessories in other media. Different sets of this evidence appear to reflect different levels within the societies that produced them. Texts offer us an idea of broad regions where there are intensive productive industries for textiles (Figure 6.63), centred in Mesopotamia and central Anatolia, and the eastern Mediterranean. The lack of texts in other regions means, however,

Figure 6.48. Distribution of quadruple-spiral beads. Inset: gold quadruple-spiral bead, Troy II, Treasure D (Aruz and Wallenfels 2003, 266, cat. 170a.). See Appendix C.1.3 for database.

129 Note that these types are exceptional within the overall corpus of Egyptian pottery assemblages of the 3rd and 2nd millennia BC, which, on the whole, were plain and undecorated (Arnold 1993) – much like their textiles.

Figure 6.50. Beads used in complex beaded 'collars' of the 18th dynasty: (a) collars as shown being worn in figural art Amarna (photo: Ägyptisches Museum und Papyrussammlung der Staatlichen Museen zu Berlin - Preußischer Kulturbesitz, ÄM 15000); (b) a reconstruction of glazed floral beaded collar Amarna, kept in the Petrie Museum; and parallel patterns on certain pottery types (Aston 2003, 154), (c) Deir el-Medina Tomb 357 (Drawing: IFAO Institut Francais d'Archeologie Orientale – Le Caire), (d) Gurob Tomb 605 (Drawing: Petrie Museum of Egyptian Archaeology, University College London); ; (e) blue-painted jar from Amarna (EA 56841, photo: © Trustees of the British Museum, #AN30996001). The 'beads' may also be durable skeuomorphs of petals or feathers.

Figure 6.49. An example of a 'beaded dress', found at Qau and dated to Dynasty 5 (i.e. between c. 2494 to 2345 BC); (photo: Petrie Museum of Egyptian Archaeology, University College London, UC17743.

Figure 6.51. A clump of tiny dark and light blue beads which represent the remains of a fragment of beaded cloth with gold thread found in the burnt remains of a 'palace' at Acemhöyük near modern Aksaray (Ankara Anadolu Medeniyetleri Müzesi).

that we may be missing many other systems. Most of the more substantial well-preserved textile remains appear to represent elite items of some kind, but each cache is relatively small, without parallels that could allow us to reconstruct the routes or networks by which such cloth may have travelled. Aside from some exceptional examples (such as the appearance of cotton in the Caucasus, the

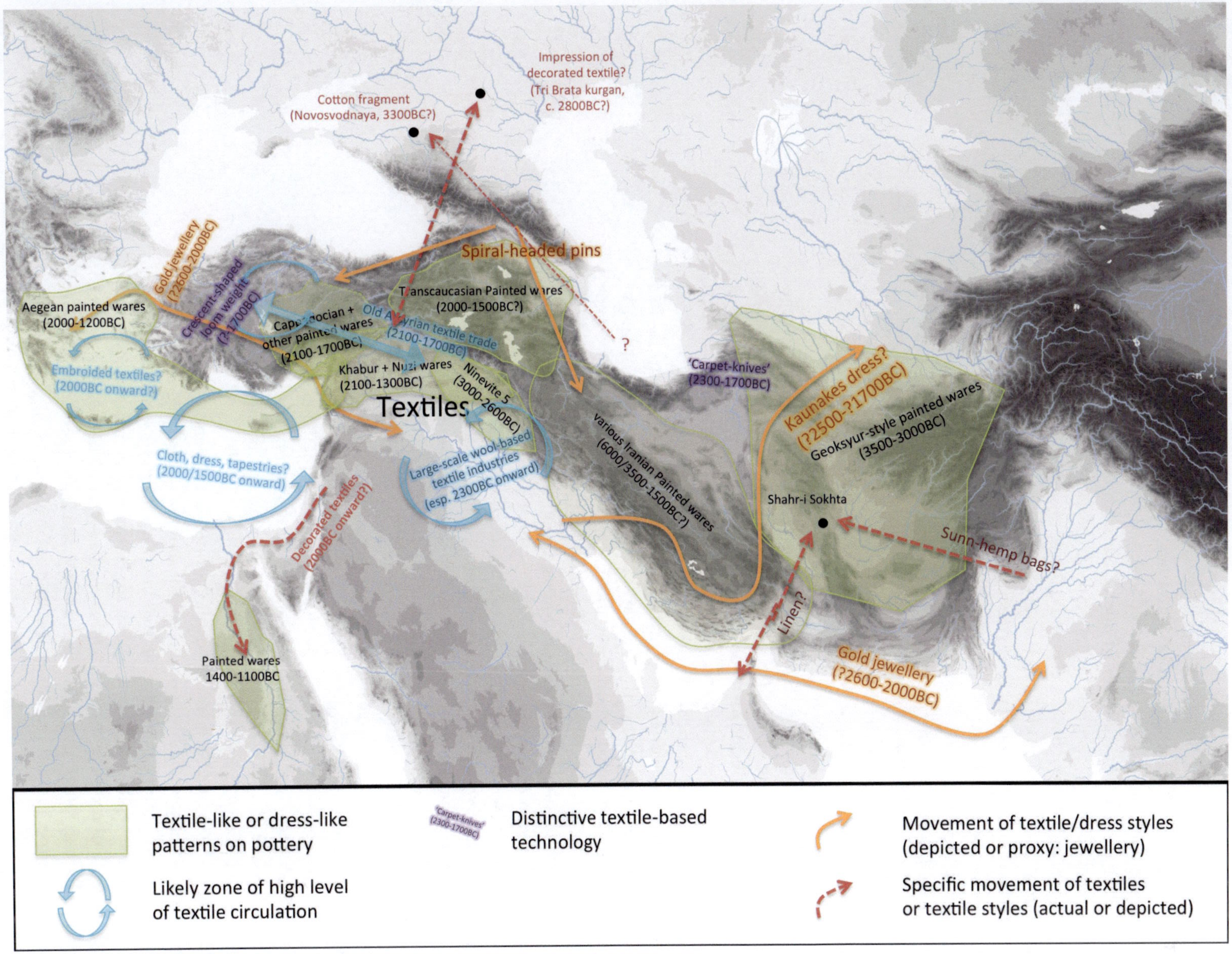

Figure 6.52. Summary of distribution data and of the directionality of textile flows (including likely areas of large-scale textile manufacture, indications for movement of fibre/cloth/dress styles, regions with textile-inspired patterns on pottery) over the entire 3000-1500BC period.

'Syrian' tunic in Egypt), this evidence is mostly too fragmentary to create a picture of the movement of dress or textiles. The pins and dress accessories may offer an easier way to access the extent of textile networks since they are more densely distributed. Again the items appear more likely to be shared international 'elite' items, especially given the full geographic distribution of some items: compare for example the northern focus of double-spiral pins (Figure 6.46) to the east-west transect running through Mesopotamia for the flat circular and quadruple-spiral beads (Figures 6.47 and 6.48). The depictions of clothing in statuary and other forms are likely to show typical 'elite' clothing, either mortal or divine and no doubt reflect values of what might be called – for lack of a better word – 'urbanity' in the adornment of bodies. Whether they represent fashions typical of our case-study regions is impossible to say, however. In Central Asia the Bactrian figurines suggest that the *kaunakes* was known, perhaps as special item of dress for certain rituals, and it seems likely that these connections were made via the south-eastern Iranian plateau. By contrast, in Transcaucasia and eastern Anatolia, there are no such direct representations by which to make comparisons between dress styles.

The distribution of distinctive technologies (such as the twill-producing crescent-shaped loom weights) or of textile-like patterns and motifs in other media offer counter-weights to the exclusive focus on elite-dress or at least provide hope that we may be able to examine small-scale networks of textile distribution amongst non-urban groups. This is particularly the case for Transcaucasian Painted Ware/TPW, which, it has been suggested above, may represent the transfer of motifs used for localized identities onto apparently mortuary containers. Their distribution and variation may provide detailed information about the geographical links between communities – although the level of data currently available is not sufficient to do this. The rise of textile-like patterns on pottery is not unique to Transcaucasia, however, and appears to document a widespread increasing popularity of painting patterns on ceramic vessels that suggests a greater degree of social integration in this region than might otherwise be suspected. The patterns on (much earlier) Namazga III and IV wares may have similar textile origins and reflect similar networks of interaction, but the chronology is very different and is not synchronized with the TPW or other eastern Mediterranean examples. Again it is pottery that provides the basis of detailed evidence about social networks, suggesting further investigation and synthesis of other data could prove productive. Much more work on detailed differences in relevant material culture (particularly through the loom weight and spindle whorl evidence) may also reveal changes and movements on both technical and cultural levels. The crescent-shaped loom weight may, for example, index the production of twill-based fabrics in Anatolia.

As is obvious from surviving texts, textiles formed a substantial part of the Mesopotamian exchange economy, and though the amounts consumed and precise forms of textiles and clothing may have differed in other regions, it seems likely – in the context of visual representations, evidence for textile production and apparent cross-craft overspill of textile patterns into other media – that textiles were as central to personal and group moral values, 'ritual-economy' and self-representation in our case-study regions too. Overall the different strands of evidence highlighted in this chapter document the spread not just of physical commodities but of intangible ideas and attitudes materialized and performed through textiles. The distribution maps thus hint – over a large scale and over a long-time period – at the flow of ethnic, regional and elite identities and in-group/out-group notions of 'urbanity' or 'civilization'.

Chapter 7

Discussion: Tying the Threads

7.1 Introduction

In the last 3 chapters, I have attempted to follow the ancient flows of three broad categories of material – stone, metal and textiles – through various forms of direct and indirect evidence. In each case, relevant data have been mapped in as much detail as possible, with the social and archaeological contexts of the evidence as well as the specific qualities and particular significance of the materials also examined. The diversity of evidential sources required to trace these materials, combined with the inherent chronological uncertainties, makes for a potentially confusing bundle of individually interesting but – at first glance – superficially disconnected data. The original goal was not only to establish the geographical locations of 'routes' in the sense of natural corridors of movement, but also to assess the evolving density of travel along such corridors – *i.e.* the shifting cargoes between cultural zones. To begin to do this, we need to weave these various threads of evidence into some kind of chronological tapestry. My two aims, then, in this 'discussion' chapter are:- first to begin to place the collected data and interim commentaries into a coherent chronological and geographical narrative; and second to highlight certain cultural patterns and trends that are revealed by this synthesis. The chronological focus of this study should not, of course, be mistaken to suggest that interregional interaction only began after 3000BC: indeed the first evidence for various material flows mentioned in this book (*e.g.* obsidian, lapis lazuli and probably copper) started well before this. The chronological summary below traces the shifting material flows (as gleaned from the evidence presented in Chapters 4 to 6), in loosely grouped blocks of 300 years from 3200 to 1400BC. This grouping has been selected as a balance between the imprecision of our chronological frameworks (as discussed in Chapter 1) and the necessity to present this synthesis in a manageable form. Any such periodization is to some extent arbitrary, but the 300 year scheme selected seems to fit the indications of moments of change more neatly than any of the alternatives.

7.2 Shifting material flows

7.2.1 3200-2900BC

Even before the beginning of the third millennium BC, members of the communities in both our case-study regions, Transcaucasia/eastern Anatolia and western Central Asia, were already involved in the flow of materials and information outside of their own local cultural boundaries. The appearance of 'Uruk' material in the Upper Euphrates region, *e.g.* at sites like Arslantepe, in the later half of the 4th millennium appears to document movement of individuals and communities from the south in search of desirable materials not available in the plains of Mesopotamia. That this was not restricted to imported items from

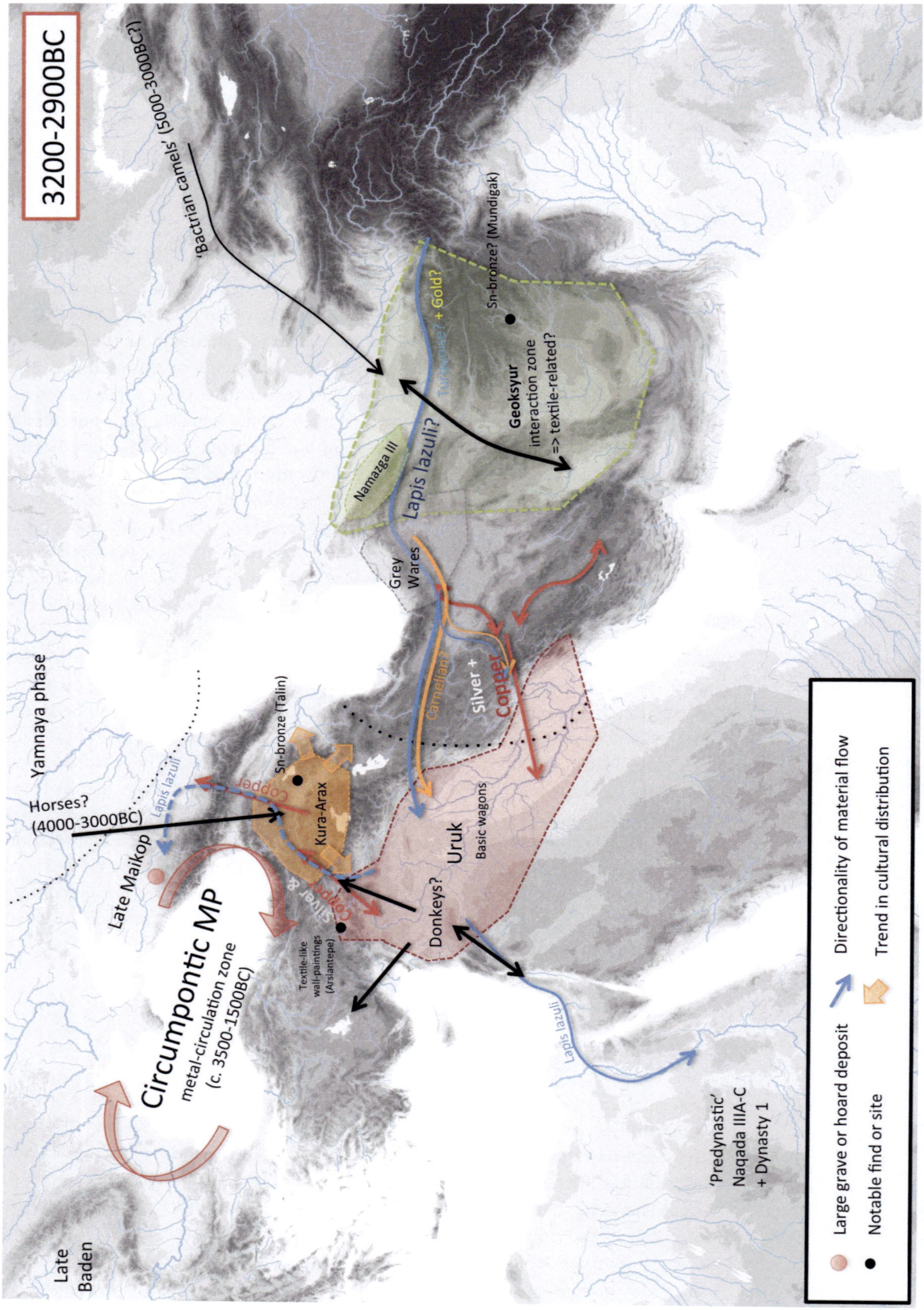

Figure 7.1. Summary of data on flows of stones, metals and textiles during the period 3200-2900BC.

the south, but also – apparently – related forms of social organization, suggests the intensity of this movement. The mobility of resources was perhaps facilitated by the use of donkeys (Section 1.4.1) to increase the quantity of heavier materials that could be transported. The main focus of this cultural 'expansion' from the southern point of view may well have been on obtaining metals: the highlands of eastern Anatolia and the Transcaucasus are, as we have seen, rich in copper sources (Section 5.2.2). On the opposite side of the Caucasus, the emergence of rich Maikop tumuli with their substantial deposits of metals and other precious goods, many with iconographic or technological connections to similar objects from Mesopotamia, signal an interrelated development to (or perhaps a distorted mirror image of) the Uruk phenomenon (Lyonnet 2000; Anthony 2005, 283-285).

The importance of metals in long-distance interaction is also demonstrated by the existence of metal parallels and similar technologies over a very wide region – called by Chernykh the 'Circumpontic Metallurgical Province' – whose beginnings appear to date to around the mid 4th millennium (Section 5.4.1). Thus the small-scale communities of highland eastern Anatolia and Transcaucasia who were, at that time, using early forms of dark or black-burnished Kura-Arax pottery (Section 5.6.5) seem very likely to have been somehow involved in the flow of metals and information between these different cultural zones north and south of highland eastern Anatolia and Transcaucasia. Between 3200 and 2900BC, however, the extent of the Kura-Arax grouping was rather restricted and it is difficult to point to concrete indicators for interaction. Obsidian, a potential source of detailed information about connections, is understudied after the Neolithic and Chalcolithic periods (Section 4.5/4.5.1). Unlike their northern neighbours, the Maikop folk, the Kura-Arax intermediaries appear to have preferred not to sacrifice large quantities of metals or other materials into the ground making it difficult to recognise their outside connections. One exceptional example is the richly-endowed 'royal tomb' at Arslantepe – constructed, it seems, in the abandoned ruins of the Late Uruk town between around 3000 and 2900BC (Section 5.5.2). The contents of this tomb show material links to both southern and northern traditions in pottery and metallurgy, hinting at otherwise invisible cultural hybridity (and implied flows of materials and ideas) in the wider region. Metals were probably not the exclusive commodity in transit: the painted wall-paintings of Late Uruk, Arslantepe (Section 6.6.2) remind us of the potential for the circulation of both images (and associated myths) and perhaps of contemporary textiles, given the 'embroidered' look of the images.

In Central Asia, the later 4th millennium can be characterized as a period of wide interconnections indicated by the common motifs found on pottery of the 'Geoksyur' period, or late Namazga III (Section 6.6.3). Parallel decorative assemblages have been found in the Kopet Dag piedmont/southern Turkmenia, Seistan, the Helmand region and in Baluchistan (Sarianidi 1983); as well as in the Zerafshan (at Sarazm). Whilst it has long been difficult to establish the exact nature of these interconnections, it is hard to see how these motifs could have travelled so far without some intermediate medium. Given the nature of the patterns, which have strong textile aesthetics (being sometimes called a 'tapestry style'), the most likely such medium is woven textiles or basketry. The pottery distribution therefore seems to document the existence of some kind of early textile trading network, though, if we are to assume that the model was baskets rather than other types of textiles, then perhaps the important factor was the association of the *contents* of such vessels. Whatever the interpretation of the patterns, the similarities at least suggest a zone of common social, economic or craft-based values.

Contemporary to this, a parallel but distinct craft-based zone can be seen developing at sites just to the west of this 'Geoksyur zone' in north-east Iran, such as Tepe Hissar (IIb) and Shah Tepe, yielding quantities of 'Early Eastern Grey Ware' pottery (Section 5.6.4.ii). The complexity of metallurgical technology revealed at Hissar and other contemporary Iranian sites at this time and the metallic aesthetic of these grey wares hints at the centrality of the circulation of metals (rather than textiles) to this zone. There appears to be relatively little in the way of direct cultural convergence in either the 'Geoksyur'/Namazga III or 'Eastern Grey Ware' zones toward Mesopotamian cultural models, but, as is the case for Transcaucasia, we have to assume that members of these Central Asian communities were nonetheless involved with flows of materials which were directed towards the Tigris-Euphrates river plains. Precious stones such as lapis lazuli (or turquoise) are the most obvious of these: their likely sources lie to the east of the 'Geoksyur' zone in Badakhshan and the Karakum respectively (Section 4.3.1), and were found in increasing numbers in graves of Mesopotamia (and also in Maikop burials) at this time. It is difficult to know exactly how such material travelled such a distance, though it seems likely to have been through sets of locally-orientated intermediaries in northern and central Iran rather than in the form of an organized long-distance trade. The 'north-south' orientation of the 'Geoksyur' and 'Early Eastern Grey Ware' cultural zones suggests that the export of precious stones to Mesopotamia was a small part of the overall economy. The need for weight-bearing animals may have been limited for commodities like lapis lazuli, turquoise and carnelian. On the other hand, a trade in textiles (if indeed that is what is represented by the Geoksyur network) or metals (if the Grey Wares represent a metallic network) would benefit considerably from the use of pack-animals: it is possible that camels were initially used from the late 4th millennium for local purposes, and that this later contributed to long-distance connections to the west, after around 2500BC, when the first clear evidence of the use of Bactrian camels for hauling wagons can be dated (Section 1.4.2).

7.2.2 2900-2600BC

Across the Near East, the period between 2900-2600BC appears to have been one in which the foundations of deeper integration were laid but still nascent. This can be most clearly seen in the example of the apparently simultaneous emergence of various weighing systems sometime around 2800BC onwards, in Mesopotamia, the Aegean, Egypt and the Indus (Section 4.6). Both the existence in the first place of such systems (presumably designed to control and check quantities of important traded commodities in circulation) and their apparent inter-calculability provide a strong index of integration and interaction. The fact that the relative denominations of these systems were interchangeable suggests that they were used to measure commodities that were flowing *between* each region, their inter-operability perhaps enforced by the standardized forms in which those commodities arrived. Such systems were suited to standardization within systems of 'quality control' for metals, precious stones and, perhaps bulk textiles, while other materials (liquids, food stuffs) may have been measured by volume rather than weight. The spread of weighing systems (and cylinder seals) not only reflects

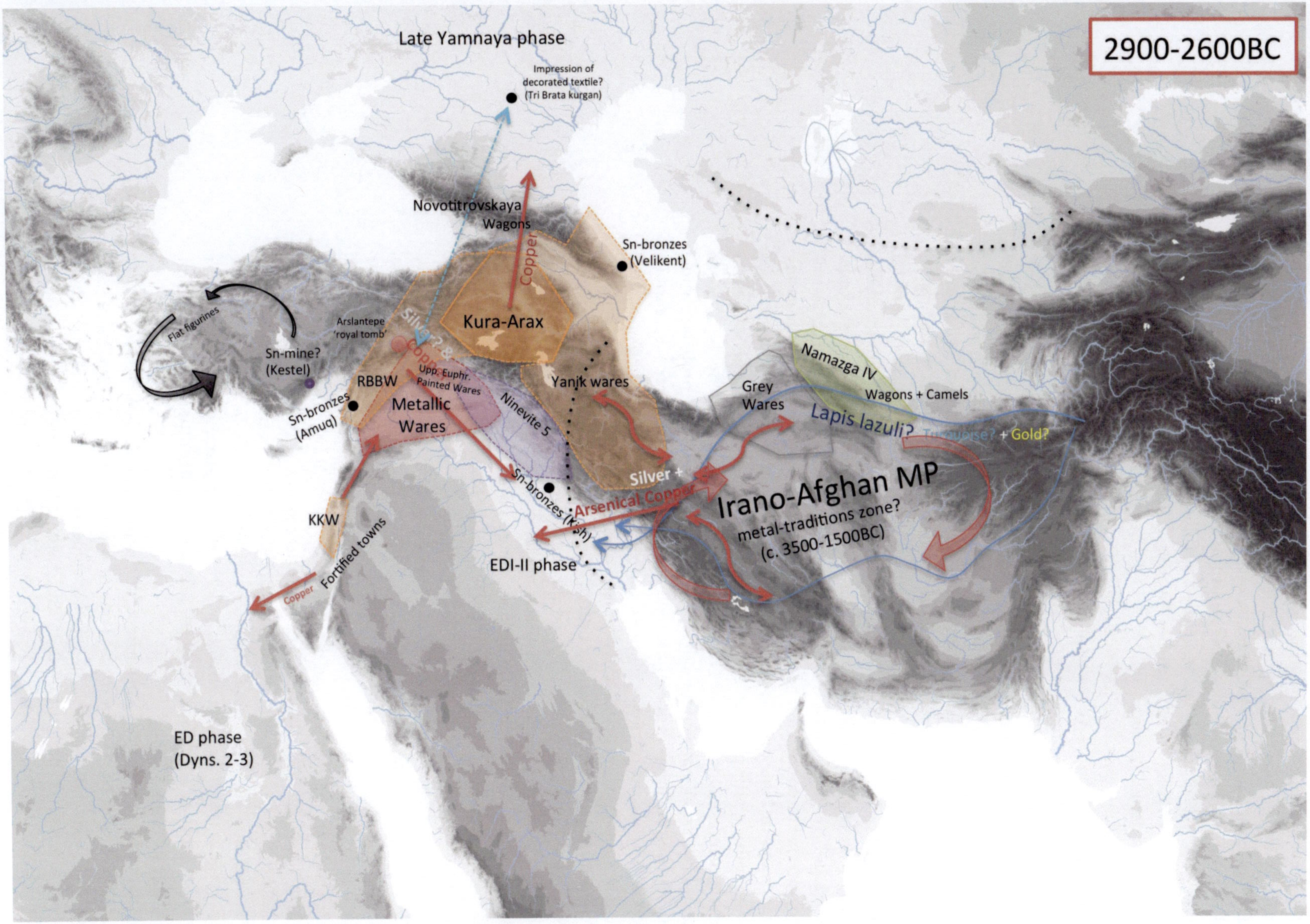

Figure 7.2. Summary of data on flows of stones, metals and textiles during the period 2900-2600BC.

the flow of materials themselves, but also of ideas about the abstract values of materials – and of forms of economic exchange. No clear evidence for weighing systems has been found in Transcaucasia or Central Asia at this time, or indeed until after the Bronze Age, unless you count the 'intercultural' stone pocketbooks (Figures 4.7c, 4.12f). If this is not simply a visibility issue (*i.e.* that weights were made in non-surviving forms) then this suggests the existence (in these regions) of alternative economic systems.

Various evidence points to the increasing variety of metals in circulation in this period (Section 5.2.5; 5.4.3), including sporadic experimentation with tin-bronze documented by the possible tin extraction at Kestel/Göltepe and Deh Hossein (Section 5.2.3) and increasing numbers of tin-bronze objects manufactured (*e.g.* the standing figurines from the Amuq, Section 6.5.4/Figure 6.30), albeit still a very minor proportion of the total (Section 5.4.3). The pottery repertoire of northern Syria and the southern part of the Upper Euphrates also begins to be dominated by undecorated assemblages of so-called 'metallic wares' from around 2700BC onwards (Section 5.6.3.i), apparently mimicking metal vessels or a metal aesthetic. It should be noted, however, that a new set of painted ware traditions also begins to emerge in the Upper Euphrates region at around the same time, after around 2800-2600BC (including Karababa, Gelinciktepe and Malatya-Elazığ wares). The connection between these wares and the perhaps slightly earlier painted wares of Ninevite 5 is unclear, though in all these cases the decorations are conceivably textile in origin (Section 6.6.1). Depictions of contemporary dress in Mesopotamia in this 'Early Dynastic period' are dominated by waist-down skirts, whose exact nature, though apparently fleece-like and probably wool-based, remains unclear – as does the extent of its adoption (Section 6.5.1). Various widely distributed fragmentary clues hint at the state of textile circulation: the change in spindle whorl styles at Arslantepe after 2800BC (connected, perhaps, to the Kura-Arax occupation at the site) has been argued to document a change to using wool-fibres (Section 6.4.3); and the fascinating but elusive impression of a mat or textile from a burial at Tri Brata, north of the Caucasus (Section 6.3.2) shows similar patterns to the Early Bronze painted pottery.

The most striking change in (and indeed beyond) Transcaucaisa/eastern Anatolia at the beginning of the 3rd millennium BC is the massive expansion of Kura-Arax (or at least Kura-Arax-inspired) material culture beyond its Transcaucasian 'homeland', first to the Upper Euphrates (Malatya-Elazığ) area and then from around 2800BC onward to the Amuq (RBBW) and southern Levant (the 'Khirbet Kerak' wares) in the south-west; to Daghestan in the north; and to north-west Iran (Yanik wares) in the east (Section 5.6.5). Earlier Kura-Arax pottery assemblages are augmented by red-black burnished styles and a greater degree of decoration by relief (Marro 2000). Meanwhile, portable fire-stands (or andirons) with possibly similar origins in Transcaucasia also appear even further afield in western Anatolia and the Aegean (see Rahmstorf 2010b, 272-277). This massive 'expansion' remains difficult to explain, but undoubtedly essential to understanding the whole region during this period.

By contrast, in western Central Asia, this period (particularly between 2800 and 2600BC) may be characterized as one of contraction or at least a far greater degree of localism on the basis of ceramic evidence. Painted pottery continues into the Namazga IV period (2700-2500BC), though of a different style (more checked 'geometric') and its motifs less widely shared than the preceding Namazga

III period. Its cross-craft origins still seem likely to lie with textiles or baskets, however (Section 6.6.3). Kirtcho (2009) has argued for a "transportation crisis" between 2750 and 2450BC in which travel was made difficult by an environmental deterioration that made cattle-driven wagons impractical. This "crisis" could only be passed once camels were instead harnessed for the job after 2600 or 2300BC. If the amount of lapis lazuli uncovered in Mesopotamian contexts between 2900 and 2600BC is any true reflection of the amount of lapis lazuli actually in circulation at the time, then the dip in finds in the early 3rd millennium noted by G. Herrmann (Section 4.3.1; Hermann 1968) fits with a picture of temporary localism or a similar 'mobility crisis' to the north-east of Mesopotamia.

It should be noted that, if current reconstructions are correct, the period between 3300 and 2700BC marked the period when the Caspian Sea was at its lowest – between c. 34-37m below sea level as opposed to today's c. 29m below (Section 1.2.2/Figure 1.6). Five to eight metres may not seem very much, but it is enough to submerge (and then destroy or hide) many archaeological sites in this region that would be very relevant to the situations in western Central Asia and eastern Anatolia/Transcaucasia – and also to any connections or boundaries between them along the north Iranian coastal plain (see Figure 1.6). It means that during the period between 3300 and 2700BC, there was a potentially very different landscape of occupation, with the coastline on the eastern side of Caspian perhaps located many hundreds of kilometres to the west.

7.2.3 2600-2300BC

Overall, the period 2600-2300 may be characterized as one of expansion and intensification of interaction across the entire Near East. Around 2600BC, in the Early Dynastic III period, the demand for a wider variety of metals (perhaps especially in Mesopotamia) seems to have increased: particularly gold and silver for ornaments and vessels (*cf.* Section 5.4.3). Silver seems to have taken on its long-term role as abstract mediator of value in exchange at around this time (Section 5.5.1). Tin-bronze also becomes increasingly popular in Mesopotamia during the period between 2600 and 2300BC, although arsenic-bronze still remained proportionately much more common in most regions (including Iran, Anatolia, the Caucasus and Central Asia) until the later second millennium (Section 5.4.3). 'Precious' stones like lapis lazuli appear to have become essential symbols of elite identity and display. The Royal Cemetery of Ur, with its extraordinary amounts of lapis lazuli and gold (including a tiara with leaves of a south Asian tree, Section 6.7) document an apparent intense connection with the east, perhaps via the Indus, since both commodities may have originated in Afghanistan (Section 5.2.3). Lapis lazuli apparently reached further west, as the no-doubt 'tip-of-the-iceberg' finds from Troy suggest (Section 4.3.1 and Figures 4.1a, *cf.* 4.2b). This period also witnesses the 'rise' of the Mature Harappan urban communities in the Indus – presumably with their own consumptive patterns – whose etched carnelian beads are also found as far west as the Aegean, though are mostly restricted to the Persian Gulf and southern Mesopotamia (Section 4.3.2).

The demand for larger amounts and greater variety of metals (and, indeed, for precious stones) may have created incentives and opportunities for opening new flows of materials. Between 2600-2300BC, we find the earliest 'intercultural style' vessels in Mesopotamia and the Gulf, whose origins seem to lie in the cultural

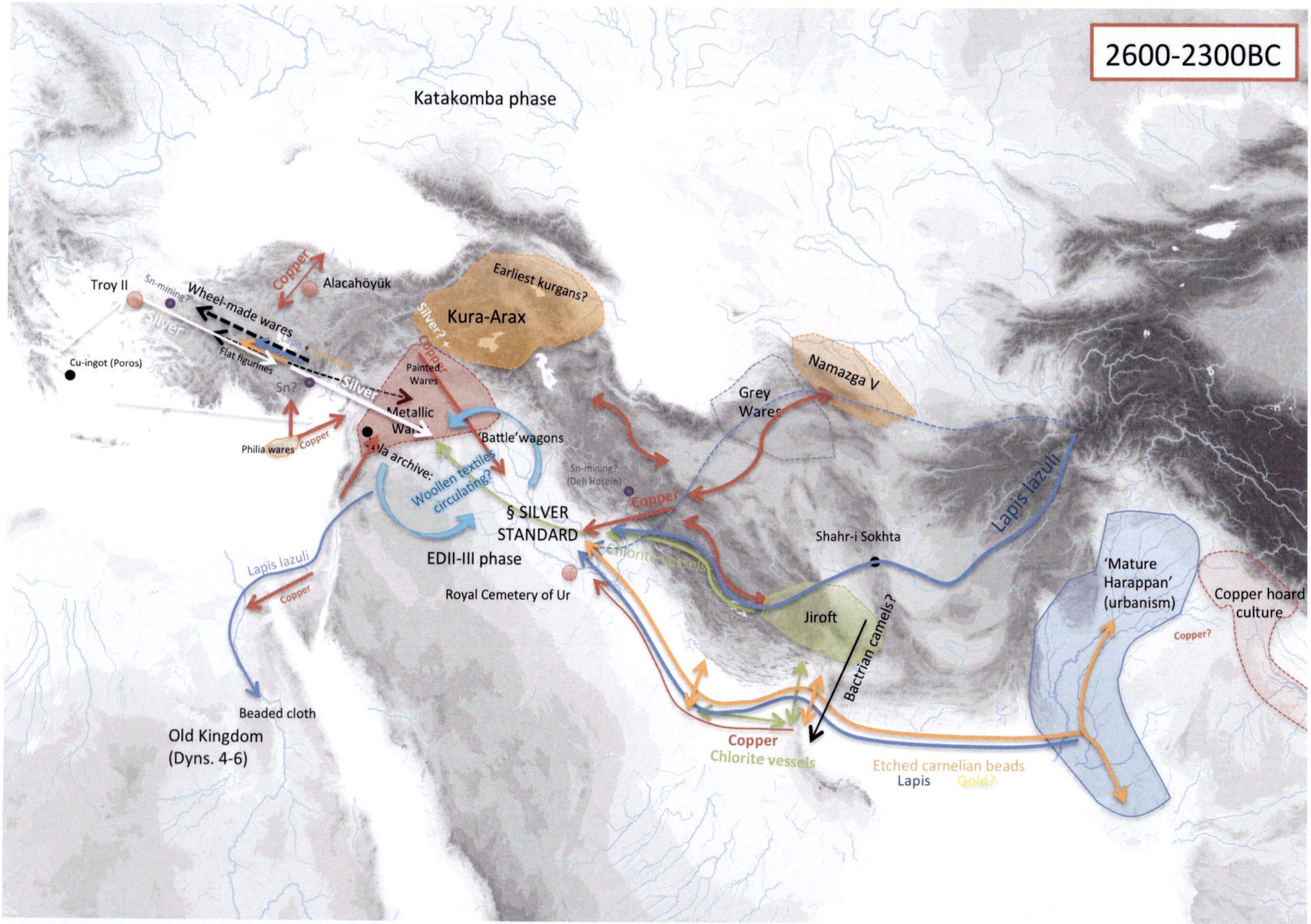

Figure 7.3. Summary of data on flows of stones, metals and textiles during the period 2600-2300BC.

milieu of south-east Iran (Section 4.4.1). Sail-based sea-transport may have played an increasing role in both the east Mediterranean and Persian Gulf, though its earliest introduction is difficult to date (Section 1.4.5). Meanwhile, in the west, the demand for silver may have been the driving force behind cultural transformations in western Anatolia, the Aegean and across the eastern Mediterranean more generally: here a much greater variety of metal objects is recorded than in previous periods, alongside the adoption of wheel-made pottery and assemblages with increasingly strong metallic aesthetics (Section 5.6.3.i). A 'Levantine' copper ingot from the site of Poros in the Aegean hints at the systematization of metal exchange networks developing during the early to mid-3rd millennium (Section 5.2.5; Day and Doonan 2007). With regards to metal skeuomorphs, it is worth noting the contents of the Alacahöyük 'royal' tombs in Central Anatolia (earlier referred to more generally in the context of patterns of metal deposition, Section 5.5.2 and Figures 5.35, 5.36). These graves contained a rich assortment of metal objects and wagon accoutrements and the black-burnished wares appearing beside similarly shaped metal vessels starkly illustrate an example of cross-craft interaction and the emerging centrality of metals. The exact dating of the Alacahöyük tombs remains unsettled, but they may date as far back as 2500BC. Meanwhile, the first 'depas amphikypellon' (the famous two-handled drinking cup) appears at a similar or slightly later time (~2500BC) in Anatolia, the Aegean and north-west Syria, suggesting the circulation not only of distinctive metal vessels (on which they are clearly based – given the metal examples, *e.g.* from Troy) but also, probably, the adoption of distinctive drinking customs perhaps involving wine (Section 5.6.3.i).

Wide interconnections based on elite dress, that cross-cut cultural boundaries, are suggested by the massive distribution and increased popularity of certain styles of metal ornaments, such as flat circular beads with central tubes and quadruple-spiral beads (Section 6.7.2). The elaboration of metal jewellery must relate to the use of lapis lazuli and carnelian beads for extravagant bodily adornment, and presumably point towards fashions in (elite) clothing or 'dress' that were highly transportable and highly desirable. The indications from textual material, for example from the Ebla archives, dating to around 2500-2300BC (Section 6.2.1) also show the rapidly increasing economic importance of sheep, wool and textile manufacture in Syro-Mesopotamia, which is no doubt related to this mobility of dress and jewellery.

From 2600BC, or perhaps a little before (~2700BC?), metals seem to start flowing through communities in western Central Asia in much larger amounts, if the greater variety of copper objects that have been uncovered from Namazga IV contexts in southern Turkmenistan provide any indication (Masson and Sarianidi 1972, 107, 120). Similarly the emergence of new pottery styles appears to represent a cross-craft signal for the rising priority of metals. First high quality undecorated wares start to be imported or produced locally more regularly, including 'grey wares' such as at Ak Depe and Kara Depe, at the western end of the Kopet Dag (Section 5.6.4). These styles appear to relate to the 'Eastern Grey Wares' that continue to dominate in north-east Iran. The 'intrusion' of grey wares and the synchrony with greater metallurgical sophistication suggest strongly that metals were now circulating differently between these two cultural zones. A large number of zoomorphic figurines have also been found at sites like Altyn Depe and Namazga Depe, and these include camels and model carts with camel heads, which supports the idea that camels were becoming (or had already become) an aid to transportation (Section 1.4.2; 1.4.3). By around 2500BC at the latest, pottery assemblages changed again, with the complete disappearance of Namazga IV style painted wares in favour of the plain wheel-made pottery of Namazga V (Section 5.6.4). That this pottery has a strong metal aesthetic seems indisputable, and it again supports the idea that metals (and possibly metal vessels) were flowing in much larger quantities through the exchange networks of this region. Though there are shared shapes, Namazga V wares differ from the 'grey wares' of north-east Iran, suggesting that, though the same or similar metal vessels may have been the model for pottery in both regions, cultural differences in the choice of metals used and in their colour treatment kept each region distinct. The extent to which other materials like lapis lazuli and other precious stones may have been circulating through this region remains unclear. Interestingly, the highest concentration of evidence for lapis lazuli working at this time – to be found further south at the site of Shahr-i Sokhta (Section 4.3.1) – does not lie on the 'cheapest' route of movement between Badakshan and Mesopotamia (see Figure 4.4).

Though poorly recorded contexts mean that it is still difficult to date precisely, 2600-2300BC seems to witness the replacement of a tradition of three-dimensional figurines ('Eneolithic types') by more schematic flat figurines ('Namazga V types') in Central Asia (Section 6.5.4). Curiously, whilst these figurines appear to have a local style of their own, the shift to similar hand-sized, more abstract or flat figurines is paralleled in other regions at around the same time or even a little earlier, including most clearly in Syria, eastern Iran and the Indus (with possible additional examples from Cyprus and western Anatolia). The mechanism by which this transition from 'three-dimensional' to 'flat' was transmitted over such large distances remains unclear but presumably it includes the movement of figurines made in materials which were either too valuable to be disposed of in the ground (metal), too physically unstable to have survived in the archaeological record (*e.g.* wood, wax) or, indeed, both. Moulds (from Sippar and Titriş Höyük) and actual metal figurines (*e.g.* from Tell Brak, Alacahöyük, Hasanoğlu, Tell al-Judeidah and İkiztepe) offer support for the idea of metal models, which were, perhaps, too symbolically charged to be placed in the ground. The manufacture and consumption of these various figurines needs closer examination, but it seems likely that their widespread popularity at this time documents some shared ritual practices or values placed on such images – itself an indication of the depth of interaction.

Meanwhile, in a reversal of the previous period, the full distribution of Kura-Arax-related wares contracts markedly after 2600BC: the distinctive red-black assemblages slowly disappear from areas such as the southern Levant, the Upper Euphrates and north-west Iran. Dark and red-black burnished wares (now with greater concentration on incised decorations) continue to be manufactured in Transcaucasia and some parts of eastern Anatolia, at sites like Sos Höyük, perhaps well into the 2nd millennium (Section 5.6.5). At some point between 2600 and 2000BC more varied decorated pottery types start to appear (including Bedeni and Martkopi types, associated mostly with kurgan burials). What the 'contraction' or disappearance of the distinctive red-black burnished pottery and associated material culture represents in social terms remains difficult to explain. Certainly, whatever mechanism was the original driving force behind the original 'expansion' must have disappeared in this time.

7.2.4 2300-2000BC

The Akkadian domination of Syro-Mesopotamia (2300-2200BC) and the subsequent emergence of the Ur III dynasty (2100-2000BC) represent periods of attempted political unification, with significant cultural consequences, but to what extent these political and cultural changes affected the flows of materials is difficult to judge with precision. It has been noted that amounts of lapis lazuli uncovered from the Akkadian period are considerably less than the previous ones (Section 4.3.1), but it is likely that this is simply an artefact of differential visibility (fewer rich burials in Mesopotamia) or indeed of uneven archaeological investigation.

The Kopet Dag sites of Namazga Depe and Altyn Depe in Turkmenistan reach their greatest size during the mid-to-late Namazga V period (the local 'Middle Bronze'), at the latest around 2400-2100BC, with monumental ('ziggurat-like') and public architecture, accompanied by considerable evidence for specialized industry (of both metal and pottery), regular seal usage and increased social differentiation documented in small but specialized burials (Section 5.6.4). All of this implies some kind of proto-'urban' status for these communities. In the past this was often attributed to rising 'Mesopotamian' influence – *ex occidente lux* – though similarities in pottery shapes between Central Asia and the Indus region (a continuation of the north-south orientation of cultural affiliation and, by implication, routes of movement), and appearance of intercultural vessels in Central Asia, indicate a much more complex situation. The period between 2300-2000BC may well be the peak of the so-called 'Middle Asian Interaction Sphere' (Possehl 2002) or the '*l'age d'échanges inter-iraniens*' (Amiet 1986). By the end of the third millennium, however, the towns along the Kopet Dag piedmont contracted in size or were, in some instances, completely abandoned. The expansion of settlement evidence from the Murghab delta at around the same time or soon after 2000BC has sometimes been taken to imply a migration into this 'new' territory, but the mechanisms and reasons for these settlement shifts remain unclear. The location of the Murghab and Bactrian regions between Western Asia and the major tin sources of the Zerafshan has been suggested as a possible reason for the emergence, here, of these new 'BMAC' communities (Section 5.6.4), though there is so far no direct evidence for tin mining at this early time, nor anything to connect later mining activities with BMAC material culture. The appearance of certain specialized knives in north-east Iran and (soon after) in the burials of the Sumbar valley may suggest the production of 'pile'-based textiles (*i.e.* pile carpet) from around 2300BC onward (Section 6.4.5).

In Transcaucasia, the widespread construction of kurgan burials of the Martkopi and Bedeni types (which may have begun earlier, perhaps even as early as 2600-2500BC) indicate zones in which considerable material wealth and presumably power were accumulated but periodically sacrificed in dramatic funerary performances which included the deposition of metal vessels and other objects (Sections 5.5.2, 5.3.1, Figure 5.13). Though Kura-Arax-like pottery continues to be produced in some locations, new monochrome painted varieties are found in the kurgans. It is also possible that early monochrome 'Transcaucasian Painted Wares' (such as the Trialeti-type) began to appear in certain areas during this period (Section 6.6.2). It is therefore worthwhile noting the emergence of other painted wares (such as the Cappadocian, Alishar III *etc.*) in Central Anatolia

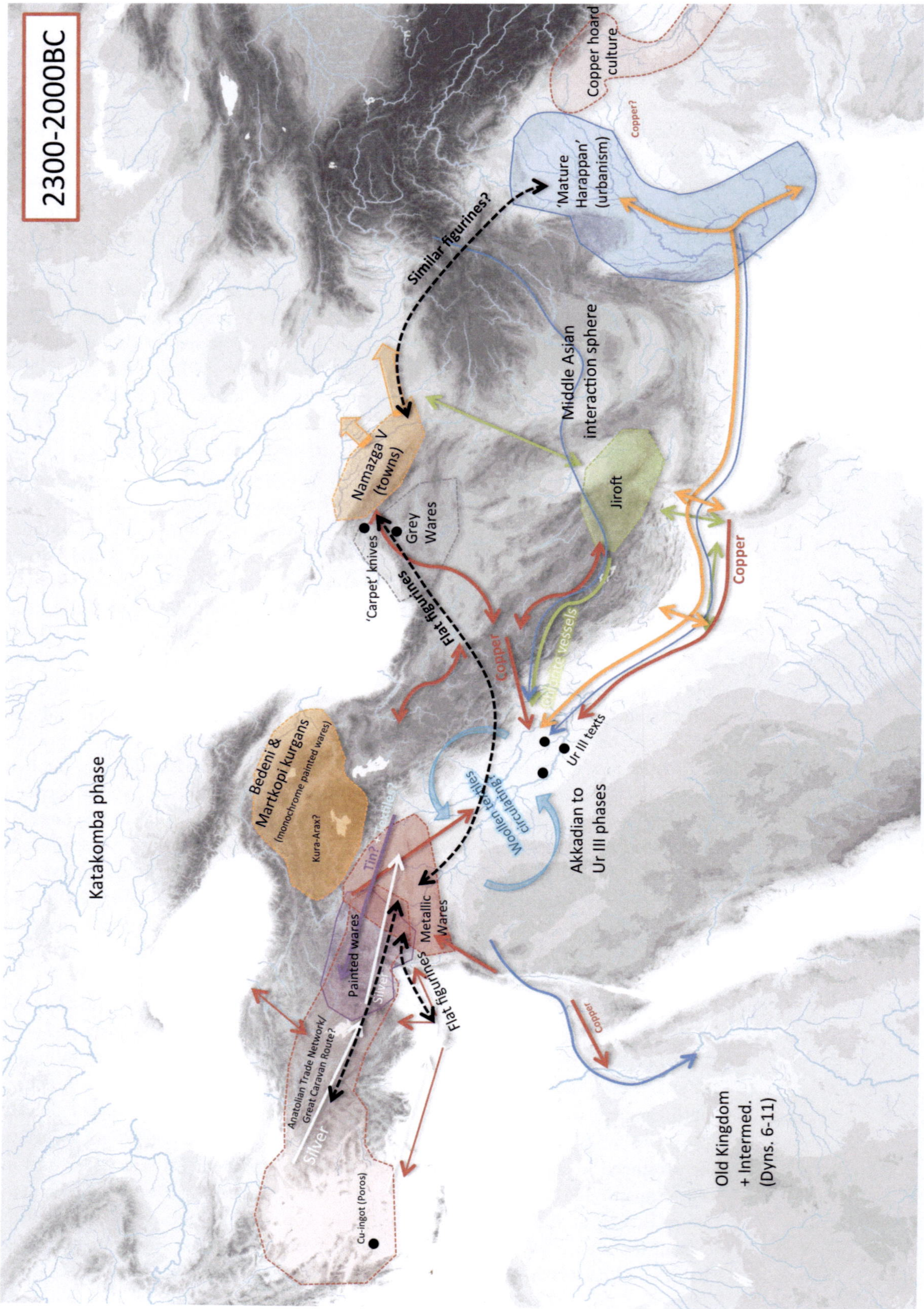

Figure 7.4. Summary of data on flows of stones, metals and textiles during the period 2300-2000BC.

around 2100-2000BC, whose balance of correspondences and variation and the nature of the patterns which seem inspired by netting or textiles (Section 6.6.1; Figure 6.38) may index a textile exchange network that predates the emergence of the 'Old Assyrian Trading Colonies' in the same area of central Anatolia. The expansion of textile production and exchange in Syro-Mesopotamia toward the end of this period is documented by the extraordinarily organized wool-textile production system recorded in the Ur III texts (Section 6.2.2). Mesopotamian figural sculpture shows clothing styles with off-the-shoulder fleeces of differing kinds, but it is difficult to say how representative such depictions are (Section 6.5.1). Even if we have little idea of what the final products described in the Ur III texts looked like, we can assume that by this time the most valuable high quality pieces were being transported over very large distances.

Meanwhile, in western Anatolia and the Aegean, dark metal-inspired pottery types predominate and the uniformity of pottery repertoire from Cilicia to the Aegean, documents what Vasıf Şahoğlu has called the 'Anatolian Trade Network' along what Turan Efe has labelled the 'Great Caravan Route' (Section 5.6.3.i). From one of the defining types of the 'Anatolian Trade Network', namely the 'Syrian bottle', both metal and pottery versions have been uncovered. Despite a distribution mainly restricted to Syria and central and western Anatolia, a fragment was also found at Sos Höyük in eastern Anatolia. The representation of a net bag around some pottery examples suggests that, besides the type's status as proxy of the movement of certain 'metal' containers, the vessels also played a role in the transportation of some valuable but currently unknown organic product (*e.g.* oil, alcohol, perfume *etc.*).

7.2.6 2000-1700BC

The period following the end of the 3rd millennium BC marks a set of apparently important transitions in both Transcaucasia/eastern Anatolia and western Central Asia. It is also the period (between 1950 and 1750BC) in which we have a remarkably detailed textual record of the workings of the workings of one large and organized trading system, that of the 'Old Assyrian Trading Colonies'. The records of this system, gleaned mostly from tablets found at Kültepe in Anatolia, provide positive evidence for the transportation and exchange of massive quantities both of textiles of diverse qualities and types (Section 6.2.3) and of metals including tin, copper and silver (Section 5.5.1) between Central Anatolia and northern Mesopotamia. More specifically merchants exchanged tin they had obtained in Aššur (apparently sourced originaly from some even more 'eastern' origin) for silver (whose ultimate origins are unclear) in trade centres across Central Anatolia. As Barjamovic (2011) has argued, this system must have required an extraordinary level of infrastructure (roads, inns, social systems protecting traders, credit *etc.*), and yet the material footprint (apart from the texts themselves) is almost unknown. This implies that similar systems may have been in place at other times, or in other places – including , perhaps, central Asia where there were similar levels of settlement complexity at this time.

In western Central Asia, in the Murghab delta and the Bactrian plain, this period witnesses the establishment of the 'Bactro-Margiana Archaeological Complex'/BMAC (Section 1.3.4; 5.6.4), much of the basic pottery material growing out of the Namazga V assemblages, but with a distinctive combination of

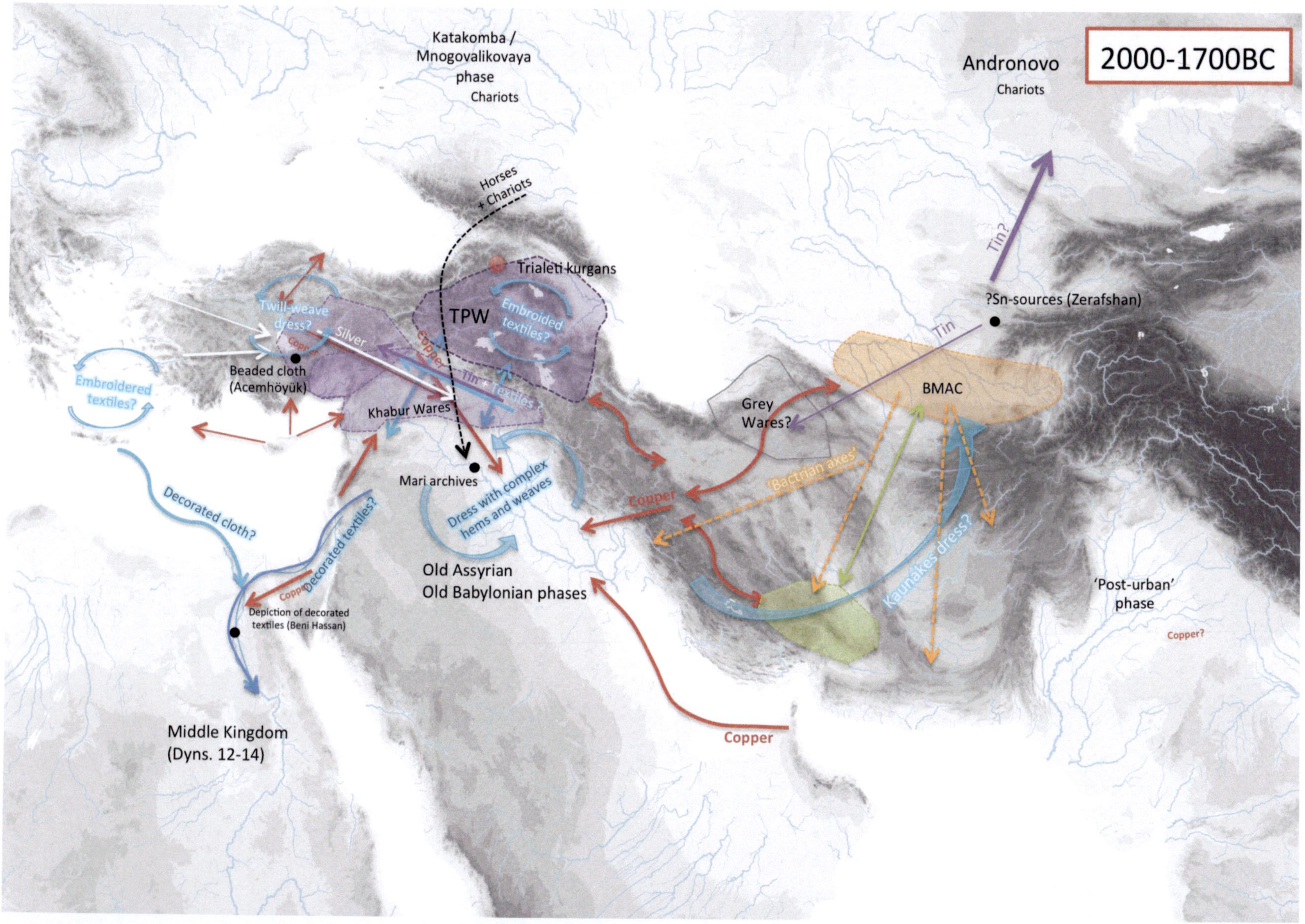

Figure 7.5. Summary of data on flows of stones, metals and textiles during the period 2000-1700BC.

iconography and crafts which suggests links to Elam and south-west Iran in some instances, and with the Indus in others. Many sites appear to have been fairly heavily fortified, or at least included strongly enclosed monumental buildings, in contrast to the earlier Namazga V towns along the Kopet Dag. The reasons for the emergence of this 'complex', much of whose material has sadly only come to light through uncontrolled or illicit excavations of burials, remain open. Greater detail about the links between local settlements within the region (*e.g.* along canals and fluvial corridors in the Murghab) might help us to unpick this process better in the future and help explain material circulation.

The direction of exchange of 'BMAC'-related objects is not always clear: BMAC miniature vials may relate to the 'intercultural-style' tradition of stone vessel manufacture in south-east Iran (Section 4.4.1; Figure 4.7f); as might the various shared iconographic schemes, such as those seen on the very similar metal vessels from 'Persepolis' and Gonur Depe (Section 6.5.3; Figure 6.27) and the various types of stone figurines (Section 4.4.2; 6.5.3). Interestingly, metal vessels decorated with figures are popular in both this 'trans-Iranian' zone and to the north-west in the Caucasus, in the form of vessels deposited in the Trialeti kurgans which also have parallels in the glyptic of Old Assyrian cylinder seals (Rubinson 2003) (Section 5.4.3). We can presume that both sets of vessels represent a tiny fraction of the original number of such items. Despite differing in exact style of execution, they nonetheless suggest the circulation of motifs and techniques designed to produce arresting drinking accoutrements for some inter-related social or ritual purposes. So-called 'Bactrian' axes – presumably manufactured in Central Asia – are found over an extremely wide area, often alongside other 'BMAC' material including, surprisingly, pottery vessels, at sites like Shahdad (Hakemi 1997) and Susa (Section 5.3.4). These items signal sharing of myths, rituals and clothing fashions, but the mechanism behind this circulation remains uncertain. The apparently intensified connections across the Iranian plateau may been balanced by a reduction in the importance of sea-routes in the Persian Gulf, if the 'collapse', or rather 'de-urbanization', of the Harappan/Indus civilization from around 1900BC can be used as any kind of indicator.

In Transcaucasia around or just before 2000BC, the kurgan traditions of the Bedeni and Martkopi are superceded by those of the Trialeti (Section 1.3.3; 5.3.1). Metal vessels from Trialeti burials have already been mentioned, but these kurgan assemblages also contain new painted pottery types and very rich metal assemblages. Next to this mortuary equipment, we have very little evidence about contemporary or related settlement. At a similar time, new ceramic assemblages emerge in Transcaucasia and eastern Anatolia, such as the polychrome 'Transcaucasian Painted wares' (Sevan-Uzerlik/Van-Urmia/Kizilvank *etc.*). The patterns on these vessels, as for many other painted wares emerging in the wider region of this time (including the Khabur and Levantine painted wares in northern Syria and the Levant respectively, and the Middle Bronze painted wares of Central Anatolia), seem to be inspired by textiles or other woven media (Section 6.6.2; 6.6.1). The diversity of textiles in circulation from Syria and beyond can be glimpsed through a range of evidence including: the variety of depictions in Assyrian seal impressions (Section 6.5.2; Figures 6.25); the fragmentary remains of a patterned, beaded dress at the palace of Acemhöyük (Section 6.7.2; Figure 6.52); the wall-paintings from Mari which may reflect tapestry or wall-hanging traditions (Section 6.6.1; Figure 6.34); and, more distantly, the colourful depictions of 'foreigners' at the Egyptian cemetery at Beni Hasan (Section 6.6.2). Aegean pottery of this period carries many painted patterns whose potential textile origins have often been noted (Section 6.6.1). Evidence for the production of twill – a weaving style that produces a distinctive visual effect – also comes in the form of specialized 'crescent-shaped' loom-weights in central and western Anatolia (Section 6.4.4; Figures 6.18; 6.19), apparently becoming particularly popular in this period. The slow rise of painted wares, and their apparent flourishing from around 2100-1900BC onward across the Near East, may reflect the increasing role of textiles as an important medium for the representation of identities: perhaps reflected in traditions of using the hems of dress as an alternative for seals in contracts which might date back to this period (Section 6.6.2). Depictions of Mesopotamian dress from this period show far fewer 'fleeces' and greater complexity in the edging of fabrics. Fleeces, of the '*kaunakes*'-type do continue to appear in the iconography of BMAC figurines and metal vessels, however (Section 6.5.3), perhaps as an archaism that indicated their special role in conservative rituals.

7.2.7 1700-1400BC

The period following 1700BC sees the apparent expansion of interaction networks in the eastern Mediterranean arena, with the growth of the Minoan and subsequent Mycenaean centres in the Aegean, the growth of the Hittite and Mitanni empires in Central and south-east Anatolia respectively, and an apparently increasing integration of Egypt into the eastern Mediterranean sphere during the 18th dynasty period (c. 1550-1300BC). This contrasts with the Persian Gulf trade, especially following the 'collapse' or decline of the Harappan settlements in the Indus, when the flow of items such as etched carnelian beads, intercultural vessels and perhaps also lapis lazuli to the west either stopped or became archaeologically invisible. To what extent this picture represents a real decrease in flows, or merely a transformation of the kinds of evidence available to us to measure these flows, is an open question. Certainly, new materials, such as faience (perhaps first invented as an artificial alternative to lapis lazuli as early as the 4th millennium BC) or amber from the Baltic (perhaps an alternative orange stone to carnelian), may have played an increasing role in substituting expensive desirable materials: but they may have augmented the availability of alternatives and increasing the complexity of social differentiation rather than lessening the actual flows of the original materials. New traditions of metallurgy and metalworking emerged around 1600-1400BC, as – according to Chernykh – the unity of the Circumpontic Metallurgical Province was broken by 3 new entities: the 'Eurasian', 'Irano-Anatolian' and 'Caucasian' Metallurgical Provinces (Section 5.4.1).

In part due to the nature of the evidence, which increasingly comes to rely on mortuary contexts (and thus the resultant uncertainty of chronologies), it becomes very difficult to accurately describe the developmental trajectories of communities in both eastern Anatolia and western Central Asia after 1700BC. In Transcaucasia, eastern Anatolia and north-west Iran, the polychrome 'Transcaucasian Painted wares' continue to be produced, although more detail from sequences such as Kyultepe-Nakhchevan need examination to understand the development and significance of these assemblages, since much of the evidence until recently has come from uncontrolled excavations. Just to the south of the area of TPW wares, the 'Nuzi' pottery assemblage, whose distribution appears to stretch across the

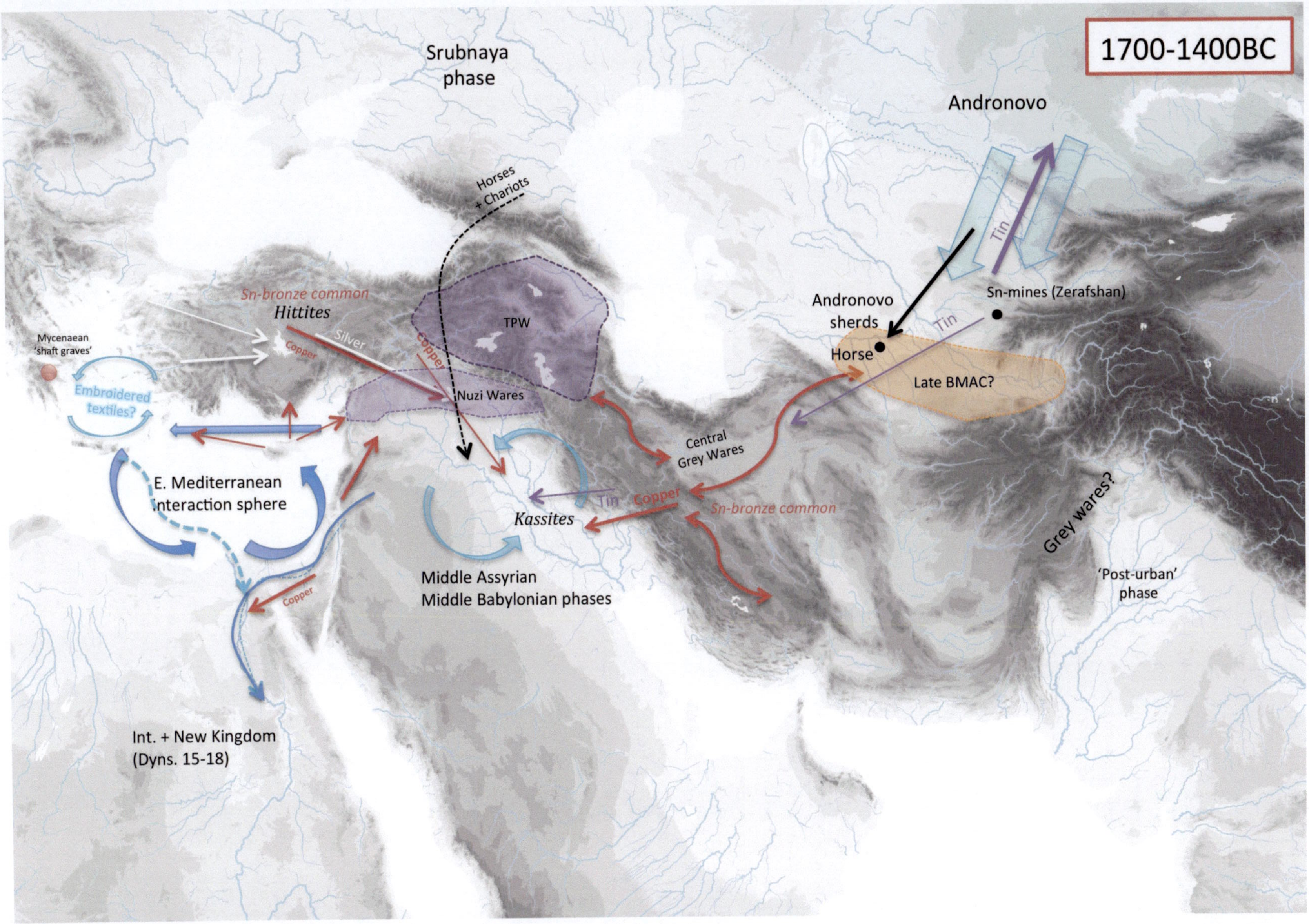

Figure 7.6. Summary of data on flows of stones, metals and textiles during the period 1700-1400BC.

plain of northern Syria, appears to continue a tradition of referencing textile designs (albeit very different ones) from the earlier Khabur wares (Section 6.6.1). Wall-paintings from Nuzi may also reflect traditions of tapestry or wall-hangings (Figure 6.34a), relating back to those of Mari. Wall-paintings have also been identified in the Aegean (for example at Akrotiri on the island of Thera), hinting at an increasing circulation of such ideas, perhaps through the circulation of tapestries themselves (Section 6.6.1). In Central Asia, in the latest occupation levels of BMAC sites, there have been a number of finds of the very rough pottery associated with Andronovo assemblages. The Andronovo culture had been developing since the beginning of the 2nd millennium in the Eurasian steppes to the north of the Karakum and Kyzylkum deserts. Though few, these fragments do seem to document new or intensifying north-south connections over gigantic distances and difficult landscapes. It is, perhaps, no coincidence that the earliest unequivocal evidence for mining in the Zerafshan region comes from deposits of Andronovo pottery between 1700 and 1400BC (Section 5.2.3). There may be a relationship between these new connections and the emergence of greater mobility on the steppes in the form of the fast chariot and horsemanship (Section 1.4.4), but though the earliest horse remains for this region come from a BMAC grave from Gonur Depe, there is very little direct evidence of horses used for transportation or to pull chariots in western Central Asia at this time, despite the appearance of both horse and chariot in contemporary Mesopotamia. The southward encroachment of Andronovo material culture may instead relate to a demand for tin from northern Eurasia, perhaps attempting to tap into Near East sources of metal in Afghanistan and beyond. It is not necessary to drag in mounted or chariot-driving marauding 'Indo-Aryans' from the steppes to 'explain' the disappearance of both the BMAC/Oxus and Indus civilizations and the emergence of certain 'grey wares' across Central Asia, Iran and the Indus; but it is possible that the transformations of this period also involved the movement of languages (and of course language-carrying agents) in the context of a new mobile social environment.

7.2.8 Subsequent trends after 1400BC

Though falling outside of the original remit of this book, it is worth briefly reviewing a few items of evidence from after 1400BC that have been mentioned as comparative material, to reflect on subsequent developments in our case-study areas. The period 1400-1200BC may be characterized as a 'high-point' of interconnection between Late Bronze eastern Mediterranean and Mesopotamian societies. Considerable evidence for interaction comes from the eastern Mediterranean in the form of the wrecks, such as the Uluburun; texts describing treaties and battles between Hittites, Egyptians and Syrians (Mitanni, Hurrians *etc.*); and various archaeological connections between constituent regions. Decorated textiles (from the Levant?) were apparently imported into Egypt during the Amarna period (1350-1320BC), as illustrated by Tutankhamun's wardrobe (Section 6.3.2; Figure 6.11); and direct evidence for richly decorated textiles also comes from Syria itself, such as the remains of purple murex-dyed cloth in the royal tomb at Qatna (Section 6.4.2). Textual evidence (from Linear B tablets) hint at textile 'factories' in the Aegean during this period, and the elaborate textiles depicted in Aegean wall-paintings, as well as similar wall-paintings at coastal sites

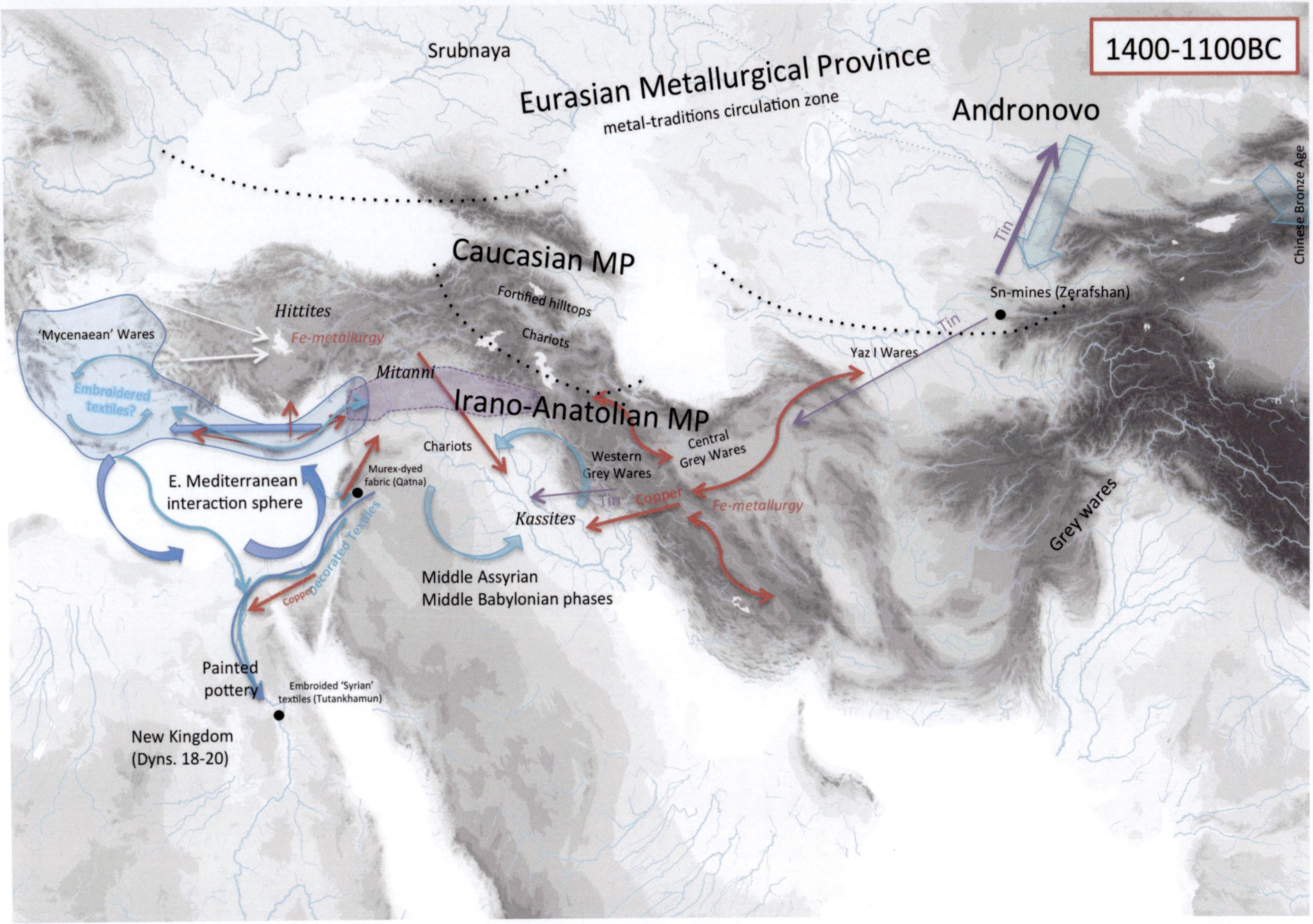

Figure 7.7. Summary of broad trends in flows of stones, metals and textiles for the period 1400-1100BC.

in the Levant and Egypt at this time suggest a textile or tapestry-based *koine* (Section 6.6.1). The Uluburun wreck (apparently sunk before 1300BC near the coast of Lycia) contained metal ingots, both copper and tin, providing a snapshot of the method of circulation of metals at this time (Section 6.2.2). The archaeology of Mesopotamia during the 'Middle Assyrian' and 'Middle Babylonian' period is rather less synthesized than its political history: the discovery of detailed texts has resulted in less archaeological effort devoted to understanding the period in material terms, including exchange relationships. This problem is not limited to Mesopotamia, of course, since the discovery of texts at Hittite sites in central Anatolia has also led to a reduced focus on archaeological material and greater reliance on textual sources.

In part a result of this archaeological blind-spot in the 'core', a similar degree of uncertainty is to be found in our case-study areas. In the centuries following 1400BC, Transcaucasia appears to provide an arena for new fortified hilltop sites (see *e.g.* Ristvet *et al.* 2012), which may have lain the foundations for social and economic structures of the 'Urartu' states in the first half of the 1st millennium BC. Aside from the impressive metal objects from graves of the Koban culture in northern Georgia (Chernykh 1992: 283-290), however, detailed material evidence from late 2nd millennium eastern Anatolia and Transcaucasia is still poorly researched and published. A parallel paucity of archaeological research on sites dating to this period in Central Asia means that we are similarly 'in the dark' about social processes and material flows on the other side of the Near East. Discussions remain as to whether there was a hiatus between the end of the BMAC/Namazga sequence, and the following Early Iron/Yaz sequences. At the very least, it seems the settlement patterns of the later period had changed dramatically (Early Iron Age sites are apparently founded newly away from Late Bronze predecessors), though the degree of cultural continuity may be higher than often assumed. The pottery assemblages in southern Turkmenistan are mostly aligned to that of the 'Yaz Depe' sequence with various plain and painted grey wares (see *e.g.* Boucharlat *et al.* 2005) .

7.3 Patterns and processes

The above chronological narrative is, of course, not a complete picture of the indicators of interaction across the entire Near East from 3200 to 1400BC. The focus has been on placing those threads of information that have been picked up in the material chapters on stones, metals and textiles into a chronological framework, with a particular focus on the two case-study regions. With more time, other potential categories of material could have been included which could contribute to this tapestry (for example foodstuffs, bitumen objects, pottery shape traditions), and greater detail could have been drawn from other regions outside of our case-studies, particularly Mesopotamia, Iran, the Indus, the Levant, the Arabian peninsula, Egypt and the Eurasian steppes. However, it should be clear that even were these additional sets of information included, the extent and quality of different datasets, and our knowledge of different regions in different periods, would remain extremely uneven. This is partly a factor of the differential survival of evidence (for example, the preponderance of mortuary evidence in the Caucasus and Central Asia between 2000 and 1400BC), but it is also a result of different research traditions for different times and places (for example, where textual records outshine the more equivocal material remains, such as during the Old Assyrian Colonies period in Anatolia). Prehistorians have been more likely to address questions of long-term, long-distance interaction than those studying historical eras, who are instead faced with the documents of detailed political changes or quantified records of exchange or production. One of the great challenges of macro-scale research into such 'cross-over' periods is precisely how to deal with these differences of evidential quality and differences of tradition. Greater focus on the material evidence is undoubtedly a large part of the answer to this, allowing more direct comparisons between 'text-less' and 'text-ful' periods. What then can we make of this selective slice through the archaeology of the period from 3000 to 1500BC presented in this book? How might it help us to understand both the trajectories of interaction in and beyond our case-study regions?

7.3.1 Transcaucasia: the Kura-Arax phenomenon

One of the most important phenomena to demand explanation in the Transcaucasian/eastern Anatolian region is, of course, the Kura-Arax or Early Trans-Caucasian cultures, and in particular its rapid expansion around 3000-2800BC and subsequent contraction a couple of centuries later (around 2600BC), with an apparently long continuation afterwards in central Transcaucasia itself. It seems very likely that the process of expansion must have involved the movement of individuals and communities, as, indeed, proponents of one of the popular explicatory paradigms have argued (Batiuk and Rothman 2007). Studies of the functional composition of the southern Levantine variant, Khirbet Kerak ware, at Tel Bet Yerah has shown an emphasis on food (or drink) consumption rather than storage in the overall repertoire (Greenberg 2007). Since food is often a central component of migrant or diaspora identity, this has been taken as positive evidence for intense migration (Paz 2009). The close similarities in technical production of the burnished wares between Transcaucasian, Upper Euphrates, Amuq and southern Levantine examples also support a very small number of technical transfers between regions (*i.e.* closer relationship between the potters) and thus increase the argument for the immigration of potters, rather than the emulation of a desirable style (Iserlis *et al.* 2010).

'Migration' on its own can only be an explanation for change, however, if one assumes that the rest of the time people are stationary. This is an unlikely scenario, and instead such change must be a matter of the intensity and nature of human traffic. It would be unrealistic to deny the importance of the flow of people in this process of Kura-Arax expansion. But the extent, motivations and mechanisms of such a movement of people remain obscure. Batiuk and Rothman have argued for a 'rippled' process of migration in which 'push' factors in Transcaucasia/eastern Anatolia – environmental deterioration, social systems of inheritance which encourage dispossessed children to set out to found new communities – are balanced by 'pull' factors in the destination zones (the Upper Euphrates, Amuq, Levant and north-west Iran) where, they argue, there was suitable/available *lebensraum* along previously extant 'information paths' (Batiuk 2005) – corridors which were presumably created through earlier metal or obsidian exchange routes. This scenario is not entirely implausible, though the evidence for environmental deterioration, suitable social systems or a recent history of Caucasian-Levantine information/material flows is rather lacking (so to the connection with north-west Iran). It also does not adequately explain the timing of the expansion: Proto and Early Kura-Arax communities were present in Transcaucasia since around the mid 4th millennium: what could have made them suddenly more mobile? The Transcaucasian expansion process has sometimes been portrayed as one of 'decline' or 'relapse' to a simpler form of life in the Upper Euphrates and the Amuq, as though the peak of urbanism and highly organized agricultural economy of the Uruk period was simply replaced (or pushed out) by the more 'primitive' non-urban pastoralist economy of the Kura-Arax culture – in order to 'fill the vacuum' left behind. This passive (de-)evolutionary perspective seems not to reflect the complex economy of 4th and 3rd millennium Kura-Arax communities. 'Pastoralism' in the sense of herding animals including sheep, goats and cattle appears to have been a successful strategy in the highlands and steppes of eastern Anatolia and Transcaucasia, since the later Neolithic to the present day, especially in those regions with limited area for crop-based agriculture and when forests may have been much more extensive. However a roaming pastoral economy seems never to have been an exclusive one (see, *e.g.* Piro 2009). There is certainly little direct evidence that the Kura-Arax communities were fully 'nomadic' in the sense of later nomadic pastoralists such as the Bedouin or Central Asian Mongols: permanent settlements are well documented until the rise of the kurgans after 2500-2300BC (and many may have continued after this time). The distinctive handles of Kura-Arax vessels (Palumbi 2010) may indicate a desire for greater mobility (*e.g.* as opposed to contemporary Syrian vessels and cups without such features, or indeed the vessels of the subsequent TPW wares), but continued analysis has supported the idea of local production of these vessels (Batiuk 2005; Iserlis *et al.* 2010). The pottery *style* and its *technique* of manufacture (particularly the surface treatment), however, was obviously highly mobile. It is possible that additional mobility was provided to Kura-Arax groups by the integration of the horse at the beginning of the 3rd millennium (Section 1.4.1) – though there is limited direct evidence of this, except for the inclusion of horse bones in certain Kura-Arax assemblages. Wagons or carts seem likely to have also played a role in the creation of a more geographically extended 'Kura-Arax' network (Sagona 2013). The limited Kura-Arax burial evidence curtails our access to relevant iconography or large items such as carts. However, suggestive is the presence of a very large number of vehicles in burials of the latest Maikop phase ('Novotitrovskaya'), in an area to just the north of Transcaucasia, around the period of the Kura-Arax expansion (Section 1.4.3).

But what were the motivations behind the desire for greater mobility? One potential hypothesis is the role of Kura-Arax communities in the circulation of metals. The 'expansion period' Kura-Arax pottery has a much more 'metallic' aesthetic than the earlier types (with use of relief and some incised decoration), particularly some variants of the KKW, Yanik and Red-Black Burnished Wares, which gives stronger weight to the idea that the (social-) economy of its users during this period increasingly relied upon metals (as suggested in Section 5.6.5). It may be that the handles and lugs of Kura-Arax wares – which might be taken to support the desirability of the pottery's mobility – might alternatively represent a cross-craft skeuomorph from another medium: for example, wood, leather or metal[130]. A connection with mining of minerals more generally is also suggested by the extensive Kura-Arax presence at the salt deposits of Duzdaği (Section 5.6.5, fn. 35), although for what purpose (*e.g.* food flavouring, preserving meat or skins, care of sheep herds or processing of metals or other materials) remains unclear.

The geographical distribution of Kura-Arax pottery is interesting: plotting this distribution onto an 'accessibility' archaeotopogram (*i.e.* type A2) of regions with substantial copper sources shows the densest distribution of sites clustering just outside those areas with the best access to mining zones (Figure 7.8). Whilst such all such maps should be treated with caution since they may be a result of various data biases (*e.g.* sites are more easily recognized in the valleys than in mountainous mining areas), if we assume they are broadly representative, then these clusters are situated at junctures between the sources and routes towards probable major markets: between Transcaucasia/eastern Anatolia and Mesopotamia (in the south) or the steppes (in the north); between eastern Mediterranean (Cyprus?) and Syro-Mesopotamia; and between the Feinan sources and Syria. It is possible then that Kura-Arax communities were expert 'middle-men' with detailed knowledge of metallurgy, perhaps built into oral lore that simultaneously enforced a strong distinctive identity, particularly rituals of craft manufacture and food traditions.

The contraction of Kura-Arax-related wares is as difficult to explain as its expansion. The disappearance of Khirbet Kerak ware in the southern Levant has recently been characterized as the assimilation of the original migrants into the local population (Paz 2009), for example. Batuik speculates whether there might have been a 'return migration process' as whatever incentives had brought Transcaucasians to the Levant and Amuq compelled them now to return. Certainly it seems likely that explanations for 'expansion' and 'contraction' are interrelated: some structured socio-economic mechanism must have encouraged relatively conservative communities developing interactively though slowly in Transcaucasia and eastern Anatolia to temporarily reach out far beyond their traditional geographical borders. If this mechanism was related to the flow of metals, it is clear that the Kura-Arax expansion appears not have been aimed at *direct* access or control to new mines: no Khirbet Kerak Ware has been found in the copper sources of Feinan in the southern Levant, and north-west Iran does not seem to have been a major source of copper ores. Control of intermediate connections, however, seems more likely. The Yanik-based expansion into north-west Iran could have been orientated precisely toward the control of central Iranian sources of copper into the north. Central Iran had been an important source of copper since at least the 5th millennium BC as documented by finds at Arisman/Veshnoveh, Sialk, Tepe Hissar *etc.* (Section 5.2.2), and perhaps served Mesopotamian consumers with arsenic-copper for millennia (Mathews and Fazeli 2004). Perhaps then, the Kura-Arax expansion was related to a short-lived shift in the density of connections (and the source of metals) between 2800 and 2600BC during which eastern Anatolian sources of metal and eastern Anatolian control over other sources became temporarily ascendent, before Mesopotamian consumers started to take advantage of Omani copper after 2600BC via new sea-based routes in the Persian Gulf.

---

130 By way of comparison, the presence of handles on EBA pottery from western Anatolia is not normally taken to represent pottery mobility here, but is more likely to be inherited from pouring and holding techniques of consumption from original metal prototypes.

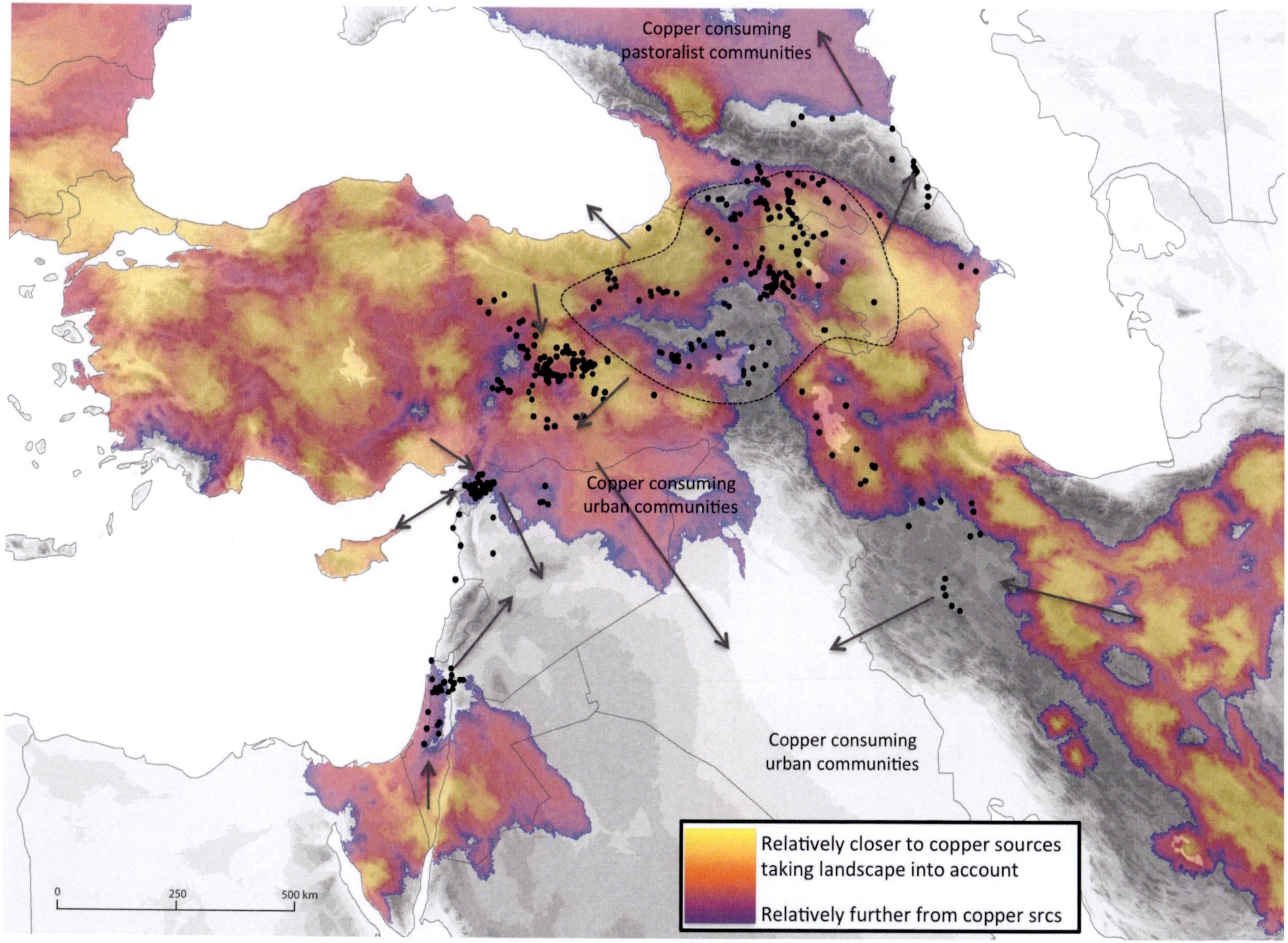

Figure 7.8. The relationship between Kura-Arax assemblages (at their greatest extent) and the accessibility to copper sources known to modern geology (as an archaeotopogram type 'A2'). Kura-Arax-related sites are located between areas of high accessibility and areas of low accessibility to copper supporting the interpretation of their role as potential 'middleman' in the exchange of copper.

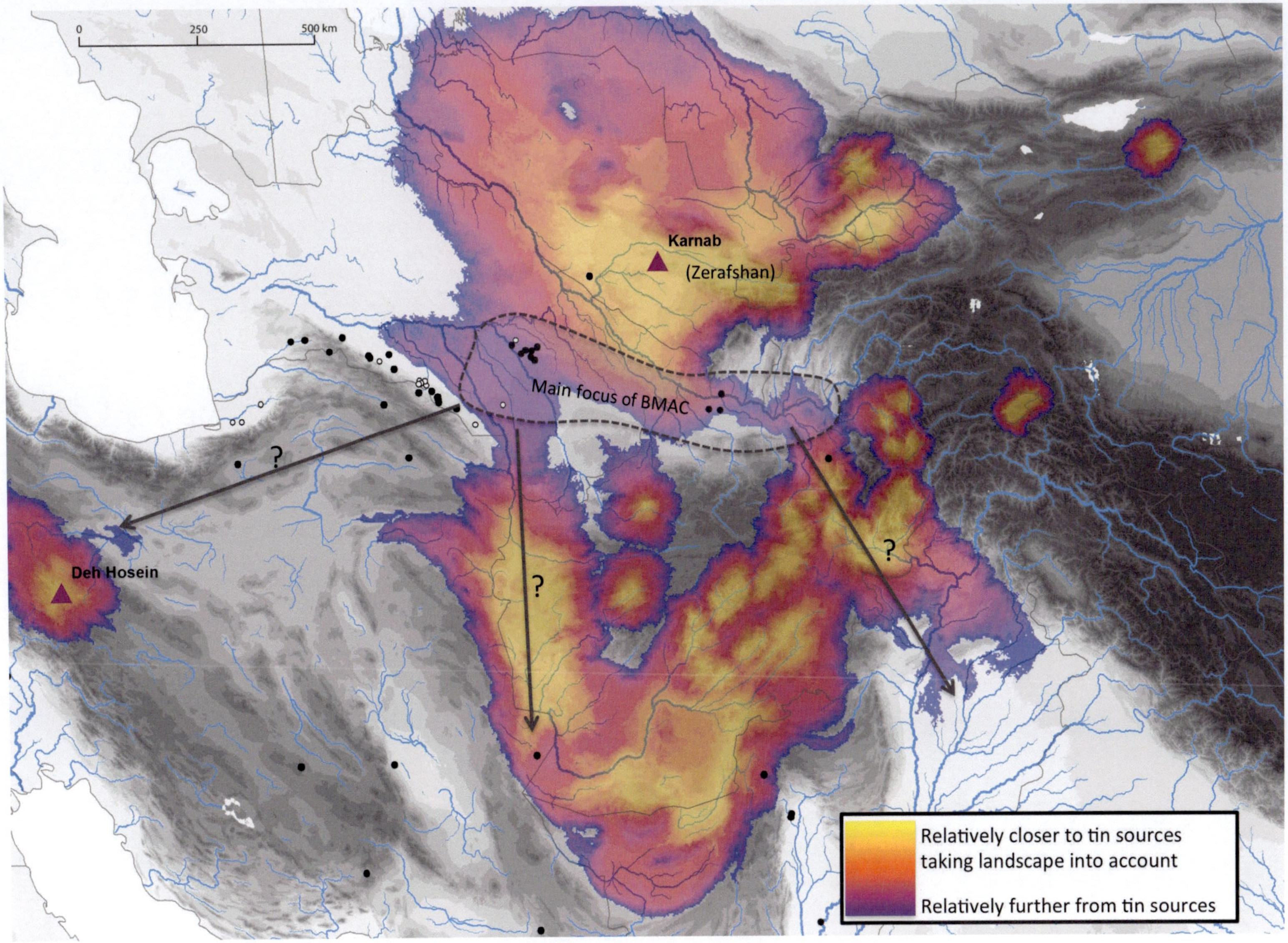

Figure 7.9. The relationship between BMAC/Namazga VI-related material culture, the central BMAC zone and areas of high accessibility to tin sources (as archaeotopogram type 'A2'). As for the Kura-Arax relationship to copper, the BMAC could be argued to have played 'middleman' in an exchange system involving tin, but seem unlikely to be involved in direct procurement (i.e. mining or panning) of the metal.

The subsequent appearance of rich kurgans in Transcaucasia (Bedeni, Martkopi, Sachkere and, perhaps after 2000BC, Trialeti) appears to document a new set of economic attitudes in which status differentiation by expensive mortuary rituals including the sacrifice of material wealth became increasingly central (a more 'sacrificial' economy, in Wengrow's terms). This must have involved a radical social revolution from the preceding less-differentiating Kura-Arax economic system, which – with the exception of the Arslantepe 'royal tomb' (arguably in a very short-lived period of turbulence and rapid change) – appears to have promoted the continual transmission or circulation of metals. To what extent this shift is also related to a possible shift away from sedentism and toward greater nomadism in eastern Anatolia and Transcaucasia is rather difficult to assess, in part because of our limited control of chronology in this period. Research at Sos Höyük, for example, appears to show continued localized 'sedentism' into the late 3rd and perhaps even early 2nd millennium BC (Sagona 2000). This transition, therefore, highlights the possibility that the Kura-Arax ritual-economy may have been closer in nature to the Uruk and later urban societies than its geographical descendants.

7.3.2 Western Central Asia: 'urbanism' and the BMAC phenomenon

Of the many changes in the Central Asian record which demand explanation, the emergence of the BMAC phenomenon is amongst the most challenging. This is partly a result of the 'eclectic' nature of BMAC material culture (most accounts including perhaps this book still paint an overly monolithic picture of the entity), and partly because of the converse but interrelated factor of an extensive distribution of BMAC objects (such as Bactrian axes, and BMAC pottery), particularly across Iran. Hiebert (1998) has proposed the idea of travelling and prosyletizing Zoroastrians as the bearers of BMAC culture in Iran, Afghanistan and Baluchistan. Potts (2008) has recently pointed to the historically-attested politico-ethnic 'Šimaški', who were a significant force in Elam between 2100-2000BC, as a warrior-group who might have originated in the BMAC milieu. Both explanations provide interesting speculative cultural roles for the BMAC culture, but they do not help us understand why communities of the BMAC should emerge in the way that they did. Some structured social or economic basis for BMAC power and the circulation of their material (and perhaps people) needs to be sought to parallel the 'historical' data.

In parallel to the scenario painted above with regards to copper in eastern Anatolia and Transcaucasia, the possible role of BMAC communities (or the related Sapalli culture) in the tin trade of the early 2nd millennium BC has already been discussed (Section 5.6.4). Lyonnet (2005) has gone so far as to paint Gonur Depe and contemporary BMAC settlements as a mirror image of Kültepe and the Old Assyrian *karum* colonies in Anatolia, with the only missing component being written texts. This is an interesting hypothesis, and one that requires further investigation. Although the earliest direct evidence for mining in the Zerafshan area so far dates a little later (*i.e.* after 1700BC) than the apparent emergence of the BMAC (2100-2000BC), and is associated with Andronovo material culture (Section 5.2.3), it is not unreasonable to assume that initial exploitation may have started earlier.

The geographical location of the Murghab delta and the Bactrian oases – on the only practical routes across or around the desert to the Zerafshan and the steppes beyond from Iran (and ultimately Mesopotamia) – is of course very suggestive. If we overlay the distribution of 'BMAC' (and related) finds onto an accessibility archaeotopogram ('type A2') of likely tin-sources in the region (Figure 7.9), including those of Afghanistan, we see a similar situation to that described for the relationship between copper and the Kura-Arax pottery. The main clusters of BMAC settlements are found at the 'edges' of the most accessible areas to the tin sources: thus in a kind of intermediary location. This could support the argument that BMAC users were 'middle-men' in a tin-trade, though with even less direct involvement with mining than for the Kura-Arax folk (given the greater distances involved). The very low numbers of tin-bronze objects from this time in Central Asia (and Iran) is not necessarily surprising if the material was more valuable in exchange than used locally. That metals (perhaps of different types) were already circulating in Central Asia is documented by the earlier 'metallification' of Namazga V pottery (Section 5.6.3) in the period before the emergence of the BMAC. Whether the Namazga V urban sites of the Kopet Dag were mostly consumers or trading agents in metals is difficult to say: it is tempting to assume that the extraction of tin, gold and perhaps silver in areas beyond the Kopet Dag could have started much earlier and in parallel to a similar process that happening contemporaneously in the west in Central Anatolia and the Aegean.

7.3.4 Linking eastern Anatolia/Transcaucasia and western Central Asia?

Back in 1988, Philip Kohl compared the early archaeological records of southern Central Asia and Transcaucasia as two sections of what he called the "Northern Frontier" of the ancient Near East.[131] "[A]n initial striking and superficially correct impression of the material culture remains from the two areas" he noted, "is that they are highly distinctive with few obvious parallels" (Kohl 1988, 593). He could identify no direct connections between them, though acknowledged that this might partly be the result of the limited knowledge of archaeology in the northern Iranian Caspian littoral (something that may change in the next few years, as studies from Azerbaijan and the Caspian littoral at sites such as Gohar Tappeh continue to be investigated, see Mahfroozi *et al.* 2009). Despite the distinctiveness of each of these cultural zones, Kohl hinted at some more obscure but perhaps no less significant structural connection:

> *"What is intriguing, however, and important for comparison with Transcaucasia is that the long-lived settlements in southern Turkmenistan seem to have been largely abandoned by the closing centuries of the third millennium B.C., roughly*

131 At that time, both regions were part of the soon-to-collapse Soviet Union, and in the context of *perestroika* and the more open atmosphere of exchange between academics on both sides of the Iron Curtain during the 1980s, this article (published in the *American Journal of Archaeology*) apparently aimed to inform the Western reader of material which was little known outside Russian academic contexts at the time. Remarkably little has changed in our basic picture since the article was published, in part because post-Soviet political and economic instability has halted or significantly lessened archaeological research in both regions. However, with the focus on Soviet material, Kohl effectively excluded eastern Turkey from the 'Transcaucasia' label (despite acknowledging its geographical continuation) and shied away from detailed consideration of north-eastern Iran (also partly due to state of research at that time) in this particular article.

> *the same time period when the Kura-Araxes settlements were abandoned, when the Transcaucasian archaeological record shifts to materials coming almost exclusively from kurgans"*
> (Kohl 1988, 594).

Kohl suggested in this article that this chronological synchrony might have something to do with mutual "qualitative advance[s] in metallurgical technology", a parallel emphasis on weaponry and the "use of specific jewellery techniques and of precious metals, such as silver and gold, for vessels" (Kohl 1988, 596). No doubt in response to the reaction created by his own applications of a 'world-systems' approach to interactions between Mesopotamia, Central Asia, the Indus and eastern Iran (*cf.* Kohl 1987b; 2011), he was at pains to displace the engine of change from any Mesopotamian centre:

> *"The barbarian 'peripheries' or northern frontiers of Transcaucasia and Central Asia, like their Aegean counterpart far to the west, did not reflect the light of civilization emanating from the ancient Near East; rather, they stimulated the later 'civilized' area and profoundly affected its course of development"*
> (Kohl 1988, 596).

Today, revisions of the respective absolute chronologies make this particular scenario of synchrony less clear-cut, although given the inherent uncertainties and resolution of the data, not necessarily baseless. The emergence of the very first kurgans in Transcaucasia may date as early as 2600BC (Edens 1995; Kavtaradze 2004). Certainly this period is also associated with a decline of Kura-Arax-inspired assemblages from the areas of its greatest extent, but in pockets of eastern Anatolia and Transcaucasia itself, detailed excavations from locations like Sos Höyük suggest there may be considerable continuity of settlement and material culture well into the late 3rd millennium, *i.e.* after the emergence of the kurgans. Meanwhile, the decline or abandonment of Namazga V settlements along the Kopet Dag may indeed belong to the last centuries of the 3rd millennium but the chronological and cultural relationship to the emergence of 'BMAC' sites in Margiana and Bactria is still unclear (*e.g.* whether there is any chronological overlap, and to what extent the populations were related).

These revisions suggest that while the parallels Kohl cited may not be exactly right, the overall evidence still suggests that we should be looking for synchronies that help us to document material flows and to explain cultural change on a grand scale. If the currently-favoured absolute chronologies are broadly right, then a better chronological match between Transcaucasia and western Central Asia might be found between the contracting Kura-Arax zone (and first kurgans in Transcaucasia) with the rise of urbanization settlements in southern Turkmenistan between 2600-2000BC. These are not culturally or formally parallelled (each phenomenon is clearly rather different in nature), but they do hint at deeper but less perceptible structural interrelationships.

To explain these periods of synchronous (but contrary) change, we need to throw the net even wider, and look at contemporary cultural processes taking place in urban centres[132] in Mesopotamia, the Indus and elsewhere from the second

132 Note the plural: it is preferable to see Mesopotamia, and indeed other regions, as being formed of constellations of 'centres' (nodes) and their interconnections (vertices), to avoid the inflexibility of centre-periphery dichotomy.

quarter of the 3rd millennium onwards. These processes include a decisive first movement towards a metallic-monetary or, as Wengrow has called it, an 'archival' economy with its links to weight metrology, sealing and record keeping in writing (see Section 5.5.2); related innovations in the shaping of precious metal objects and in the desirability of different metals (silver, gold and tin); and at the same time or soon after (and perhaps not coincidentally) the intensification, or perhaps better extensification (McCorriston *et al.* 1997), of textile production, associated with the rise in the use of woollen fibres for textiles, and new forms of widely shared bodily adornment connected to them: common beads and jewellery. The material remains of these processes are strikingly illustrated in the eclectic assemblages from the Royal Cemetery of Ur dated to around 2600-2500BC which are echoed in the distributions of objects described in the last 3 chapters. In this economic context, there appears to have been an increasing demand for silver, gold and metals generally. As silver in particular becomes more valuable (both as object of desire in itself, and later in its role as mediator of other exchanges), the increased demand probably resulted in ever greater amounts of goods being taken to ever more distant locations in order to elicit ever greater amounts of silver from source regions. This may have been the engine behind the first evidence of connections between the eastern Mediterranean and the Aegean (S. Sherratt 2010, 91-93). Similar demands were probably also felt in tin-yielding regions. If initial flows of tin may have come from smaller sources in the Taurus, Zagros or Caucasus in the early or mid-3rd millennium, larger sources in distant Afghanistan and Zerafshan may have come to play a bigger part in the late 3rd or early 2nd millennium. The result of all this was probably many local 'imbalances' – perhaps better described as 'eddies' or 'sinkholes' in the flow of metals – in the accumulation of commodities, and this provided ever greater power to those who could obtain them.

If this sounds dangerously close to the *ex oriente lux* paradigm that Kohl was so keen to reject, then the two-way (or rather multi-directional) nature of these processes should instead be emphasized. The sources of many of the materials imported by Mesopotamia during the 3rd and 2nd millennia BC remained many hundreds if not thousands of kilometres outside of their political control (*cf.* Stein 1999; 2002; Kohl 2011). Actors in the 'source regions' were clearly motivated to exchange by obtaining materials travelling in the opposite direction that were desirable for their own local cultural purposes – without the need for the direct political manipulation and 'exploitation' by actors from the 'core' by which modern colonialism is usually characterized. Instead the localized networks had their own cultural logic designed to take advantage of resource flows created by external and internal demand. Of course, all of these economic and cultural transformations were affected, in unknown ways, by contemporary environmental changes about which we still remain rather ignorant. The rise of the Caspian Sea at this time is only one example that may well have affected local food production and settlement possibilities as well as ease of transportation in adjacent regions.

7.3.5 Systemic (im)balances: the centrality of metals in the Bronze Age?

Kohl indicated metallurgical techniques and use of new metal types (especially silver and gold) as one of the potential indicators of deeply-embedded connectivity between Transcausia and Central Asia, and metal technology is something he has emphasized again more recently in a recent textbook on the Eurasian Bronze Age (Kohl 2011, 244-260). The idea of metals as central to Bronze Age economies was discussed in terms of 'metalification' in Chapter 5. Metal ore deposits are known primarily from mountainous regions, where tectonic and geological processes are more likely to coincide to expose or make mineral sources accessible. In contrast, the earliest large-scale urban societies (including the Mesopotamian, Egyptian and Indus) developed in lowland and relatively metal-poor regions, despite extensive use of metals and their incorporation into urban social systems. This polarity or 'potential difference' between metal-rich highlands and early social complexity in the alluvial plains forms the foundation of metal flows, and has long been recognized as significant in the creation of urbanization societies. This uneven resource distribution is what originally led Kohl to characterize the economy of the 3rd millennium in terms of an unequal balance of trade and direct dependency – in which the mineral-poor but agriculturally-rich peoples from the plains of Mesopotamia 'exploited' the mineral wealth of highland communities (*e.g.* of Anatolia or Iran) by exporting their agricultural surplus in exchange for metals and precious stones (Kohl 1978). This picture soon breaks down, however, once you start to look in more detail or further afield – something which Kohl acknowledged in later revisions of his 'world-systems' approach (*cf.* Kohl 1987a; 1987b; 2011). For example, on the basis of cumulative analytical results, Chernykh (1992) argues that vast amounts of metals were being exported from Caucasia to the massive area around the Black Sea and Pontic steppes, and not just to Mesopotamia. He calculates that only 10% of copper deposited in the northern area of his Circumpontic Metallurgical Province came from the Urals during the EBA, falling to 5% during the MBA – the majority was thus arriving from the south, *i.e.* the Caucasus and Anatolia (Chernykh 1992, 159). If these statistics are representative, then the notions of 'centre' and 'periphery' as often applied in a world-systems perspective (and even Andrew Sherratt's notion of a 'margin', A. Sherratt 1994) seem insufficient: the highland communities of Anatolia and the Caucasus were not simply part of a passive 'periphery' but independent cultural entities with their own social, political and economic imperatives – who also had economic relationships of interdependency (but not dependency) with a variety of different partners in different regions (S. Sherratt 2010). While it is clear that the flow of metals (raw materials, metal objects and assemblages/techniques) somehow linked distant communities over a massive area of Eurasia by the 3rd millennium, such that the economies of each were closely intertwined, the manner in which such metals were deployed indicates a range of very different systems of value. Importantly, however, and unlike in modern colonialism, the power over these structural connections was not held by any one area (*e.g.* by a Mesopotamian 'centre' – indeed Mesopotamia itself was politically fractured), and any imbalances came about primarily as 'emergent' qualities of economic interrelationships and changing consumer demands.

The use of metals (particularly silver) as currency from the 3rd millennium onwards (Section 5.5.1) makes it easy to place metals at the centre of ancient economies, and in a tacit acceptance of Three-Age evolutionary functionalism, to place the flow and development of metal technologies at the centre. Whilst undoubtedly an important substance to the ancient societies in which we are interested, we should remain cautious that metals were always the central commodity. Indeed, as I have attempted to indicate through the study of proxy evidence of various kinds, many important commodity flows were less durable (including textile or food products) and some simply more subtle (such as aesthetic or colour 'tastes'), but may have been equally 'central'.

7.3.6 Cross-craft indicators and the flow of metals and textiles

A common thread in these reconstructions described above is the evidence from cross-craft indicators for the flows of other materials: *i.e.* 'inter-media aesthetic traditions' as revealed through skeuomorphism, particularly in the pottery assemblages. It is striking that much of the pottery from both our case-study regions (and well beyond) can be argued to show strong evidence of formal or 'aesthetic' mimicry of either metals or textiles and these patterns in turn require further examination.

Whilst the aesthetic basis of the earliest 'Kura-Arax' pottery – dark/black and shiny – could recall either metals, obsidian, or perhaps leather, later examples, including the red-black burnished vessels (with their inverse patterning) and certain shapes with incised and ribbed decorations seem to show such a strong orientation towards metal that it would be surprising that metal prototypes have not been recovered – were it not for the general lack of metalwork in Kura-Arax contexts. The pattern of expansion and contraction of related wares makes a uniform picture of what happens next extremely difficult to create: but when the Kura-Arax wares disappear in the Upper Euphrates (after 2600BC), new painted pottery styles with basic netting or textile patterns become increasingly dominant. In Transcaucasia, different pottery assemblages with first monochrome and then later by the beginning of the 2nd millennium, polychrome painted patterns – also with strong textile associations – replace the earlier Kura-Arax repertoire. From a wider perspective, these sets of painted ware traditions – each very distinctive in details though both sharing abstract geometric motifs whose textile origin, though difficult to prove, seems extremely likely – are part of a wider trend for painted pottery styles, starting in the late 3rd millennium, that displace earlier plain (and apparently 'metallic' inspired) assemblages from the Aegean to Syria.

Meanwhile, in western Central Asia, the record seems to show precisely the opposite process on the surface of local pottery objects, *i.e.* from textiles to metals, at around the same time. Before the second quarter of the 3rd millennium, painted pottery with suggestively textile-inspired patterns (to the extent that some earlier Namazga III period pottery has been described as 'tapestry ware') dominates the record. These decorations completely disappear between 2700 and 2500BC in favour of the wheel-made, mass-produced plain wares of the Namazga V/VI and the following (or overlapping?) BMAC periods. The idea that painting on ceramics disappeared because mass production left no time for such decoration seems unconvincing since they remain 'high quality' wares. A more congruent explanation is, of course, that these plain wares are emulations of metal vessels (or again, a metal aesthetic) – and, though we find relatively little in the way of metal vessels during the early Namazga V period probably due to the predominance of settlement evidence for this time, the idea is strongly supported by the parallels between pottery and metal vessels found in later BMAC contexts such as those at Gonur Depe (see, especially, Figure 5.46). Early inspiration for this shift may have come from the new connections with the innovative metallurgists of north-eastern Iran, as documented by the 'intrusion' of 'Eastern Grey Wares' (*e.g.* at Ak Depe and Kara Depe).

How can we explain these shifts, and is there any connection between them other than the superficial mirror image reversal of metal and textile 'skeuomorphism' on pottery? The exact meaning of such pottery skeuomorphism is unclear and unlocking it in detail awaits further research into the processes behind such effects. Tentatively, both represent what we might call *Kulturschock*[133] – a cross-craft response to the (?rapid) rise in the *importance* of some particular medium. It does not document the first arrival of that particular medium (or material); *i.e.* it does not mean that there were no textiles or that such textiles were not important to early 3rd millennium communities in eastern Anatolia, or that metals were not valuable to 4th and early 3rd millennium communities in western Central Asia. The evidence in both instances is against this. Rather such *Kulturschock* appears to document the moment at which such materials become central to local cultural discourse on value (aesthetic, economic, social, moral or religious) and thus desirable to emulate or refer to in other materials. Since pottery is generally a 'low-value' but highly plastic material, the aesthetic it mimics is likely to reflect that of the 'most-valuable' material(s) of the moment. If consumption is generally related to the creation of relative identities, metal and textile skeuomorphism may just provide a glimpse into the basis of the system of identity in each respective area. Thus the more-or-less contemporary material changes in eastern Anatolia and western Central Asia indicate more-or-less synchronous periods of change in the conception of identity, value and the performance of status, albeit substantively different, which would seem likely to be a result of both 'internal' social dynamics and shifts in supra-regional flows.

What were those internal dynamics and supra-regional shifts of material flows? The Mesopotamian demand for metals (copper, silver, gold and tin) in the mid-3rd millennium has already been discussed above, and the role of this demand for metals (and its contagion to other regions whose communities started to demand metals in greater amounts themselves). In this sense, the communities of the Namazga V phase parallel those of central and western Anatolia (and the Aegean) more closely than those of Transcaucasia or eastern Anatolia, as being areas with new pottery repertoires that emphasize the metal aesthetic and suggesting that this skeuomorphism represents the importance of metal procurement to the societies in question. However, though the Namazga V vessels share many similarities with Indus ones (and hence we could set the unit of analysis as a much larger area to include the mines of Afghanistan), the oases of southern Turkmenistan themselves are, in fact, relatively metal-poor (*unlike* the Aegean and Anatolia). The metallic aesthetic is not, then, a direct 'reflection' of the *amount* of, say, silver (or tin or gold) in circulation, but of their role in culturally significant rituals and practices involving high-value materials: *i.e.* their desirability and symbolic capital. Such capital probably included the trade in metals itself in some regions. Those involved in the trade must have had access to greater resources generally and higher status leading to emulation of their practices. Simultaneously, new sets of social rituals (including drinking and eating) could spread because of that trade. Further work on exactly which vessel types use which colours and which are 'metalized' first

133 Coinage modelled on the meaning of *Metallschock* (Renfrew 1972, 338). One might also propose the less catchy '*Wolleschock*' for the impact of woollen textiles.

might also help reveal differences in the relative connectivity of different groups *within* Namazga communities. In other regions, like the Kopet Dag, such capital may have related to local strategies of consumption, often based on practices from elsewhere, but with distinctively local aims[134].

In eastern Anatolia and Transcaucasia, the shift to textile patterns on pottery (which seems to happen in two different phases in the valleys and then in the highlands) appears to mark different responses to shifting material flows in a region where metals had long been a key component of exchange. The contraction of Kura-Arax networks may index the decline of copper in value relative to other metals in the mid to late 3rd millennium. On the above model of *Metallschock*, the emergence of painted pottery with textile patterns appears to indicate the rise of textiles as central mediators of status and value that, because of the poor preservation-qualities of textiles, presents itself only in its mimicry in pottery patterns. The patterns themselves may of course have had various specific meanings, in common or divergent across large areas, but in a sense this is less relevant than the overall trend toward imitating such patterns on a medium which does not have very similar functional objectives: metal vessels are understandable models for pots, but associating textiles with vessels makes for less obvious metaphors. Inspiration from basketry could provide the 'container' model required, and perhaps that is what is mimicked by some of the early painted 3rd millennium painted wares in the Upper Euphrates[135]. The later patterns of Transcaucasian Painted wares, in common with the Cappadocian wares, involve net patterns (which could also refer to baskets or similarly mobile vessels), but the patterns are often arranged more like abstract friezes (more akin to miniature geometric tapestries than baskets).

In ethnographic accounts of pottery manufacture and the role of pottery in social metaphors (see *e.g.* Gosselain 1999), pots are frequently likened to human bodies[136], and it is therefore tempting in the light of the role of clay for human origins which is present in Mesopotamian mythology and that is fixed in textual forms in the late 3rd or early 2nd millennium (*e.g.* in the myth of Enki and Ninmah), to consider whether these pots are in some sense being moulded and decorated in a manner which reinforces this metaphor. Ceramic pots were perhaps being 'clothed' like human bodies, and possibly in similar ways (*i.e.* mostly plain but with small but expensive coloured banded features, hems, tassels *etc.*). Indeed, pots in other materials (*e.g.* metal vessels) may also have been 'clothed' with real textiles, at the same time that metal objects of other shapes, such as weapons, were being 'clothed' with tassels and sheaths that so rarely survive in the record (except as tiny scraps preserved by chemicals in the metals). That clothing on human bodies is associated with 'identity' and 'status' seems unsurprising to the modern view, but – as argued before (Section 6.6.2) – the use of hems as an alternative to seals in order to 'sign' (as recorded in early 2nd millennium Old Babylonian texts) argues for a systematization of this relationship in this period, perhaps even for the first time. Of course further archaeological and philological research needs to be undertaken to develop this argument further, but it is very tempting to see

134 The fact that the shapes of the new pottery are primarily table wares – designed for eating or drinking rather than storage *etc.* – suggests that identity was also linked to the performance of eating and drinking rituals, perhaps in communal contexts.

135 For example the Gelinciktepe-style painted wares – *i.e.* baskets with small coloured bands of fabric (Marro 1997).

136 The archaeological naming of pot parts also sometimes references this: *e.g.* mouth, body, foot.

the combination of emerging painted pottery styles of the late 3rd millennium, the evidence for highly organized textile production from texts (and a massive increase in textile production equipment in the 3rd millennium across Anatolia and Mesopotamia), as part of an emergence of a new system of identity focussed on patterned coloured textiles.

7.3.7 Documenting inequality? The mobility of value

It seems fair to argue that much of the material in circulation over long-distances in the 3rd and 2nd millennia BC was destined for the consumption by an elite – or rather sets of elites. The main purpose of this consumption appears to be similar to the 'tournaments of value' as discussed by Arjun Appadurai (1986): namely regular and irregular ritualized events (feasts, festivals, tournaments, marriages, real or simulated battles, funerals) in which values and status are negotiated by the deployment of resources (weapons, body adornment, food and drink) in a competitive but controlled manner. Of course, only a few of these events have left clear material traces. The obvious examples include the monumentalized remains of elaborate funerary rituals: at Alacahöyük, Ur, Umm al-Mara, Maikop, Trialeti, Gonur Depe, where international connections are very clear. A common factor is the general focus on the 'portability' of wealth and a concomitant mobility that this gave to its owners and would-be owners. Metals were often portable both in object form (as weapons or as jewellery) but also in the flexible nature of their materiality, since they could be melted down and remade to fit new social requirements depending on the context in which they ended up. Textiles of various kinds (clothing, decorations) were also portable physically, and because of this they facilitated the spread of motifs and fashions over much greater distances than other more weighty products, so that their 'value' could be understood in many different cultural contexts.

The deployment of these particular material resources can be argued to signal a desire by elite groups to free themselves of direct constraints of status based on concrete and regular social relations within a community, towards an abstract concept of wealth based on material premises. Both metals and textiles facilitated new conceptions of status and power that did not rely on locally embedded ritual traditions. Gordon Childe associated the introduction of metalworking with the rise of cities and argued that the former required the social complexity of the latter to facilitate it. With greater chronological resolution and more detailed contextual information, we now know this was not exactly the case – since many early 'non-urban' groups (including those of the Caucasus) developed sophisticated metal production without 'cities' *per se*. However, a link between urbanism and the portability of wealth (facilitated through the 'liquid' conception of economy which metals provide) still seems relevant. It is ironic then that 'urbanism' is often seen to represent a process of settlement and sedentism, when in fact it appears to have involved a much greater degree of mobility (in the movement of people and goods) and a focus on the increase of 'portability' of wealth and abstraction of social relations. It is perhaps no surprise that an important feature of many of elite funerary assemblages are items relating to transportation technology (carts, wagon accoutrements), a direct manifestation of the symbolic importance of mobility to power.

7.3.8 Flows and boundaries

Of course, whilst the main focus of this work and the evidence discussed above has been on interconnections and interaction between regions, these same cultural zones and object distributions also reveal cultural boundaries and barriers to flows of material or ideas. Thus, even in the face of massive social and economic transformations in certain fields, we can also hazily identify cultural resistance to change or rather active maintenance of traditions which do not follow the same trajectory as those of the mainstream 'world-system' or 'centre(s)'. The most obvious example of this is the lack of lapis lazuli and weighing systems in Transcaucasia/ eastern Anatolia during the Kura-Arax era. The geographical proximity of these communities and their likely role in the circulation of metals from Anatolia and the Caucasus makes it difficult to believe that such communities were truly isolated from the expanding Mesopotamian economic system. However, the lack of lapis lazuli, obvious weighing or sealing practices and perhaps even the lack of wheel-made pottery all point towards a strong cultural resistance to a set of values which were apparently more easily adopted in, say, the Aegean or western Anatolia. This may reflect a kind of socially-embedded conservativism within the Kura-Arax system which enforced these boundaries.

Perhaps less obviously there are differences in the colour schemes of metallic-inspired pottery assemblages which might be best explained by active cultural colour preferences. In Central Asia, for example, it is possible that we see the desirability of polished silver (and gold): the paste used to manufacture Namazga IV and V vessels produces almost exclusively a pale white to yellow shade (in contrast to the reds of earlier Namazga pottery styles). This scheme contrasts markedly with those 'metallic wares' from Anatolia and the Aegean, which in general tend to be shaded darkly – blacks, greys and bright or dark rusty reds. This suggests that metal vessels in Central Asia were preferred when lightly coloured and untarnished though not overly shiny, whilst in Anatolia, metal vessels were often dark and unpolished or even actively tarnished. Whilst local availability of clays may sometimes play a role in traditions, given the high-level of skills that crafts people of both regions demonstrate (and the existence in each region of alternative but much less-well used colour schemes), it seems more likely that cultural preferences for particular clays is more likely. The explanation for these preferences, I believe, may lie in low-level but long-lived traditions of 'cross-craft' aesthetic preferences that frequently force new materials to conform to the aesthetics of others because of the values assigned to certain material qualities. In Anatolia, the likely candidate for an aesthetic model for metals (and pottery) is of course obsidian – a dark and shiny material which was clearly very highly valued. It is therefore curious to note that the eastern extent of obsidian circulation during the Neolithic and Chalcolithic does not reach as far east as Central Asia, and therefore was not used there as a basis for signalling new 'high-value' materials. This boundary – to be drawn somewhere in central Iran – also divides the 'metallurgical provinces' of the Circumpontic and Irano-Afghan zones during the 3rd millennium.

Another source of evidence for boundaries can be seen in our knowledge of textile circulation, for which we must mostly rely on textual sources given our limited archaeological database. It is obvious from the Old Assyrian texts, for example, that whilst textiles were manufactured in both northern Syria and central Anatolia, there were clear differences in quality and type between different regions that enabled the exchange. Kohl has argued (*e.g.* 2011) that in the period of early state formation, technologies were easily transferable and this militates against the kind of 'unequal exchanges' which he had originally envisaged as being a part of the economic 'world-system' of the 3rd millennium Near East. These differences in textile production (and the 'aesthetics' of metals hinted at in pottery skeuomorphs) suggest that technologies were not always easily partible from their social contexts and the social organization that had created them. The manufacture of woollen textiles in Syro-Mesopotamia in the later 3rd and early 2nd millennium in particular seems to have involved a massive social re-organization which was either not possible, or not desirable in other regions until later. Long periods of material 'monopoly' were therefore possible in this era because of clear socio-cultural barriers to certain types of movement and exchange, and it is this fact, along with changing consumer demands, which must have had such a massive effect on shifting economic fortunes. This suggests that further research must thus be directed not only at tracing *flows* but also *boundaries* between cultural and economic zones – of technologies, of materials and of ideas.

7.4 Summary of discussion

This chapter has pulled together the various strands of data about stones, metals and textiles in the previous 3 chapters into a more coherent chronological framework. Using this data, explanations have been put forward for the expansion of the Kura-Arax phenomenon and emergence of the BMAC respectively on the basis of international metal flows, and tested using the visualizations (archaeotopograms) of the geographical relationship between the metals and cultural zones in question. Periods of synchrony are considered as ways of connecting the trajectories of Transcaucasian/eastern Anatolian and western Central Asian communities more directly and the evidence of cross-craft mimicry of metals and textiles in pottery form explored as a potential indicator for the basis of these interconnections. It was noted that many of the indicators for exchange are essentially 'elite' based and appear to relate to the long-term consequences of attempts to widen control through making common values of wealth and power more mobile and more widely accepted. Finally it was argued that whilst it is on 'flows' that the data in this book has emphasized, absences of certain features in certain regions remind us that it is equally important to search for 'boundaries' between regions in our explanations.

Chapter 8

Conclusions

8.1 Introduction

The research presented in this book aimed to map the geographical routes of exchange and the nature of interaction across Eurasia between 3000 and 1500BC, synthesizing archaeological and other data, and using database and GIS technology to map the distribution and flow of materials. Various themes have informed analysis in this work, and several questions were designed to focus these themes. In this final chapter, I return to the original research questions set out in Chapter 1, and concisely assess to what extent the data presented were able to answer them, and to offer ideas for how these questions might be taken forward in the future.

8.2 Reconstructing and re-visualizing routes

Research Question: ***What methods can we use to reconstruct and understand the physical routes of movement and interaction during the 3rd and early 2nd millennium BC in the Near East?***

Two main methods have been outlined:

1. The traditional method: based on the principle of 'route inertia' in which major routes are assumed to have remained the same through long periods of time and thus data from better documented (normally later) periods can be used to project likely frameworks for the period of interest.
2. A novel computerized method: based on the principle of landscape continuity, in which the traversability of terrain is modelled and visualized using cost-surface GIS techniques, and this can then be used in association with period-specific distribution data to suggest the density of travel across this terrain.

8.2.1 Route inertia and route dynamism

Research Question: ***To what extent is the normally unspoken reliance on the concept of 'route inertia' a help or a hindrance to the understanding of past interaction networks?***

Chapter 2 was devoted to the outline of the principle of 'route inertia' and of collecting suitable data about routes (from historical sources and archaeological indicators) from relevant areas of the Near East to inform the study of routes within our case-study regions. This review highlighted a number of things:

1. First, our knowledge of the location of roads (or routes) of later periods is often much less certain than is sometimes assumed.
2. Our knowledge of the hierarchy of routes, or the actual density of traffic along these roads or routes is extremely low.

3. The use of the principle of 'route inertia' relies on a default mythological approach to routes in the archaeological imagination, and requires a greater deal of critique, testing and case-by-case demonstration.

4. Many famous routes, such as the 'silk road' are more mythic or literary tropes than empirical phenomena or useful analytical concepts.

Most importantly, it was argued that an over-reliance on the concept of route inertia makes it was impossible to find periods of route dynamism. It is important therefore to find new ways to document either inertia or dynamism within the physical location of interregional networks.

Chapter 3 argued that part of the problem with the 'route inertia' approach is the way in which we tend to visualize ancient networks of interconnection, *i.e.* in the anachronistic form of 'pseudo-road-map', often on a topography-free background. A closer examination of the concept of 'the route' led to a more flexible definition as a 'corridor' of movement, which requires different forms of visualization. It was argued that landscape continuity (topography, climate) provides a more helpful basis for the mechanism of route inertia than historical routes. However, the physical aspects of routes are only one part of the whole, and to better characterize routes and interaction, we need to be able to incorporate the flows of material (and people) along these corridors into our models and visualizations.

8.2.2 Visualizing threads: future directions for GIS and the study of ancient interactions

Research Question: ***Can GIS-based cost-surface analyses help us to visualize routes and the spatiality of exchange better than traditional methodologies, especially where no physical roads have survived?***

In Chapter 3 a number of GIS-based techniques devoted to the modelling of movement were reviewed, and a novel methodology based on cost-surface techniques was set out. This approach was designed to re-visualize routes and flows of materials, in particular to avoid the 'road-map' reflex. The archaeological material presented in subsequent parts of this study formed a sample data set to test the usefulness of such an approach in as far as can be achieved with standard GIS hardware and software. Several 'archaeotopograms', based on these techniques, were presented in this study (for example, Figures 4.2a and c, 4.4, 4.6, 5.2, 5.3, 5.11) and from these it is possible to make some general comments about the utility of archaeotopograms in general, and how, if at all, they might be taken forward in the future.

1. The first thing to note is that different data sets do not benefit equally from being analysed in this way and that the results from different data sets will provide information about interaction on different scales.

It is important to select appropriate entities for which there is a distributional logic, a relatively 'representative' sample and sufficient density of distribution to create a meaningful archaeotopogram. No archaeotopogram for the distribution of all direct textile remains was produced, for example, because their distribution is dominated by the random effects of preservation and is not likely to reflect any ancient cultural patterns. Aspects of some materials may be better visualized than others. For example, it is possible to produce an informative 'type A2' archaeotopogram of raw material sources, such as lapis lazuli (since we are reasonably confident that we are not missing too many alternative sources). Such a map provides a guide to how accessible such sources are to the entire region, and thus indicates the relative 'cost' implied by the movement of such material to each location. Since we know that vast amounts of lapis lazuli ended up at Ur, it is possible to focus on this one site and look at its geographical relation with the sources and hence produce a least-cost corridor or even (a slightly more arbitrary) least-cost-path (as per Figure 4.4). On the other hand, the distribution of lapis lazuli *objects* in the archaeological record is not very representative of their circulation (or that of the raw material) because such objects were rarely consigned to the ground. The distribution of finds in the current database is certainly not dense enough to provide much of an indication or confirmation of the small-scale detailed 'routes' or corridors by which lapis actually travelled.

A similar argument could be made for ore sources and metal objects. The accessibility of copper and tin sources (as we currently know them) can be modelled easily using archaeotopograms ('type A2') – and, taking the analysis a step further ('type B2'), the regions with 'overlapping' accessibility highlighted to show the likely regions of early tin-bronze experimentation as seen in Figure 5.11. If sufficiently large databases of metal objects could be assembled and made easily available electronically, these could also be compared to these maps and also, perhaps, used to create maps of 'corridors' ('type C') based on distributions of distinctive shapes, metallic composition or other features.

2. The most productive category of evidence for visualization in archaeotopogram form for the purposes of characterizing route networks turn out to be pottery types. (Samples of these were shown in Figures 5.50b and 6.43).

This is because databases of pottery types with sufficient density to characterize regional 'zones' or 'corridors' are much easier to assemble than for other materials. That similar pottery types tend to indicate very close interaction between constituent communities suggests that archaeotopograms derived from these distributions show the location of dense exchange networks or 'pathways' (Section 3.2.1). Where the pottery can be seen to mimic other more valuable materials (metals, textiles) we can assume that it reveals at least one of the major materials that was being transported along such networks.

3. The limitations of current software and hardware platforms that were available for this study made it ultimately impossible to produce 'type C' corridor-based archaeotopograms because of the sheer setup, programming and processing time required.

This is a shame, because such visualizations could produce results which more closely match the 'corridor'-like definition of routes that is much more appropriate than the route-as-'road' definition. The question is whether, computationally, processing these models is 'solvable' within realistic human time-scales (*cf.* the 'travelling salesperson problem' and other problems which become exponentially harder as the data sets grow). More research into resistance-based approaches may provide a productive way to achieve such results efficiently in the near future.

4. The resulting visualizations are most useful if treated as heuristic tools of interpretation, and not as 'objective' maps of past exchange networks.

The results of cost-based analyses as shown are, of course, contingent on the model and on the detail of the archaeological data involved, and it is worth emphasizing again – as I have tried to show in their deployment here – that they should be treated as tools of interpretation rather than 'objective' maps of past exchange networks.

5. The expanding size of the archaeological data available to us means that future applications of these approaches would benefit from better ways of managing archaeological data 'from the ground up'.

The utility of such analyses becomes greater as the number of 'networks' that are to be traced is increased and as the quality of the source data improves. In this book it has only been possible to assemble a small range of suitable databases for analysis that provide tantalizing but equivocal windows onto the dynamics of ancient exchange. It is to be hoped that in future years, as far more archaeological data becomes available digitally, it will be possible to refine, augment and improve on interpretations through comparing different data sets. For this to be possible (and also for researchers to be able to handle the ever-growing products of archaeological excavation, survey and scientific analysis), the archaeological community needs to come up with standardized ways of presenting, distributing and querying such data. To a large degree, the future of synthetic approaches in archaeology must lie, therefore, in the digital management of data.

8.3 Direct and indirect evidence for interaction

Research Question: ***How much can we reconstruct about the flows of the three selected material foci (stones, metals and textiles) in this period, using direct and indirect evidence?***

Chapters 4, 5 and 6 were devoted to the detailed review of significant indicators of interaction of these important classes of material, stones, metals and textiles. Besides the individual conclusions of individual categories of evidence, some general conclusions can be made:

1. Direct evidence of individual categories of material are rarely dense enough to be able to recover the detailed location of networks of their movement. This is particularly true for metals and textiles.
2. Indirect evidence is often more helpful to reconstructing the location of networks in more detail, though we should take care about what it is these proxy indicators actually indicate.
3. Overall, the unevenness of the data currently limits the degree to which we can quantify the relevant flows of materials and information during our period of interest, and we must therefore often rely on inferences and impressions to create a coherent account.

8.3.1 Materiality, mobility and the performance of power and identity

Research Question: ***What were the consequences of the specific 'materialities' of these categories on social values and practice, and what significance did these have for interaction between regions?***

Chapters 4, 5 and 6 all attempt to highlight the consequences of the specific 'materialities' or material qualities of the stones, metals and textiles traced. Some general comments can be made:

1. Precious stones like lapis lazuli and carnelian depended on their colour and visual aesthetic, but also their stability for their value.

These values were probably also associated with moral, magic and symbolic power which are difficult to reconstruct in the archaeological record, but are hinted at by later texts.

2. Metals relied partly upon their shiny aesthetics and hardness for social value, but perhaps their liquidity and recyclability had greater social impact.

Like precious stones, metal's aesthetic properties and technical requirements no doubt had magical associations. Additionally, however, the liquidity of metals allowed them (particularly silver) to become an abstract mediator of value, and to increase the mobility of 'wealth' and transfer of power by its portability. Different moral or ritual codes responded differently to this liquidity and recyclability resulting in different 'economic' systems. This is documented most clearly by the appearance of hoards or richly-endowed burials in certain regions at certain times, as such dangerously mobile and recyclable symbols of power or wealth were sacrificed in elaborate social rituals.

3. The almost infinite possibilities for weaving patterns or alternative modes of dressing allowed textiles (perhaps particularly in the form of clothing) to play an increasing role in the performance of both group and individual identities.

Special textiles may have become one of the most valuable items of elite property, though because of the perishability of textile materials, it is difficult to record how this might have developed. The flexibility of textile fibres to be woven into complex and, as dyeing technology improved after around 2600-2300BC, extremely colourful adornments for the body allowed the performance of both 'exceptionality' (kingly dress) and 'commonality' (*e.g.* ethnic dress) at different times and places.

4. In the creation of relative value systems, all of the materials relied (to differing extents) on the value attributed to performance of difficult technical tasks to transform them into social objects: whether this meant carving images, casting weapons, or the weaving of complex patterns in cloth. These technical tasks required detailed knowledge of the materials in question.

Whilst material qualities may have been important to social function, the context of consumption and deposition may be of greater importance to the overall understanding of interaction. As noted, many of the material indicators for exchange during the 3rd and 2nd millennium BC are essentially designed for 'elite' consumption.

8.3.2 Cross-craft interaction: 'epiaesthetic' skeuomorphism?

Research Question: ***To what extent can evidence of cross-craft-interaction help us to understand the flow of materials that are poorly represented or otherwise invisible in the archaeological record?***

Throughout this book, the 'unboundness' of material categories has been a continuous theme: both in the intentional appearance of materially different but functionally similar items in the 'wrong' chapters, and in the deployment of skeuomorphs and apparent 'cross-craft' aesthetics as proxy evidence for other materials.

1. Pottery is the most common and most useful material to show evidence for cross-craft interaction.

This is presumably in part because of pottery's extreme degree of 'plasticity', which allows it to be formed into many different shapes, but also through various manipulations allows it to take on the visual (and sometimes tonal) qualities of other materials such as metal and stone. Polychrome painted patterns on pottery (or on plaster) can also closely mimic woven or embroidered patterns from textiles.

2. The surface of pottery vessels – what might be labelled their 'epiaesthetic' qualities – seems also to reflect some aspect of the material priorities of the producers. In turn this may be used as a proxy for zones of circulation for certain forms of the emulated material.

Though we cannot be sure of the exact significance of cross-craft indicators, it seems highly likely, as has been argued here, that the decision to make pottery vessels that emulate metal vessels directly, or a metal aesthetic, or else paint patterns on pottery vessels with textile patterns results from either (a) a functional overlap, but one in which the non-pottery material is seen as the primary medium; and/or (b) an attempt to draw on the 'aura' or material values of the mimicked material. In both cases it seems likely that the distribution of such pottery in some sense indexes the zone of circulation of certain forms of the emulated material, and thus reflects the otherwise invisible flow of these materials.

Cross-craft indicators are very rarely given serious attention – in part because of the difficulty of 'proving' the skeuomorphic relations. The approach offers many insights into the economy of materials which are otherwise inaccessible, however, and also may help to explain the conformity of pottery production over large distances if more mobile materials (metals, textiles) can be assumed to be the 'missing link' in the transmission of shapes and patterns.

8.4 Cross-cultural flows and local development

Research Question: ***What effects did the flow of these materials have on the trajectories of communities and cultures in our case-study regions?***

Research Question: ***Can we explain the expansion of the Kura-Arax phenomenon or the emergence of the BMAC, on the basis of information about the flow of materials and techniques, or the lack of it, in and beyond these regions?***

In Chapter 7, the various threads of evidence discussed in early chapters were placed into a chronological frame, and an attempt made to make sense of the different flows of materials and their role in the trajectories of communities in the case-study regions, Transcaucasia /eastern Anatolia and western Central Asia.

1. In both cases, the international flow of metals has been argued to form an important role in local development.

It is suggested, for example, that the expanded distribution of the Kura-Arax phenomenon between 2800 and 2600BC may be the result of shifting emphasis of metal flows, and that Kura-Arax communities may have played a role as expert middle-men in the copper exchange networks of the period. Meanwhile, the emergence of the BMAC (2100-1700BC) might be explained in a similar way, as the result of an increased demand for tin, and the role of Central Asian communities as middle-men to the tin sources to the north, in the Zerafshan or beyond.

2. Evidence of absence may be as important as evidence of flows.

Whilst the apparent lack of lapis lazuli and definite weighing systems in Transcaucasia/eastern Anatolia could be a factor of archaeological visibility, if real then it appears to suggest that these communities were following a very different socio-economic path to their southern neighbours in the Uruk and Early Dynastic periods. As argued earlier (Section 7.3.7), there are indicators for various cultural boundaries (economic/ritual systems, colour preferences or craft techniques) which if investigated more fully, may offer greater insights into the motivations and long-term changes of each of these local foci.

8.5 Final note: whither a Bronze Age 'Silk Road'?

Research Question: ***To what extent does the evidence point towards the existence of prehistoric precursors to the famed 'Silk Road(s)'?***

When this research began, I, somewhat naively, hoped that it would be possible to compare and contrast historical routes (like the 'silk roads') with prehistoric ones in a way which would reveal the slow evolution of long-distance routes and, hopefully, some of the engines behind such an evolution. The analysis of various flows of material in their geographic context does enable a kind of synthesis, as demonstrated by the chronological and thematic summaries in Chapter 7. However, as I argued in Chapter 2, it is almost impossible to fix the 'silk roads' to any geographic or chronological specificity: there was no road which could have been described as the official or authentic 'silk road' and even the multiple diverging routes across Eurasia that might be traced through historical records and material goods of the 1st and early 2nd millennia AD – for which the plural term 'silk routes' might in any case be more appropriate – are much less studied or understood than we would hope. This major barrier to comparison is of course compounded by the traditional imagination of a route as a 'road' rather than a 'corridor'. A more productive definition for such 'route constellations' was provided by Andrew Sherratt: "a directional chain of preferentially orientated transactions, which allow[s] a complementary flow of products" (A. Sherratt 2003, 241). In this way, such routes can be seen as structural rather than literally geographic, and this transforms the process of comparison between periods into a characterization of alternative structures rather than assumption of geographic continuity. Sherratt's account aims to explain the "dialectics of change" between 4500 and 1500BC in the Circumpontic region through inter-connected technological and cultural transformations in both urban and non-urban systems. He argues, to cite in full:

> *"It was the growth of indigenous cultural foci, like the Kura-Araks culture in Transcaucasia or Altyn Depe and the Namazga culture in Turkmenistan, which provided a continuity of partners interested in*

> *exchange; and (like the Silk Route, along which materials flowed, although no individual travelled its entire length) a directional chain of preferentially orientated transactions, which allowed a complementary flow of products"* (A. Sherratt 2003, 241).

Coincidentally, it is precisely these particular "indigenous cultural foci" that we have been examining in the course of our investigations. Does this definition therefore allow us to compare the 'directional chains of preferentially orientated transactions' of the 3rd and early 2nd millennia BC with those of the 1st and early 2nd millennia AD? – Perhaps, though this task requires far more information about the structural relationships of the later routes before a serious comparison can be made. What we can suggest is, however, that the "continuity of partners" was probably far from continuous, and that it was precisely the constant transformations of partners (or rather the transformation of their preferences of consumption) that drove the evolution of routes. A full assessment of the extent to which such evolution can be said to have been either a cumulative process (with particular routes gaining momentum through time) or a matter of oscillating intensities (with routes going in and out of use) will have to await future research, but the small sample of material flows explored in this book suggests some combination of both. If this is the case, the 'silk routes' may well have had their 'roots' in the exchange networks of the 3rd and 2nd millennia BC...

Bibliography

Adovasio, J. M., Soffer, O., and Klima, B. 1996. Upper Palaeolithic fibre technology: interlaced woven finds from Pavlov I. *Antiquity* 70: 526-534.

Agrawal, D. P. 2000. *Ancient Metal Technology and Archaeology of South Asia: a Pan-Asian Perspective.* New Delhi: Aryan Books International.

Agrawal, D. P. and Seshadri, R. 1998. The Metallurgical Tradition of the Harappans. In *Archaeometallurgy in India.* V. Tripathi (ed). Delhi: Sharada Publishing House. 10-16.

Albenda, P. 1978. Assyrian Carpets in Stone. *The Journal of the Ancient Near East Society* 10: 1-34.

Algaze, G. 1989. The Uruk expansion: cross-cultural exchange in early Mesopotamian civilization. *Current Anthropology* 30 (5): 571-608.

Algaze, G. 1993. *The Uruk World System. The Dynamics of Expansion of Early Mesopotamian Civilization.* Chicago: University of Chicago Press.

Alimov, K., Boroffka, N., Bubnova, B., Burjakov, J., Cierny, J., Jakubov, J., Lutz, J., Parzinger, H., Pernicka, E., Radililovskij, V., Ruzanov, V., Širinov, T., Staršinin, S. and Weisgerber, G. 1998. Prähistorischer Zinnbergbau in Mittelasien: Vorbericht der Kampagne 1997. *Eurasia Antiqua* 4: 137-199.

Alvarez-Mon, J. 2005. The Introduction of Cotton in the Near East: A View from Elam. *Achemenet*, http://www.achemenet.com/ (accessed 15th September, 2011).

Amiet, P. 1986. *L'âge des échanges inter-iraniens, 3500-1700 avant J. C.* Paris: Éditions de la Réunion des musées nationaux.

Anthony, D. W. 1998. The Opening of the Eurasian Steppe at 2000 BCE. In *The Bronze Age and Early Iron Age Peoples of Eastern Central Asia.* V. H. Mair (ed). The Institute for the Study of Man. 94-113.

Anthony, D. W. 2007. *The Horse, the Wheel, and Language: How Bronze-Age Riders from the Eurasian Steppes shaped the Modern World.* Princeton: Princeton University Press.

Anthony, D. W. and Brown, D. R. 2011. The Secondary Products Revolution, Horse-Riding, and Mounted Warfare. *Journal of World Prehistory* 24: 1-30.

Anthony, D. W., Brown, D. R., and George, C. 2006. Early horseback riding and warfare: the importance of the magpie around the neck. In *Horses and Humans: The Evolution of the Equine-Human Relationship.* S. Olsen, S. Grant, A. Choyke, and L. Bartosiewicz (eds.). British Archaeological Reports International Series, S1560. Oxford: Archaeopress. 137–156.

Appadurai, A. 1986. Introduction: commodities and the politics of value. In *The social life of things: commodities in cultural perspective.* A. Appadurai (ed). Cambridge: Cambridge University Press. 3-63.

Arne, T. J. 1945. *Excavations at Shah Tepé, Iran.* The Sino-Swedish Expedition - Publication 27. Stockholm: Elanders Boktryckeri Artibolag.

Arnold, D. 1993. Decoration. In *An Introduction to Ancient Egyptian Pottery: Fascicle 1, Techniques and Traditions in the Pottery of Ancient Egypt.* D. Arnold (ed). Mainz: Verlag Philipp von Zabern. 88-102.

Arpe, K. and Leroy, S. A. G. 2007. The Caspian Sea Level forced by the atmospheric circulation, as observed and modelled. *Quaternary International* 173/174: 144-152.

Artioli, G., Giardino, G., Guida, A., Lazzari, A., and Vidale, M. 2005. On the exploitation of copper ores at Shahr-i Sokhta (Sistan, Iran) in the 3rd millennium BC. In *South Asian Archaeology 2003*. U. Franke-Vogt and H.-J. Weisshaar (eds.). Bonn: Verlag Linden Soft. 179-184.

Aruz, J. 2003. Art and Interconnections in the Third Millennium B.C. In *Art of the First Cities: The Third Millennium B.C. from the Mediterranean to the Indus*. J. Aruz and R. Wallenfels (eds.). New Haven: Yale University Press. 239-250.

Aruz, J., Benzel, K., and Evans, J. M. (eds.). 2008. *Beyond Babylon: Art, Trade and Diplomacy in the Second Millennium B.C.* New Haven/London: The Metropolitan Museum of Art/Yale University Press.

Aruz, J. and Wallenfels, R. (eds.). 2003. *Art of the First Cities: The Third Millennium B.C. from the Mediterranean to the Indus*. New Haven: Yale University Press.

Asher-Greve, J. M. and Sweeney, D. 2006. On Nakedness, Nudity, and Gender in Egyptian and Mesopotamian Art. In *Images and Gender: Contributions to the Hermeneutics of Reading Ancient Art*. S. Schroer (ed). OBO 220. 125-176.

Askarov, A. 1973. *Сапаллитепа* [Sapallitepe]. Tashkent: Izdatel'stvo "F.A.N." UzSSR.

Askarov, A. 1992. The Beginning of the Iron Age in Transoxania. In *History of civilizations of Central Asia. Volume I: The dawn of civilization: earliest times to 700BC*. A. H. Dani and V. M. Masson (eds.). Paris: UNESCO Publishing. 441-458.

Askarov, A. and Abdullaev, B. N. 1983. *Джаркутан* [Djarkutan]. Tashkent: Izdatel'stvo "F.A.N." UzSSR.

Aston, D. A. 2003. New Kingdom Pottery Phases as revealed through Well-dated Tomb Contexts. In *The Synchronisation of Civilisations in the Eastern Mediterranean in the Second Millennium B.C. II. Proceedings of the SCIEM 2000 - EuroConference, Haindorf 2nd May - 7th May 2001*. M. Bietak (ed). Wien: Österreichische Akademie der Wissenschaften. 135-162.

Astour, M. C. 2000. Overland Trade Routes in Ancient Western Asia. In *Civilizations of the Ancient Near East*. J. M. Sasson, J. Baines, G. Beckman, and K. S. Rubinson (eds.). Peabody: Hendrickson. 1401-1420.

Avilova, L. I. 2008. Regional Models of Metal Production in Western Asia in the Chalcolithic, Early and Middle Bronze Ages. *Trabajos De Prehistoria* 65 (1): 73-91.

———. 2009. Models of metal production in the Near East (Chalcolithic - Middle Bronze Age). *Archaeology, Ethnology and Anthropology of Eurasia* 37 (3): 50-58.

Aydıngün, Ş. 2005. Tunç Çağı'nın Gizemli Kadınları / Mysterious Women of the Bronze Age. [Catalogue of an exhibition held at Vedat Nedim Tör Museum, Istanbul, 13 Oct 2005 - 30 Mar 2006] ed. Istanbul: Yapı Kredi Kültür Sanat Yayınları.

Bachhuber, C. 2011. Negotiating Metal and the Metal Form in the Royal Tombs of Alacahöyük in north-central Anatolia. In *Interweaving Worlds: Systemic Interactions in Eurasia, 7th to 1st millennia BC*. T. C. Wilkinson, S. Sherratt, and J. Bennet (eds.). Oxford: Oxbow Books. 158-174.

Badalyan, R., Chataigner, C., and Kohl, P. 2004. Trans-Caucasian Obsidian: the Exploitation of the Sources and their Distribution. In *A View from the Highlands: archaeological studies in honour of Charles Burney*. A. G. Sagona (ed). Herent: Peeters Press. 437-465.

Badalyan, R. S. and Avetisyan, P. S. 2007. *Bronze and Early Iron Age Archaeological Sites in Armenia.* Oxford: Archaeopress.

Badre, L. 1980. *Les Figurines Anthropomorphes en Terre Cuite à l'Age de Bronze en Syrie.* Paris: Librarie Orientaliste Paul Geuthner.

Bagh, T. 2003. The Relationship between Levantine Painted Ware, Syro/Cilician Ware and Khabur Ware and the Chronological Implications. In *The Synchronisation of Civilisations in the Eastern Mediterranean in the Second Millennium B.C. II. Proceedings of the SCIEM 2000 - EuroConference, Haindorf 2nd May - 7th May 2001.* M. Bietak (ed). Wien: Österreichische Akademie der Wissenschaften. 219-237.

Bakhshaliyev, V. and Marro, C. 2009. The Archaeology of Nakhichevan: Ten Years of Discoveries. Istanbul: Ege Yayınları.

Ball, W. 1998. Following the Mythical Road. *Geographical* 70 (3): 18-23.

Ball, W. and Gardin, J. -C. 1982. *Archaeological gazetteer of Afghanistan: Catalogue des sites archéologique d'Afghanistan.* Paris: Editions Recherche sur les Civilisations.

Bar-Adon, P. 1971. רמשמ לחנמ םיאצממ ,ןומטמה תרעמ. [The Cave of the Treasure: The Finds from the Caves of Nahal Mishmar]. Jerusalem: The Bialik Institute and the Israel Exploration Society. [non vidi].

Barber, E. J. W. 1991. *Prehistoric Textiles.* Princeton: Princeton University Press.

———. 1994. *Women's Work, the first 20,000 years: Women, Cloth, and Society in Early Times.* New York: Norton.

———. 1999. *The Mummies of Ürümchi.* New York: Norton.

Barjamovic, G. 2011. *A Historical Geography of Ancient Anatolia in the Assyrian Colony Period.* Carsten Niebuhr Institute Publications 38. Copenhagen: Museum Tusculanum Press.

Bass, G. F. 2000. Sea and River Craft in the Ancient Near East. In *Civilizations of the Ancient Near East.* J. M. Sasson, J. Baines, G. Beckman, and K. S. Rubinson (eds.). Peabody: Hendrickson. 1421-1431.

Batiuk, S. D. 2005. *Migration theory and the distribution of the Early Transcaucasian Culture.* PhD thesis, University of Toronto. [Unpublished thesis].

Batiuk, S. D. and Rothman, M. S. 2007. Early Transcaucasian Cultures and Their Neighbors: Unraveling migration, trade, and assimilation. *Expedition* 49 (1): 7-17.

Bayburtoğlu, B. and Yıldırım, S. 2008. Gold and Silver in Anatolia. In Anatolian Metal IV. Ü. Yalçın (ed). Bochum: Deutsches Bergbau-Museum. 43-54.

Baykal-Seeher, A. and Obladen-Kauder, J. 1996. *Demircihüyük. Band IV. Die Ergebnisse der Ausgrabungen 1975 - 1978: Die Kleinfunde.* Mainz am Rhein: Verlag von Philipp von Zabern.

Beaugeard, A.-C. 2010. Les textiles du Moyen-Euphrate à l'époque paléo-babylonienne d'après un ouvrage récent. In *Textile Terminologies in the Ancient Near East and Mediterranean from the Third to the First Millennia BC.* C. Michel and M.-L. Nosch (eds.). Ancient Textiles 8. Oxford: Oxbow Books. 283-289.

Beck, H. C. 1933. Etched carnelian beads. *The Antiquaries Journal* 13: 384-398.

Bedianashvili, G. 2008. The Pattern Burnished Ornament in Georgia During the Bronze Age. In *Ceramics in Transition: Chalcolithic Through Iron Age in the Highlands of the Southern Caucasus and Anatolia.* K. S. Rubinson and A. G. Sagona (eds.). Louvain: Peeters Press. 87-100.

Begemann, F., Haerinck, E., Overlaet, B., Schmitt-Strecker, S., and Tallon, F. 2008. An Archaeo-Metallurgical Study of the Early and Middle Bronze Age in Luristan, Iran. *Iranica Antiqua* 43: 1-66.

Begemann, F. and Schmitt-Strecker, S. 2009. Über das Frühe Kupfer Mesopotamiens. *Iranica Antiqua* 44: 1-45.

Bell, T., Wilson, A., and Wickham, A. 2002. Tracking the Samnites: landscape and communications routes in the Sangro Valley, Italy. *American Journal of Archaeology* 106 (2): 169-186.

Bellelli, G. M. 1989. Reflections on an 'unexcavated' precious vase with zoomorphic decoration. *Iranica Antiqua* 24: 87-108.

Belli, O. and Bahşaliyev, V. 2001. Nahçıvan Bölgesi'nde Orta ve Son Tunç Çağı Boya Bezemeli Çanak Çömlek Kültürü. Istanbul: Arkeoloji ve Sanat Yayınları.

Belli, O. and Sevin, V. 1999. Nahçıvan'da Arkeolojik Araştırmalar / Archaeological Survey in Nakhichevan. Istanbul: Arkeoloji ve Sanat Yayınları.

Berezkin, Y. E., Charles, M. P., Gosden, C., Harris, R. D., Hillman, C. G., Kasparov, K. A., Korobkova, F. G., Kurbansakhatov, K., Legge, J. A., Limbrey, S., and Masson, V. M. 1993. Investigating early agriculture in Central Asia: new research at Jeitun. *"Antiquity"* 67: 324-338.

Berger, J. 1969. *The World's Major Fibre Crops: Their Cultivation and Manuring.* Zurich: Centre d'Etude de l'Azote.

Berthoud, T., Cleuziou, S., Hurtel, L. P., Menu, M., and Volfovsky, C. 1982. Cuivres et alliages en Iran, Afghanistan, Oman au cours des IVe et IIIe millénaires. *Paléorient* 8 (2): 39-54.

Berthoud, T., Craddock, P. T., Hauptmann, A., Maddin, R., Muhly, J., Pigott, V. C., Stech-Wheeler, T., and Weisgerber, G. 1980. Production, échange et utilisation des métaux: bilan et perspectives des recherches archéométriques récentes dans le domaine oriental. *Paléorient* 6 (1): 99-127.

Betts, A., van der Bord, K., de Jong, A., McClintock, C., and van Stryndonck, M. 1994. Early Cotton in North Arabia. *Journal of Archaeological Science* 21: 489-499.

Beyer, D. 1989. Un nouveau témoin des relations entre Mari et le monde iranien au IIIème millénaire. *Iranica Antiqua* 24: 109-120.

Biga, M. G. 2010. Textiles in the Administrative Texts of the Royal Archives of Ebla (Syria, 24th century BC) with Particular Emphasis on Coloured Textiles. In *Textile Terminologies in the Ancient Near East and Mediterranean from the Third to the First Millennia BC.* C. Michel and M.-L. Nosch (eds.). Ancient Textiles 8. Oxford: Oxbow Books. 146-172.

Bilgi, Ö. 1984. Metal Objects from İkiztepe-Turkey. *"Beiträge zur Allgemeinen und Vergleichenden Archäologie"* 6: 31-96.

———. 2001. *"Metallurgists of the Central Black Sea Region: a new perspective on the question of the Indo-Europeans' Original Homeland / Orta Karadeniz Bölgesi Madencileri: Hind-Avrupalıların Anavatanı Sorununa Yeni Bir Yaklaşım".* Istanbul: TASK Yayınları.

Bloch, M. and Parry, J. 1989. Introduction: Money and the Morality of Exchange. In *Money and the Morality of Exchange.* M. Bloch and J. Parry (eds.). Cambridge: Cambridge University Press. 1-32.

Bobokhyan, A. 2008. *Kommunikation und Austausch im Hochland zwischen Kaukasus und Taurus, ca. 2500-1500 v. Chr.* British Archaeological Reports International Series 1853. Oxford: John and Erica Hedges Ltd.

Bokovenki, N. A. 2000. The Origins of Horse Riding and the Development of Ancient Central Asian Nomadic Riding Harnesses. In *Kurgans, Ritual Sites, and Settlements: Eurasian Bronze and Iron Age.* J. Davis-Kimball, E. M. Murphy, L. Koryakova, and L. T. Yablonsky (eds.). British Archaeological Reports International Series S890. Oxford: Archaeopress. 304-310.

Bolger, D. 1996. Figurines, fertility, and the emergence of complex society in prehistoric Cyprus. *Current Anthropology* 37 (2): 365-373.

Bonora, G. L. and Vidale, M. 2008. An Aspect of the Early Iron Age (Yaz I) Period in Margiana: Ceramic Production at Site No. 999. In *The Archaeological Map of the Murghab Delta. Volume II. The Bronze Age and Early Iron Age in the Margiana Lowlands.* S. Salvatori and M. Tosi (eds.). British Archaeological Reports International Series S1806. Oxford: BAR. 155-195.

Booth, W. J. 1994. Household and Market: On the Origins of Moral Economic Philosophy. *The Review of Politics* 56 (2): 207-235.

Boroffka, N., Cierny, J., Lutz, J., Parzinger, H., Pernicka, E., and Weisgerber, G. 2002. Bronze Age Tin from Central Asia: Preliminary notes. In *Ancient Interactions: East and West in Eurasia.* K. Boyle, C. Renfrew, and M. Levine (eds.). Cambridge: MacDonald Institute for Archaeological Research. 135-160.

Boucharlat, R. 2002. Pasargadae. *Iran* 40: 279-282.

Boucharlat, R., Francfort, H.-P. and Lecomte, O. 2005. The Citadel of Ulug Depe and the Iron Age Archaeological Sequence in Southern Central Asia. *Iranica Antiqua* 40: 479-514.

Braidwood, R. J. and Braidwood, L. S. 1960. *Excavations in the Plain of Antioch I: The Earlier Assemblages Phases A-J.* Oriental Institute Publications 61. Chicago: University of Chicago Oriental Institute Publications.

Braudel, F. 1996. *The Mediterranean and the Mediterranean World in the Age of Philip II.* Berkeley: University of California Press.

Briant, P., Henkelman, W., and Stolper, M. (eds.). 2008. *L'archive des fortifications de Persépolis. État des questions et perspectives de recherches.* Persika 12. Paris: Éditions de Boccard.

Broekmans, T. 2006. Insights Into North Mesopotamian 'Metallic Ware'. *Archaeometry* 48 (2): 219-227.

Brovender, Y. M. 2009. Copper Ores of the Northern Pontic Region as Raw Materials for Production Activity in the Paleometal Age (Based on the Study of the Kartamysh Ore Mining and Metallurgy Complex). *Baltic-Pontic Studies* 14: 103-123.

Buchanan, B. 1966. *Catalogue of Ancient Near Eastern Seals in the Ashmolean Museum I, Cylinder Seals.* Oxford: Clarendon Press.

Budd, P. and Taylor, T. 1995. The faerie smith meets the bronze industry: magic versus science in the interpretation of prehistoric metal-making. *World Archaeology* 27 (1): 133-143.

Burney, C. A. 1961. Excavations at Yanik Tepe, North-West Iran. *Iraq* 23 (2): 138-153.

———. 1962. The Excavations at Yanik Tepe, Azerbaijan, 1961 Second Preliminary Report. *Iraq* 24 (2): 134-152.

———. 1964. The Excavations at Yanik Tepe, Azerbaijan, 1962: Third Preliminary Report. *Iraq* 26 (1): 54-61.

———. 1972. Excavations at Haftavan Tepe 1969: Second Preliminary Report. *Iran* 10: 127-142.

———. 1989. The Khirbet Kerak Question and the Early Transcaucasian Background. In *L'urbanisation de la Palestine à l'âge du Bronze ancient: Bilan et perspectives des recherches actuelles*. P. de Miroschedji (ed). British Archaeological Reports International Series S527. Oxford: BAR. 331-340.

Burney, C. and Lang, D. M. 2001 [1971]. *The Peoples of the Hills: Ancient Ararat and Caucasus*. London: Wiedenfeld and Nicolson.

Cann, J. R. and Renfrew, C. 1964. The characterization of obsidian and its application to the Mediterranean region. *Proceedings of the Prehistoric Society* 30: 11-33.

Carter, R. A. 2006. Boat remains and maritime trade in the Persian Gulf during sixth and fifth millennia BC. *Antiquity* 80: 52-63.

Carter, T. 2008. The consumption of obsidian in the Early Bronze Age Cyclades. In *Horizons: A Colloquium on the Prehistory of the Cyclades*. N. Brodie, J. Doole, G. Gavalas, and C. Renfrew (eds.). McDonald Institute Monographs. Cambridge: McDonald Institute for Archaeological Research. 225-235.

Casanova, M. 1992. The Sources of the Lapis-lazuli found in Iran. In *South Asian archaeology, 1989: papers from the Tenth International Conference of South Asian Archaeologists in Western Europe, Musée national des arts asiatiques-Guimet, Paris, France, 3-7 July 1989*. C. Jarrige (ed). Madison: Prehistory Press. 49-56.

———. 1993. Lapis lazuli beads in Susa and Central Asia: a preliminary study. *Annales Academiae Scientiarum Fennicae. Series B* 271 (1): 137-145.

———. 1998. *Le lapis-lazuli dans l'Orient ancien: gisements, production, circulation, des origines au début du 2ème millénaire avant J.-C.*. Doctorat, Université Paris I, Sorbonne. [Unpublished thesis].

———. 2000. Le lapis-lazuli de l'Asie centrale à la Syrie au Chalcolithique et à l'âge du Bronze: traits communs et particularités régionales. In *Proceedings of the First International Congress on the Archaeology of the Ancient Near East. Rome, May 18th-23rd 1998, vol. I*. P. Matthiae, A. Enea, L. Peyronel, and F. Pinnock (eds.). Roma: Dipartimento di Scienze Storiche, Archaeologiche e Antropologiche Dell'Antichità. 171-183.

———. 2001. Le lapis-lazuli, la pierre précieuse de l'Orient ancien. *Dialogues d'histoire ancienne* 27 (2): 149-170.

———. 2008. Lapis lazuli. In *Beyond Babylon: Art, Trade and Diplomacy in the Second Millennium B.C.* J. Aruz, K. Benzel, and J. M. Evans (eds.). New Haven/London: The Metropolitan Museum of Art/Yale University Press. 71-72.

———. 2013. *Le lapis-lazuli dans l'Orient ancien: production et circulation du Néolithique au IIe millénaire av. J.-C.* Documents préhistoriques 27. Paris, Éditions du Comité des travaux historiques et scientifiques.

Cattani, M. 2008. The Final Phase of the Bronze Age and the "Andronovo Question" in Margiana. In *The Archaeological Map of the Murghab Delta. Volume II. The Bronze Age and Early Iron Age in the Margiana Lowlands*. S. Salvatori and M. Tosi (eds.). British Archaeological Reports International Series S1806. Oxford: BAR. 133-151 .

Cauvin, M.-C., Gourgaud, A., Gratuze, B., Arnaud, N., Poupeau, G., Poidevan, J. -P., and Chataigner, C. (eds.). 1998. *L'obsidienne au Proche et Moyen Orient: du volcan à l'outil.* British Archaeological Reports International Series 738. Oxford: British Archaeological Reports.

Cecchini, S. M. 1965. *La Ceramica Di Nuzi.* Studi Semitici 15. Roma: Istituto di Studi del Vicino Oriente.

Charles, J. A. 1985. Determinative Mineralogy and the Origins of Metallurgy. In *Furnaces and Smelting Technology in Antiquity.* P. T. Craddock and M. J. Hughes (eds.). London: British Museum Publications. 21-28.

Chase-Dunn, C. K. and Hall, T. D. 1997. *Rise and Demise. Comparing World-Systems.* Boulder, Colorado: Westview Press.

Chegini, N. N., Helwing, B., Parzinger, H., and Vatandoust, A. 2004. Eine prähistorische Industriesiedlung auf dem iranischen Plateau: Forschungen in Arisman. In *Persiens Antike Pracht.* T. Stöllner, R. Slotta, and A. Vatandoust (eds.). Katalog der Ausstellung des Deutschen Bergbau-Museums Bochum vom 28. November 2004 bis 29. Mai 2005. Bochum: Deutsches Bergbau-Museum. 210-216.

Chegini, N. N., Momenzadeh, M., Parzinger, H., Pernicka, E., Stöllner, T., Vatandoust, A., and Weisgerber, G. 2000. Preliminary report on the archaeometallurgical investigations around the prehistoric site of Arisman near Kashan western Central Iran. *Archäologische Mitteilungen aus Iran und Turan* 32: 281-318.

Chernykh, E. N. 1992. *Ancient Metallurgy in the USSR: the Early Metal Age.* S. Wright (trans.). Cambridge: Cambridge University Press.

———. 2009. *Степной Пояс Евразии Феномен Кочевых Культур* [The phenomenon of Nomadic cultures in the Eurasian Steppe Belt]. Moskva: Rukopisnye Pamyatniki Drevnei Rusi.

Chernykh, E. N., Avilova, L. I., and Orlovskaya, L. B. 2002. *Metallurgy of the Circumpontic Area: From Unity to Disintegration. In Anatolian Metal II.* Ü. Yalçın (ed). Bochum: Deutsches Bergbau-Museum. 83-100.

Childe, V. G. 1915. On the date and origin of Minyan ware. *Journal of Hellenic Studies* 35: 196-207.

Chlopin, I. N. 1986. *Jungbronzezeitliche Gräberfelder im Sumbar-Tal, Südwest-Turkmenistan.* München: C. H. Beck.

Chorley, R. J. and Haggett, P. 1968. *Socio-economic models in geography.* London: Methuen.

Çilingiroğlu, A. 1993. Van-Dilkaya Höyüğü Kazıları: Kapanış. In *XIV. Kazı Sonucları Toplantısı.* Ankara: Ankara Üniversitesi Basımevi. 29-38.

Çilingiroğlu, A. and Dedoğlu, F. 2007. Ulucak Höyük Kazıları 2005 yılı Çalışmaları. In *28. Kazı Sonuçları Toplantısı*, 1. Cilt. Kültür ve Turizm Bakanliği/Dösimm Basimevi: Ankara. 137-146.

Clark, S. R. 2012. *The Social Lives of Figurines: Recontextualizing the Third Millennium BC Terracotta Figurines from Harappa (Pakistan).* Oxford: Oxbow Books.

Clark, S. R. 2003. Representing the Indus body: sex, gender, sexuality, and the anthropomorphic terracotta figurines from Harappa. *Asian Perspectives* 42 (2): 304-328.

Clarke, D. 1977. *Spatial archaeology.* I. Hodder (ed). London: Academic Press.

Cleuziou, S. and Berthoud, T. 1982. Early Tin in the Near East: A Reassessment in the Light of New Evidence from Western Afghanistan. *Expedition* 25: 14-19.

Clutton Brock, J., Mittre, V., and Gulati, A. N. (eds.). 1961. *Technical reports on Archaeological Remains.* Deccan College Post-Graduate and Research Institute, Department of Archaeology and Ancient History Publication 2. Poona: Deccan College.

Collischonn, W. and Pilar, J. V. 2000. A direction dependent least-cost-path algorithm for roads and canals. *International Journal of Geographical Information Science* 14: 397-406.

Compagnoni, B. and Tosi, M. 1978. The Camel: its Distribution and State of Domestication in the Middle East During the 3rd Millennium BC in Light of Finds from Shahr-i Sokhta. In *Approaches to Faunal Analysis in the Middle East.* R. Meadow and M. Zeder (eds.). Cambridge, MA: Harvard University Press. 91-103.

Conolly, J. and Lake, M. W. 2006. *Geographical Information Systems in Archaeology.* Cambridge: Cambridge University Press.

Conti, A. M. and Persiani, A. C. 1993. When Worlds Collide: Cultural Developements in Eastern Anatolia in the Early Bronze Age. In *Between the Rivers and Over the Mountains: archaeologica anatolica et mesopotamica Alba Palmieri dedicata.* M. Frangipane, H. Hauptmann, M. Liverani, P. Matthiae, and M. Mellink (eds.). Roma: Dipartimento di Scienze Storiche, Archaeologiche e Antropologiche Dell'Antichità. 361-413.

Cook, R. M. 1987. 'Artful Crafts': A Commentary. *The Journal of Hellenic Studies* 107: 169-171.

Coqueuegniot. 1998. L'obsidienne en Méditerranée orientale aux époques post-néolithiques. In *L'obsidienne au Proche et Moyen Orient: du volcan à l'outil.* M.-C. Cauvin, A. Gourgaud, B. Gratuze, N. Arnaud, G. Poupeau, J. -P. Poidevan, and C. Chataigner (eds.). British Archaeological Reports International Series 738. British Archaeological Reports. 351-362.

Courcier, A. 2007. La métallurgie dans les pays du Caucase au Chalcolithique et au début de l'âge du Bronze: bilan des études et perspectives nouvelles. In *Les cultures du Caucase (VIème-IIIème millénaires avant notre ère).* B. Lyonnet (ed). Paris: CNRS éditions. 199-232.

Craddock, P. T. (ed.). 1980. *Scientific studies in early mining and extractive metallurgy.* London: British Museum.

Cremaschi, M. 1998. Palaeohydrography and Middle Holocene Desertification in the Northern Fringe of the Murghab Delta. In *The archaeological map of the Murghab Delta: Preliminary reports 1990-1995.* A. Gubaev, G. Koshelenko, and M. Tosi (eds.). Rome: Istituto Italiano per l'Africa et l'Oriente. 15-25.

Crowfoot, E. 1960. Appendix A: Textiles, Matting and Basketry. In *Excavations at Jericho, I: Tombs Excavated in 1952-4.* K. M. Kenyon (ed). London. 519-526.

Crowfoot, G. M. and Davies, N. 1941. The Tunic of Tut'ankamūn. *Journal of Egyptian Archaeology* 27: 113-30.

Crowfoot Payne, J. 1968. Lapis Lazuli in Early Egypt. *Iraq* 30: 58-61.

Cunliffe, B. 1971. Some aspects of hill-forts and their cultural environments. In *The Iron Age and its Hillforts.* M. Jesson and D. Hill (eds.). Southampton: Southampton University Archaeology Society. 53-70.

Dalley, S. 1984. *Mari and Karana.* London: Longman.

———. 1991. Ancient Assyrian Textiles and the Origins of Carpet Design. *Iran* 29: 117-135. http://www.jstor.org/stable/4299853.

Dankoff, R. 2006. *An Ottoman Mentality: the World of Evliya Çelebi.* Leiden: Brill.

Davis, J. 1986. *Keos V: Ayia Irini: Period V.* Mainz: Philipp von Zabern.

Day, P. M. and Doonan, R. C. (eds.). 2007. *Metallurgy in the Early Bronze Age Aegean.* Oxford: Oxbow Books.

Del, A. and Tavernari, C. 2009. Utilisation d'un réseau de polygones de Thiessen pour la géolocalisation robuste de caravansérails décrits dans les récits anciens de voyageurs. http://www.esrifrance.fr/sig2009/evcaucaravan.htm (accessed August 9, 2011).

Delmas, A. B. and Casanova, M. 1990. The lapis lazuli sources in the Ancient Near East. In *South Asian archaeology 1987 - Proceedings of the 9th International conference of the Association of South Asian archaeologists in Western Europe, Rome.* M. Taddei (ed). 493-505.

Delougaz, P. 1940. *The Temple Oval at Khafajah.* Oriental Institute Publications 53.

Dercksen, J. G. 1996. *The Old Assyrian Copper Trade in Anatolia.* Istanbul: Nederlands Historisch-Archaeologisch Instituut.

———. 1999. *Trade and Finance in Ancient Mesopotamia: Proceedings of the 1st MOS Symposium, Leiden, December 19-20.* PIHANS (Publications de l'Institut historique-archeologique neerlandais de Stamboul) 84. Istanbul: Nederlands Historisch-Archaeologisch Instituut.

———. 2005. Metals According to Documents from Kültepe-Kanish Dating to the Old Assyrian Colony Period. In *Anatolian Metal III.* Ü. Yalçın (ed). Bochum: Deutsches Bergbau-Museum. 17-34.

Deshayes, J. 1967. Ceramiques peintes de Tureng Tepe. *Iran* 5: 123-131.

———. 1972. Tureng Tépé et la Période Hissar IIIC. In *The Memorial Volume of the VIth International Congress of Iranian Art and Archaeology. Tehran - Isfahan - Shiraz, 11th-18th April 1968.* Tehran: Iranian Center for Archaeological Research. 34-38.

Dickinson, O. 2002. *The Aegean Bronze Age.* Cambridge: Cambridge University Press.

Dietz, U. L. 1992. Zur Frage vorbronzezeitlicher Trensenbelege in Europa. *Germanica* 70 (1): 17-36.

Donahue, C. J. 2006. *The Importance of Cloth: Aegean Textile Representation in Neopalatial Wall Painting.* Unpublished MA thesis, The Florida State University College of Visual Arts, Theatre and Dance. [Unpublished thesis].

Doonan, R. C., Day, P. M., and Dimopoulou-Rethemiotaki, N. 2007. Lame Excuses for Emerging Complexity in Early Bronze Age Crete: the Metallurgical FInds from Poros Katsambas and their Context. In *Metallurgy in the Early Bronze Age Aegean.* P. M. Day and R. C. Doonan (eds.). Oxford: Oxbow Books. 98-122.

Dothan, T. K. (ed.). 1979. *Excavations at the Cemetery of Deir El-Balah. Qedem 10.* Jerusalem.

Dupree, L., Gouin, P., and Omer, N. 1971. The Khosh Tapa Hoard from North Afghanistan. *Archaeology* 24 (1): 28-34.

Durand, J.-M. 2009. *La nomenclature des habits et des textiles dans les textes de Mari.* Archives Royales de Mari 30. Paris: CNRS éditions.

During Caspers, E. C. L. 1972. Etched cornelian beads. *Bulletin of the Institute of Archaeology* 8/9: 83-98.

Dyson, R. H. and Howard, S. M. (eds.). 1989. *Tappeh Hesar: Reports of the Restudy Project, 1976.* Firenze: Case Editrice le Lettere.

Easton, D. F. 1976. Towards a Chronology for the Anatolian Early Bronze Age. *Anatolian Studies* 26: 145-173.

———. 1995. The Troy treasures in Russia. *Antiquity* 68: 11-14.

Edens, C. 1995. Transcaucasia at the End of the Early Bronze Age. *BASOR* 299/300: 53-64.

Edgü, F. (ed.). 1983. *The Anatolian Civilisations, I: Prehistoric / Hittite / Early Iron Age.* Istanbul: Turkish Ministry of Culture and Tourism.

Edwards, M. 1983. *Excavations in Azerbaijan (North-western Iran) 1 : Haftavan, Period VI.* British Archaeological Reports International Series S182. Oxford: BAR.

Efe, T. 2002. The Interaction Between Cultural/Political Entities and Metalworking in Western Anatolia during the Chalcolithic and Early Bronze Ages. In *Anatolian Metal II.* Ü. Yalçın (ed). Bochum: Deutsches Bergbau-Museum. 49-66.

———. 2007. The theories of the 'Great Caravan Route' between Cilicia and Troy: the Early Bronze Age III period in inland western Anatolia. *Anatolian Studies* 57: 47-64.

Efe, T. and Fidan, M. E. 2006. Pre-Middle Bronze Age Metal Objects from Inland Western Anatolia: A Typological and Chronological Evaluation. *Anatolia Antiqua* 14: 15-43.

Ehrich, R. W. (ed.). 1992. *Chronologies in Old World archaeology.* Chicago: University of Chicago Press.

Ellis, R. 1989. Note on a Textile Sample from the Main Mound at Tappeh Hesar. In *Tappeh Hesar: Reports of the Restudy Project, 1976.* R. H. Dyson and S. M. Howard (eds.). Firenze: Case Editrice le Lettere. 199-220.

Emre, K. 1966. The pottery from Acemhoyuk. *Anadolu* 10: 99-143.

———. 1971. *Anadolu Kurşun Figurinleri ve Taş Kalıpları.* Ankara: Türk Tarih Kurumu.

Engels, D. A. 1978. *Alexander the Great and the logistics of the Macedonian army.* Berkeley: University of California Press.

Erdmann, K. and Erdmann, H. 1961. *Das anatolische Karavansaray des 13. Jahrhunderts.* Berlin: Mann, 1966.

Esin, U. 1969. *Kuantitatif Spektral Analiz Yardımıyla Başlangıcından Asur Kolonileri Çağina Kadar Bakır ve Tunç Madenciliği (1967).* İstanbul Üniversitesi Edebiyat Fakültesi Publication 1427. Istanbul: İstanbul Üniversitesi.

Evliya Çelebi. 1996. *Evliya Çelebi Seyahatnâmesi.* 10 volumes. İstanbul: Yapı Kredi Yayınları.

Fairbairn, A. 2004. Archaeobotany at Kamen-Kalehöyük 2003. *Kamen-Kalehöyük: Anatolian Archaeological Studies* 13: 107-120.

Farmer, S., Sproat, R., and Witzel, M. 2004. The collapse of the Indus-script thesis: The myth of a literate Harappan civilization. *Electronic Journal of Vedic Studies* 11 (2): 19-57.

Faroqhi, S. 1994. *Pilgrims and Sultans: The Hajj under the Ottomans 1517-1683.* London: Tauris & Co.

Feininger, A. 1960. *Frauen und Goettinnen*. Köln: M. DuMont Schauberg.

Finley, M. I. 1970. Aristotle and Economic Analysis. *Past & Present*, 47: 3-25.

Fleming, S. J., Pigott, V. C., Swann, C. P., and Nash, S. K. 2005. Bronze in Luristan: Preliminary analytical evidence from copper/bronze artifacts excavated by the Belgian Mission in Iran. *Iranica Antiqua* 40: 35-64.

Forlanini, M. 2008. The historical geography of Anatolia and the transition from the karum-period to the Early Hittite Period. In *Anatolia and the Jazira during the Old Assyrian Period*. J. G. Dercksen (ed). PIHANS (Publications de l'Institut historique-archeologique neerlandais de Stamboul) 111. Leiden: Nederlands Instituut voor het Nabije Oosten. 57–86.

Forsdyke, E. J. 1914. The pottery called Minyan Ware. *Journal of Hellenic Studies* 34: 126-156.

Frahm, E. E. 2010. *The Bronze-Age obsidian industry at Tell Mozan (Ancient Urkesh), Syria: redeveloping electron microprobe analysis for 21st-Century sourcing research and the implications for obsidian use and exchange in Northern Mesopotamia after the Neolithic*. PhD dissertation, University of Minnesota. [Unpublished thesis].

Frahm, E. and Feinberg, J. M. 2013a. Empires and resources: Central Anatolian obsidian at Urkesh (Tell Mozan, Syria) during the Akkadian period. *Journal of Archaeological Science* 40 (2): 1122-1135.

Frahm, E. and Feinberg, J. M. 2013b. Environment and collapse: Eastern Anatolian obsidians at Urkesh (Tell Mozan, Syria) and the third-millennium Mesopotamian urban crisis. *Journal of Archaeological Science* 40 (4): 1866-1878.

Frame, L. 2010. Metallurgical investigations at Godin Tepe, Iran, Part I: The metal finds. *Journal of Archaeological Science* 37 (7): 1700-1715.

Francfort, H.-P. (ed.). 1989. *Fouilles de Shortugaï. Recherches sur l'Asie Centrale Protohistorique*. Paris: Diffusion de Boccard.

———. 1994. The Central Asian dimension of the symbolic system in Bactria and Margiana. *Antiquity* 68: 406-418.

———. 2003. La civilisation de l'Asie Centrale à l'âge du Bronze et à l'âge du Fer. In *De l'Indus à l'Oxus: Archéologie de l'Asie Centrale (Catalogue de l'Exposition)*. O. Bopearachchi, C. Landes, and C. Sachs (eds.). Lattes: Association IMAGO, Musée de Lattes.

Frangipane, M. 1997. A 4th-millennium temple/palace complex at Arslantepe-Malatya. North-South relations and the formation of early state societies in the Northern regions of Greater Mesopotamia. *Paléorient* 23 (1): 45-73.

Frangipane, M., Andersson Strand, E., Laurito, R., Möller-Wiering, S., Nosch, M.-L., Rast-Eicher, A., and Wisti Lassen, A. 2009. Arslantepe, Malatya (Turkey): Textiles, Tools and Imprints of Fabrics from the 4th to the 2nd Millennium BCE. *Paléorient* 35 (1). 5-30.

Frangipane, M., Di Nocera, G. M., Hauptmann, A., Morbidelli, P., Palmieri, A., Sadori, L., Schultz, M., and Schmidt-Schultz, T. 2001. New Symbols of a New Power in a "Royal" Tomb from 3000BC Arslantepe, Malatya (Turkey). *Paléorient* 27 (2): 105-139.

Frangipane, M. and Palumbi, G. 2007. Red-Black ware, pastoralism, trade, and Anatolian-Transcaucasian interactions in the 4th-3rd millennium bc. In *Les cultures du Caucase (VIème-IIIème millénaires avant notre ère).* B. Lyonnet (ed). Paris: CNRS éditions. 233-256.

Frank, A. G. 1970 [1966]. The Development of Underdevelopment. In *Imperialism and Underdevelopment.* R. I. Rhodes (ed). New York: Monthly Review Press. 4-17.

———. 1993. Bronze Age World System Cycles. *Current Anthropology* 34 (4): 383-429.

Frank, A. G. and Gills, B. K. (eds.). 1993. *The World System: Five hundred or five thousand?* Abingdon: Routledge.

Franke-Vogt, U. 2003. Sohr Damb/Nal, Balučistan, Pakistan: Ergebnisse der Grabunden 2001, 2002 und 2004. *Archäologische Mitteilungen aus Iran und Turan* 35/36: 83-141.

Frei, K. M., Frei, R., Mannering, U., Gleba, M., Nosch, M. -L., and Lyngstrøm, H. 2009. Provenance of Ancient textiles - a pilot study evaluating the strontium isotope system. *Archaeometry* 51 (2): 252-276.

French, D. 1988. *Roman roads and milestones of Asia Minor.* British Archaeological Reports International Series S392. Oxford: British Archaeological Reports.

———. 1998. Pre- and Early- Roman Roads of Asia Minor: The Persian Royal Road. *Iran* 36: 15-43.

Friedman, E. 1998. Technological Style in Early Bronze Age Anatolia: A Dissertation Proposal Presented to The Department of Near Eastern Languages and Civilizations. http://oi.uchicago.edu/OI/DEPT/RA/DISPROP/Friedman_diss.html (accessed November 24, 2009).

———. 2000. *Technological Style in Early Bronze Age Anatolia: the Interrelationship between Ceramic and Metal Production at Göltepe.* PhD dissertation, University of Chicago. [Unpublished thesis].

Frifelt, K. 1991. *The Island of Umm an-Nar, 1. Third Millennium Graves.* Jutland Archaeological Society Publication 26.1. Aarhus: Jutland Archaeological Society.

Fuller, D. Q. 2006. Agricultural Origins and Frontiers in South Asia: A Working Synthesis. *Journal of World Prehistory* 20 (1): 1-86.

———. 2008. The spread of textile production and textile crops in India beyond the Harappan zone: an aspect of the emergence of craft specialization and systematic trade. Ed. Toshiki Osada and Akinori Uesugi. *Linguistics, Archaeology and the Human Past. Indus Project.* 1-26.

Fuller, D. and Rowlands, M. 2011. Ingestion and Food Technologies: Maintaining Differences over the Long-term in West, South and East Asia. In *Interweaving Worlds: Systemic Interactions in Eurasia, 7th to 1st millennia BC.* T. C. Wilkinson, S. Sherratt, and J. Bennet (eds.). Oxford: Oxbow Books. 37-60.

Gale, N. H. and Stos-Gale, Z. 1981. Lead and silver in the ancient Aegean. *Scientific American* 244 (6): 176-177.

———. 1999. Copper Oxhide Ingots and the Aegean Metals Trade. New perspectives. In *MELETEMATE. Studies in Aegean Archaeology Presented to Malcolm H. Wiener I.* P. Betancourt, V. Karageorghis, R. Lafineur, and W.-D. Niemeier (eds.). Aegeaum 20. Liège: University of Liège. 269-277.

GBNID. 1942. *Turkey, Vol. 1.* London: Naval Intelligence Division.

———. 1945. *Persia.* London: Naval Intelligence Division.

Gei, A. N. 2000. *Новотиторовская культура [Novotitorovskaya Kul'tura].* Moscow: Institut Arkheologii.

Gell, A. 1992. The Technology of Enchantment and the Enchantment of Technology. In *Anthropology, Art and Aesthetics.* J. Coote and A. Shelton (eds.). Oxford: Clarendon. 40–66.

———. 1998. *Art and Agency.* Oxford: Oxford University Press.

de Genouillac, H. 1934. *Fouilles de Telloh I.* Paris: Geuthner.

Ghirshman, R. (ed.). 1938. *Fouilles de Sialk, prés de Kashan, volume I.* Musée du Louvre, Série archéologique. Paris: Paul Geuthner.

———. 1939. *Fouilles de Sialk, prés de Kashan, volume II.* Musée du Louvre, Série archéologique. Paris: Paul Geuthner.

Gietl, R., Doneus, M., and Fera, M. 2008. Cost Distance Analysis in an Alpine Environment: Comparison of Different Cost Surface Models. In *Layers of Perception. Proceedings of the 35th International Conference on Computer Applications and Quantitative Methods in Archaeology (CAA), Berlin, Germany, April 2–6, 2007.* A. Poluschny, K. Lambers, and I. Herzog (eds.). Kolloquien zur Vor- und Frühgeschichte 10. Bonn: Habelt. 342-350.

Gilbert, A. S. 1991. Equid remains from Godin Tepe, western Iran: an interim summary and interpretation, with notes on the introduction of the horse into Southwest Asia. In *Equids in the Ancient World, vol. 2.* R. H. Meadow and H.-P. Uerpmann (eds.). Beihefte zum Tübinger Atlas des Vorderen Orients (Reihe A Naturwissenschaften) 19. Wiesbaden: Reichert Verlag. 75-122.

Gillis, C., Clayton, R., Pernicka, E., and Gale, N. 2003. Tin in the Aegean Bronze Age. In *METRON: Measuring the Aegean Bronze Age. Proceedings of the 9th International Aegean Conference. New Haven, Yale University 18-21 April 2002.* K. P. Foster and R. Laffineur (eds.). Aegeaum 24. 103-110.

Gimbutas, M. 1956. *The Prehistory of Eastern Europe: Part I, Mesolithic, Neolithic and Copper Age Cultures in Russia and the Baltic Area.* Cambridge, MA: Peabody Museum.

Goedicke, H. 1984. Abi-Sha(i)'s Representation in Beni Hasan. *Journal of the American Research Center in Egypt* 21: 203-210.

Good, I. L. 1995a. Notes on a Bronze Age textile fragment from Hami, Xinjiang with comments on the significance of twill. *Journal of Indo-European Studies* 23 (3/4): 319-346.

———. 1995b. On the question of silk in pre-Han Eurasia. *Antiquity* 69: 959-968.

———. 1998. Bronze Age Cloth and Clothing of the Tarim Basin: The Chǎchǎn Evidence. In *The Bronze Age and Early Iron Age Peoples of Eastern Central Asia.* V. H. Mair (ed). Washington, DC: The Institute for the Study of Man. 656-670.

———. 1999. *The ecology of exchange: Textiles from Shahr-i Sokhta, eastern Iran.* PhD, University of Pennsylvania. [Unpublished thesis].

———. 2006a. Textiles as a Medium of Exchange in Third Millennium BCE Western Asia. In *Contact and Exchange in the Ancient World.* V. H. Mair (ed). Honolulu: University of Hawai'i Press. 191-214.

———. 2006b. Exploring Inner Asia's High Alpine Frontier: High Alpine Transhumant Pastoralism, Vertical Cultivation and Environmental Archaeology in the Lower Vakhsh-Panj Confluence and Gorno-Badakhshan Regions, Southern Tajikistan [Unpublished work].

———. 2007. Cloth in the Babylonian World. In *The Babylonian World.* G. Leick (ed). London: Routledge.

———. 2011. Strands of Connectivity: Assessing the Evidence for Long Distance Exchange of Silk in Later Prehistoric Eurasia. In *Interweaving Worlds: Systemic Interactions in Eurasia, 7th to 1st millennia BC.* T. C. Wilkinson, S. Sherratt, and J. Bennet (eds.). Oxford: Oxbow Books. 218-230.

———. 2012. Changes in Fiber use and Spinning Technologies on the Iranian Plateau: A comparative and diachronic study of spindle whorls ca 4500-2500 BCE. *Paléorient* 36 (1): 111-126.

Good, I. L., Kenoyer, J. M., and Meadow, R. H. 2009. New Evidence For Early Silk in the Indus Civilization. *Archaeometry* 51 (3): 457-466.

Gopal, L. 1961. Textiles in ancient India. *Journal of the Economic and Social History of the Orient* 4 (1): 53-69.

Gosselain, O. P. 1999. In Pots We Trust: The Processing of Clay and Symbols in Sub-Saharan Africa. *Journal of Material Culture* 4: 205-230.

Goulder, J. 2010. Administrators' bread: an experiment-based re-assessment of the functional and cultural role of the Uruk bevel-rim bowl. *Antiquity* 84: 351-362.

Götze, A. 1902. Die Kleingeräte aus Metall, Stein, Knochen, u.s.w. In *Troja und Ilion.* W. Dörpfield (ed). Athens: Beck and Barth. 320-423.

Götzelt, T. 1996. *Ansichten der Archäologie Süd-Turkmenistans bei der Erforschung der 'mittleren Bronzezeit' ('Periode' 'Namazga V').* Archäologie in Eurasien 2. Espelkamp: Marie Leidorf.

Grajetzki, W. and Quirke, S. 2000. *Digital Egypt* http://www.digitalegypt.ucl.ac.uk/Welcome.html (accessed September 30, 2009).

Gratuze, B. 1999. Obsidian characterization by laser ablation ICP-MS and its application to prehistoric trade in the Mediterranean and the Near East: sources and distribution of obsidian within the Aegean and Anatolia. *Journal of Archaeological Science* 26 (8): 869-881.

Greenberg, R. 2007. Transcaucasian Colors: Khirbet Kerak Ware at Khirbet Kerak (Tel Bet Yerah). In *Les cultures du Caucase (VIème-IIIème millénaires avant notre ère).* B. Lyonnet (ed). Paris: CNRS éditions. 257-268.

Greenberg, R., Eisenberg, E., Paz, S., and Paz, Y. 2006. *Bet Yerah - The Early Bronze Age Mound: Vol. I - Excavation Reports 1933-1986.* Israel Antiquities Authority Reports 30. Jerusalem: Israel Antiquities Authority.

Greenberg, R. and Goren, Y. 2009. Introduction: Migrating Technologies at the Cusp of the Early Bronze Age III. *Tel Aviv* 36: 129-134.

Greenberg, R. and Porat, N. 1996. A Third Millennium Levantine Pottery Production Center: Typology, Petrography, and Provenance of the Metallic Ware of Northern Israel and Adjacent Regions. *BASOR* 301: 5-24.

Greenhalgh, M. 2006. Iraq Heritage Program: Image: 210 (Image). http://opencontext.org/media/GHF1RES0000005815 (accessed June 10, 2011).

Gulati, A. N. 1961. A note on the early history of silk in India. In *Technical reports on Archaeological Remains.* J. Clutton Brock, V. Mittre, and A. N. Gulati (eds.). Deccan College Post-Graduate and Research Institute, Department of Archaeology and Ancient History Publication 2. Poona: Deccan College.

Gulati, A. N. and Turner, A. J. 1929. A note on the early history of cotton. *Journal of the Textile Institute* 20: 1-9.

Günel, S. 1999. *Panaztepe 2: M.Ö. 2. bine tarihlendirilen Panaztepe seramiğinin Batı Anadolu ve Ege arkeolojisindeki yeri ve önemi.* Ankara: Türk Tarim Kurumu.

Gupta, S. P. 1979. *Archaeology of Soviet Central Asia, and the Indian borderlands.* Delhi: B. R. Publishing Corporation.

Haerinck, E. 2002. Textile remains from Eastern Arabia and new finds from Shakhoura (Bahrain) and ed-Dur (Umm al-Qaiwain, UAE). *Arabian Archaeology and Epigraphy* 13 (2): 246-254.

———. 2011. Painted Pottery of the first half of the Early Bronze Age (Late 4th - first centuries of the 3rd millennium BC) in Luristan, W-Iran. *Iranica Antiqua* 46, 55-106.

Haines, R. C. 1956. Where a Goddess of Love and War Was Worshipped 4000 Years Ago. *Illustrated London News* 229: 266-269.

Hakemi, A. 1972. *Catalogue de l'exposition: Lut. Shahdad "Xabis".* Tehran.

———. 1997. *Shahdad: Archaeological Excavations of a Bronze Age Center in Iran.* S. M. S. Sajjadi (ed). Rome: Istituto Italiano per il Medio ed Estremo Oriente/Centro Scavi e Ricerche Archeologiche.

Hall, M. and Steadman, H. 1991. Anatolia and Tin: Another Look. *Journal of Mediterranean Archaeology* 4 (1): 77-90.

Hall, R. 1986. *Egyptian textiles.* Shire Egyptology 4. Princes Risborough: Shire, 2001.

Halstead, P. and Isaakidou, V. 2011. Revolutionary Secondary Products: the Development and Significance of Milking, Animal-Traction and Wool-Gathering in Later Prehistoric Europe and the Near East. In *Interweaving Worlds: Systemic Interactions in Eurasia, 7th to 1st millennia BC.* T. C. Wilkinson, S. Sherratt, and J. Bennet (eds.). Oxford: Oxbow Books. 61-76.

Hansen, D. P. 1970. A Proto-Elamite Silver Figurine in the Metropolitan Museum of Art. *Metropolitan Museum Journal* 3: 5-26.

Harley, J. B. and Woodward, D. 1987. *The History of Cartography. Volume 1: Cartography in Prehistoric, Ancient, and Medieval Europe and the Mediterranean.* Chicago: University of Chicago Press.

Harley, J. B. and Woodward, D. (eds.). 1992. *The History of Cartography. Volume 2, Book 1: Cartography in the Traditional Islamic and South Asian Societies.* Chicago: University of Chicago Press.

Harmankaya, S. and Erdoğlu, B. 2002. *Türkiye Arkeolojik Yerleşmeleri - 4 - Ilk Tunç* [Archaeological Sites of Turkey - 4 - Early Bronze]. Istanbul: TASK Yayınları.

Harris, S. 2008. Textiles, Cloth and Skins: The Problem of Terminology and Relationship. *Textiles the Journal of Cloth and Culture* 6 (3): 222-237.

Hauptmann, A., Rehren, T., and Schmitt-Strecker. 2003. Early Bronze Age copper metallurgy at Shahr-i Sokhta (Iran), reconsidered. In *Man and Mining—Mensch und Bergbau: Studies in Honour of Gerd Weisgerber: Der Anschnitt, Beiheft 16.* T. Stöllner, G. Koerlin, G. Steffens, and J. Cierny (eds.). Bochum: Deutsches Bergbau-Museum. 197-213.

Hauptmann, A. and Palmieri, A. M. 2000. Metal Production in the Eastern Mediterranean during the 4th/3rd millennium: case studies from Arslantepe. In *Anatolian Metal I.* Ü. Yalçın (ed). Bochum: Deutsches Bergbau-Museum. 75-82.

Hauptmann, A. and Weisgerber, G. 1980. The Early Bronze Age copper metallurgy of Shahr-i Sokhta (Iran). *Paléorient* 6: 120-127.

Haustein, M., Gillis, C., and Pernicka, E. 2010. Tin isotopy - a new method for solving old questions. *Archaeometry* 52 (5): 816-832.

Healey, E. 2007. Obsidian as an Indicator of Inter-Regional Contacts and Exchange: Three Case-Studies from the Halaf Period. *Anatolian Studies* 57: 171-189.

Heinsch, M. and Vandiver, P. 2006. Recent Xeroradiographic Analysis of Kura-Araxes Ceramics. In *Beyond the Steppe and the Sown: Proceedings of the 2002 University of Chicago Conference on Eurasian Archaeology.* D. L. Peterson, L. M. Popova, and A. T. Smith (eds.). Leiden: Brill Academic Publishers. 382-394.

Helwing, B. 2005. Early mining and metallurgy on the western Iranian Plateau: First results of the Iranian-German archaeological research at Arisman, 2000-2004. *Archäologische Mitteilungen aus Iran und Turan* 37: 425-434.

———. 2009. Rethinking the Tin Mountains: Patterns of Usage and Circulation of Tin in Greater Iran from the 4th to the 1st Millennium BC. *TÜBA-AR* 12: 209-221.

Herrmann, G. 1968. Lapis Lazuli: the early phases of its trade. *Iraq* 30: 21-57.

Hiebert, F. T. 1994a. *Origins of the Bronze Age Oasis Civilization in Central Asia.* Cambridge, MA: Peabody Museum of Archaeology and Ethnology, Harvard University.

———. 1994b. Production evidence for the origins of the Oxus Civilization. *Antiquity* 68: 372-387.

———. 1998. Central Asians on the Iranian Plateau: A Model for Indo-Iranian Expansionism. In *The Bronze Age and Early Iron Age Peoples of Eastern Central Asia.* V. H. Mair (ed). The Institute for the Study of Man. 148-161.

———. 2002. Bronze Age Interaction between the Eurasia Steppe and Central Asia. In *Ancient Interactions: East and West in Eurasia.* K. Boyle, C. Renfrew, and M. Levine (eds.). Cambridge: MacDonald Institute for Archaeological Research. 237-248.

———. 2002. The Kopet Dag Sequence of Early Villages in Central Asia. *Paléorient* 28 (2): 25-41.

Hiebert, F. T. and Dyson, R. H. 2002. Prehistoric Nishapur and the Frontier between Central Asia and Iran. *Iranica Antiqua* 37: 113-149.

Hiebert, F. T. and Kurbansakhatov, K. 2003. *A Central Asian village at the dawn of civilization : excavations at Anau, Turkmenistan.* Philadelphia: University of Pennsylvania Museum.

Hiebert, F. T. and Moore, K. M. 1993. New stratigraphic excavations at Gonur Depe (North). In *Information Bulletin,* 19. 96-108.

Hijmans, R. J., Cameron, S. E., Parra, J. P., Jones, P. G., and Jarvis, A. 2005. Very high resolution interpolated climate surfaces for global land areas. *International Journal of Climatology* 25: 1965-1978.

Hirth, F. 1917. The Story of Chang Ki'en, China's Pioneer in Western Asia. *Journal of the American Oriental Society* 37: 89-116.

Hodder, I. and Orton, C. 1976. *Spatial Analysis in Archaeology.* Cambridge: Cambridge University Press.

Hoffman, B. C. and Miller, H. M. L. 2009. Production and Consumption of Copper-base Metals in the Indus Civilization. *Journal of World Prehistory* 22 (3): 237-264.

Holzer, H. F. and Momenzadeh, M. 1971. Ancient Copper Mines in the Veshnoveh Area, Kuhestan-E-Qom, West Central Iran. *Archaeologia Austriaca* 49: 1-22.

Højlund, F. and Andersen, H. H. 1994. *Qala'at al-Bahrain. Vol. 1. The Northern City Wall and the Islamic Fortress.* Aarhus: JASP.

Hoogendoorn, R. M., Boels, J. F., Kroonenberg, S. B., Simmons, M. D., Aliyeva, E., Babazadeh, A. D., and Huseynov, D. 2005. Development of the Kura delta, Azerbaijan; a record of Holocene Caspian sea-level changes. *Marine Geology* 222: 359-380.

Hopkins, L. 2003. *Archaeology at the North-East Anatolian Frontier, VI: An Ethnographical Study of Sos Höyük and Yiğittaşı Village.* Ancient Near Eastern Studies Supplement 14. Louvain: Peeters Press.

Hopkirk, P. 1992. *The Great Game: The Struggle for Empire in Central Asia.* Kodansha International.

Hours, F. (ed.). 1994. *Atlas des sites du Proche Orient (14000-5700BP).* Lyon: Maison de l'Orient méditerranéen.

Howes, D. 2003. *Sensual Relations: Engaging the Senses in Culture and Social Theory.* Ann Arbor: University of Michigan Press.

Huot, J. -L. 1969. La diffusion des épingles à tête à double enroulement. *Syria* 46: 57-98.

———. 2009. Quelques réflexions sur les épingles à double spirale. *Syria* 86: 183-202.

Hutchinson, J. B., Silow, R. A., and Stephens, S. G. 1954. New evidence on the origin of the Old World cottons. *Heredity* 8 (2): 225-241.

Ingold, T. 2000. To journey along a way of life: Maps, wayfaring and navigation. In *The Perception of the Environment: essays in livelihood, dwelling and skill.* London: Routledge. 219-242.

Inizan, M.-L. 1995. Cornaline et agates: Production et circulation de la préhistoire à nos jours. In *Les pierres précieuses de l'Orient ancien des Sumériens auz Sassanides.* F. Tallon (ed). Paris: Éditions de la Réunion des musées nationaux. 21-29.

Isakov, A. H. 1991. *Саразм* [Sarazm]. Dushanbe: Donish.

Isakov, A. and Lyonnet, B. 1988. Céramiques de Sarazm (Tadjikistan, URSS): problèmes d'échanges et de peuplement à la fin du Chalcolithique et au début de l'Age du Bronze. *Paléorient* 14 (1): 31-47.

Iserlis, M. 2009. Khirbet Kerak Ware at Bet Yera: Segregation and Integration through Technology. Tel Aviv 36: 181-195.

Iserlis, M., Greenberg, R., Badalyan, R., and Goren, Y. 2010. Bet Yerah, Aparan III and Karnut I: Preliminary Observations on Kura-Araxes Homeland and Diaspora Ceramic Technologies. *TÜBA-AR* 13: 245-262.

James, M. A., Reifarth, N., Mukherjee, A. J., Crump, M. P., Gates, P. J., Sandor, P., Robertson, F., Pfälzner, P., and Evershed, R. P. 2009. High prestige Royal Purple dyed textiles from the Bronze Age royal tomb at Qatna, Syria. *Antiquity* 83: 1109-1118.

Janaway, R. C. 1989. Corrosion Preserved Textile Evidence: Mechanism, Bias and Interpretation. In *Evidence Preserved in Corrosion Products.* R. C. Janaway and B. Scott (eds.). UKIC Occasional Papers 8. UKIC. 21-29.

Jeffreys, M. D. W. 1951. Some Notes on the Fon of Bikom. *African Affairs* 50 (200): 241-249.

de Jesus, P. S. 1978. Metal Resources in Ancient Anatolia. *Anatolian Studies* 28: 97-102.

———. 1980. *The development of prehistoric mining and metallurgy in Anatolia.* British Archaeological Reports International Series S74. Oxford: BAR.

Ibn Jubayr. 1952. *The Travels of Ibn Jubayr.* R. J. C. Broadhurst (ed). London: Cape.

Kadish, B. 1969. Excavations of Prehistoric Remains at Aphrodisias, 1967. *American Journal of Archaeology* 73: 49-65.

Kaniuth, K. 2006. *Metallobjekte der Bronzezeit aus Nordbaktrien.* Mainz: Verlag Philipp von Zabern.

———. 2007. The Metallurgy of the Late Bronze Age Sapalli Culture (Southern Uzbekistan) and its implications for the 'tin question'. *Iranica Antiqua* 42: 23-40.

Kantner, J. 2004. Geographical Approaches for Reconstructing Past Human Behavior from Prehistoric Roadways. In *Spatially Integrated Social Science: Examples in Best Practice.* M. Goodchild and D. G. Janelle (eds.). Oxford: Oxford University Press. 323-344.

Kamberi, D. 1994. The Three Thousand Year Old Chārchān Man Preserved at Zaghunluq: Abstract Account of a Tomb Excavation in Chārchān County of Uyghuristan. *Sino-Platonic Papers* 44: 1-12.

Karaulašvili, T. 1979. *Vestmik Gosudarstvennogo Muzeja Gruzii* [Peasant Methods of Preparation of Cloth at Kakheti]. [non vidi].

Kavtaradze, G. L. 1999. The importance of metallurgical data for the formation of Central Transcaucasian chronology. In *The Beginnings of Metallurgy: Proceedings of the International Conference "The Beginnings of Metallurgy", Bochum 1995.* A. Hauptmann, E. Pernicka, T. Rehren, and Ü. Yalçın (eds.). Bochum: Deutsches Bergbau-Museum. 67-99.

———. 2004. The Chronology of the Caucasus during the Early Metal Age: Observations from Central Trans-Caucasus. In *A View from the Highlands: archaeological studies in honour of Charles Burney.* A. G. Sagona (ed). Herent: Peeters Press. 539-556.

Kawami, T. S. 1992. Archaeological evidence for textiles in pre-Islamic Iran. *Iranian Studies* 25 (1): 7-18.

Kelly-Buccellati, M. 1990. Trade in Metals in the Third Millennium: Northeastern Syria and Eastern Anatolia. In *Resurrecting the Past: a Joint Tribute to Adnan Bounni.* P. Matthiae, M. van Loon, and H. Weiss (eds.). Istanbul: Nederlands Historisch-Archaeologisch Instituut te Istanbul. 117-131.

———. 1996. Nuzi Viewed from Urkesh, Urkesh Viewed from Nuzi: Stock Elements and Framing Devices in Northern Syro-Mesopotamia. In *Studies on the Civilization and Culture of Nuzi and the Hurrians, volume 8.* D. I. Owen and G. Wilhelm (eds.). Richard F. S. Starr Memorial Volume ed. Bethesda, MD: CDL Press. 247-268.

———. 2004. Andirons at Urkesh: New Evidence for the Hurrian Identity of the Early Trans-Caucasian Culture. In *A View from the Highlands: archaeological studies in honour of Charles Burney.* A. G. Sagona (ed). Herent: Peeters Press. 67-89.

Kendall, A. I. 1937. Chalcolithic Textile Fragments. In *The Alishar Hüyük: Seasons of 1930-32, Part III.* H. Von der Osten (ed). Chicago: University of Chicago Press. 334-335.

Kenoyer, J. M. 1997. Trade and technology of the Indus Valley: new insights from Harappa, Pakistan. *World Archaeology* 29 (2): 262-280.

———. 2003. Stone Beads and Pendant Making Techniques. In *A Bead Timeline: Vol. 1 Prehistory to 1200 CE.* J. W. Lankton. Washington DC: The Bead Museum. 14-19.

Kenoyer, J. M. and Miller, H. M. -L. 1999. Metal Technologies of the Indus Valley Tradition in Pakistan and Western India. In *The Archaeometallurgy of the Asian Old World.* V. C. Pigott (ed). Philadelphia: University of Pennsylvania Museum. 107-151.

Khachikian, L. 1966. The Ledge of the Merchant Hovhannes Joughayetsi. *Journal of the Asiatic Society* 8 (3): 153-186.

Khachatryan, T. S. 1975. *Drevnyaya Kul'tura Skiraka III-I tys. do n.e.* Yerevan. [non vidi].

Khlopin, I. N. 1981. The Early Bronze Age Cemetery of Parkhai II: The First Two Seasons of Excavations, 1977-78. In *The Bronze Age Civilization of Central Asia: Recent Soviet Discoveries.* P. L. Kohl (ed). New York: M. E. Sharpe. 3-34.

———. 1982. The Manufacture of Pile Carpets in the Bronze Age Central Asi. *Halı* 5 (2): 116-118.

———. 1983. *Юго-Западная Туркмения в зпоху поздней Ъронзы* [South-West Turkmenia in the Late Bronze Age]. Leningrad: Leningradskoe Otdelenie.

———. 2002. *Эпоха Бронзы Юго-Западного Туркменистана* [Bronze Age of South-West Turkmenistan]. Saint-Petersburg: Russian Academy of Sciences.

Khlopina, L. I. 1981. Namazga-depe and the Late Bronze Age of Southern Turkmenia. In *The Bronze Age Civilization of Central Asia: Recent Soviet Discoveries.* P. L. Kohl (ed). New York: M. E. Sharpe. 35-60.

Kiguradze, T. and Sagona, A. G. 2003. On the Origins of the Kura-Araxes Cultural Complex. In *Archaeology in the borderlands: investigations in Caucasia and beyond.* A. T. Smith and K. S. Rubinson (eds.). Los Angeles: Cotsen Institute of Archaeology, UCLA. 38-94.

Killick, D. 2009. Cairo to Cape: The Spread of Metallurgy Through Eastern and Southern Africa. *Journal of World Prehistory* 22: 399-414.

Kircho, L. B., Korobkova, G. F., and Masson, V. M. 2008. *Технико-Технологический Потенция Алтын-Депе как Основа Становления Раннегородской Цивилизации* [The Technical and Technological Potential of the Eneolithic Population of Altyn-Depe as the basis of the rise of an Early Urban Civilization]. St Petersburg: European House.

Kirtcho, L. B. 2009. The Earliest Wheeled Transport in Southwestern Central Asia: New Finds from Altyn-Depe. *Archaeology, Ethnology and Anthropology of Eurasia* 37 (1): 25-33.

Knappett, C. 2002. Photographs, Skeuomorphs and Marionettes: Some Thoughts on Mind, Agency and Object. *Journal of Material Culture*, 7: 97-117.

———. 2011. *An archaeology of interaction : network perspectives on material culture and society*. Oxford ; New York: Oxford University Press.

Knappett, C., Evans, T., and Rivers, R. 2011. The Theran eruption and Minoan palatial collapse: new interpretations gained from modelling the maritime network. *Antiquity* 85: 1008-1023.

Kohl, P. L. 1978. *Seeds of upheaval : the production of chlorite at Tepe Yahya and an analysis of commodity production and trade in Southwest Asia in the mid-third millenium.* PhD dissertation, Harvard University. [Unpublished thesis].

———. 1981. The Namazga Civilization: An Overview. In *The Bronze Age Civilization of Central Asia: Recent Soviet Discoveries.* P. L. Kohl (ed). New York: M. E. Sharpe. vii-xl.

———. 1984. *Central Asia: Palaeolithic beginnings to the Iron Age / L'Asie centrale: des origines á l'âge du fer.* Paris: Editions Recherche sur les Civilisations.

———. 1987a. The ancient economy, transferable technologies and the Bronze Age world-system: a view from the northeastern frontier of the Ancient Near East. In *Centre and Periphery in the Ancient World.* M. Rowlands, M. Larsen, and K. Kristiansen (eds.). Cambridge: Cambridge University Press. 13-24.

———. 1987b. The Use and Abuse of World Systems Theory: The Case of the Pristine West Asian State. *Advances in Archaeological Method and Theory* 11: 1-35.

———. 1988. The Northern "Frontier" of the Ancient Near East: Transcaucasia and Central Asia Compared. *American Journal of Archaeology* 92 (4): 591-596.

———. 1992. Central Asia (Western Turkestan): Neolithic to the Early Iron Age. In *Chronologies in Old World archaeology.* R. W. Ehrich (ed). Chicago: University of Chicago Press. 179-195.

———. 2001. Migrations and Cultural Diffusions in the Later Prehistory of the Caucasus. In *Migration und Kulturtransfer: Der Wandel vorder- und zentralasiatischer Kulturen im Umbruch vom 2. zum 1. vorchristlichen Jahrtausend.* R. Eichmann and H. Parzinger (eds.). Akten des Internationalen Kollquiums Berlin, 23. bis 26. November 1999. Bonn: Dr. Rudolf Habelt. 313-328.

———. 2006. The Early Integration of the Eurasian Steppes with the Ancient Near East: Movements and Transformations in the Caucasus and Central Asia. In *Beyond the Steppe And the Sown: Proceedings of the 2002 University of Chicago Conference on Eurasian Archaeology.* D. L. Peterson, L. M. Popova, and A. T. Smith (eds.). Leiden: Brill Academic Publishers. 3-39.

———. 2007. *The Making of Bronze Age Eurasia.* Cambridge: Cambridge University Press.

———. 2009a. The Maikop Singularity: the Unequal Accumulation of Wealth on the Bronze Age Eurasian Steppe? In *Social Complexity in Prehistoric Eurasia: Monuments, Metals and Mobility.* B. K. Hanks and K. M. Linduff (eds.). Cambridge: Cambridge University Press. 91-103.

———. 2009b. Origins, Homelands and Migrations: Situating the Kura-Araxes Early Transcaucasian 'Culture' within the History of Bronze Age Eurasia. *Tel Aviv: Journal of the Institute of Archaeology of Tel Aviv University* 36 (2): 241-265.

———. 2011. World-Systems and Modelling Macro-Historical Processes in Later Prehistory: an Examination of Old and a Search for New Perspectives. In *Interweaving Worlds: Systemic Interactions in Eurasia, 7th to 1st millennia BC.* T. C. Wilkinson, S. Sherratt, and J. Bennet (eds.). Oxford: Oxbow Books. 77-86.

Kökten, K., Özgüç, N., and Özgüç, T. 1945. Türk Tarih Kurumu Asına Yapılan Samsun Bölgesi Kazıları Hakkinda İlk Kısa Rapor. *Belleten* 9 (35): 361-400.

Korenevsky, S. N. 1995. *Galugay I — The settlement of Maikop culture : the archaeological source on the problem of ancient farmers and cattle-breeders on the Caucasus's boundary of the Near East and the East Europa.* Biblioteka Rossiĭskogo ėtnografa. Moskva: In-t arkheologii RAN.

Korfmann, M. 1982. *Tilkitepe: Die ersten Ansätze prähistorischer Forschung in der östlichen Türkei.* Istanbuler Mitteilungen, Beiheft 26. Tübingen: Verlag Ernst Wasmuth.

Korobkova, G. F. 1981. Ancient Reaping Tools and Their Productivity in the Light of Experimental Tracewear Analysis. In *The Bronze Age Civilization of Central Asia: Recent Soviet Discoveries.* P. L. Kohl (ed). New York: M. E. Sharpe. 325-349.

Koryakova, L. and Epimakhov, A. 2007. *The Urals and Western Siberia in the Bronze and Iron Ages.* Cambridge: Cambridge University Press.

Kosarev, A. N. and Yablonskaya, E. A. 1994 [1987]. *The Caspian Sea.* The Hague: SPB Academic Publishing.

Košak, S. 1985. The gospel of iron. In *Kaniššuwar. A Tribute to H. G. Giiterbock on His 75th Birthday.* Chicago: University of Chicago Press. 125-135.

Koşay, H. Z. 1951. *Alaca Höyük Kazısı 1937-1939'daki çalışmalara ve keşiflere ait ilk rapor.* Ankara: Türk Tarim Kurumu.

Koşay, H. Z. and Turfan, K. 1959. Erzurum Karaz kazısı raporu. *Belleten* 23: 349-413.

Köhler-Rollefson, I. 1993. Camels and camel pastoralism in Arabia. *The Biblical Archaeologist* 56 (4): 180-188.

Kroll, S. E. 2002. Metalle und Metallurgie in Nordwest-Iran und Armenien im Chalkolithikum und in der Frühen Bronzezeit. In *Anatolian Metal II.* Ü. Yalçın (ed). Bochum: Deutsches Bergbau-Museum. 131-136.

Kuftin, B. A. 1950. *Arkheologicheskie izyskanija v Rionskoj Nizmenosti.* Tbilisi.

Kunze, R., Bobokhyan, A., Pernicka, E., and Meliksetian, K. 2013. Project Ushkiani. Investigations of the cultural landscape around the Sotk prehistoric gold-working region. In *Archäologie in Armenien II.* H. Meller and P. Avetisyan (eds.). Veröffentlichungen des Landesamtes für Denkmalpflege und Archäologie Sachsen-Anhalt. Halle: Landesmuseum für Vorgeschichte. 49-88.

Kushnareva, K. 1993. *Южный Кавказ в IX-II тыс. до н.э.* [South Caucasus IX-II mil. BC (Stages of cultural and social-economical evolution)]. Sankt-Peterburg: Russian Academy of Sciences.

———. 1997. *The Southern Caucasus in prehistory. Stages of cultural and socioeconomic development from the 8th to the 2nd millennium B.C.* Philadelphia: University of Pennsylvania Museum.

Kushnareva, K. and Chubinishvili, T. 1970. *Древнце Культуры Южного Кавказа (V-III тыс. до Н. 3.)* [Ancient Cultures of Southern Caucasus]. Izdatel'stvo Nauka; USSR Academy of Sciences, Institute of Archaeology ed. Leningrad: Leningradskoe Otdelenie.

Kushnareva, K. and Markovin, V. I. 1994. *Epokha Bronzy Kavkaza i Srednei Azii: Rannyay i srednyaya bronza Kavkaza* [The Bronze Age in Caucasia and Central Asia: Early and Middle Bronze Caucasia]. Izdatel'stvo Nauka; USSR Academy of Sciences, Institute of Archaeology ed. Moscow: Nauka.

Kuzmina, E. E. 2007. *The Prehistory of the Silk Road.* V. Mair (trans.) Philadelphia: University of Pennsylvania Press.

Kuznetsov, P. F. 2006. The emergence of Bronze Age chariots in eastern Europe. *Antiquity* 80: 638-645.

Kühne, H. 1976. *Die Keramik vom Tell Chuera und ihre Beziehungen zu Funden aus Syrien-Palästina, der Türkei und dem Iraq.* Berlin: Gebr. Mann Verlag.

Kühne, H. and Schneider, G. 1988. Neue Untersuchungen zur Metallischen Ware. *Damaszener Mitteilungen* 3: 83-139.

Kvachidze, V. 2006. First Expeditions Beneath the Caspian. *Azerbaijan International* 14 (4): 58-65. http://azer.com/aiweb/categories/magazine/ai144_folder/144_articles/144_viktor_index.html (accessed October 5, 2011).

Lacaisne, Z. 1912. Notes sur des tissus recouvrant des haches de cuivre. *Mémoires de la Délégation en Perse* 13: 163-64.

Lacy, A. D. 1967. *Greek pottery in the Bronze Age.* London: Methuen.

Lal, B. B. 1992. The Painted Grey Ware Culture of the Iron Age. In *History of civilizations of Central Asia. Volume I: The dawn of civilization: earliest times to 700BC.* A. H. Dani and V. M. Masson (eds.). Paris: UNESCO Publishing. 421-440.

Lamberg-Karlovsky, C. C. 1975. Third Millennium Modes of Exchange and Modes of Production. In *Ancient Civilization and Trade.* J. Sabloff and C. C. Lamberg-Karlovsky (eds.). Santa Fe: University of New Mexico Press. 341-368.

———. 1985. The Longue Durée of the Ancient Near East. In *De l'Indus aux Balkans, Recueil à la mémoire de Jean Deshayes.* J. -L. Huot, M. Yon, and Y. Calvet (eds.). Paris: Editions Recherche sur les Civilisations. 55-72.

———. 1989. Mesopotamia, Central Asia and the Indus Valley: so the kings were killed. In *Archaeological Thought in America.* C. C. Lamberg-Karlovsky (ed). Cambridge: Cambridge University Press. 241-267.

Larsen, M. T. 1967. *Old Assyrian Caravan Procedures.* Istanbul: Nederlands Historisch-Archaeologisch Instituut.

Latour, B. 2005. *Reassembling the Social: an Introduction to Actor-Network-Theory.* Oxford: Oxford University Press.

Law, R. W. 2011. *Inter-regional interaction and urbanism in the ancient Indus Valley: a geologic provenience study of Harappa's rock and mineral assemblage.* Kyoto: Research Institute for Humanity and Nature.

Lawler, A. 2003. Jiroft discovery stuns archaeologists. *Science* 302 (5647): 973-974.

———. 2004. The Indus Script—Write or Wrong? *Science* 306 (5704): 2026.

Le Bigre, P. 1996. Les caravanserails routiers du Moyen-Orient (IXème-XIXème siècles), Construction d'un regard – Découverte d'un langage. *Les Annales Archéologiques Arabes Syriennes* 42: 217-231.

———. 2004. L'inventaire des Caravanserails en Asie centrale. UNESCO (ed.). http://www.unesco.org/culture/dialogue/eastwest/pdf/brochure-en.pdf (accessed 6th February, 2008).

Lehmann, G. 2002. *Bibliographie der archäologischen Fundstellen und Surveys in Syrien und Libanon.* Orient-Archäologie 9. Rahden: Marie Leidorf.

Lehner, B., Verdin, K., and Jarvis, A. 2008. HydroSHEDS: Technical Documentation, version 1.1. http://gisdata.usgs.net/HydroSHEDS/downloads/HydroSHEDS_TechDoc_v11.pdf (accessed July 24, 2009).

Leroi-Gourhan, A. 1964. *Le Geste et la parole. I, Technique et langage.* Paris: Albin Michel.

Levine, M. 1999. The Origins of Horse Husbandry on the Eurasian Steppe. In *Late prehistoric exploitation of the Eurasian steppe.* M. Levine, Y. Rassamakin, A. Kislenko, and N. Tatarintseva (eds.). Cambridge: McDonald Institute for Archaeological Research. 5-58.

Lévi-Strauss, C. 1983 [1963]. *The Raw and the Cooked: Mythologiques, Volume 1.* Chicago: University of Chicago Press, 1983.

Levy, T. E., Adams, R. B., Hauptmann, A., Prange, M., Schmitt-Strecker, S., and Najjar, M. 2002. Early Bronze Age metallurgy: a newly discovered copper manufactory in southern Jordan. *Antiquity* 76: 425-437.

Ligabue, G. and Salvatori, S. 1990. *Bactria: An ancient oasis civilization from the sands of Afghanistan.* Venice: Erizzo Editrice.

Lindsay, I., Smith, A. T., and Badalyan, R. 2010. Magnetic survey in the investigation of sociopolitical change at a Late Bronze age fortress settlement in northwestern Armenia. *Archaeological Prospection* 17 (1): 15-27.

Littauer, M. A. and Crouwel, J. H. 1979. *Wheeled Vehicles and Ridden Animals in the Ancient Near East.* Leiden: E. J. Brill.

Lloyd, S. and Mellaart, J. 1962. *Beycesultan, Vol. I: the Chalcolithic and Early Bronze Age Levels.* London: British Institute of Archaeology at Ankara.

Lo Giudice, A., Re, A., Calusi, S., Giuntini, L., Massi, M., Olivero, P., Pratesi, G., Albonico, M., and Conz, E. 2009. Multitechnique characterization of lapis lazuli for provenance study. *Analytical and Bioanalytical Chemistry* 395 (7): 2211-2217.

Lombard, P. 1986. Une coupe à boire en argent et ses accessoires à Tell Khazneh. In *Failaka. Fouilles françaises 1984–1985.* Y. Calvet and J. -F. Salles (eds.). Lyon: TMO. 281-290.

Lyonnet, B. 2000. La Mésopotamie et le Caucase du Nord au IVe et au début du IIIe millenaires av. N. E.: Chronologiques de la culture de Majkop. État de la question et nouvelles propositions. In *Chronologies des Pays du Caucase et de l'Euphrate aux IVe-IIIe Millenaires.* C. Marro and H. Hauptmann (eds.). Istanbul: Institut Français d'Etudes Anatoliennes. 299-320.

———. 2005. Another Possible Interpretation of the Bactro-Margiana Culture (BMAC) of Central Asia: The Tin Trade. In *South Asian Archaeology 2001.* C. Jarrige and V. Lefevre (eds.). Paris: Editions Recherche sur les Civilisations. 191-200.

———. 2007. La culture de Maïkop, la Transcaucasie, l'Anatolie orientale et le Proche-Orient: relations et chronologie. In *Les cultures du Caucase (VIème-IIIème millénaires avant notre ère)*. B. Lyonnet (ed). Paris: CNRS éditions. 133-162.

MacDermot, B. C. and Schippman, K. 1999. Alexander's march from Susa to Persepolis. *Iranica Antiqua* 34: 283-308.

Macdonald Kinneir, J. 1818. *Journey through Asia Minor, Armenia, and Koordistan in the years 1813 and 1814; with remarks on the marches of Alexander and retreat of the ten thousand.* London: John Murray.

Madjidzadeh, Y. 2003. *Jiroft: the Earliest Oriental Civilization.* Tehran: Iranian Cultural Heritage Organization.

Magee, P., Petrie, C., Knox, R., Khan, F., and Thomas, K. 2005. The Achaemenid Empire in South Asia and Recent Excavations in Akra in Northwest Pakistan. *American Journal of Archaeology* 109 (4): 711-741.

Magomedov, R. 2006. The Kura-Araxes 'Culture' in the North-Eastern Caucasus: Problems in its Identification and Chronology. In *Beyond the Steppe And the Sown: Proceedings of the 2002 University of Chicago Conference on Eurasian Archaeology.* D. L. Peterson, L. M. Popova, and A. T. Smith (eds.). Leiden: Brill Academic Publishers. 142-159.

Mahfroozi, A., Piller, C. K., Ghasemi, S., Bagherpour, N., Neumann, T., and Ögüt, B. 2009. First preliminary report on the joint Iranian-German excavations at Gohar Tappe, Māzandarān, Iran. *Archäologische Mitteilungen aus Iran und Turan* 41: 177-209.

Majidzadeh, Y. 1982. Lapis lazuli and the Great Khorasan Road. *Paléorient* 8 (1): 59-69.

———. 2008. Excavations at Konar Sandal in the region of Jiroft in the Halil Basin: First preliminary report (2002-2008). *Iran* 46: 69-104.

Mallory, J. P. and Mair, V. H. 2000. *The Tarim Mummies: Ancient China and the Mystery of the Earliest Peoples from the West.* London: Thames and Hudson.

Manhart, C. 2001. The Afghan Cultural Heritage Crisis: UNESCO's Response to the Destruction of Statues in Afghanistan. *American Journal of Archaeology* 105 (3): 387-388.

Marchetti, N. 2000. Clay Figurines of the Middle Bronze Age from Northern Inner Syria: Chronology, Symbolic Meaning and Historical Relations. In *Proceedings of the First International Congress on the Archaeology of the Ancient Near East. Rome, May 18th-23rd 1998, vol. I.* P. Matthiae, A. Enea, L. Peyronel, and F. Pinnock (eds.). Roma: Dipartimento di Scienze Storiche, Archaeologiche e Antropologiche dell'Antichità. 839-867.

Marcolongo, B. 2002. Geoarchaeology and Palaeoenvironment Evolution of the Merv Plain. In *Margiana: Gonur-depe Necropolis. 10 years of excavations by Ligabue Study and Research Centre.* G. Rossi Osmida (ed). Padua: Il Punto Edizione. 57-67.

Marcolongo, B. and Mozzi, P 1998. Outline of Recent Geological History of the Kopet-Dagh Mountains and the Southern Kara-Kum. In *The archaeological map of the Murghab Delta: Preliminary reports 1990-1995.* A. Gubaev, G. Koshelenko, and M. Tosi (eds.). Rome: Istituto Italiano per l'Africa et l'Oriente. 1-13.

Margueron, J. -C. (ed.). 2004. *Mari: Métropole de l'Euphrate au IIIe et au début du IIe millénaire av. J.-C.* Paris: Picard.

Markofsky, S. 2010. *Illuminating the Black Sands: Survey and Settlement in the Bronze Age Murghab Delta, Turkmenistan*. PhD Thesis, University College London. [Unpublished thesis].

Marom, N. and Bar-Oz, G. 2009. Culling profiles: the indeterminacy of archaeozoological data to survivorship curve modelling of sheep and goat herd maintenance strategies. *Journal of Archaeological Science* 36 (5): 1184-1187.

Marro, C. 1997. *La Culture du Haut-Euphrate au Bronze Ancien: Essai d'Interpretation a partir de la Ceramique Peinte de Keban (Turquie)*. Paris: Institude Français de l'Anatolie.

———. 2004. Upper Mesopotamia and the Caucasus: an Essay on the Evolution of Routes and Road Networks from the Old Assyrian Kingdom to the Ottoman Empire. In *A View from the Highlands: archaeological studies in honour of Charles Burney*. A. G. Sagona (ed). Herent: Peeters Press. 91-120.

Marro, C., Bakhshaliyev, V., and Sanz, S. 2010. Archaeological investigations on the salt mine of Duzdağı (Nakhchivan, Azerbaijan). Türkiye Bilimler Akademisi Arkeoloji Dergisi (*TÜBA-AR*) 13: 229-244.

Marshall, J. 1931. *Mohenjo-daro and the Indus Civilization*. London: Arthur Probsthain.

Massa, M. 2011. Anatolian travels: analysing communication routes in the late prehistoric of Asia Minor. *Heritage Turkey* 1: 39.

Masson, V. M. 1968. The Urban Revolution in South Turkmenia. *Antiquity* 42: 178-187.

———. 1988. *Altyn-Depe*. H. Michael (trans.) Philadelphia: The University Museum, Pennsylvania.

———. 1992. The Bronze Age in Khorasan and Transoxania. In *History of civilizations of Central Asia. Volume I: The dawn of civilization: earliest times to 700BC*. A. H. Dani and V. M. Masson (eds.). Paris: UNESCO Publishing. 225-245.

Masson, V. M. and Merpert, N. J. 1982. *Энеолит СССР* [Eneolithic USSR]. Moskva: Nauka.

Masson, V. M. and Sarianidi, V. I. 1972. *Central Asia: Turkmenia Before the Achaemenids*. London: Thames & Hudson.

———. 1973. *Среднеазиатская Терракота Эпохи Бронзы* [Central Asian Terracotta Figurines of the Bronze Age]. Moskva: Izdatelstvo «Nauka».

Masuda, S. 1974. Tepe Sang-e Čaxamaq. *Iran* 12: 222-223.

Matheson, S. 1976. *Persia: An Archaeological Guide*. 2nd ed. London.

Matthews, R. and Fazeli, H. 2004. Copper and Complexity: Iran and Mesopotamia in the Fourth Millennium B.C. *Iran* 42: 61-75.

Maunsell, F. R. 1894. *The Geography of Eastern Turkey in Asia*. Aldershot: Gale & Polden.

Mauss, M. 1935. Les techniques du corps. *Journal de Psychologie* 32: 271-293.

Maxwell-Hyslop, K. R. 1974. Assyrian sources of iron. A preliminary survey of the historical and geographical evidence. *Iraq* 36 (1/2): 139-154.

McCorriston, J., Evans, S. T., Huot, J. L., Liverani, M., Morrison, K., Potts, D., Sherratt, A., Wright, R. P., and Zettler, R. L. 1997. The Fiber Revolution: Textile Extensification, Alienation, and Social Stratification in Ancient Mesopotamia [and Comments and Reply]. *Current Anthropology* 38 (4): 517-549.

McGeehan-Liritzis, V. 1987. Yugoslavian Tin Deposits and the Early Bronze Age Industries of the Aegean Region. *Oxford Journal of Archaeology* 6 (3): 287-300.

McGovern, P. and Notis, M. D. (eds.). 1989. *Cross-craft and Cross-cultural Interactions in Ceramics*. Ceramics and Civilization 4. Ohio: Westerville.

McNaughton, P. R. 1988. *The Mande Blacksmiths: Knowledge, Power and Art in West Africa*. Bloomington: Indiana University Press.

Mei, J. and Shell, C. 1998. Copper and Bronze Metallurgy in Later Prehistoric Xinjiang. In *The Bronze Age and Early Iron Age Peoples of Eastern Central Asia*. V. H. Mair (ed). Washington, DC: The Institute for the Study of Man.

Meiggs, D. C. 2009. *Investigation of neolithic ovicaprine herding practices by multiple isotope analysis: A case study at PPNB grittile, southeastern Turkey*. PhD thesis, University of Wisconsin-Madison. [Unpublished thesis].

Mellaart, J. 1954. Preliminary report on a survey of pre-Classical remains in southern Turkey. *Anatolian Studies* 4: 175-240.

———. 1958. Excavations at Hacılar. First Preliminary Report. *Anatolian Studies 8*: 127-156.

———. 1966. *The Chalcolithic & Early Bronze Ages in the Near East & Anatolia*. Beirut: Khayats.

———. 1968. Anatolian trade with Europe and Anatolian geography and culture provinces in the Late Bronze Age. *Anatolian Studies* 18: 187-202.

Mellink, M. J. 1969. Excavations at Karataş-Semayük in Lycia, 1968. *American Journal of Archaeology* 73 (3): 319-331.

———. 1982. Archaeology in Asia Minor. *American Journal of Archaeology* 86 (4): 557-576.

———. 1990. Archaeology in Anatolia. *American Journal of Archaeology* 94 (1): 125-151.

———. 1992. Anatolia. In *Chronologies in Old World archaeology*. R. W. Ehrich (ed). Chicago: University of Chicago Press. 207-221.

———. 1993. The Anatolian South Coast in the Early Bronze Age: the Cilician Perspective. In *Between the Rivers and Over the Mountains: archaeologica anatolica et mesopotamica Alba Palmieri dedicata*. M. Frangipane, H. Hauptmann, M. Liverani, P. Matthiae, and M. Mellink (eds.). Roma: Dipartimento di Scienze Storiche, Archaeologiche e Antropologiche Dell'Antichità. 495-508.

———. 1998. Anatolia and the Bridge From East to West in the Early Bronze Age. *TÜBA-AR* 1: 1-8.

Menze, B. H. and Ur, J. A. 2012. Mapping patterns of long-term settlement in Northern Mesopotamia at a large scale. *Proceedings of the National Academy of Sciences* 109 (14): E778-E787.

Meyer, K. and Brysac, S. 1999. *Tournament of Shadows: The Great Game and the Race for Empire in Asia*. Berkeley: Counterpoint.

Michailidou, A. 2010. Measuring by weight in the Late Bronze Age Aegean: The people behind the measuring tools. In *The Archaeology of Early Quantification and Cosmology: comprehending Heaven, Earth and time in ancient societies.* I. Morley and C. Renfrew (eds.). Cambridge: Cambridge University Press. 71-87.

Michel, C. 1992. Transporteurs, responsables et propiétaries de convois dans les tablettes paléo-assyriennes. In *La circulation des biens, des personnes et des idées dans le Proche-Orient, XXXVIII R.A.I.* D. Charpin and F. Joannas (eds.). Paris: Editions Recherche sur les Civilisations. 137-156.

———. 2001. *Correspondance des marchands de Kaniš au début du IIe millénaire avant J.-C.* Littératures anciennes du Proche-Orient 19. Paris: Les Éditions du Cerf.

———. 2003. *Old Assyrian Bibliography of Cuneiform Texts, Bullae, Seals and the Results of the Excavations at Aššur, Kültepe/Kaniš, Acemhöyük, Alişar and Boğazköy.* Old Assyrian Archives Studies 1. Leiden.

———. 2005. Old Assyrian Bibliography 1. *Archiv Für Orientforschung* 51: 436-449.

Michel, C. and Nosch, M.-L. (eds.). 2010. *Textile Terminologies in the Ancient Near East and Mediterranean from the Third to the First Millennia BC.* Ancient Textiles 8. Oxford: Oxbow Books.

Michel, C. and Veenhof, K. R. 2010. The Textiles Traded by the Assyrians in Anatolia (19th-18th centuries BC). In *Textile Terminologies in the Ancient Near East and Mediterranean from the Third to the First Millennia BC.* C. Michel and M.-L. Nosch (eds.). Ancient Textiles 8. Oxford: Oxbow Books. 210-271.

Migne, J. P. 1864. *Patrologia Graeca.* Paris.

de Miroschedji, P. 2000a. An Early Bronze Age III Pottery Sequence for Southern Israel. In *Ceramics and change in the early Bronze Age of the southern Levant.* G. Philip and D. Baird (eds.). London: Continuum International Publishing Group. 315-346.

———. 2000b. La céramique de Khirbet Kerak en Syro-Palestine: état de la question. In *Chronologies des Pays du Caucase et de l'Euphrate aux IVe-IIIe Millenaires.* C. Marro and H. Hauptmann (eds.). Istanbul: Institut Français d'Etudes Anatoliennes.

Mirtskhulava, G. 2011. Problems of the Kura-Araxes Culture. *Studies of the Society of Assyriologists, Biblical Studies and Caucasiiologists* 5. Tbilisi.

Mohammadifar, Y., Motarjem, A., and Khorasani, T. 2009. Tepe Pissa: new investigations at a Kura-Araxes site in central western Iran. *Antiquity (Project Galley).* http://antiquity.ac.uk/ProjGall/mohammadifar1/ .

Momenzadeh, M., Hajisoltan, A., and Momenzadeh, M. 2004. Metallische Bodenschätz in Iran in antiker Zeit. Ein kurzer Überblick. In *Persiens Antike Pracht.* T. Stöllner, R. Slotta, and A. Vatandoust (eds.). Katalog der Ausstellung des Deutschen Bergbau-Museums Bochum vom 28. November 2004 bis 29. Mai 2005. Bochum: Deutsches Bergbau-Museum. 8-21.

Momenzadeh, M. and Sadighi, T. 1989. Places Names. A Guide in Detecting Ancient Gold Mines in Iran. In *Archäometallurgie der Alten Welt: Beiträge zum Internaionalen Symposium "Old World Archaeometallurgy" Heidelberg 1987.* A. Hauptmann, E. Pernicka, and G. A. Wagner (eds.). Bochum: Deutsches Bergbau-Museum. 307-318.

Moorey, P. R. S. 1982. Archaeology and Pre-Achaemenid Metalworking in Iran: a Fifteen Year Retrospective. *Iran* 20: 81-101.

———. 1986. The Emergence of the Light, Horse-Drawn Chariot in the Near-East c. 2000-1500 BC. *World Archaeology* 18 (2): 196-215.

———. 1994. *Ancient Mesopotamian Materials and Industries: the Archaeological Evidence.* Oxford: Clarendon Press.

———. 1999. *Ancient mesopotamian materials and industries: the archaeological evidence.* Winona Lake, IA: Eisenbrauns.

———. 2003. *Idols of the People: Miniature Images of Clay in the Ancient Near East.* The Schweich Lectures of the British Academy 2001. Oxford: Oxford University Press.

Moran, W. L. 1992. *The Amarna Letters.* Baltimore: Johns Hopkins University Press.

Morgan, L. 1995. Minoan Painting and Egypt: The Case of Tell el-Dab'a. In *Egypt, the Aegean and the Levant: Interconnections in the Second Millennium BC.* W. V. Davies and L. Schofield (eds.). London: British Museum Press.

Motzenbäcker, I. 1996. *Sammlung Kossnierska: der Digorische Formenkreis der Kaukasischen Bronzezeit.* Bestandskataloge 3. Berlin: Museum für Vor- und Frühgeschichte.

Moulherat, C., Tengberg, M., Haquet, J. F., and Mille, B. 2002. First evidence of cotton at Neolithic Mehrgarh, Pakistan: Analysis of mineralized fibres from a copper bead. *Journal of Archaeological Science* 29 (12): 1393-1401.

Muhly, J. D. 1973. *Copper and Tin. The Distribution of Mineral Resources and the Nature of the Metals Trade in the Bronze Age.* Transactions of the Connecticut Academy of Arts and Sciences 46. Hamden: Archib Books.

———. 1985. Sources of Tin and the Beginnings of Bronze Metallurgy. *American Journal of Archaeology* 85 (2): 275-291.

———. 1989. Çayönü Tepesi and the Beginnings of Metallurgy in the Old World. In *Archäometallurgie der Alten Welt: Beiträge zum Internaionalen Symposium "Old World Archaeometallurgy" Heidelberg 1987.* A. Hauptmann, E. Pernicka, and G. A. Wagner (eds.). Bochum: Deutsches Bergbau-Museum. 1-12.

———. 1993. Early Bronze Age Tin and the Taurus. *American Journal of Archaeology* 97 (2): 239-253.

———. 1999. Copper and Bronze in Cyprus and the Eastern Mediterranean. In *The Archaeometallurgy of the Asian Old World.* V. C. Pigott (ed). Philadelphia: University of Pennsylvania Museum. 15-25.

———. 2005. Cyprus and Copper for the World. In *Anatolian Metal III.* Ü. Yalçın (ed). Bochum: Deutsches Bergbau-Museum. 137-142.

Muhly, J. D., Begemann, F., Öztunalı, Ö., Pernicka, E., Schmitt-Strecker, S., and Wagner, G. 1991. The Bronze Age Metallurgy of Anatolia and the Question of Local Tin Sources. In *Archaeometry '90.* E. Pernicka and G. A. Wagner (eds.). Switzerland: Birkhauser Verlag. 209-220.

Muhly, J. D., Maddin, R., and Karageorghis, V. (eds.). 1982. *Early Metallurgy in Cyprus 4000-500BC.* International Archaeological Symposium (1981 Larnaca, Cyprus). Nicosia: Pierides Foundation.

Muhly, J. D., Maddin, R., Stech, T., and Özgen, E. 1985. Iron in Anatolia and the Nature of the Hittite Iron Industry. *Anatolian Studies* 35: 67-84.

Munchaev, R. M. 2005. Месопотамия, Кавказ и Циркумпонтийская Металлургическая Провинция [Mesopotamia, the Caucasus and the Circumpontic metallurgical province]. *Rossiyskaya Arkheologiya* 4: 13-24.

Muscarella, O. W. 2001. Jiroft and Jiroft-Aratta a review article of Yousef Madjidzadeh, Jiroft: The earliest oriental civilization. *Bulletin of the Asia Institute* 15: 173-198.

Nakou, G. 2007. Absent Presences: Metal Vessels in the Aegean at the End of the Third Millennium. In *Metallurgy in the Early Bronze Age Aegean.* P. M. Day and R. C. Doonan (eds.). Oxford: Oxbow Books. 225-244.

Narimanov, I. 2004. Archaeological Sites of the Early Bronze Age in North Azerbaijan: A Gazetteer. In *A View from the Highlands: archaeological studies in honour of Charles Burney.* A. G. Sagona (ed). Herent: Peeters Press. 467-473.

Negahban, E. O. 1996. *Marlik, the Complete Excavation Report.* University Museum monograph 87. Philadelphia: University of Pennsylvania Museum.

Newhard, J. M. L., Levine, N., and Rutherford, A. 2008. Least-cost pathway analysis and inter-regional interaction in the Göksu valley, Turkey. *Anatolian Studies* 58: 87-102.

Nezafati, N., Pernicka, E., and Momenzadeh. 2006. Ancient tin: Old question and a new Answer. *Antiquity (Project Gallery).* http://antiquity.ac.uk/ProjGall.

Niemeier, B. and Niemeier, W. -D. 2000. Aegean Frescoes in Syria-Palestine: Alalakh and Tel Kabri. In *The Wall Paintings of Thera, Proceedings of the First International Symposium.* S. Sherratt (ed). Athens: Thera Foundation. 763-802.

Northover, J. P. 1989. Properties and Use of Arsenic-Copper Alloys. In *Archäometallurgie der Alten Welt: Beiträge zum Internaionalen Symposium "Old World Archaeometallurgy" Heidelberg 1987.* A. Hauptmann, E. Pernicka, and G. A. Wagner (eds.). Bochum: Deutsches Bergbau-Museum. 111-118.

Oates, D., Oates, J., and McDonald, H. (eds.). 2001. *Excavations at Tell Brak. Vol. 2: Nagar in the third millenium BC.* McDonald Institute of Archaeological Research and British School of Archaeology in Iraq. Oates, J. 2003. A Note on the Early Evidence for Horse and the Riding of Equids in Western Asia. In *Prehistoric steppe adaptation and the horse.* M. Levine, C. Renfrew, and K. Boyle (eds.). Cambridge: McDonald Institute for Archaeological Research. 115-125.

Oates, J., McMahon, A., Karsgaard, P., Quntar, S. A., and Ur, J. 2007. Early Mesopotamian urbanism: a new view from the north. *Antiquity* 81: 585-600.

Ökse, A. T. 2005. Kızılırmak ve Fırat Havzalarını Birbirine Bağlayan Eski Kervan Yolları. *Bilig* 34: 15-32.

———. 2007. Ancient mountain routes connecting central Anatolia to the upper Euphrates region. *Anatolian Studies* 57: 35-45.

Öktü, A. G. 1973. *Die Intermediate-Keramik in Kleinasien.* München: W. & J. M. Salzer.

Orphanides, A. G. 1983. *Bronze Age Anthropomorphic Figurines in the Cesnola Collection at the Metropolitan Museum of Art.* Göteburg: Paul Aströms Förlag.

Özfırat, A. 2001. *Doğu Anadolu Yayla Kültürleri.* Istanbul: Arkeoloji ve Sanat Yayınları.

———. 2005. Transhumance on the Eastern Anatolian High Plateau in the 2nd mill. B.C. *Archäologische Mitteilungen aus Iran und Turan* 37: 139-152.

Özgüç, N. 1966. Acemhöyük Kazıları/Excavations at Acemhöyük. *Anadolu* 10: 1-52.

———. 2006. *Kültepe-Kaniš/Neša: Yerli Peruwa ve Aššur-ımittī'nin oğlu Assur'lu Tüccar Usur-ša-Ištar'ın Arşivlerine ait Kil Zarfların Mühür Baskıları / Seal Impressions on the Clay Envelopes from the Archives of the Native Peruwa and Assyrian Trade Usur-ša-Ištar son of Aššur-ımittī.* Ankara: AKDTYK Türk Tarih Kurumu.

Özgüç, T. 1948. *Die Bestattungsbräuche im vorgeschichtlichen Anatolien.* Ankara: Türk Tarih Korumu.

Özgüç, T. and Akok, M. 1958. *Horoztepe.* Ankara: Türk Tarih Kurumu.

Palmieri, A., Begemann, F., Schmitt-Strecker, S., and Hauptmann, A. 2002. Chemical Composition and Lead Isotopy of Metal Objects from the "Royal" Tomb and Other Related Finds at Arslantepe, Eastern Anatolia. *Paléorient* 28 (2): 43-69.

Palmieri, A. M., Sertok, K., and Chernykh, E. 1993. From Arslantepe Metalwork to Arsenical Copper Technology in Eastern Anatolia. In *Between the Rivers and Over the Mountains: archaeologica anatolica et mesopotamica Alba Palmieri dedicata.* M. Frangipane, H. Hauptmann, M. Liverani, P. Matthiae, and M. Mellink (eds.). Roma: Dipartimento di Scienze Storiche, Archaeologiche e Antropologiche Dell'Antichità. 573-599.

Palumbi, G. 2003. Red-Black Pottery: Eastern Anatolian and Transcaucasian Relationships around the Mid-Fourth Millennium BC. *Ancient Near Eastern Studies* 40: 80-134.

———. 2008. Mid-Fourth Millennium Red-Black Burnished Wares from Anatolia: A Cross-Comparison. In *Ceramics in Transition: Chalcolithic Through Iron Age in the Highlands of the Southern Caucasus and Anatolia.* K. S. Rubinson and A. G. Sagona (eds.). Louvain: Peeters Press. 39-58.

———. 2008. *The Red and Black: Social and Cultural Interaction between the Upper Euphrates and the Southern Caucasus Communities in the Fourth and Third Millennium BC.* Studi di Preistoria Orientale 2. Roma: Sapienza Università di Roma.

———. 2010. Pastoral Models and Centralised Animal Husbandry. The Case of Arslantepe. In *Economic Centralisation in Formative States. The Archaeological Reconstruction of the Economic System in 4th Millennium Arslantepe.* M. Frangipane (ed). Studi di Preistoria Orientale 3. Roma: Sapienza Università di Roma. 149-163.

Parpola, S. and Porter, M. 2001. *The Helsinki Atlas of the Near East in the Neo-Assyrian Period.* Helsinki: Casco Bay Assyriological Institute.

Parrot, A. 1958. *Mission archéologique de Mari: II. Le Palais, Peintures murales.* Paris: Librarie Orientaliste Paul Geuthner.

———. 1959. *Mission archéologique de Mari: II. Le Palais, Documents et Monuments.* Paris: Librarie Orientaliste Paul Geuthner.

Parzinger, H. 2002. Das Zinn in der Bronzezeit Eurasiens. In *Anatolian Metal II.* Bochum: Deutsches Bergbau-Museum. 159-178.

Parzinger, H. and Boroffka, N. (eds.). 2003. *Das Zinn der Bronzezeit in Mittelasien I: Die siedlungsarchäologischen Forschungen im Umfeld der Zinnlagerstätten.* Mainz: Philipp von Zabern.

Paz, S. 2009. A Home Away from Home? *The Settlement of Early Transcuacasian Migrants at Tel Bet Yerah.* Tel Aviv 36: 196-216.

Pegolotti, F. B. 1936. *La pratica della mercatura.* A. Evans (ed). Cambridge: Mediaeval Academy of America.

Pelfer, G. 2007. The Cost Surface Analysis as a Predictive Model for the Reconstruction of the Ancient Road Network in the Territory of the Protohistorical Tarquinia. In *The World is in your Eyes. Computer Applications and Quantitative Methods in Archaeology, Proceedings of the 33rd CAA Conference, Tomar, March 2005.* A. Figueredo and G. Velho (eds.). Tomar: CAAPortugal. 329-334.

Pernicka, E. 2004. Kupfer und Silber in Arisman und Tappeh Sialk und die frühe Metallurgie in Iran. In *Persiens Antike Pracht.* T. Stöllner, R. Slotta, and A. Vatandoust (eds.). Katalog der Ausstellung des Deutschen Bergbau-Museums Bochum vom 28. November 2004 bis 29. Mai 2005. Bochum: Deutsches Bergbau-Museum. 232-239.

Perrot, J. and Madjidzadeh, Y. 2005. L'iconographie des vases et objets en chlorite de Jiroft (Iran). *Paléorient* 31 (2): 123-152.

———. 2006. À travers l'ornamentation des vases et objets en chlorite de Jiroft. *Paléorient* 32 (1): 99-112.

Peyronel, L. 2007. Spinning and Weaving at Tell Mardikh-Ebla (Syria): Some Observations on Spindle-Whorls and Loom-Weights from the Bronze and Iron Ages. In *Ancient Textiles: Production, Craft and Society.* C. Gillis and M.-L. Nosch (eds.). Oxford: Oxbow. 26-35.

Pfister, R. 1939. Traces de tissu sur un miroir de cuivre provenant d'une tombe proto-elamite de Sialk. In *Fouilles de Sialk, prés de Kashan, vol. II.* R. Ghirshman (ed). Paris. 201-202.

Philip, G. 1999. Complexity and Diversity in the Southern Levant during the Third Millennium BC: The Evidence of Khirbet Kerak Ware. *Journal of Mediterranean Archaeology* 12 (1): 26-57.

———. 2000. Khirbet Kerak Ware in the Levant: the Implications of Radiocarbon Chronology and Spatial Distribution. In *Chronologies des Pays du Caucase et de l'Euphrate aux IVe-IIIe Millenaires.* C. Marro and H. Hauptmann (eds.). Istanbul: Institut Français d'Etudes Anatoliennes. 279-291.

Pigott, V. C. 1989. Archaeo-metallurgical investigations at Bronze Age Tappeh Hesar, 1976. In *Tappeh Hesar: Reports of the Restudy Project, 1976.* R. Dyson and S. M. Howard (eds.). Firenze: Case Editrice le Lettere. 25-34.

———. 1999a. The Development of Metal Production on the Iranian Plateau: an Archaeometallurgical Perspective. In *The Archaeometallurgy of the Asian Old World.* V. C. Pigott (ed). Philadelphia: University of Pennsylvania Museum. 73-106.

———. 1999b. A heartland of metallurgy: Neolithic/Chalcolithic metallurgical origins on the Iranian Plateau. In *The Beginnings of Metallurgy: Proceedings of the International Conference "The Beginnings of Metallurgy", Bochum 1995.* A. Hauptmann, E. Pernicka, T. Rehren, and Ü. Yalçın (eds.). Bochum: Deutsches Bergbau-Museum.

———. 2004c. Bedeutung Irans für die Erforschung prähistorischer Kupfermetallurgie. In *Persiens Antike Pracht.* T. Stöllner, R. Slotta, and A. Vatandoust (eds.). Katalog der Ausstellung des Deutschen Bergbau-Museums Bochum vom 28. November 2004 bis 29. Mai 2005 Bochum: Deutsches Bergbau-Museum. 28-43.

Piller, C. K. 2003. Zur Mittelbronzezeit im nördlichen Zentraliran - Die Zentraliranische Graue Ware (Central Grey Ware) als mögliche Verbindung zwischen Eastern und Western Grey Ware. *Archäologische Mitteilungen aus Iran und Turan* 35/36: 143-173.

Pinhasi, R., Gasparian, B., Areshian, G., Zardaryan, D., Smith, A., Bar-Oz, G., and Higham, T. 2010. First Direct Evidence of Chalcolithic Footwear from the Near Eastern Highlands. *Plos One* 5 (6): e10984.

Pinner, R. 1982. Decorative Designs on Prehistoric Turkmenian Ceramics. *Halı* 5 (2): 118-119.

Piro, J. J. 2009. *Pastoralism in the Early Transcaucasian Culture: The Faunal Remains from Sos Höyük.* PhD, New York University. [Unpublished thesis].

Pittman, H. 2003. Reconsidering the *Trouvaille de la Statuette d'Or.* In *Yeki bud, yeki nabud: essays on the Archaeology of Iran in Honor of William M. Sumner.* N. F. Miller and K. Abdi (eds.). Los Angeles: Cotsen Institute of Archaeology, UCLA. 177-192.

Pogrebova, M. 2003. The Emergence of Chariots and Riding in the South Caucasus. *Oxford Journal of Archaeology* 22 (4): 397-409.

Pomponio, F. 2010. New Texts Regarding the Neo-Sumerian Textiles. In *Textile Terminologies in the Ancient Near East and Mediterranean from the Third to the First Millennia BC.* C. Michel and M.-L. Nosch (eds.). Ancient Textiles 8. Oxford: Oxbow Books. 186-200.

Possehl, G. L. 1996. Meluhha. In *The Indian Ocean in Antiquity.* J. Reade (ed). London: British Museum.

———. 2002a. *The Indus civilization: a contemporary perspective.* Altamira Press.

———. 2002b. Indus-Mesopotamian trade. *Iranica Antiqua* 37: 325-342.

Pottier, M. -H. 1984. *Matériel Funéraire de la Bactriane Méridionale de l'Age du Bronze.* Paris: Editions Recherche sur les Civilisations.

Potts, D. T. 1997. *Mesopotamian Civilization: the Material Foundations.* Egyptology and Ancient Near Eastern Studies. London: The Athlone Press.

———. 1999. *The Archaeology of Elam: Formation and Transformation of an Ancient Iranian State.* Cambridge: Cambridge University Press.

———. 2004a. Camel Hybridization and the Role of Camelus bactrianus in the Ancient Near East. *Journal of the Economic and Social History of the Orient* 47 (2): 143-165.

———. 2004b. Exit Aratta: Southeastern Iran and the Land of Marhashi. *Nāme-ye Irān-e Bāstān* 4 (1): 1-11.

———. 2008a. Puzur-Inšušinak and the Oxus Civilization (BMAC): Reflections on Šimaški and the geo-political landscape of Iran and Central Asia in the Ur III period. *Zeitschrift Für Assyriologie* 98: 165-194.

———. 2008b. The Persepolis Fortification Texts (PFTs) and the Royal Road: Another look at the Fahliyan area. In *L'archive des fortifications de Persépolis. État des questions et perspectives de recherches.* P. Briant, W. Henkelman, and M. Stolper (eds.). Persika 12. Paris: Éditions de Boccard. 275-302.

Potts, D. T., Roustaei, K., Petrie, C. A., and Weeks, L. R. (eds.). 2009. *The Mamasani Archaeological Project Stage One: A report on the first two seasons of the ICAR - University of Sydney Expedition to the Mamasani District, Fars Province, Iran.* Oxford: Archaeopress.

Potts, T. F. 1993. Patterns of Trade in Third-Millennium BC Mesopotamia and Iran. *World Archaeology* 24 (3): 379-402.

———. 1994. *Mesopotamia and the East: an archaeological and historical study of foreign relations ca. 3400-2000 B.C.* Oxford: Oxford University Committe for Archaeology.

Pruß, A. 2000. The Metallic Ware of Upper Mesopotamia: Definition, Chronology and Distribution. In *Chronologies des Pays du Caucase et de l'Euphrate aux IVe-IIIe Millenaires.* C. Marro and H. Hauptmann (eds.). Istanbul: Institut Français d'Etudes Anatoliennes. 193-203.

Pulak, C. 1998. The Uluburun Shipwreck: An Overview. *The International Journal of Nautical Archaeology* 27 (3): 188-224.

Puturidze, M. 2003. Social and Economic Shifts in the South Caucasian Middle Bronze Age. In *Archaeology in the borderlands: investigations in Caucasia and beyond.* A. T. Smith and K. S. Rubinson (eds.). Los Angeles: Cotsen Institute of Archaeology, UCLA. 111-127.

P'yankova, L. 1994. Central Asia in the Bronze Age: sedentary and nomadic cultures. Trans. F Hiebert. *Antiquity* 68: 355-72.

Rahmstorf, L. 2006. Zur Ausbreitung vordasiatischer Innovationen in die frühbronzezeitliche Ägäis. *Prähistorische Zeitschrift* 81: 49-96.

———. 2010a. The concept of weighing during the Bronze Age in the Aegean, the Near East and Europe. In *The Archaeology of Early Quantification and Cosmology: comprehending Heaven, Earth and time in ancient societies.* I. Morley and C. Renfrew (eds.). Cambridge: Cambridge University Press. 88-105.

———. 2010b. Indications of Aegean-Caucasian relations during the third millennium BC. In *Von Majkop bis Trialeti: Gewinnung und Vernreitung von Metallen und Obsidian in Kaukasien im 4.-2. Jt. v. Chr.* S. Hansen, A. Hauptmann, I. Motzenbäcker, and E. Pernicka (eds.). Bonn: Dr. Rudolf Habelt. 263-295.

———. 2011. Re-integrating 'Diffusion': the Spread of Innovations among the Neolithic and Bronze Age societies of Europe and the Near East. In *Interweaving Worlds: Systemic Interactions in Eurasia, 7th to 1st millennia BC.* T. C. Wilkinson, S. Sherratt, and J. Bennet (eds.). Oxford: Oxbow Books. 100-119.

Ramsay, W. M. 1890. *The History Geography of Asia Minor.* Royal Geographical Society, Supplementary Papers IV. London: John Murray.

Ratnagar, S. 2001. The Bronze Age: Unique Instances of a Pre-Industrial World System? *Current Anthropology* 42 (3): 351-379.

———. 2006. *Trading Encounters: From the Euphrates to the Indus in the Bronze Age.* Delhi: Oxford University Press India.

Reade, J. 1979. *Early etched beads and the Indus-Mesopotamia trade.* British Museum Occasional Paper 2. London: British Museum.

Reade, W. J. and Potts, D. T. 1993. New evidence for late third millennium linen from Tell Abraq, Umm al-Qaiwain, UAE. *Paléorient* 19 (2): 99-106.

Reese, D. S. 1987. Palaikastro shells and Bronze Age purple-dye production in the Mediterranean basin. *The Annual of the British School at Athens* 82: 201-206.

Renfrew, C. 1968. Megaliths, territories and populations. In *Acculturation and continuity in Atlantic Europe mainly during the Neolithic period and the Bronze Age : papers presented at the IV Atlantic Colloquium.* S. J. de Laet (ed). Bruge: De Tempel. 198-220.

———. 1969. The Development and Chronology of the Early Cycladic Figurines. *American Journal of Archaeology* 73 (1): 1-32.

———. 1972. *The Emergence of Civilisation: The Cyclades and the Aegean in The Third Millennium BC.* London: Routledge.

———. 1975. Trade as Action at a Distance: Questions of Integration and Communication. In *Ancient Civilization and Trade.* J. Sabloff and C. C. Lamberg-Karlovsky (eds.). Santa Fe: University of New Mexico Press. 3-59.

———. 1990. *Archaeology and Language: The Puzzle of Indo-European Origins.* London: Routledge.

Riefstahl, E. 1944. *Patterned Textiles in Pharaonic Egypt.* New York: Brooklyn Museum, Brooklyn Institute of Arts and Sciences.

Ristvet, L., Gopnik, H., Bakhshaliyev, V., Lau, H., Ashurov, S., and Bryant, R. 2012. On the Edge of Empire: 2008 and 2009 Excavations at Oğlanqala, Azerbaijan. *American Journal of Archaeology* 116 (2): 321-362.

Roaf, M. 2000. *Cultural Atlas of Mesopotamia and the Near East.* New York: Facts on File.

Roberts, B. W., Thornton, C. P., and Pigott, V. C. 2009. Development of metallurgy in Eurasia. *Antiquity* 83: 1012-1022.

Robson, E. 2001. Technology in society: three textual case studies from Late Bronze Age Mesopotamia. In *The social contexts of technological change: Egypt and the Near East, 1650-1550 BC.* A. Shortland (ed). Proceedings of a conference held at St. Edmund Hall, Oxford 12-14 September 2000. Oxford: Oxbow. 39–57.

von Rosen, L. 1988. *Lapis Lazuli in Geological Contexts & in Ancient Written Sources.* Partille, Paul Astroms Forlag.

Rossel, S., Marshall, F., Peters, J., Pilgram, T., Adams, M. D., and O'Connor, D. 2008. Domestication of the donkey: Timing, processes, and indicators. *Proceedings of the National Academy of Sciences* 105 (10): 3715.

Rossi Osmida, G. (ed.). 2002a. *Margiana: Gonur-depe Necropolis. 10 years of excavations by Ligabue Study and Research Centre.* Padua: Il Punto Edizione.

———. 2002b. Considerations on the Necropolis at Gonur-Depe. In *Margiana: Gonur-depe Necropolis. 10 years of excavations by Ligabue Study and Research Centre.* G. Rossi Osmida (ed). Padua: Il Punto Edizione. 69-119.

———. 2007. Le Figurine Anthropomorfe: The Anthropomophic Figurines. In *Adji Kui Oasis III-II mill. BC, Vol. I : La cittadella delle Stauette. The Citadel of the Figurines. Heykellerin sitadeli.* G. Rossi Osmida (ed). A report of the figurines found at Gonur Depe - useful summaries (some taken from Masson & Sarianidi's work), but also some classification of the Gonur Depe stuff. Venezia: Il Punto, Centro Studi Ricerche Ligabue. 146-197.

Rothe, R. and Rapp, G. 1995. Trace-Element Analyses of Egyptian Eastern Desert Tin and Its Importance to Egyptian Archaeology. In *Proceedings of the Egyptian-Italian Seminar on the Geosciences and Archaeology in the Mediterranean Countries.* A. A. A. Hussein, M. Miele, and S. Riad (eds.). Special Publication 70. Cairo: Geological Survey of Egypt. 229-244.

Rothman, M. S. 2003a. "Ripples in the Stream": Transcaucasia-Anatolian Interaction in the Murat/Euphrates Basin at the Beginning of the Third Millennium BC. In *Archaeology in the borderlands: investigations in Caucasia and beyond.* A. T. Smith and K. S. Rubinson (eds.). Los Angeles: Cotsen Institute of Archaeology, UCLA. 95-110.

———. 2003b. Style Zones and Adaptations Along the Turkish-Iranian Borderland. In *Yeki bud, yeki nabud: essays on the Archaeology of Iran in Honor of William M. Sumner.* N. F. Miller and K. Abdi (eds.). Los Angeles: Cotsen Institute of Archaeology, UCLA. 207-216.

———. 2006. Transcaucasians: Settlement, migration, and trade in the Kura-Araxes periods. *Archäologische Mitteilungen aus Iran und Turan* 37: 53-62.

Rothman, M. S. and Kozbe, G. 1997. Interpreting Muş in the Early Bronze Age. *Anatolian Studies* 47: 105-126.

Rova, E. 1991. Die Keramik aus dem Gräberfeld. In *Gräber des 3. Jahrtausends v. Chr. im syrischen Euphrattal 2. Ausgrabunden in Wreide.* W. Orthmann and E. Rova (eds.). Saarbrücken. 71-171.

Rova, E., Puturidze, M., and Makharadze, Z. 2010. The Georgian-Italian Shida Kartli Archaeological Project: A Report on the First Two Field Seasons 2009 and 2010. *Rivista di Archeologia* 34: 5-30.

Rowlands, M., Larsen, M., and Kristiansen, K. (eds.). 1987. *Centre and Periphery in the Ancient World.* Cambridge: Cambridge University Press.

Rowlands, M. and Warnier, J.-P. 1993. The magical production of iron in the Cameroon Grassfields. In *The Archaeology of Africa: Food, metals and towns.* T. Shaw, P. Sinclair, B. Andah, and A. Okpoko (eds.). London: Routledge. 512-550.

Rubinson, K. S. 2003. The Hasanlu Gold "Bowl": a View from Transcaucasia. In *Yeki bud, yeki nabud: essays on the Archaeology of Iran in Honor of William M. Sumner.* N. F. Miller and K. Abdi (eds.). Los Angeles: Cotsen Institute of Archaeology, UCLA. 237-242.

Rubinstein, J. B. and Barsky, L. 2002. *Non-ferrous metal ores: deposits, minerals and plants.* Mineral Deposits and Resources of the former Soviet Union. London: Taylor and Francis.

Ibn Rustah. 1892. *Kitāb al-A'lāk an-Nafīsa.* M. J. De Goeje (ed). Bibliotheca Geographorum Arabicorum. Leiden: Brill.

Sagona, A. G. 1981. Spiral-headed Pins: a Further Note. *Tel Aviv* 8: 152-159.

———. 1984. *The Caucasian Region in the Early Bronze Age. (Parts 1, 2 & 3).* British Archaeological Reports International Series S214. Oxford: BAR.

———. 1993. Settlement and Society in Late Prehistoric Trans-Caucasus. In *Between the Rivers and Over the Mountains: archaeologica anatolica et mesopotamica Alba Palmieri dedicata.* Roma: Dipartimento di Scienze Storiche, Archaeologiche e Antropologiche Dell'Antichità. 453-474.

———. 1999. The Bronze Age-Iron Age transition in northeast Anatolia: a view from Sos Höyük. *Anatolian Studies* 49: 153-157.

———. 2000. Sos Höyük and the Erzurum Region in Late Prehistory: a Provisional Chronology for Northeast Anatolia. In *Chronologies des Pays du Caucase et de l'Euphrate aux IVe-IIIe Millenaires.* C. Marro and H. Hauptmann (eds.). Istanbul: Institut Français d'Etudes Anatoliennes. 329-373.

———. 2004. Social Boundaries and Ritual Landscapes in Late Prehistoric Trans-Caucasis and Highland Anatolia. In *A View from the Highlands: archaeological studies in honour of Charles Burney.* A. G. Sagona (ed). Herent: Peeters Press. 475-538.

———. 2006. *The Heritage of Eastern Turkey: from Earliest Settlements to Islam.* Victoria: Macmillan Art Publishing.

———. 2013. Wagons and Carts of the Trans-Caucasus. In *Tarham Armağanı: M. Taner Tarhan'a Sunulan Makaleler / Essays in Honour of M. Taner Tarhan.* O. Tekin, M. H. Sayar, and E. Konyar (eds.). İstanbul: Ege Yayınları. 277-297.

Sagona, A. G., Erkmen, M., Sagona, C., and Thomas, I. 1996. Excavations at Sos Höyük, 1995: second preliminary report. *Anatolian Studies* 46: 27-52.

Sagona, A. G., Nikolaishvili, V., Sagona, C., Ogleby, C., Pilbrow, V., Briggs, C., Giunashvili, G., and Manjegaladze, G. 2010. Excavations at Samtavro, 2008-2009: An Interim Report. *Ancient Near Eastern Studies* 47: 1-136.

Sagona, A. G., Sagona, C., and Özkorucuklu, H. 1995. Excavations at Sos Höyük 1994: first preliminary report. *Anatolian Studies* 45: 193-218.

Sagona, A. G. and Sagona, C. 2004. *Archaeology at the North-Eastern Anatolian frontier 1: An Historical Geography and a Field Survey of the Bayburt Province.* Louvain: Peeters Press.

Sagona, C. 1999. An archaeological survey of the Erzurum province, 1999: The region of Pasinler. *Ancient Near Eastern Studies* 36: 108-131.

———. 2004. Did Xenophon take the Aras High Road? Observations on the Historical Geography of North-East Anatolia. In *A View from the Highlands: archaeological studies in honour of Charles Burney.* A. G. Sagona (ed). Herent: Peeters Press. 299-333.

Salvatori, S. 1998. Margiana Archaeological Map: the Bronze Age Settlement Pattern. In *The archaeological map of the Murghab Delta: Preliminary reports 1990–1995.* A. Gubaev, G. Koshelenko, and M. Tosi (eds.). Rome: Istituto Italiano per l'Africa et l'Oriente. 57-65.

———. 2003. Pots and peoples: the "Pandora's Jar" of Central Asian archaeological research: on two recent books on Gonur graveyard excavations. *Rivista Di Archeologia* 27: 5-20.

———. 2007. About recent excavations at a Bronze Age site in Margiana (Turkmenistan). *Rivista Di Archeologia* 31: 11-28.

———. 2008. Cultural Variability in the Bronze Age Oxus Civilisation and its Relations with the Surrounding Regions of Central Asia and Iran. In *The Archaeological Map of the Murghab Delta. Volume II. The Bronze Age and Early Iron Age in the Margiana Lowlands.* S. Salvatori and M. Tosi (eds.). British Archaeological Reports International Series S1806. Oxford: BAR. 77-100.

Salvatori, S. and Tosi, M. (eds.). 2008. *The Archaeological Map of the Murghab Delta. Volume II. The Bronze Age and Early Iron Age in the Margiana Lowlands.* British Archaeological Reports International Series S1806. Oxford: BAR.

Sarianidi, V. I. 1971. The Lapis Lazuli Route in the Ancient East. *Archaeology* 24 (1): 12-15.

———. 1981. Margiana in the Bronze Age. In *The Bronze Age Civilization of Central Asia: Recent Soviet Discoveries.* P. L. Kohl (ed). New York: M. E. Sharpe. 165-193.

———. 1983. The Pottery of Shahr-i Sokhta I and its Southern Turkmenian Connections. In *Prehistoric Sīstān 1.* M. Tosi (ed). Rome: Istituto Italiano per il Medio ed Estremo Oriente. 183-198.

———. 1990a. Soviet Excavations in Bactria: the Bronze Age. In *Bactria: An ancient oasis civilization from the sands of Afghanistan.* G. Ligabue and S. Salvatori (eds.). Venice: Erizzo Editrice. 109-123.

———. 1990b. *Древности Страны Маргуш* [Antiquities of the country of Margush]. Ashgabat: Ylym.

———. 2001. *Некрополь Гонура и Иранское Язычесмво* [Necropolis of Gonur and Iranian Paganism]. Moscow: World Media.

———. 2002a. Chamber Graves of the Gonur Necropolis. In *Complex Societies of Central Eurasia from the 3rd to the 1st Millennium BC: Regional Specifics in Light of Global Models.* K. Jones-Bley and D. G. Zdanovich (eds.). Journal of Indo-European Studies Monograph Series (No. 46). Washington, D.C.: Institute for the Study of Man. 574-593.

———. 2002b. *Marguş* [Маргуш. Древневосточное царство в старой дельте реки Мурғаб / Margush. Ancient Oriental Kingdom in the Old Delta of the Murghab river]. Ashgabat: Turkmen Dowlet Habarlary.

———. 2006. *Goňurdepe, Türkmenistan: City of kings and gods.* Aşgabat: Miras.

Schachermeyr, F. 1955. *Die ältesten Kulturen Griechenlands.* Stuttgart: W. Kohlhammer.

Schachner, A. 2002. Zur Entwicklung der Metallurgie im östlichen Transkaukasien (Azerbaycan und Nahçevan) während des 4. und 3. Jahrtausends v. Chr. In *Anatolian Metal II.* Ü. Yalçın (ed). Bochum: Deutsches Bergbau-Museum. 115-130.

Schafer, E. 1963. *The Golden Peaches of Samarkand: A Study of T'ang Exotics.* Berkeley: University of California Press.

Schiaparelli, E. 1927. *La Tomba intatta dell' Architetto Cha. Relazione sui Lavori della Missione archeologica in Egitto 2.* Turin.

Schiltberger, J. 1879. *The bondage and travels of Johann Schiltberger: a native of Bavaria, in Europe, Asia, and Africa, 1396-1427.* Hakluyt Society.

Schliemann, H. 1880. *Ilios: The City and Country of the Trojans.* London.

Schmidt, A. and Fazeli, H. 2007. Tepe Ghabristan: A Chalcolithic Tell Buried in Alluvium. *Archaeological Prospection* 14 (1): 38-46.

Schmidt, E. F. 1937. *Excavations at Tepe Hissar, Damghan.* Philadelphia: University of Pennsylvania Press.

Schmidt, H. 1902. *Heinrich Schliemanns Sammlung trojanischer Altertümer.* Berlin.

Schneider, G. 1989. A Technological Study of North-Mesopotamian Stone Ware. *World Archaeology* 21 (1): 30-50.

Schneider, J. 2011. Anticipating the Silk Road: Some Thoughts on the Wool-Murex Connection in Tyre. In *Interweaving Worlds: Systemic Interactions in Eurasia, 7th to 1st millennia BC.* T. C. Wilkinson, S. Sherratt, and J. Bennet (eds.). Oxford: Oxbow Books. 295-302.

Schoop, U.-D. 2006. Dating the Hittites with Statistics: Ten Pottery Assemblages from Boğazköy-Hattuša. In *Strukturierung und Datierung in der hethitischen Archäologie : Voraussetzungen, Probleme, neue Ansätze/Structuring and dating in Hittite archaeology: requirements, problems, new approaches; Internationaler Workshop, Istanbul, 26-27. November 2004.* D. P. Mielke, U.-D. Schoop, and J. Seeher (eds.). Istanbul: Ege Yayınları. 215-240.

Seeden, H. 1980. *The Standing Armed Figurines in the Levant.* Prähistorische Bronzefunde I:1. München: C. H. Beck.

Sevin, V. 2004. Pastoral Tribes and Early Settlements of the Van Region, Eastern Anatolia. In *A View from the Highlands: archaeological studies in honour of Charles Burney.* A. G. Sagona (ed). Herent: Peeters Press. 179-203.

Shaffer, J. G. 1978. The Later Prehistoric Periods. In *The Archaeology of Afghanistan: from earliest times to the Timurid period.* F. Allchin and N. Hammond (eds.). London: Academic Press. 71-186.

Shah, V. B. and McRae, B. 2008. Circuitscape: a tool for landscape ecology. In P*roceedings of the 7th Python in Science Conference (SciPy 2008).* G. Varoquaux, T. Vaught, and J. Millman (eds.). 62-66.

Shatzmiller, J. 1998. Jews, Pilgrimage, and the Christian Cult of Saints: Benjamin of Tudela and His Contemporaries. In *After Rome's Fall: Narrators and Sources of Early Medieval History.* W. A. Goffart and A. C. Murray (eds.). Toronto: University of Toronto. 337-347.

Shaw, I. (ed.). 2000. *The Oxford History of Ancient Egypt.* Oxford: Oxford University Press.

Shaw, M. C. 1970. Ceiling Patterns from the Tomb of Hepzefa. *American Journal of Archaeology* 74 (1): 25-30.

Shaw, M. C. and Laxton, K. 2002. Minoan and Mycenaean Wall Hangings. New light from a wall painting at Ayia Triada. *Creta Antica* 3: 93-104.

Sherratt, A. G. 1976. Resources, technology and trade; an essay in early metallurgy. In *Problems in Social and Economic Archaeology.* I. Longworth, G. Sieveking, and K. Wilson (eds.). London: Duckworth. 557-581.

———. 1981. Plough and pastoralism: aspects of the secondary products revolution. In *Pattern of the Past: Studies in Honour of David Clarke.* I. Hodder, G. Isaac, and N. Hammond (eds.). Cambridge: Cambridge University Press. 261–305.

———. 1983. The secondary exploitation of animals in the Old World. *World Archaeology* 15 (1): 90-104.

———. 1985. *Ancient Times: an archaeological map and timescale for Europe, Western Asia and Egypt.* Oxford: Ashmolean Museum.

———. 1993. What would a Bronze-Age World System Look Like? Relations between Temperate Europe and the Mediterranean in Later Prehistory. *Journal of European Archaeology* 1 (2): 1-57.

———. 1994. Core, periphery and margin: perspectives on the Bronze Age. In *Development and Decline in the Mediterranean Bronze Age.* C. Mathers and S. Stoddart (eds.). Sheffield Archaeological Monographs 8. Sheffield: Sheffield Academic Press. 335-345.

———. 1996. Why Wessex? The Avon Route and River Transport in Later British Prehistory. *Oxford Journal of Archaeology* 15 (2): 211-234.

———. 1997. Troy, Maikop, Altyn Depe: Early Bronze Age Urbanism and its Periphery. In *Economy and Society in Prehistoric Europe.* A. G. Sherratt (ed). Edinburgh: Edinburgh University Press. 457-470.

———. 2003. The Horse and the Wheel: the Dialectics of Change in the Circum-Pontic Region and Adjacent Areas, 4500-1500BC. In *Prehistoric steppe adaptation and the horse.* M. Levine, C. Renfrew, and K. Boyle (eds.). Cambridge: McDonald Institute for Archaeological Research. 233-252.

———. 2004. Trade Routes: Growth of Global Trade. *ArchAtlas.* http://www.archatlas.org/Trade/Trade.php (accessed November 8, 2011).

———. 2005. Obsidian Trade in the Near East, 14,000 to 6500 BC. *ArchAtlas.* http://www.archatlas.org/ObsidianRoutes/ObsidianRoutes.php (accessed July 19, 2011).

———. 2006. The Trans-Eurasian Exchange: the Prehistory of Chinese Relations with the West. In *Contact and Exchange in the Ancient World.* V. H. Mair (ed). Honolulu: University of Hawai'i Press. 30-61.

Sherratt, A. G. and Sherratt, E. S. 1997. The Archaeology of Indo-European: An Alternative View. In *Economy and Society in Prehistoric Europe.* A. G. Sherratt (ed). Edinburgh: Edinburgh University Press. 471-485.

Sherratt, [E.] S. 1994a. Comment on Ora Negbi, The 'Libyan Landscape' from Thera: A Review of Aegean Enterprises Overseas in the Late Minoan IA Period. *Journal of Mediterranean Archaeology* 7: 237-240.

———. 1994b. Commerce, iron and Ideology: metallurgical innovation in 12th-11th Century Cyprus. In *Cyprus in the 11th Century BC.* V. Karageorghis (ed). Proceedings of the International Symposium of the Archaeological Research Unit of the University of Cyprus and the Anastasios G. Leventis Foundation (30-31 October 1993). Nicosia: Anastasios G. Leventis Foundation. 59-107.

———. 2000. The Circulation of metals and the end of the Bronze Age in the Eastern Mediterranean. In *Metals Make the World Go Round: The Supply and Circulation of Metals in Bronze Age Europe.* C. Pare (ed). Oxford: Oxbow Books. 82-95.

———. 2007. The Archaeology of Metal Use in the Early Bronze Age Aegean - a review. In *Metallurgy in the Early Bronze Age Aegean.* P. M. Day and R. C. P. Doonan (eds.). Oxford: Oxbow Books. 245-263.

———. 2008a. Representation and hidden technologies. Lecture presented at the Sheffield Centre for Aegean Archaeology Roundtable 2008. [Unpublished work].

———. 2008b. Vitreous Materials in the Bronze and Early Iron Ages: Some Questions of Values. In *Vitreous Materials in the Late Bronze Age Aegean.* Oxford: Oxbow Books. 210-232.

———. 2010. The Aegean and the Wider World: Some Thoughts on a World-Systems Perspective. In *Archaic State Interaction: The Eastern Mediterranean in the Bronze Age. School for Advanced Research Advanced Seminar Series.* W. A. Parkinson and M. L. Galaty (eds.). Santa Fe, NM: School for Advanced Research Press. 81-106.

Shimelmitz, R. 2003. A Glance at the Early Trans-Caucasian Culture. *Tel Aviv* 30: 204-221.

Shirazi, R. 2007. Figurines anthropomorphes du Bronze ancien de Shahr-i Sokhta période II (Séistan, Sud-Est de l'Iran): approche typologique. *Paléorient* 33 (2): 147-162.

Shishlina, N. (ed.). 1999. *Текстиль Эпохи Ъронзы Евразыйских Степей* [Bronze Age Textiles of the Eurasian Steppes]. Papers of the State Historical Museum 109. Moskva: State Historical Museum.

———. 2008. *Reconstruction of the Bronze Age of the Caspian Steppes: Life styles and life ways of pastoral nomads.* British Archaeological Reports International Series S1876. Oxford: Archaeopress.

Shishlina, N. I. and Hiebert, F. T. 1998. The Steppe and the Sown: Interaction between Bronze Age Eurasian Nomads and Agriculturalists. In *The Bronze Age and Early Iron Age Peoples of Eastern Central Asia.* V. H. Mair (ed). The Institute for the Study of Man. 222-237.

Shishlina, N. I., Golikov, V. P., and Orfinskaya, O. V. 2000. Bronze Age Textiles of the Caspian Sea Maritime Steppes. In *Kurgans, Ritual Sites, and Settlements : Eurasian Bronze and Iron Age.* J. Davis-Kimball, E. M. Murphy, L. Koryakova, and L. T. Yablonsky (eds.). British Archaeological Reports International Series S890. Oxford: Archaeopress. 109-117.

Shishlina, N. I., Orfinskaya, O. V., and Golikov, V. P. 2003. Bronze Age Textiles from the North Caucasus: New Evidence of Fourth Millennium BC Fibres and Fabrics. *Oxford Journal of Archaeology* 22 (4): 331-344.

Simon, E. and Hirmer, M. 1985. *Die Götter der Griechen.* München: Hirmer.

Simonyan, H. *Shengavit: an Early Transcaucasian Site in Yerevan on the Ararat Plain, Republic of Armenia.* http://muse.widener.edu/~msrothma/shengavitweb2.html (accessed April 20, 2011).

Sinitsyn, I. V. 1948. Pamjatniki predskifskoj epokhi v stepjakh nizhnego povolzh'ja [Finds of the pre-Scythian period in the area of the lower Volga]. *Sovetskaja Arkheologija* 10: 143-160.

Smith, A. T. 2005. Prometheus unbound: Southern Caucasia in prehistory. *Journal of World Prehistory* 19 (4): 229-279.

Smith, A. T., Badalian, R. S., and Avetisyan, P. 2009. *The Archaeology and Geography of Ancient Transcaucasian Societies: Foundations of research and regional survey in the Tsaghkahovit Plain, Armenia.* Chicago: Oriental Institute of the University of Chicago.

Soressi, M. and Geneste, J. -M. 2011. The History and Efficacy of the Chaîne Opératoire Approach to Lithic Analysis: Studying Techniques to Reveal Past Societies in an Evolutionary Perspective. *PaleoAnthropology* 334: 334-350.

Sprenger, A. 1864. *Post- und Reiserouten des Orients* . Leipzig: F.A. Brockhaus.

Stacul, G. 1987. *Prehistoric and Protohistoric Swāt, Pakistan (c. 3000 - 1400BC).* Rome: Istituto Italiano per il Medio ed Estremo Oriente.

Stark, F. 1958. *Alexander's path: from Caria to Cilicia.* London: Murray.

———. 1966. *Rome on the Euphrates : the story of a frontier.* London: Murray.

Starr, R. F. S. (ed.). 1939. *Nuzi: Report on the Excavations at Yorgan Tepa Near Kirkuk, Iraq.* Cambridge, MA: Harvard University Press.

Stech, T. and Pigott, V. C. 1986. The Metals Trade in Southwest Asia in the Third Millennium B.C. *Iraq* 48: 39-64.

Stein, D. L. 1984. Khabur Ware and Nuzi Ware: Their Origin, Relationship, and Significance. *Assur (Monographic Journals of the Near East)* 4 (1): 1-65.

Stein, G. J. 1999. *Rethinking World-Systems: Diasporas, Colonies, and Interaction in Uruk Mesopotamia.* Tucson: University of Arizona Press.

———. 2002. From Passive Periphery to Active Agents: Emerging Perspectives in the Archaeology of Interregional Interaction. *American Anthropologist* 104 (3): 903-916.

Stein, M. A. 1929. *On Alexander's Track to the Indus: Personal Narrative of Explorations on the North-West Frontier of India.* London: Macmillan & Co.

———. 1937. *Archaeological reconnaissances in north-western India and south-eastern Iran.* London.

Steinkeller, P. 1982. On Editing Ur III Economic Texts. *Journal of the American Oriental Society* 102 (4): 639-644.

Stöllner, T. 2005. Early Mining and Metallurgy on the Iranian Plateau. In *Anatolian Metal III.* Ü. Yalçın (ed). Bochum: Deutsches Bergbau-Museum. 191-208.

Stöllner, T., Doll, M., Eskanderi, M. M., Momenzadeh, M., Pasternak, R., and Steffens, G. 2004. Bronzezeitliche Kupferezgewinnung in Veshnāveh [Bronze Age copper exploitation at Veshnovah]. In *Persiens Antike Pracht.* T. Stöllner, R. Slotta, and A. Vatandoust (eds.). Katalog der Ausstellung des Deutschen Bergbau-Museums Bochum vom 28. November 2004 bis 29. Mai 2005. Bochum: Deutsches Bergbau-Museum. 240-257.

Strathern, M. 1990. *The Gender of the Gift: Problems with Women and Problems with Society in Melanesia.* Berkeley: University of California Press.

Strommenger, E. 1962. *Fünf Jahrtausende Mesopotamien.* München: Hirmer Verlag.

Stronach, D. B. 1957. The Development and Diffusion of Metal Types in Early Bronze Age Anatolia. *Anatolian Studies* 7: 89-125.

Süel, A. and Süel, M. 1997. 1995 yılı Ortaköy-Şapinuva Kazı Çalışmaları. In *XVIII. Kazı Sonuçları Toplantısı, I.* Ankara: Kültür Bakanliği Milli Kütüphane Basimevi. 339-351.

Summers, G. 2004. Yanik Tepe and the Early Trans-Caucasian Culture: Problems and Perspectives. In *A View from the Highlands: archaeological studies in honour of Charles Burney.* A. G. Sagona (ed). Herent: Peeters Press. 617-643.

Şahoğlu, V. 2005. The Anatolian Trade Network and the Izmir Region During the Early Bronze Age. *Oxford Journal of Archaeology* 24 (4): 339-361.

Taddeucci, A., Palmieri, A. M., Malpieri, L., and Fornaseri, M. 1975. Analyses of Obsidians from the late Chalcolithic Levels of Arslantepe (Malatya). *Paléorient* 3 (1): 231-246.

Talbert, R. J. A. 2000. *Barrington Atlas of the Greek and Roman world.* Princeton NJ: Princeton University Press.

Tavernari, C. 2009. The CIERA program and activities: focus on the roads and wayside caravanserais in medieval Syria. *ArchAtlas.* http://www.archatlas.org/workshop09/works09-tavernari.php (accessed August 9, 2011).

Tengberg, M., Potts, D. T., and Francfort, H. P. 2008. The golden leaves of Ur. *Antiquity* 82: 925-936.

Terekhova, I. I. 1990. Обработка металлов в Древней Маргиане [Processing of Metals in Ancient Margiana]. In *Древности Страны Маргуш.* V. I. Sarianidi (ed). Ashgabat: Ylym.

Thornton, C. P. 2009. The emergence of complex metallurgy on the Iranian Plateau: Escaping the Levantine Paradigm. *Journal of World Prehistory* 22: 301-327.

———. 2010. The Rise of Arsenical Copper in Southeastern Iran. *Iranica Antiqua* 45: 31-50.

Thornton, C. P., Rehren, T., and Pigott, V. C. 2009. The production of speiss (iron arsenide) during the Early Bronze Age in Iran. *Journal of Archaeological Science* 36 (2): 308-316.

Tolstikow, W. P. and Trejster, M. J. 1996. *Der Schatz aus Troja: Schliemann und der Mythos des Priamos-Goldes.* C. König (trans.). Stuttgart/Zürich: Belser Verlag.

Tosi, M. 1974. The lapis lazuli trade across the Iranian Plateau in the 3rd millennium B.C. In *Gururājamañjarikā: Studi in onore di Giseppe Tucci.* A. Forte, L. Polese Remaggi, and M. Taddei (eds.). Napoli: Istituto Universitario Orientale. 3-22.

———. 1983a. A Bronze Female Statuette from Shahr-i Sokhta: Chronological Problems and Stylistic Connections. In *Prehistoric Sīstān 1.* M. Tosi (ed). Rome: Istituto Italiano per il Medio ed Estremo Oriente.

———. 1983b. Development, Continuity and Cultural Change in the Stratigraphical Sequence of Shahr-i Sokhta. In *Prehistoric Sīstān 1.* M. Tosi (ed). Rome: Istituto Italiano per il Medio ed Estremo Oriente. 127-182.

Tosi, M. and Wardak, R. 1972. The Fullol Hoard: A New Find from Bronze-Age Afghanistan. *East and West* 22: 9-18.

Trifonov, V. A. 1994. The Caucasus and the Near East in the Early Bronze Age (Fourth & Third Millennia BC). *Oxford Journal of Archaeology* 13 (3): 357-360.

Tsareva, E. 2007. Gonur City Linens of the second millennium BC. In *Necropolis of Gonur.* V. I. Sarianidi (ed). Athens: Kapon Editions. 330-334.

Tuck, A. 2006. Singing the Rug: Patterned Textiles and the Origins of Indo-European Metrical Poetry. *American Journal of Archaeology* 110 (4): 539-550.

Tufnell, O. 1958. *Lachish IV: The Bronze Age.* London.

Türker, L. and Demirci, Ş. 1998. Şapinuva, Ortaköy (Çorum) Kazısından Elde Edilen bir Tekstil Örneğinin Analizi. In *XIII. Arkeometri Sonuçları Toplantısı.* Ankara: Kültür Bakanliği Milli Kütüphane Basimevi. Udemuradov, B. 2002. Ceramic Material from Gonur-Depe Necropolis. In *Margiana: Gonur-depe Necropolis. 10 years of excavations by Ligabue Study and Research Centre.* G. Rossi Osmida (ed). Padua: Il Punto Edizione. 133-143.

Uerpmann, H.-P. 1990. Die Domestikation des Pferdes im Chalcolithikum West- und Mitteleuropas. *Madrider Mitteilungen* 31: 109-153.

Ullmann, L. Z. 2011. *Movement and the Making of Place in the Hittite Landscape.* PhD dissertation. Columbia University. [Unpublished thesis].

Ur, J. A. 2003. CORONA satellite photography and ancient road networks: a northern Mesopotamian case study. *Antiquity* 77 (295): 102-115.

———. 2009. Emergent landscapes of movement in Early Bronze age northern Mesopotamia. In *Landscapes of Movement: Paths, Trails, and Roads in Anthropological Perspective.* J. E. Snead, C. Erickson, and W. A. Darling (eds.). Philadelphia: University of Pennsylvania Press. 180-203.

Van Loon, M. N. 2001. *Selenkahiye: Final Report on the University of Chicago and University of Amsterdam Excavations in the Tabqa Reservoir, Northern Syria, 1967 - 1975.* Leiden: Nederlands Historisch-Archaeologisch Instituut te Istanbul.

Vatandoust, A., Parzinger, H., and Helwing, B. 2011. *Early Mining and Metallurgy on the Western Central Iranian Plateau.* Archäologie in Iran und Turan. Mainz: Philipp von Zabern.

Veenhof, K. R. 1972. *Aspects of Old Assyrian Trade and its Terminology.* Leiden: E. J. Brill.

Veselovskij, N. I. 1898. Kubanskaja oblast. *Russkaja Arkheologicheskaja Kommissija: Otchët.*

Vickers, M. 1985. Artful crafts: the influence of metal work on Athenian painted pottery. *Journal of Hellenic Studies* 105: 108-128.

———. 1989. The Cultural Context of Ancient Greek Ceramics: an Essay in Skeuomorphism. In *Cross-craft and Cross-cultural Interactions in Ceramics.* P. McGovern and M. D. Notis (eds.). Ohio: Westerville. 45-64.

Vickers, M. J. and Gill, D. W. J. 1994. *Artful crafts: ancient Greek silverware and pottery.* Oxford: Clarendon.

Vogelsang-Eastwood, G. M. 1993. *Pharaonic Egyptian Clothing.* Leiden: Brill.

———. 1993. The Çanönü Textile. [non vidi].

———. 1999. *Tutankhamun's Wardrobe: Garments from the tomb of Tutankhamun.* Rotterdam: Barjesteh van Waalwijk van Doorn & Co.

Völling, E. 2008. *Textiltechnik im Alten Orient: Rohstoffe und Herstellung.* Würzburg: Ergon.

De Waele, A. and Haerinck, E. 2006. Etched (carnelian) beads from northeast and southeast Arabia. *Arabian Archaeology and Epigraphy* 17 (1): 31-40.

Waetzoldt, H. 1983. Leinen (Flachs). In *Reallexikon der Assyriologie.* Berlin: Walter de Gruyter. 583-594.

———. 2010. The Colours and Variety of Fabrics from Mesopotamia during the Ur III Period (2050BC). In *Textile Terminologies in the Ancient Near East and Mediterranean from the Third to the First Millennia BC.* C. Michel and M.-L. Nosch (eds.). Ancient Textiles 8. Oxford: Oxbow Books. 201-209.

Waghorne, I. P. 1994. *The Raja's Magic Clothes, Re-visioning Kingship and Divinity in England's India.* Pennsylvania: Pennsylvania State University Press.

Wagner, G. A., Gentner, W., and Gropengiesser, H. 1979. Evidence for third millenium lead-silver mining on Siphnos Island (Cyclades). *Naturwissenschaften* 66 (3): 157-158.

Wagner, G. A., Öztunalı, Ö., and Eibner, C. 1989. Early Copper in Anatolia. Archaeometallurgical Field Evidence. In *Archäometallurgie der Alten Welt: Beiträge zum Internaionalen Symposium "Old World Archaeometallurgy" Heidelberg 1987.* A. Hauptmann, E. Pernicka, and G. A. Wagner (eds.). Bochum: Deutsches Bergbau-Museums. 299-306.

Wagner, G. A. and Öztunalı, Ö. 2000. Prehistoric Copper Sources in Turkey. In *Anatolian Metal I.* Ü. Yalçın (ed). Bochum: Deutsches Bergbau-Museum. 31-68.

Waldbaum, J. C. 1978. *From Bronze to Iron.* Studies in Mediterranean Archaeology 54. Göteborg: Paul Astroms Forlag.

———. 1999. The coming of iron in the eastern Mediterranean: Thirty years of archaeological and technological research. In *The Archaeometallurgy of the Asian Old World.* V. C. Pigott (ed). Philadelphia: University of Pennsylvania Museum. 27-57.

Wallerstein, I. M. 1974. *The Modern World-System, vol. I: Capitalist Agriculture and the Origins of the European World-Economy in the Sixteenth Century.* New York/London: Academic Press.

Warnier, J.-P. 2007. *The pot-king : the body and technologies of power.* African Social Studies Series 17. Leiden: Brill.

Waugh, D. 2007. Richthofen's "Silk Roads": Toward the Archaeology of a Concept. *The Silk Road Newsletter* 5 (1): 1-10. http://www.silk-road.com/newsletter/vol5num1/srjournal_v5n1.pdf.

Waugh, D. and Lee, A. 2000. Ancient Silk Road Travellers. http://www.silk-road.com/artl/srtravelmain.shtml (accessed June 1, 2011).

Webb, J. M., Frankel, D., Stos, Z. A., and Gale, N. 2006. Early Bronze Age Metal Trade in the Eastern Mediterranean. New Compositional and Lead Isotope Evidence from Cyprus. *Oxford Journal of Archaeology* 25 (3): 261-288.

Weeks, L. R. 2003. *Early Metallurgy of the Persian Gulf: Technology, Trade, and the Bronze Age World.* Boston: Brill Academic Publishers.

Weiner, A. B. and Schneider, J. (eds.). 1991. *Cloth and Human experience.* Washington: Smithsonian Institution.

Weisgerber, G. 1990. Montanarchäologische Forschungen in Nordwest-Iran 1978. *Archäologische Mitteilungen aus Iran* 23: 73-84.

Weisgerber, G., Kroll, S. E., Gropp, G., and Hauptmann, A. 1990. Das Bergbaurevier von Sungun bei Kighal in Azarbaidjan (Iran). *Archäologische Mitteilungen aus Iran* 23: 85-103.

Weisgerber, G. and Yule, P. 2003. Al-Aqir near Bahlā' – an Early Bronze Age dam site with planoconvex 'copper' ingots. *Arabian Archaeology and Epigraphy* 14 (1): 24-53.

Weiss, B. L. 2003. *Sacred Trees, Bitter Harvests: Globalizing Coffee in Northwest Tanzania.* Portsmouth NH: Heineman.

Wengrow, D. 2001. The evolution of simplicity: aesthetic labour and social change in the Neolithic Near East. *World Archaeology* 33 (2): 168-188.

———. 2011. 'Archival' and 'Sacrificial' Economies in Bronze Age Eurasia: an Interactionist Approach to the Hoarding of Metals. In *Interweaving Worlds: Systemic Interactions in Eurasia, 7th to 1st millennia BC.* T. C. Wilkinson, S. Sherratt, and J. Bennet (eds.). Oxford: Oxbow Books. 135-144.

Wertime, T. A. 1973. The beginnings of metallurgy: a new look. *Science* 182: 875-87.

Whitehouse, D. 1975. Carnelian in the Persian Gulf. *Antiquity* 49: 129-130.

Whitfield, S. 2001. *Life Along the Silk Road.* University of California Press.

Wilkinson, K. N., Gasparian, B., Pinhasi, R., Avetisyan, P., Hovsepyan, R., Zardaryan, D., Areshian, G. E., Bar-Oz, G., and Smith, A. 2012. Areni-1 Cave, Armenia: A Chalcolithic-Early Bronze Age settlement and ritual site in the southern Caucasus. *Journal of Field Archaeology* 37 (1): 20-33.

Wilkinson, T. C. 2009. Pathways and highways: routes in Bronze Age Eurasia. *ArchAtlas* http://www.archatlas.org/workshop09/works09-wilkinson.php. (accessed 15th September, 2011).

Wilkinson, T. C., Sherratt, S., and Bennet, J. (eds.). 2011. *Interweaving Worlds: Systemic Interactions in Eurasia, 7th to 1st millennia BC.* Oxford: Oxbow Books.

Wilkinson, T. J. 1993. Linear hollows in the Jazira, upper Mesopotamia. *Antiquity* 67: 548-548.

Wilkinson, T. J., Miller, N., Reichel, C., and Whitcomb, D. 2004. *On the margin of the Euphrates: settlement and land use at Tell es-Sweyhat and in the upper Lake Assad area, Syria.* Chicago: Oriental Institute of the University of Chicago.

Williams, J. 1829. *Two essays on the geography of ancient Asia; intended partly to illustrate the campaigns of Alexander, and the Anabasis of Xenophon.* London: John Murray.

Williams-Thorpe, O. 1995. Obsidian in the Mediterranean and the Near East: a Provenancing Success Story. *Archaeometry* 37 (2): 217-248.

Winter, I. J. 2000. Thera Paintings and the Ancient Near East: The Private and Public Domains of Wall Decoration. In *The Wall Paintings of Thera, Proceedings of the First International Symposium.* S. Sherratt (ed). Athens: Thera Foundation. 745-762.

Wisti Lassen, A. 2010a. Tools, Procedures and Professions: A review of the Akkadian textile terminology. In *Textile Terminologies in the Ancient Near East and Mediterranean from the Third to the First Millennia BC.* C. Michel and M.-L. Nosch (eds.). Ancient Textiles 8. Oxford: Oxbow Books. 272-282.

———. 2010b. The Trade in Wool in Old Assyrian Anatolia. *Jaarbericht Ex Oriente Lux* 42: 159-179.

———. 2013a. Technology and Palace Economy in Middle Bronze Age Anatolia: the Case of the Crescent Shaped Loom Weight. In *Textile Production and Consumption in the Ancient Near East: archaeology, epigraphy, iconography*. M. -L. Nosch, H. Koefoed, and E. Andersson Strand (eds.). Ancient Textiles Series 12. Oxford: Oxbow. 78-92.

———. 2013b. Weaving with Crescent Shaped Loom Weights: an investigation of a special kind of loom weight. *Tools and Textiles - Texts and Context.* E. Andersson and M.-L. Nosch (eds.). Oxford: Oxbow.

Woolley, L. 1934. *Ur Excavations 2.* London and Philadelphia: British Museum and Museum of the University of Pennsylvania.

Wordsworth, P. D. 2010. Traversing the Karakum: Approaches to defining trade networks through the desert landscapes of Medieval Central Asia. *ArchAtlas*. http://www.archatlas.org/workshop09/works09-wordsworth.php (accessed 15th September 2011).

Yakar, J. 1984. Regional and Local Schools of Metalwork in Early Bronze Age Anatolia: Part I. *Anatolian Studies* 34: 59-86.

———. 1985. *The Later Prehistory of Anatolia: The Late Chalcolithic & Early Bronze Age: Part i.* British Archaeological Reports International Series. Oxford: BAR.

———. 2002. East Anatolian Metallurgy in the Fourth and Third Millennia BC: Some Remarks. In *Anatolian Metal II.* Ü. Yalçın (ed). Bochum: Deutsches Bergbau-Museum. 15-26.

Yener, K. A. 1986. The Archaeometry of Silver in Anatolia: The Bolkardağ Mining District. *American Journal of Archaeology* 90 (4): 469-472.

———. 2000. *The Domestication of Metals: The rise of complex metal industries in Anatolia.* Leiden: Brill.

Yener, K. A. and Vandiver, P. 1993a. Tin Processing at Göltepe, an Early Bronze Age Site in Anatolia. *American Journal of Archaeology* 97 (2): 207-238.

———. 1993b. Reply to J.D. Muhly, "Early Bronze Age Tin and the Taurus", with an appendix by Lynn Willies: Early Bronze Age Tin Working at Kestel. *American Journal of Archaeology* 97 (2): 255-264.

Yerasimos, S. 1991. *Les voyageurs dans l'empire Ottoman (XIVe - XVIe siècles). Bibliographie, itinéraires et inventaire des lieux habités.* Publications de la Société Turque d'Histoire VII. Ankara: Imprimerie de la Société Turque d'Histoire.

Yule, H. 1863. *Jordanus, a version of the Mirabilia with a commentary.* Hakluyt Society.

Yule, P. 1985. *Figuren, Schmuckformen und Täfelchen der Harappa-Kultur.* Prähistorische Bronzefunde I:6. München: C. H. Beck.

Zarins, J. 1978. The domesticated equidae of third millennium BC Mesopotamia. *Journal of Cuneiform Studies* 30 (1): 3-17.

Zhang, X., Good, I., and Laursen, R. 2008. Characterization of dyestuffs in ancient textiles from Xinjiang. *Journal of Archaeological Science* 35 (4): 1095-1103.

Zindorf, A., Horowitz, S., and Blum, R. 1971. Textile Remains of Nahal Mishmar [The Cave of the Treasure]. In *The Cave of the Treasure.* P. Bar-Adon (ed). 248-50.

Zimmermann, T. 2005. Perfumes and Policies - A 'Syrian bottle' from Kinet Höyük and Anatolian Trade Patterns in the Advanced Third Millennium BC. *Anatolica* 31: 161-165.

———. 2006. Bottles and Netbags - Some additional notes on the article about 'Syrian bottles' in Anatolica 31, 2005. *Anatolica* 32: 229-231.

———. 2007. Anatolia and the Balkans, Once Again: Ring-Shaped Idols from Western Asia and a Critical Reassessment of some 'Early Bronze Age' items from İkiztepe, Turkey. *Oxford Journal of Archaeology* 26 (1): 25-33.

Zischow, A. S. 2004. *Die Bronzezeitlichen Fundplätze des Tsalka-Plateaus In Trialeti/ Georgien im Kontext der Bronzezeit Transkaukasiens.* Magisterarbeit, Üniversität Tübingen. [Unpublished thesis].

Zohary, D. and Hopf, M. 1988. *Domestication of Plants in the Old World.* Oxford: Clarendon Press.

Zöldföldi, J. and Kasztovsky, Z. S. 2009. Provenance Study of Lapis Lazuli. Non destructive Prompt Gamma Activation Analysis (PGAA). In *ASMOSIA VII. Proceedings of the 7th International Conference of Association for the Study of Marble and Other Stones in Antiquity.* Y. Maniatis (ed). Athens: École française d'Athènes. 677-691.

Appendix A

Cost-of-surface models and Archaeotopograms

Selected parts of the processed GIS models described here will be available online at:

http://www.archatlas.org/databank/2014/wilkinson.tc/a/index
http://tobywilkinson.co.uk/threadsofeurasia/a/index

A.1 Foundations for a basic model of 'cost-of-passage'

Many different factors could be modelled by 'cost-surface' / 'friction' concepts, limited only by the availability and accessibility of data in suitable forms to be included. Until recently, representative coverage (or even reasonably precise interpolated data) has been difficult to obtain or else data sizes too big for average computers to allow the combination of factors outlined below. Three major factors in the cost of travel have been identified here: topography, water distribution and climatic extremes. Relevant data sources were then sought to provide a realistic empirical basis for their effects.

The following sections outline a 'basic model' that was used to create the archaeotopograms shown in this book. This description includes the source data upon which each factor is based; any processing it has received to make it usable (including the software used and processing parameters), and any relevant nuances or potential problems with the data or processing. The factors and final model are all stored in raster image formats. Raster images are effectively two-dimensional grids, which for mapping purposes are more usefully displayed through a coloured symbology (for example, on a colour ramp from green through yellow to red to represent increasing cost values). As a result of the diverse sources of the data, each factor may have its own internal relative measure of cost. Wherever possible the cost value is measured as a multiple, so that a cell with the value of 4 should be twice as difficult to traverse as a cell with the value of 2. In the absence of true experimental data on such relative costs (which cannot always be easily measured in the same 'energy expenditure' measurements that, for example, topography can), some of these value systems are of course hypothetical. In future applications it would be worth rethinking them, but in general terms they still have analytical value for the purposes of this work. The cell size is based on a resolution of around 1km, or rather 900m – small enough to show differences over the regional scale but large enough to facilitate realistic computer processing with current available hardware. Hence each raster cell is represented in mapping by a 900 x 900 m square. Again the diverse sources of data mean that exact resolution of each factor may be different in origin, and are interpolated (usually downgraded) at the combination stage or earlier. The individual factor cost values must also ultimately be *standardized* before being combined into the complete model. This standardization and combination process again requires a decision on the relative importance of each factor, the process and logic of which are described in the last section.

Note on Processing: Much of the actual processing of the data to form these factors was done with widely available tools that form part of the ArcGIS 9.2 software suite, with additional help from open source tools including geotiff utilities and QGIS.

A.1.1 Factor 1: Topography

F1. Slope

The most important aspect of topography on routes is the amount of effort expended on ascending and descending inclined ground surface, *i.e.* slope. There is a substantial literature on the effects of slope on travel (Conolly and Lake 2006, 217-221), and there are numerous applications of slope-based cost-surface rasters (*e.g.* Bell, Wilson and Wickham 2002). The slope raster used in this model is derived from SRTM (version 2) data[137] which was downgraded from a resolution of 90m to 900m in order to make the processing manageable. Small gaps in the SRTM data were filled in by the GlobalMapper export function. The slope raster was generated from the elevation data using a 'Slope' function[138]. A correction formula as described in Bell *et al.* (2002) converted the 'slope' value (measured in degrees) into a more realistic value of relative effort: tan(slope)/tan(1deg).

A secondary aspect of topography and slope is the effect of anisotropic slope on later cost-distance analyses: *i.e.* the fact that ultimate diagonal distance travelled is greater for a horizontal cell where the slope is steeper (though of course on the cost surface/friction map, the slope cost may function to impede cost distance), but also that the relationship between uphill and downhill effort expenditure when walking does not follow a lineal or a normal curve. To take account of this, the slope is also included as a 'vertical factor' (see below Section A.1.5)

The downgraded horizontal resolution of 900m, combined with the source data's vertical resolution of a maximum 10-16m, could miss small areas of high-resistance. For example relatively small but significant topographical features that could obstruct movement in a local region may not necessarily appear here as the data is smoothed for interpolation. The vertical resolution is simply the limitation of current SRTM data (which may or may not improve in the future), and the horizontal resolution is unfortunately a necessary sacrifice for the ability to process the data on current hardware. These problems, which may be solved by greater processing power and better data in the future, only really affect the small scale however, and over longer distances, the overall effect is probably less significant.

Data category	Filename	Range	Ratio
(Topography - DEM)	dem.tif		n/a
(Slope)	f1-slopes.tif	0-66.0438	n/a
F1. Adjusted slope	f1-slopeffort.tif	0-128.94	n/a

Table A.1. Factor 1, Topography.

137 SRTM-v2, was downloaded from the NASA website.

138 Using the standard ArcGIS Spatial Analyst 'Slope' function, and a z-value of 1 (since the SRTM is in metres, the z-value can also be metres)

A.1.2 Factor 2: Proxies for water availability

The availability of water is an important factor in routes for two reasons. First the difficulty in locating water sources for travellers is a serious restriction on the ability to move over certain landscapes. Second, the absence of adequate water sources (for drinking, irrigation *etc.*) more generally makes the habitation and settlement extremely difficult, resulting in low settlement density, and thus a low incentive for certain travellers (whether for comfort of for exchange purposes). Whilst it is true that even in the harshest deserts, it is possible to find evidence for human exploitation (normally through access to deep groundwater wells), the density of occupation and potential 'connectedness' across it is directly correlated to water availability.

Compared to topography, direct data on water availability is difficult to collect because the sources of water are somewhat diverse. In the model used here, 3 main data sources are used as proxies for water availability or restriction: precipitation patterns; proximity to major rivers (those for which the flow 'catchment' is much greater than others); and the general zones of groundwater. Technological adaptations (such as the great *buzhane* or "ice-houses" of the Islamic cities of Khorasan, localized cisterns or the Iranian *qalat* irrigation systems) are not taken into account, though given that the data is treated at a macro-regional scale, this may not be too significant because such adaptations still require precipitation. It must also be borne in mind that precipitation patterns or river courses may have been different in the past: large-scale irrigation projects (such as those of the Euphrates and Tigris rivers, or the Karakum canal) have had dramatic effects on the distribution of water, levels of humidity and availability of groundwater. Likewise, fluctuating global and regional temperatures have had particular effects on glacial-fed streams and rivers (particularly those fed by the Pamirs and Hindu Kush). Nonetheless, the crude model presented here still provides a useful relative indicator of the potential effect of water distribution on travel for a pre-modern era (more specifically pre-motorized transport).

F2a. Low precipitation

Rainfall data was taken from the 'WorldClim' data set, a synthetic environmental data set based on cumulative interpolated data from the period 1965-1978, described in Hijmans *et al.* (2005)[139]. The resolution of this data is 30 seconds (or around 900m), which is sufficient for the macro-scale of this model. The annual rainfall raster[140] was reclassified to exclude area of high precipitation, and then invert the values in order to transform the values into a measure of relative friction: in other words, higher values indicate less water therefore greater difficulty in traversal. Though strictly it is the relationship between precipitation and evaporation which is important to defining a "desert" and semi-arid regions, it is not unreasonable to use a maximum annual precipitation of 500mm as a broad definition for low precipitation (with less 250mm indicating desert conditions). Hence, regions with annual precipitation of higher than 500mm were assumed to

139 WorldClim data, downloaded from the WorldClim website, http://www.worldclim.org/
140 'bio_12' in the WorldClim dataset.

have an unproblematic base level of water, and hence excluded from the friction surface (*i.e.* set to a base value of '1'). Regions with 481-500mm were set to a value of 2, 461-480 to 3 and so on following the table set out below[141].

Annual precipitation (in mm)	Assigned value of cost
>500mm (*i.e.* outside desert region)	1
481-500	2
461-480	3
441-460	4
421-440	5
401-420	6
381-400	7
361-380	8
341-360	9
321-340	10
301-320	11
281-300	12
261-280	13
241-260	14
221-240	15
201-220	16
181-200	17
161-180	18
141-160	19
121-140	20
101-120	21
81-100	22
61-80	23
41-60	24
21-40	25
0-20	26
[BASE]	1

Table A.2. Factor 2a, Reclassification of annual precipitation to 'cost'.

F2b. River water

Since rivers may often flow through areas with relatively low precipitation (the great rivers of the Nile, Euphrates, Tigris, Indus and Oxus all flow through regions of very low precipitation at some point in their course), they are a significant alternative source for water both directly for drinking or irrigation[142], or indirectly by increasing underground reserves. The data for river watercourses was derived from the WWF's HYDROsheds vector database (Lehner, Verdin and Jarvis 2008)[143]. The smallest watercourses (*i.e.* whose uphill drainage is less than 2000 units in the HYDROsheds database) were excluded.

The modern flow pattern directly from HYDROsheds has was preserved for the purposes of the model with two exceptions, which were manually removed: (1) massive Karakum/Lenin irrigation canal and (2) the (generally dry) *wadis* of central Arabia.

Some potential anachronisms should be highlighted: the courses of major rivers may have changed over the last few millennia, especially in Mesopotamia where both natural and man-made processes have resulted in course changes. In Central Asia, it has also been suggested that in certain periods the Amu Darya (the ancient Oxus) flowed on a different route across the Karakum region into the Caspian (rather than to the north into the Aral Sea basin, as it does now), perhaps a result of tectonic action in the region. The understanding of this remains undeveloped, however.

Because of the limitations of the data set, no attempt was made to take into account the seasonality or reliability of watercourses, and the 'flow accumulation' units were simply used as a proxy for amount of water flow. This could have the effect of over-emphasizing some flow basins that in fact have fairly little reliable water in them (such as dry *wadi* valleys or basins).

A 'Straight Line Distance' analysis[144] was used to create a 30km buffer zone in each direction around the remaining watercourses, to indicate zones proximal to these large watercourses and hence a high likelihood of available water. Areas within the 60km zone of the river were assigned a cost value of 1 and all other areas were assigned a cost value of 20. Since the purpose of including this factor is to highlight rivers in areas that normally have low rainfall, the raster was then filtered via a Con (conditional) function. The previously calculated values were

141 The assigned values which result are in some sense arbitrarily inter-related, since they are not measured against any particular model of relationship between travel and rainfall, but they do add in indicator of increasing difficult associated with increasing aridity below the 250mm level. Whilst ideally it would be better to establish a relationship function experimentally, this is not possible within the scope of this project, and would require an entirely separate programme of its own. Instead the 'lack of water' value calculated was squared to exaggerate the effect of decreasing water exponentially.

142 Though rivers themselves may often be problematic for drinking purposes without further treatment such as boiling (for teas or infusions), fermentation (to alcohols), or indirectly via animal products (such as milk).

143 HYDROsheds data is available from USGS website, http://hydrosheds.cr.usgs.gov/

144 Using the standard ArcGIS 'Straight Line' tool in Spatial Analyst.

Table A.3. *Factor 2b, Reclassification of river proximity to a value of 'cost'.*

Relationship to	Assigned value of cost
Within 30km of major river or lake (within areas with precipitation of less than 240mm per year)	1
All other land surface	20

preserved for those areas with less than 240mm of rainfall per year, but replaced by 1 in all other cases (the f2ai-annualB.tif was used to test this condition).

F2c. Poor groundwater

In theory, at least, groundwater accessibility provides one of the most direct indicators of the distribution of water for the purposes of settlement and travel, on both a local and regional level. Groundwater is distributed almost everywhere across the Earth but access and quantity remains highly differential depending on local geology. Hydro-geological models tend to be rather complex, with many different variables contributing to the ability of humans to gain access to aquifers, whether by springs or wells. Unfortunately suitable data that could be used to predict a realistic relationship between groundwater and possible routes or roadways remains currently very limited, so the model presented here will be very basic.

The limited data for groundwater in this model was taken from the pioneering BGR WHYMAP database[145], which categorizes the Earth's surface by the presence or absence of large-scale aquifers (large aquifers, complex mixes of aquifers close to non-aquifers, and shallow or insignificant aquifers) and the relative 'recharge' rate (*i.e.* how quickly the aquifers are recharged by precipitation or other means). Though this is a fascinating resource, its main purpose is as a predictor of future exploitability of aquifers (mainly through deep wells *etc.*), hence the focus on recharge rates and not on the accessibility of groundwater or springs. For our purposes, the distribution of areas of saline groundwater content were taken as representing a cost value of 20 as compared to areas with non-saline groundwater. All other areas were assigned a cost value of 1.

This is unfortunately a very crude proxy for true groundwater availability, and substantial improvements could be made to this model if detailed data becomes available in the future. Detailed data could particularly aid the study of pre-modern desert routes.

Table A.4. *Factor 2c, Reclassification of groundwater salinity to a value of 'cost'.*

Groundwater types	Assigned value of cost
Saline groundwater	20
All other land surface	1

F2. Combination procedure - water availability proxy.

This river (F2b) corridor layer was combined with the saline groundwater layer (F2c) layer by using a 'conditional' function on rasters. Where the river raster had data (between 1 and 5), then this value was used; in all other locations, the value

145 A German hydrogeological project design to map worldwide groundwater supplies. WHYMAP has been recently made available as a web-mapping service, for use GIS systems with suitable capabilities, see http://www.whymap.org/

Data category	Filename	Range	Ratio
F2a. Low precipitation zones	f2ai-annualB.tif	1-26	A= 60
F2b. River water zones	f2b-riverzones-dryareas.tif	1-20	B= 10
F2c. Poor groundwater	f2c-saline2.tif	1-20	C= 5
F2. Combined cost	f2combined.tif	1.01923-25	n/a

Table A.5. Factor 2, Water availability: combination of constituent factors.

from the groundwater layer was used. The resultant raster was then combined with the precipitation (F2a) layer by means of a weighted average where the temporary raster has a weight of 1 and the precipitation a weight of 1. The resultant raster surface functions as a very broad graded measure of cost (with plenty of water as low cost, and no water as highest cost).

Raster calculator:

((((60 * ((Float([f2ai-annualB.tif])) / 26)) + (10 * ((Float([f2b-riverzones-dryareas.tif])) / 20)) + (5 * ((Float([f2c-saline2.tif])) / 20))) / 3)

A.1.3 Factor 3: Weather Extremes

F3a. Snow resistance

Snow provides a considerably impedance to travel in winter, and in high-altitude zones. An annual mean duration of snow cover, in days, was derived from ESRI's Geography Network Service's ESRI_Snow dataset[146], in turn derived from meteorological and snow-measuring stations. The resolution of the derived raster is about 1km, but the true resolution of the information is slightly cruder since it is based on interpolated values which were subsequently simplified into discrete 'zones' rather than a continuous surface. This means that small areas are not really visible. Despite these limitations, the average length of snow cover provides a useful relative cost for broad regions which are widely and heavily effected by lengthy snow cover. The zonal values were reclassified by dividing by 365 and multiplying by 36, before rounding up to the nearest integer:

Annual number of days of snow cover	Assigned value of cost
0	0
(1-)10	1
(11-)25	3
(26-)50	5
(51-)100	10
(101-)150	15
(151-)200	20
(201-)250	25
(251-)300	30
(301-)365	36

Table A.6. Factor 3a, Reclassification of number of days of snow cover to value of 'cost'.

146 Derived from the ESRI description of the snow cover layer: Map Server: http://www.geographynetwork.com Map Service: ESRI_Snow

F3b. Extreme maximum and minimum temperatures

Areas which have more than one month of very high or very low temperatures (those over 30 and under 5 degrees) present costs to travel, though to my knowledge these have also never been experimentally calculated. Some of these costs may be offset by, for example, night-travel in hot climates, or investment in warm clothing and fuel in cold climates, but these are, of course, difficult to model social-technological factors. For the purposes of this model: the annual maximum temperatures in the hottest month, and annual minimum temperatures from the coldest month were extracted from the WorldClim.org data set (bio_5 & bio_6 respectively) and then reclassified using the tables below:

Maximum temp (°C)	Assigned value of cost
< 31	0
31	1
32	2
33	3
34	4
35	5
36	6
37	7
38	8
39	9
40	10
41	11
42	12
43	13
...	...
48	18

Table A.7. Factor 3b, Reclassification of maximum temperatures to a value of 'cost'.

Minimum temp (°C)	Assigned value of cost
> -5	0
-6	1
-7	2
-8	3
-9	4
-10	5
-11	6
-12	7
-13	8
-14	9
-15	10
-16	11
-17	12
-18	13
-19	14
-20	15
...	...
-42	37

Table A.8. Factor 3b, Reclassification of minimum temperatures to a value of 'cost'.

F3. Combination procedure

Again there is no experimental data to suggest likely ratios for each factor, so the values selected are essentially an interpretative prediction of the likely relevance of each. Snowcover and minimum temperatures are inter-related, hence their ratios are lower than the maximal temperatures.

Raster calculator:

(((45 * ((Float([f3b-iii-mintemp-5])) / 37)) + (35 * ((Float([f3a-snow36.tif])) / 36)) + (40 * ((Float([f3b-i-maxtemp31])) / 17))) / 3)

Data category	Filename	Range	Ratio
F3a. Annual snow cover	f3a-snow36.tif	0-25	B= 35
F3b-i. Maximum temperatures	f3b-i-maxtemp31.tif	(0)-17	C= 40
F3b-iii. Minimum temperatures	f3b-iii-mintemp-5.tif	(0)-37	A= 45
F3. Combined cost	f3combined.tif	1.18972-23.0754	n/a

Table A.9. Factor 3, Extreme climate: combination of constituent factors.

A.1.4 Combined basic 'cost-of-passage' models

The base model friction map (which will be used as the default in later analyses) was created by weighted addition. Each factor was standardized, and then given a relative weight (per mille ‰). Again the relative weight value for each factor is difficult to assign as there is no a priori or experimental knowledge from which this can be currently deduced: *instead the values are subjective estimates of the relative importance of each factor to the others*[147].

It must also be borne in mind, that the effects of each factor can be manipulated by the transportation technologies: for example, carts or camels can be used to carry water through desert terrain which would be otherwise difficult to traverse (see below Section A.3). Though subjective and endlessly open to revision, I believe this model still serves to better visualize already existing assumptions.

These models must be treated as *heuristic models of assumptions* rather than empirically-devised models of the actual time, energy or other potential measures of 'cost'.

Model 1 = Terrain-reduced

Raster calculator:

model1nn.tif = (((727 * (([f1-slopeffort.tif] + 1) / 128.94)) + (136 * (([f2combined.tif]) / 25)) + (136 * (([f3combined.tif]) / 23.0754))))

Data category	Filename	Range	Ratio
F1 – Terrain	f1-slopeffort.tif	0-128.94	A= 727‰
F2 – Water	f2combined.tif	1.01923-25	B= 136‰
F3 – Climatic extremes	f3combined.tif	1.18972-23.0754	C= 136‰
Basic model			n/a

Table A.10. Combination procedure for 'cost-of-passage' raster, model 1 (see also Figure 3.2).

Model 2 = Base model

Raster calculator:

model2nn.tif = (((818 * (([f1-slopeffort.tif] + 1) / 129.94)) + (91 * ((([f2combined.tif])) / (25))) + (91 * ((([f3combined.tif])) / (23.0754)))))

Data category	Filename	Range	Ratio
F1 – Terrain	f1-slopeffort.tif	0-128.94	A= 818‰
F2 – Water	f2combined.tif	1.01923-25	B= 91‰
F3 – Climatic extremes	f3combined.tif	1.18972-23.0754	C= 91‰
Basic model			n/a

Table A.11. Combination procedure for 'cost-of-passage' raster, model 2 (see also Figure 3.1).

147 See, for example, http://webhelp.esri.com/arcgisdesktop/9.2/index.cfm?TopicName=Path%20Distance:%20adding%20more%20cost%20complexity

Model 3 = Water-biased

Raster calculator:

model3nn.tif = (((682 * (([f1-slopeffort.tif] + 1) / 129.94)) + (273 * (([f2combined.tif]) / 25)) + (45 * (([f3combined.tif]) / 23.0754))))

Data category	Filename	Range	Ratio
F1 – Terrain	f1-slopeffort.tif	0-128.94	A= 682‰
F2 – Water	f2combined.tif	1.01923-25	B= 273‰
F3 – Climatic extremes	f3combined.tif	1.18972-23.0754	C= 45‰
Basic model			n/a

Table A.12. Combination procedure for 'cost-of-passage' raster, model 3 (see also Figure 3.3).

	Input File	Comments
Source – points from which cost is calculated	<Variable>	This is the variable input value as single or multiple locations: it may include the location of sites, objects or other features
Cost (distance) raster	base-models/	
model2nn.tif	See SECTION A.1.4 above	
Surface raster	base-models/	
elevation.tif	Derived from the SRTM data. This layer allows the algorithm to calculate the additional distance created by non-flat surfaces	
Horizontal factor raster	-	-
Horizontal factor parameter	-	See SECTION A.3 below
Vertical factor raster	base-models/	
elevation.tif	See SECTION A.1.5 above	
Vertical factor parameter	base-algorithms/	
vfToblerAway.txt	See SECTION A.1.5 above	
Maximum distance	-	

Table A.13. Standard input parameters values for 'PathDistance' function.

A.1.5 Anisotropic vertical factors

The model here assumes that the 'traveller' is a pedestrian, and thus uses a 'hiking' function in its assessment of the additional time costs of walking up and down slopes. This is clearly a simplification of the available means of transport in the ancient world, though so far cost algorithms have not been designed to take multiple transport techniques into account. In its favour, however, is the fact that most ancient travellers would have had to go by foot in a time where the wheel and horse, and perhaps even the donkey, were restricted to only a few individuals.

Slope/hiker function

The table used to adjust the slope costs is based on the empirically-derived (though cleaned) function originally put forward by Tobler 1993[148].

148 http://www.geodyssey.com/papers/tobler93.html
Transformation of this function to suitable .txt file following example from: http://mapaspects.org/courses/gis-and-anthropology/weekly-class-exercises/week-9-anisotropic-cost-surfaces-and-least-cost-

vfToblerAway.txt vfToblerTowards.txt

A.2 Standard parameters for archaeotopograms

In order to make archaeotopograms inter-comparable standard parameters should be used and standard symbology, otherwise the resultant maps can be misleading.

A.2.1 Standard PathDistance parameters

A.3 Future work: additional or alternative anisotropic factors

A series of factors were not possible to include in the current model because of the additional time it would take to produce meaningful working models. These include travel on rivers, lakes and seas, and variations associated with different modes of transport (*e.g.* horse, cart, camel *etc.*). The problem is that these factors are heavily reliant on technology for their costs which needs empirically-derived functions to calculate their time (or other) costs, and they are also heavily directional (where sea travel is dependant on winds and currents, and rivers on the direction of stream), which considerably adds to the complexity of the models. These factors could be built into the current model by using additional 'horizontal factors'. For example a river transport factor could be added in which a lower cost is included when travelling downstream and higher cost travelling upstream, relative to the general cost of being on a river cell. There are many complications however: travel on a boat does not have the same 'vertical factors' as hiking on land; different boats have different qualities (*e.g.* wind, rowing); and it is difficult to measure the relative costs between land and sea traversal. For now then, these more advanced models must remain unrealized.

Appendix B

Database of Sites

Selected and updated parts of the database described here will be available online at:

http://www.archatlas.org/databank/2014/wilkinson.tc/b/index
http://tobywilkinson.co.uk/threadsofeurasia/b/index

B.1 Aims and methodology of site database construction

The main aim of this database was to create a list of sites with reasonably accurate co-ordinates, ideally to within a kilometre, that could be used for map creation and to form the basis of any GIS analyses. This task, central from the beginning of the research was particularly time-consuming and problematic, mainly because it is not common practice to publish GPS or other co-ordinates with excavation reports – more common is to show a 'location map' with a dot showing site location, which at the scale of the map may be very many kilometres in diameter. That few co-ordinates are ever published is due, in part, to the relatively recent introduction of GPS to archaeological study, but also a reticence to publish 'hard-earned' co-ordinates freely remains in-built in the discipline. There are a number of archaeological gazetteers, which have aimed to overcome this shortcoming and aide the study of regional settlement patterns. However, the increased quality and quantity of freely available georeferenced satellite imagery (as seen in Google Earth), has made the low level of precision of most of these gazetteers very obvious. Many of the older gazetteers were designed around particular map sets with regional co-ordinate systems which also need time-consuming conversion to 'universal' co-ordinates (such as UTM or latitude/longitude). Few gazetteers provide any indication of the likely accuracy or precision of co-ordinates – so that it is difficult to know how confident we can be even where we do have co-ordinates without a process of verification.

The methodological procedure adopted for creating this database involved the following steps:

- Collection of co-ordinates from published gazetteers and databases;
- Digitization and georeferencing of published maps with relevant sites to find co-ordinates;
- Searching for sites using online databases (including Wikipedia and Google Earth's own place-name, photograph and relative 'point-of-interest' databases);
- 'Eyeballing' co-ordinates for those sites whose approximate location (in 100kms) can be deduced from maps or descriptions but which could not be easily verified.

Locations were verified in the following ways:

- Taking the co-ordinates derived from published gazetteers and maps, and attempting to confirm visually on Google Earth,
- Ground-truthing (by visiting sites on the ground and taking GPS readings),
- Cross-checking with original primary and secondary publications for location information.
- Most importantly, a value of 'confidence' was assigned to each co-ordinate included in the database to indicate how much such co-ordinates should be trusted.

Various issues were raised by this process including: the problems associated with variant spellings (a result, often, of different conventions for languages and transliteration of non-Latin scripts rather than spelling errors as such); the many sites with similar or identical names; and the question of what exactly constitutes a site (a single co-ordinate, though the easiest for the purposes of a macro-scale database cannot take account of the full geometry of real sites and their different occupation layers, or solve the problem of where one site 'begins' and another 'ends'). The last of these was not necessary to solve for the purpose of this book, but I attempted to take account of the first by including spelling variants and transliterations into the database whenever I came across them.

The process revealed a very uneven distribution of sites which, it seems likely, indicates the 'distribution of archaeologists' and the amount of work done in certain regions rather than giving any kind of representative picture of settlement (or burial) patterns.

B.2 Sources: Transcaucasia and eastern Anatolia

There are few comprehensive gazetteers for the area of Transcaucasia during our period of interest but two sources played the major starting point: the first is A. Sagona's gazetteer of Early Bronze age sites created as part of his doctoral thesis (Sagona 1984), and updated recently by S. Batiuk (2005). The second, covering wider eastern and central Anatolia, was the gazetteer/inventory of sites of the TAY project (*Türkiye Arkeolojik Yerleşmeleri* / Archaeological Sites of Turkey). Several volumes of documentation have been produced by this project, including an Early Bronze Age (Harmankaya and Erdoğlu 2002) and an Iron Age catalogue, though sadly Middle and Late Bronze sections have not yet been completed. Only sites within the modern borders of Turkey are included in the catalogues, but many have detailed descriptions of the location of the sites, often visited by the TAY team. The print version of the Early Bronze Age volume includes co-ordinates – though these are not always very precise. Other general sources of site co-ordinates or general locations consulted include: Belli and Sevin 1999; Lehmann 2002; Burney and Lang 1971; Kushnareva 1997; Özfırat 2001; Barjamovic 2011.

The distribution of these sites (Figure B.1) clearly demonstrates the current large gaps in knowledge for many key regions: Cilicia, the Taurus, the Kahramanmaraş, Sivas, the Upper Euphrates, the eastern Black Sea, the upper Tigris, Bingöl, Muş. Those areas that have undergone greatest research, particularly along the Euphrates dam construction zones, are already, or soon will be, under water to be covered by alluvial deposits from the dam. The main exceptions, are the Hittite heartland of

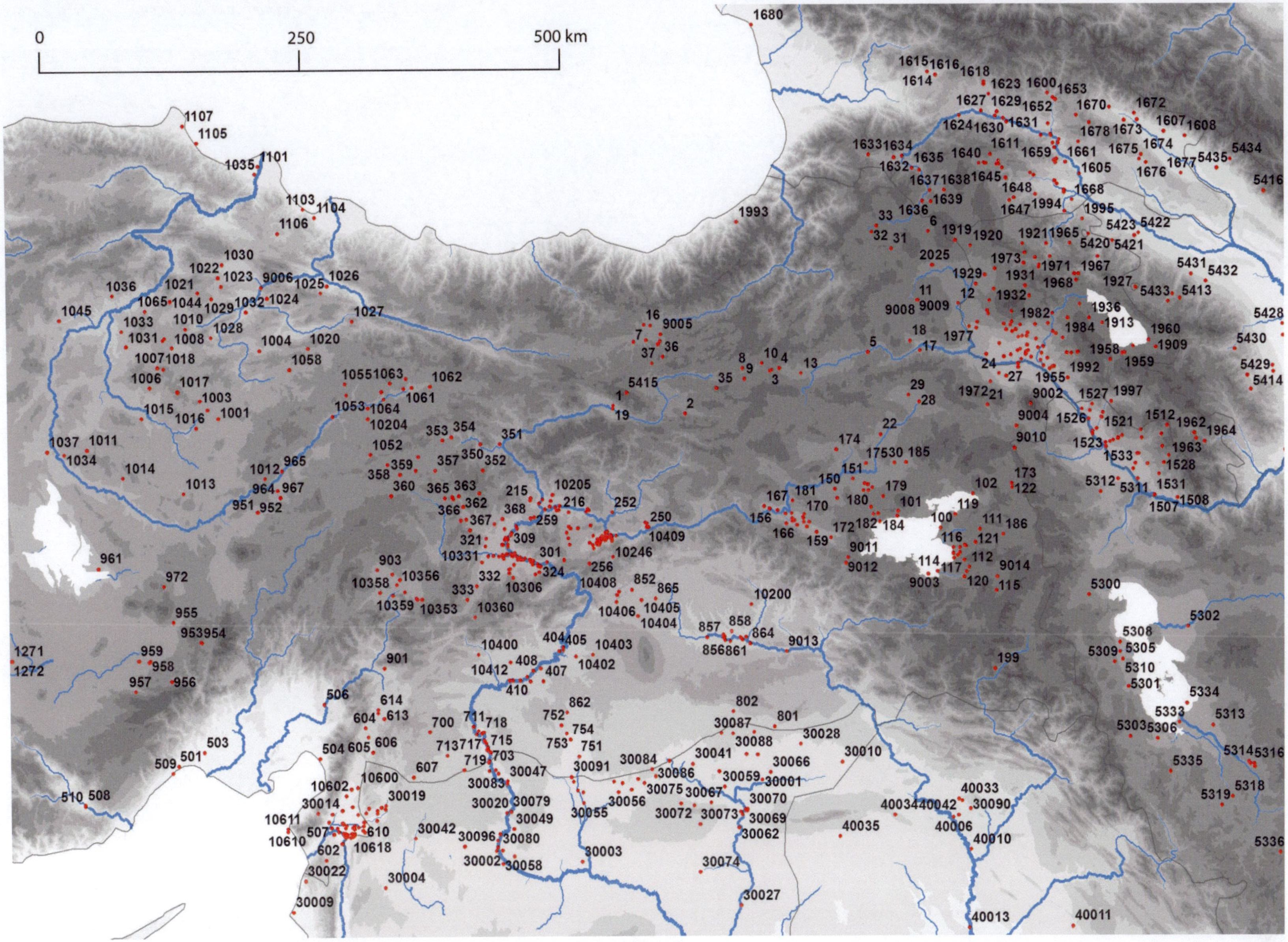

Figure B.1. All sites of eastern Anatolia in the database.

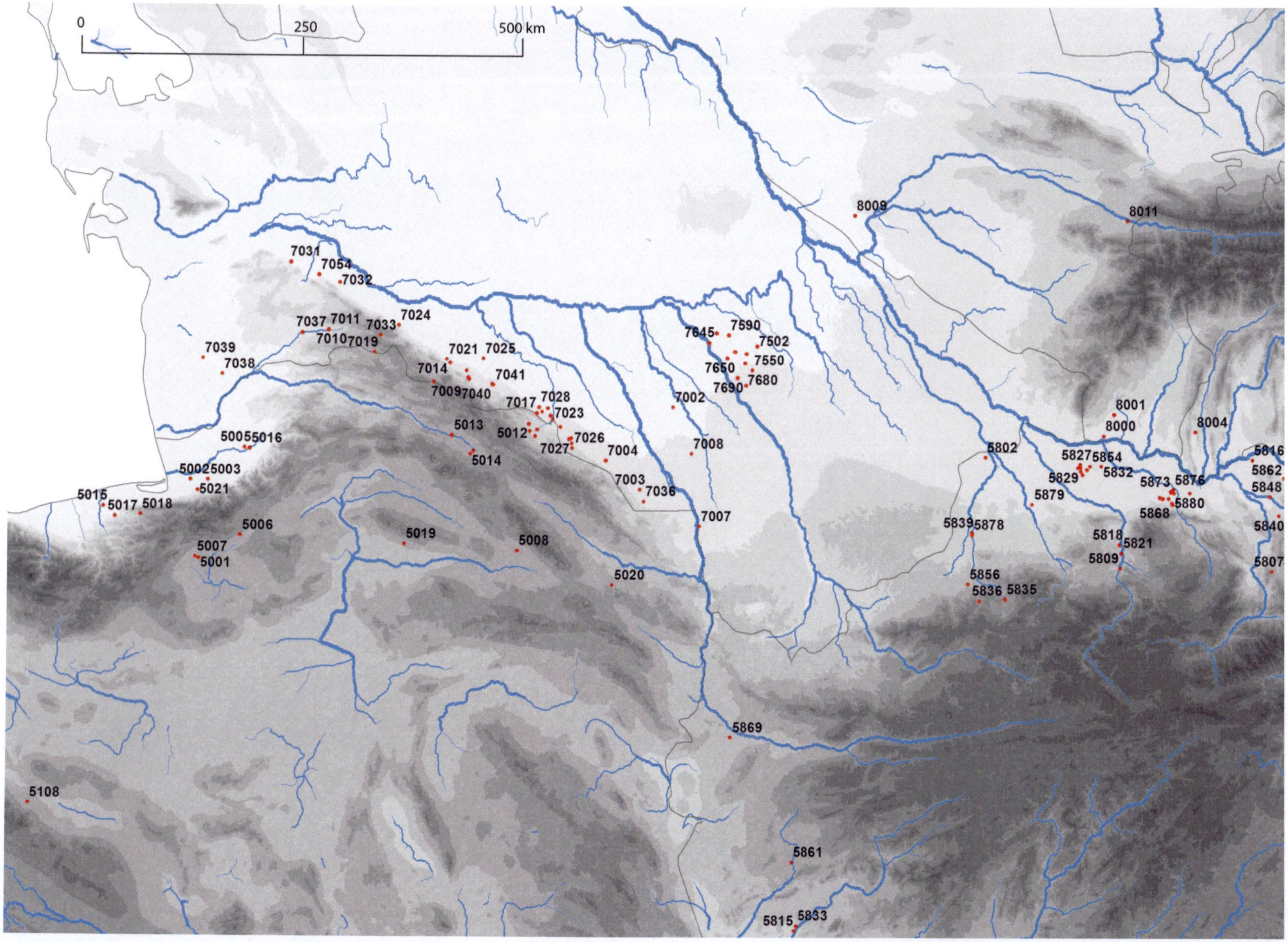

Figure B.2. All sites of western Central Asia in the database.

North-East Central Anatolia, and the area around Lake Van. Sites discovered by surface surveys, or recorded in local archaeological records may to some extent fill in the gaps, but of course less information for these locations is available, either to pinpoint the co-ordinates or assign reliable dates and cultural information. The publication of the many current surveys and excavations from the Caucasus will no doubt improve the quality of locational information over the next few years.

B.3 Sources: western Central Asia

Parts of Central Asia have been intensively surveyed by Soviet researchers (*e.g.* as part of the Yu project) and more recently by international teams such as the University of Bologna's 'Archaeological Map' projects (*The Archaeological Map of the Murghab Delta. Volume II. The Bronze Age and Early Iron Age in the Margiana Lowlands* 2008) – both were focussed on the desert regions of southern Turkmenistan. Other synthetic sources have provided useful foundation for the Central Asian section of the database, including: Ball and Gardin 1982; Kohl 1984.

The distribution of sites in the database (Figure B.2) again demonstrates an uneven distribution that is likely to be the result of uneven investigation. North-east Iran, in particular, is currently poorly represented in the literature, though improving. Similarly, archaeological research in Afghanistan essentially stopped after the 1980s (although has restarted recently) and is so poorly understood.

B.4 Special site codes

There are a few special site codes that have been used in the object databases to indicate vague location/find-spot information: *e.g.* vague, untrustworthy or unknown provenance; or 'regions' and 'countries' (for when the exact site is unknown). They are as follows:

SiteCode	CC	SiteName	AlternativeName	LatitudeDECn	LongitudeDECe
-999586	PK	Pakistan			
-999368	IQ	Iraq			
-999364	IR	Iran			
-105	IR	Eastern Iran			
-104	EG	Fayum	Faiyum	29.3084	30.84
-103		Bactria			
-102	IR	Iranian Azerbaijan		38.037	46.334
-101	TR	Amuq	Amık	36.262	36.476
-100	IR	Luristan	Lorestan	34.026	47.76
-3		(Unprovenanced - art market)			
-2		(Site location unknown)			
-1		(Site unknown)			

B.5 Key to site table

The following abbreviations are used in the row headers of the site database.

SiteCode	This is a unique identifier for each site, which is used to link object data to the site in which the object was found (see APPENDIX C).
CC	Country code (following the ISO-3166 A2 two-letter code). Country of current administration is assumed, with no commentary on the 'correct' political status of the location in question.
SiteName	Site name as normally provided by the first source in which the site was found, though occasionally adjusted to the most common spelling or term given to the site in English publications.
AlternativeSiteName	A list of alternative spellings or names for the same site, including alternative ancient names.
AncientName	The ancient name, if known.
LatitudeDECn	Latitude (given in decimal degrees north).
LongitudeDECe	Longitude (given in decimal degrees east).
Cn	Coordinate 'confidence' – a relative value of confidence from 1-10, where 10 is verified by GPS or good published source, and 1 is highly imprecise (or inaccurate). 0 means that a value of confidence has not been assigned.
V	Visited by author.
D	Destroyed – site has been destroyed (*e.g.* by construction or erosion).
CoordinateSource	The source of the co-ordinate, normally a reference to a publication in which the co-ordinate itself was published, or a map with the site location that could be digitized. There are various abbreviations used here: GE= site identified visually or by nearest village in Google Earth TCW, AGS, BCC= GPS co-ordinate obtained by the author, Andrew Sherratt or Ben Claasz Coockson. ANE.kml= an online collection of ancient places as published online by Olof Pederson. http://www.lingfil.uu.se/staff/olof_pedersen/Google_Earth/Livius.org= an online collection of ancient place locations.
T	Site Type (where known). S=Settlement G=Grave or mortuary site
E	Excavated?
S	Surveyed?

B.6 Co-ordinate confidence

The confidence (a subjective measure of accuracy and precision) for each co-ordinate is shown graphically in Figure B.3. This was considered an extremely important part of the database, since too often (unverified) co-ordinates are given with little indication of the extent to which they can be relied upon.

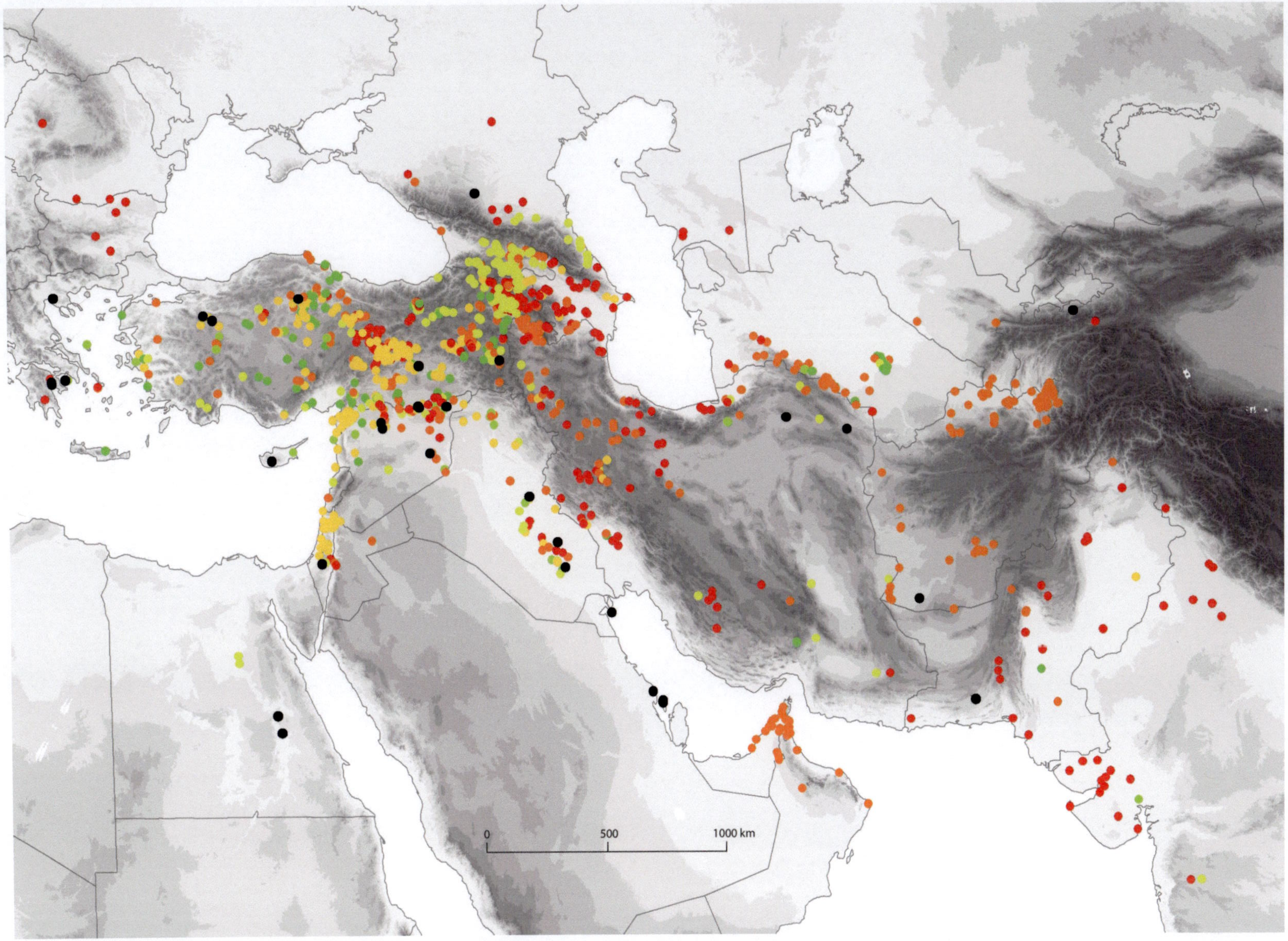

Figure B.3. Graphical representation of the 'confidence' value for each site in the database. Green indicates high confidence, orange medium and red low confidence. Dark green refers to GPS or Google Earth verified co-ordinates. Black indicates that no value of confidence was assigned yet

Appendix C

Databases of Object Types

Selected parts of the databases will be available online at:
http://www.archatlas.org/databank/2014/wilkinson.tc/c/index
http://tobywilkinson.co.uk/threadsofeurasia/c/index

C.1. Databases and file names

C.1.1 Objects discussed in Chapter 4 - Stones

Lapis objects
See file: TABC-1_lapislazuli (75 items)

Etched carnelian objects
See file: TABC-2_carnelian (19 items)

'Intercultural style' objects
See file: TABC-3_intercultural (341 items)

C.1.2 Objects discussed in Chapter 5 - Metals

BMAC objects (including Bactrian axes)
See file: TABC-4_bmac (42 items)

Decorated metal vessels
See file: TABC-5_metalvessels (5 items)

Depas amphikypellon objects
See file: TABC-6_depas (86 items)

Syrian flasks
See file: TABC-7_syrianflasks (69 items)

Metallische Ware
See file: TABC-8_metallische (57 items)

Metallic ware
See file: TABC-9_metallic (32 items)

Namazga V pottery
See file: TABC-10_nmgV (20 items)

Hissar IIB burnished grey ware pottery
See file: TABC-11_hissarIIb (5 items)

Kura-Arax and related wares
See file: TABC-12_kuraarax (471 items)

C.1.3 Objects discussed in Chapter 6 - Textiles

Textile objects
See file: TABC-13_textiles (145 items)

Crescent-shaped loom weights
See file: TABC-14_crescentloom (24 items)

Transcaucasian Painted Wares ('Aras')
See file: TABC-15_tpw (137 items)

Namazga IV-style pottery
See file: TABC-16_nmgIV (29 items)

Double spiral-headed pins
See file: TABC-17_dblsprlpin (134 items)

Flat-circular beads with tubes
See file: TABC-18_fcbeads (18 items)

Quadruple-spiral beads
See file: TABC-19_qsbeads (10 items)

Appendix D

Databases of Raw Material Sources

Selected and updated parts of the databases will be also available online at:
http://www.archatlas.org/databank/2014/wilkinson.tc/d/index
http://tobywilkinson.co.uk/threadsofeurasia/d/index

D.1 Databases and file names

D.1.1 Sources discussed in Chapter 4 - Stones

Lapis lazuli deposits
See following pages.
Also as file: TABD-1_lapislazulisrcs

Obsidian deposits
See following pages.
Also as file: TABD-2_obsidiansrcs

D.1.2 Sources discussed in Chapter 5 - Metals

Copper ore deposits
See following pages.
Also as file: TABD-3_coppersources

Tin deposits
See following pages.
Also as file: TABD-4_tinsources

Table D.1. Lapis lazuli sources.

Id	Name	Source
1	Sar-i Sang, Badakhshan	Herrmann (1968)
2	Bi-Bi-Dik, Chagai	Berthoud et al. (1982, 41 n. 21) Unlikely to exist: Law 2011, 528-543
3	Lyadzhuar Dara, Pamirs	Lo Giudice et al. (2009) Unlikely to have been exploited
4	Sirjan	Potts (1994, 210) Unconfirmed

Table D.2. Obsidian sources.

Id	Name	Source
1	Melos (Sta Nychia)	Carter (2008)
2	Melos (Dhemeneghaki)	Carter (2008)
3	Hasan Dag	Frahm (2010)
4	Nenezi Dag	Frahm (2010)
5	Gollu Dag	Frahm (2010)
6	Acigol	Frahm (2010)
7	Bingol A	Frahm (2010)
8	Bingol B	Frahm (2010)
9	Nemrut Dag	Frahm (2010)
10	Mus	Frahm (2010)
11	Suphan Dag	Frahm (2010)
12	Meydan Dag	Frahm (2010)
13	Tendurek Dag/Dogabeyazit	Frahm (2010)
14	Erzincan	Frahm (2010)
15	Erzurum	Frahm (2010)
16	Pasinler	Frahm (2010)
18	Ikizdere	Frahm (2010)
19	Sarikamis	Frahm (2010)
20	Kars (Akbab Dag	Frahm (2010)
21	Kars (Digor)	Frahm (2010)
22	Kars (Arpacay)	Frahm (2010)
23	Chikiani	Badalyan, Chataigner and Kohl (2004)
24	Sizavet	Badalyan, Chataigner and Kohl (2004)
25	Agvorik	Badalyan, Chataigner and Kohl (2004)
26	Arteni	Badalyan, Chataigner and Kohl (2004)
27	Tsakhkunyats	Badalyan, Chataigner and Kohl (2004)
28	Alapars	Badalyan, Chataigner and Kohl (2004)
29	Gutaansar	Badalyan, Chataigner and Kohl (2004)
30	Hati	Badalyan, Chataigner and Kohl (2004)
31	Geghasar	Badalyan, Chataigner and Kohl (2004)
32	Khorapor	Badalyan, Chataigner and Kohl (2004)
33	Kechaldag	Badalyan, Chataigner and Kohl (2004)
34	Spitakasar	Badalyan, Chataigner and Kohl (2004)
35	Rokr Satanakar	Badalyan, Chataigner and Kohl (2004)
36	Mets Satanakar	Badalyan, Chataigner and Kohl (2004)
37	Sevkar	Badalyan, Chataigner and Kohl (2004)
38	Bazenk	Badalyan, Chataigner and Kohl (2004)
39	Michnek Satanakar	Badalyan, Chataigner and Kohl (2004)
40	Bartsratumb	Badalyan, Chataigner and Kohl (2004)

Table D.3. Copper ore deposits.

No.	Name of region of mining area	Reference
1	Kure Group	de Jesus (1978)
2	Yaprakli Group	de Jesus (1978)
3	Giresun-Trabzon Group	de Jesus (1978)
4	Murgul-Kuvarshan Group	de Jesus (1978)
5	Ergani Group	de Jesus (1978)
6	Pontic Group	de Jesus (1978)
7	Alaverdi district	Chernykh (1992, 6-7)
8	Zangezur district	Chernykh (1992, 6-7); Schachner (2002)
9	Dari-dag	Chernykh (1992, 6-7); Schachner (2002)
10	Bursa region?	(Efe 2002)
11	Jiroft region?	Potts (1994, 146)
12	Baft region?	Potts (1994, 146)
13	Oman	Potts (1994, 146)
14	Veshnoveh	Potts (1994, 146); Stöllner (2005)
15	Kerman region?	Potts (1994, 146); Pigott (1999, 83)
16		Potts (1994, 146)
17	Anarak	Potts (1994, 146)
18	Khash?	Potts (1994, 146)
19	Birjand region?	Potts (1994, 146)
20-26		Potts (1994, 146)
27	Taknar	Pigott (1999, 75, 83)
28	Qazvin?	Algaze (1993, fig. 35)
28	Elenovka and Ush-Katta	Chernykh (1992, 6-7)
29	Qazvin?	Algaze (1993, fig. 35)
29	Bakr-Uzyak	Chernykh (1992, 6-7)
30	Qazvin?	Algaze (1993, fig. 35)
30	Tash-Kazgan and Nikolskoe	Chernykh (1992, 6-7)
31	Kalmakyr	Rubinstein and Barsky (2002, 126)
32	Kuru-Tegerek	Rubinstein and Barsky (2002, 130)
33	Mitronovsky	Rubinstein and Barsky (2002, 130)
34	Cyprus	Moorey (1994, 247)
35	Feinan	Moorey (1994, 247)
36	Wadi Arabah	Moorey (1994, 247)
37	Timna	Grajetzki and Quirke (2000, 'copper')
38	Serabit el-Khadin, Sinai	Moorey (1994, 247)
39	Wadi Sitra	Grajetzki and Quirke (2000, 'copper')
40	Hamash	Grajetzki and Quirke (2000, 'copper')
41	Buhen	Grajetzki and Quirke (2000, 'copper')
42	Kenkazgan	Chernykh (1992, 6-7)
43	Altyn-Tyube	Chernykh (1992, 6-7)
44	Dzhezkazgan	Chernykh (1992, 6-7)
45	Bukantau and Sultan-Uiz-dag district	Chernykh (1992, 6-7)
46	Kugitangtau district	Chernykh (1992, 6-7)
47	Chatkal-Talass-Kuram district	Chernykh (1992, 6-7)
48	Ferghana district	Chernykh (1992, 6-7)
49	Rudny or Western Altai	Chernykh (1992, 6-7)
50	Shoriya	Chernykh (1992, 6-7)

Table D.3. Copper ore deposits (continued).

No.	Name of region of mining area	Reference
51	Minusinsk basin	Chernykh (1992, 6-7)
52	Tuva	Chernykh (1992, 6-7)
53	Verkhnaya Racha district	Chernykh (1992, 6-7)
54	North Balkan mining centre	Chernykh (1992, 6-7)
55	Northeast Carpathian mining centre	Chernykh (1992, 6-7)
56	Banat, Bor and Vidin mining centres	Chernykh (1992, 6-7)
57	Apuseni mining centre	Chernykh (1992, 6-7)
58	Balkan-Vratsa mining centre	Chernykh (1992, 6-7)
59	Strandja mining centre	Chernykh (1992, 6-7)
60	North Thracian mining centre	Chernykh (1992, 6-7)
61	Mines near Lake Onega	Chernykh (1992, 6-7)
62	Donets Cu-bearing sandstone	Chernykh (1992, 6-7)
63	Rajasthan sources	Kenoyer and Miller (1999, 108)
64-103		Pigott (1999, 83)

Table D.4. Tin deposits.

Id	Name of region or mining area	Reference
1	Kestel	Yener (2000)
2	Madenbelenitepe, Bursa	Yakar (1984, 80)
3	Karnab group	Alimov et al. (1998)
4	Pendzikent region	Alimov et al. (1998)
5		Potts (1994, 146)
6	Kadamdzhay	Rubinstein and Barsky (2002, 126)
7	Kassansay?	Rubinstein and Barsky (2002, 130)
8	Uchkoshkon?	Rubinstein and Barsky (2002, 130)
9	Srebrenica	McGeehan-Liritzis (1987, 291)
10	Cer	McGeehan-Liritzis (1987, 291)
11	Bukulja	McGeehan-Liritzis (1987, 291)
12	Kalba and Narym	Chernykh (1992, 6-7)
13	Shilka and Argun rier basins	Chernykh (1992, 6-7)
14	Zirabulak-Ziyaetdin mountains	Chernykh (1992, 6-7)
15	Deh Hosein region	Momenzadeh, Hajisoltan and Momenzadeh (2004); Nezafati, Pernicka and Momenzadeh (2006)
16-50		Pigott (1999, 82)

Appendix E

Summary Statistics on Metal Trends

Selected parts of the summary data described here will be available online at:
http://www.archatlas.org/databank/2014/wilkinson.tc/e/index
http://tobywilkinson.co.uk/threadsofeurasia/e/index

The graphical figures of metal trends (Figures 5.23-5.32) were created based data from a number of publications. In the original publications, the statistics were presented in number of different ways:

- raw values which have been reproduced here,
- graphical representations, for which estimated percentages or values were taken in order to redraw graphs.

The combined data was then placed in unified tables, and supplemented by less systematic data (quantitative representation of qualitative descriptions: *e.g.* 'most' would be translated to 75%) from the same or other sources.

E.1 Chernykh *et al.*'s data recast by category and publication

Tables E.1 to E5 (shown on following pages) represent the assembled 'raw' data from Chernykh (1992), Chernykh, Avilova and Orlovskaya (2002) and Avilova (2008) as a basis for assessing trends in metal deposition, form and composition from objects in the Chernykh database. Where the original publication showed a graphical presentation (but gave no numbers), numbers have been estimated from the graph.

E.2 Combined tables of value, including additional data from other sources

Table E6 represent an attempt to combine all the available data provided by the Chernykh database on metal composition (based on Tables E.4 to E5) into a single comprehensive table, and supplement the gaps with data from other sources. This means that many of the figures are not systematic or directly comparable (note, for example, that different authors use different definitions: some use a value of 4.5% or above to indicate intentional tin, whilst others use more than 1% to denote a 'tin-bronze', see Kenoyer and Miller 1999, 114-116). However, in the absence of more systematic information, these provide an 'impressionistic' idea of the changing metal consumption patterns over a wide geographic and chronological span.

Metal composition statistics

See file (online): TABE-6_metalcomposition

TheRegionDescribed	EBA	MBA						
	S	C	H	I	S	C	H	I
Asia Minor	90	?	5	5	60	25	5	5
Transcaucasia	40	40		20		80	5	10
North Caucasus	2	95		3		70	5	20
S. of E. Europe	10	80	5	5	5	80	7	8
Carpatho-Balkans	65	8	7	20	10	10	35	35

Table E.1. Statistics describing the contexts of deposition for metal objects (Source: Chernykh 1992, 164). Approximate percentages from graphs; S=settlement, C=cemetery, H=hoards, I=isolated finds.

TheRegionDescribed	EBA	MBA				
	av	tw	or	av	tw	or
Asia Minor	4		60	21		40
Transcaucasia	9		50	26		30
North Caucasus	15		25	190		90
Azov & Caspian Sea	1.5		30	37		60
N. Black Sea region	7		30	9		30
Volga & Urals	1		0	3		40
Carpatho-Balkans	5		20	11		30

Table E.2. Statistics describing the average number of copper/copper-alloy objects per square kilometre in each region (Source: Chernykh 1992, 161); av= Average Cu/bronze finds per 10k sq. km; tw=% tools weapons (estimated from graph), or=% ornaments.

RegionDescribed	Total	Chalco	EBA	MBA			
		tw	or	tw	or	tw	or
Anatolia	37017	28	41	145	188	1115	34999
Mesopotamia	14893	4	2	53	439	469	13434
Levant	5500	3		366	181	1313	1990
Iran	3286	114	34	86	846	1105	750

Table E.3a. Statistics defining metal objects as tools/weapons or ornaments (Source: Chernykh 1992, 150).

RegionDescribed	EBA	MBA	LBA						
	tw	or	…	tw	or	…	tw	or	…
Asia Minor	35	50	5	50	30	10			
Transcaucasia	35	40	15	50	25				
North Caucasus	80	5	10	25	70	5			
S. of E. Europe	55	40	5	55	40	5			
Carpatho-Balkans	90	7	3	65	20	5			

Table E.3b. Statistics defining metal objects as tools/weapons or ornaments (Source: Avilova 2008, 77). Approximate percentages from graphs; tw=tools/weapons, or=ornaments, …=other.

TheRegionDescribed		Chalco	EBA	MBA			
	Total	Cu	Au	Cu	Au	Cu	Au
S. of E. Europe		1000		500		1500	
Carpatho-Balkans		3500	3200	250		1000	
N. Caucasus		100	3500	400	3000	3000	
Transcaucasia		100		200		600	2500
Asia Minor		100		400		1500	16500

Table E.4a. Statistics on metal material of objects in database. (Source: Chernykh 1992, 142). Values are approximate based on graphic original.

TheRegionDescribed	**Chalcolithic**	**EBA**	**MBA**									
	Tot	**Cu+**	**Au**	**Ag**	**Tot**	**Cu+**	**Au**	**Ag**	**Tot**	**Cu+**	**Au**	**Ag**
Aegean					1100	70	6	24	3700	44.6	53.6	1.8
Anatolia					370	95	0.5	4.5	36500	5.8	93	1.2
Levant					700	76	2.5	21.5	4700	80	17.3	2.7
N. Caucasus					8700	4.5	84	11.5	4500	98.4	0	0.5
S. Caucasus					320	98	0	2	3560	28	70.4	1.6
Mesopotamia					550	48.5	50.8	0.7	14300	25.8	46.3	27.9
Susiana					180	99	0.5	0.5	1500	90	6.7	3.3
Balkan-Carpathian					320	97	1.5	1.5	900	98.7	1	0.3
E. (Vost) Europe					880	88	1.5	10.5	3700	80	17.3	2.7

Table E.4b. Statistics on metal material of objects in database. (Source: Chernykh, Avilova and Orlovskaya 2002, 15).

TheRegionDescribed		**Chalco**	**EBA**	**MBA**									
	Total	**Cu+**	**Au**	**Ag**	**Pb**	**Cu+**	**Au**	**Ag**	**Pb**	**Cu+**	**Au**	**Ag**	**Pb**
Anatolia	37017	70		1		333	1	16	4	2125	33997	396	33
Mesopotamia	14893	5	1			262	275	4	39	3694	6619	3984	
Levant	5500	3			1	532	18	151		3759	814	124	8
Iran	3286	160				254	669	22	7	1825	180	111	21

Table E.4c. Statistics on metal material of objects in database. (Source: Avilova 2008, 77).

TheRegionDescribed	EBA	MBA								
	Tot	Cu	+As	+Sn	...	Tot	Cu	+As	+Sn	...
Asia Minor		10	70	10			5	50	30	5
Transcaucasia		5	90	5			5	60	20	5
North Caucasus		5	95	0			1	80	1	1
S of E. Europe		40	60				20	75	5	0
Carpatho-Balkans		60	30	5	5		45	15	25	5
Syria/ N. Meso										

Table E.5a. Statistics defining copper-alloy broad types for objects in database (Source: Chernykh 1992, 146). Based on approximate percentage from graph. Note: not all added up to 100% in original.

TheRegionDescribed	EBA	MBA								
	Tot	Cu	+As	+Sn	...	Tot	Cu	+As	+Sn	...
Anatolia	136	26	98	8	4	489	63	295	110	21
Balkans-Carpathians	207	90	99	4	14	724	234	130	270	90
Aegean	217	35	83	44	55	311	42	137	109	23
Levant	90	35	51	1	3	188	39	71	75	3
Mesopotamia	85	12	65	7	1	129	21	65	42	1
Susiana	30	5	25	0	0	339	0	288	43	8
South Caucasus	139	2	136	0	1	401	3	310	61	27
North Caucasus	307	15	288	0	4	1519	21	1389	76	33
Eastern Europe	471	178	292	1	0	1146	243	882	16	5

Table E.5b. Statistics defining copper-alloy broad types for objects in database (Source: Chernykh, Avilova and Orlovskaya 2002, 14). Updated no. of objects (total EBA: 1682, MBA 5246).

TheRegionDisc.	Chalcolithic	EBA	MBA									
	Tot	Cu	+As	+Sn	Tot	Cu	+As	+Sn	Tot	Cu	+As	+Sn
Anatolia	42	26	13	1	127	19	95	10	489	63	250	155
Mesopotamia	3	3			85	12	63	7	129	21	49	58
Levant	0				90	35	51	1	189	39	71	75
Iran	69	30	37		76	22	53	1	373	5	258	102

Table E.5c. Statistics defining copper-alloy broad types for objects in database (Source: Avilova 2008). Only includes analysed objects.